CW01171544

Functional Auxiliary Materials in Batteries

Functional Auxiliary Materials in Batteries

Synthesis, Properties, and Applications

Wei Hu

WILEY VCH

Author

Dr. Wei Hu
University of Science and
Technology Beijing
No. 30 Xueyuan Road
Haidian District
Beijing 100083
China

Cover Image: © VectorMine/Adobe Stock

All books published by **WILEY-VCH** are carefully produced. Nevertheless, authors, editors, and publisher do not warrant the information contained in these books, including this book, to be free of errors. Readers are advised to keep in mind that statements, data, illustrations, procedural details or other items may inadvertently be inaccurate.

Library of Congress Card No.: applied for

British Library Cataloguing-in-Publication Data
A catalogue record for this book is available from the British Library.

Bibliographic information published by the Deutsche Nationalbibliothek
The Deutsche Nationalbibliothek lists this publication in the Deutsche Nationalbibliografie; detailed bibliographic data are available on the Internet at <http://dnb.d-nb.de>.

© 2025 WILEY-VCH GmbH, Boschstraße 12, 69469 Weinheim, Germany

The manufacturer's authorized representative according to the EU General Product Safety Regulation is WILEY-VCH GmbH, Boschstraße 12, 69469 Weinheim, Germany, e-mail: Product_Safety@wiley.com.

All rights reserved (including those of translation into other languages, text and data mining and training of artificial technologies or similar technologies). No part of this book may be reproduced in any form – by photoprinting, microfilm, or any other means – nor transmitted or translated into a machine language without written permission from the publishers. Registered names, trademarks, etc. used in this book, even when not specifically marked as such, are not to be considered unprotected by law.

Print ISBN: 978-3-527-35529-7
ePDF ISBN: 978-3-527-85277-2
ePub ISBN: 978-3-527-85276-5
oBook ISBN: 978-3-527-85278-9

Typesetting Straive, Chennai, India
Druck und Bindung: CPI books GmbH Leck, Germany

Contents

Preface *xiii*

1 Application of Organic Functional Additives in Batteries *1*
1.1 Introduction *1*
1.2 Fluorinated Additives *2*
1.2.1 Functions of Fluorinated Additives *2*
1.2.1.1 Improvement of Safety Performance *2*
1.2.1.2 SEI-Forming Additives *2*
1.2.1.3 High Oxidation Stability *4*
1.2.1.4 Promotion of the Formation of Anion-Rich Solvation Structure *5*
1.2.1.5 Reduction of Desolvation Barrier *6*
1.2.2 Synergies of Fluoroethylene Carbonate with Other Compounds *6*
1.2.2.1 Fluoroethylene Carbonate and Other Fluorinated Electrolytes *6*
1.2.2.2 Fluoroethylene Carbonate and Lewis Base *7*
1.2.2.3 Fluoroethylene Carbonate and Glyme *7*
1.2.3 Drawbacks of Fluoroethylene Carbonate *8*
1.2.3.1 Generation of HF Gas *8*
1.2.3.2 Increase of Impedance and Loss of Impedance *8*
1.2.3.3 Incompatibility with Other Electrodes *8*
1.2.3.4 Recycling Issues *9*
1.3 Nitro Additive *9*
1.3.1 Functions of Nitro (NO_3^-) *9*
1.3.1.1 Participation in Solvation and Desolvation Structures *9*
1.3.1.2 Formation of Inorganic-Rich SEI *10*
1.3.1.3 CEI-Forming Additives *12*
1.3.1.4 Functions in Lithium–Sulfur Batteries *13*
1.3.1.5 Stabilization of Water Molecules *14*
1.3.2 Organic Nitro Additive *14*
1.3.2.1 Complex Nitrate-Based Additives *14*
1.3.2.2 Complex Nitro-Based Additives *15*
1.3.3 Drawbacks and Solutions of Nitro Additives *15*
1.3.3.1 Low Solubility *15*
1.3.3.2 Sacrificial Additives *17*

1.3.3.3	High Decomposition Activation Energy of LiNO$_3$	18
1.4	Nitrile Additives	19
1.4.1	Functions of Nitrile Additives	19
1.4.1.1	Plasticization	19
1.4.1.2	Facilitation of Ion Transport	20
1.4.1.3	Promotion of Lithium Salt Dissolution	22
1.4.1.4	Widening of the Electrochemical Window	22
1.4.1.5	Inhibiting the Decomposition of the Electrolyte	22
1.4.1.6	Low Flammability	22
1.4.1.7	Improvement of Polymer Flexibility	23
1.4.1.8	Modification of the Cathode Interface	23
1.4.1.9	Involvement in the Solvation Structure of Zn^{2+}	24
1.4.1.10	Weakening of Ionic Association	24
1.4.1.11	Contribution to the Formation of SEI	24
1.4.2	Compatibility Analysis of Nitrile and Lithium Metal	25
1.4.2.1	Incompatibility of Nitrile and Lithium Metal	25
1.4.2.2	Improvement of the Compatibility of Nitrile and Lithium Metal	25
1.4.3	Other Drawbacks of Nitrile Additives	28
1.4.3.1	Low Mechanical Strength	28
1.4.3.2	Prone to Polymerization	28
1.4.3.3	Crystallinity	28
1.5	Phosphate Ester Additives	29
1.5.1	Functions of Phosphate Ester Additives	29
1.5.1.1	Flame Retardant	29
1.5.1.2	Stabilization of Cathodes and Anodes	30
1.5.1.3	Involvement in Solvation Structure Regulation	31
1.5.2	Drawbacks of Phosphate Ester	32
1.5.2.1	Incompatibility with Anodes	32
1.5.2.2	Improvement of the Compatibility of Phosphate Ester and Lithium Metal	32
1.6	Sulfate Ester Additives	34
1.6.1	Functions of Sulfate Ester Additives	34
1.6.1.1	SEI-Forming Additives	34
1.6.1.2	CEI-Forming Additives	36
1.7	Conclusion and Outlook	39
	References	40
2	**Application of Biopolymers in Batteries**	**51**
2.1	Introduction	51
2.2	Overview of Biopolymers	53
2.2.1	Carboxymethyl Cellulose (CMC)	53
2.2.2	Chitosan (CS)	54
2.2.3	Sodium Alginate (SA)	56
2.2.4	Lignin	57
2.2.5	Gum Arabic (GA)	57

2.2.6	Guar Gum (GG)	*59*
2.2.7	Xanthan Gum (XG)	*59*
2.2.8	Starch	*60*
2.2.9	Gelatin	*61*
2.2.10	Tragacanth Gum (TG)	*62*
2.2.11	Cellulose (CLS)	*63*
2.2.12	Trehalose (THL)	*64*
2.2.13	Citrulline (Cit)	*64*
2.2.14	Pectin	*65*
2.2.15	Carrageenan	*66*
2.3	Application of Biopolymers in Binders	*66*
2.3.1	Carboxymethyl Cellulose	*67*
2.3.2	Chitosan	*68*
2.3.3	Sodium Alginate	*70*
2.3.4	Lignin	*72*
2.3.5	Gum Arabic	*74*
2.3.6	Guar Gum and Xanthan Gum	*75*
2.3.7	Starch	*75*
2.3.8	Gelatin	*76*
2.3.9	Tragacanth Gum (TG)	*78*
2.4	Application of Biopolymers in Electrolytes	*79*
2.4.1	Cellulose	*79*
2.4.2	Chitosan	*81*
2.4.3	Lignin	*82*
2.4.4	Gelatin	*85*
2.5	Application of Biopolymers in Electrolyte Additives	*87*
2.5.1	Cellulose	*88*
2.5.2	Trehalose	*88*
2.5.3	Citrulline	*89*
2.5.4	Pectin	*90*
2.6	Application of Biopolymers in Separators	*90*
2.6.1	Cellulose	*91*
2.6.2	Starch	*94*
2.6.3	Carrageenan	*95*
2.7	Application of Biopolymers in Anode Functional Layers	*95*
2.7.1	Cellulose	*96*
2.7.2	Chitosan and Sodium Alginate	*96*
2.8	Conclusion and Outlook	*98*
	References	*100*
3A	**Application of Synthetic Polymers in Batteries: Carbon-chain Polymers** *107*	
3A.1	Introduction	*107*
3A.2	Overview of Synthetic Polymers Materials	*107*
3A.2.1	Polyvinylidene Difluoride (PVDF)	*108*

3A.2.2	Polytetrafluoroethylene (PTFE)	*109*
3A.2.3	Styrene-Butadiene Rubber (SBR)	*110*
3A.2.4	Polyvinyl Alcohol (PVA)	*111*
3A.2.5	Polyacrylics (PA)	*111*
3A.2.6	Polyacrylonitrile (PAN)	*112*
3A.2.7	Polyvinyl Pyrrolidone (PVP)	*113*
3A.2.8	Polyolefin (PO)	*114*
3A.3	Application of Synthetic Polymers in Binders	*115*
3A.3.1	Polyvinylidene Difluoride	*115*
3A.3.2	Polytetrafluoroethylene	*117*
3A.3.3	Styrene-Butadiene Rubber	*118*
3A.3.4	Polyvinyl Alcohol	*119*
3A.3.5	Polyacrylics	*122*
3A.4	Application of Synthetic Polymers in Electrolytes	*124*
3A.4.1	Polyvinylidene Difluoride	*125*
3A.4.2	Polyacrylonitrile	*129*
3A.4.3	Polyacrylics	*131*
3A.4.4	Polyvinyl Alcohol	*133*
3A.5	Application of Synthetic Polymers in Battery Separators	*135*
3A.5.1	Polyolefin	*136*
3A.5.2	Polyvinylidene Difluoride	*138*
3A.5.3	Polyacrylonitrile	*139*
3A.5.4	Polyvinyl Alcohol	*140*
3A.6	Application of Synthetic Polymers in Anodes	*142*
3A.6.1	Polyacrylonitrile	*142*
3A.6.2	Polyacrylics	*143*
3A.7	Conclusions and Outlook	*143*
	References	*145*
3B	**Application of Synthetic Polymers in Batteries: Hetero-chain Polymers**	***155***
3B.1	Introduction	*155*
3B.2	Overview of Synthetic Polymers Materials	*155*
3B.2.1	Epoxy Resin (EPR)	*156*
3B.2.2	Polyethylenimine (PEI)	*157*
3B.2.3	Polyurethane (PU)	*158*
3B.2.4	Polyethylene Oxide (PEO)	*158*
3B.2.5	Polyethylene Terephthalate (PET)	*159*
3B.2.6	Polyimide (PI)	*160*
3B.3	Application of Synthetic Polymers in Binders	*161*
3B.3.1	Epoxy Resin	*161*
3B.3.2	Polyethylenimine	*162*
3B.3.3	Polyurethane	*164*
3B.3.4	Polyimide	*166*
3B.4	Application of Synthetic Polymers in Electrolytes	*167*

3B.4.1	Epoxy Resin	*167*
3B.4.2	Polyurethane	*170*
3B.4.3	Polyethylene Oxide	*173*
3B.4.4	Polyimide	*176*
3B.5	Application of Synthetic Polymers in Battery Separators	*178*
3B.5.1	Polyethylene Terephthalate	*178*
3B.5.2	Polyimide	*179*
3B.6	Conclusions and Outlook	*180*
	References	*182*

4 Application of Nontraditional Organic Ionic Conductors in Batteries *189*

4.1	Ionic Liquids	*189*
4.1.1	Introduction of Ionic Liquids	*189*
4.1.2	Development of Ionic Liquids	*190*
4.1.3	Catalog of Ionic Liquids	*191*
4.1.4	Advantages of Ionic Liquids for Batteries	*193*
4.1.5	Synthesis and Characterization Method of Ionic Liquids	*193*
4.1.6	Application of Ionic Liquids	*194*
4.2	Application of ILs in Batteries	*196*
4.2.1	Ionic Liquid Electrolyte	*198*
4.2.2	Ionic Liquid/Organic Solvent Electrolyte	*206*
4.2.3	Organic–Inorganic Composite Ionic Liquid Electrolyte	*210*
4.3	Single-Ion Conductive	*215*
4.3.1	Introduction of Single-Ion Conductive	*215*
4.3.2	Catalog of Single-Ion Conductive	*216*
4.4	Application of Single-Ion Conductive in Batteries	*217*
4.4.1	Organic Single-Ion Conductor Electrolyte	*217*
4.4.2	Organic–Inorganic Composite Single-Ion Conductor Electrolyte	*225*
4.5	Conclusions and Outlook	*228*
	References	*230*

5 Application of Self-Healing Materials in Batteries *239*

5.1	Introduction	*239*
5.1.1	The Need for Battery Innovation	*239*
5.1.2	Overview of Self-Healing Materials	*239*
5.1.3	Benefits of Self-Healing Technologies in Batteries	*240*
5.1.4	Challenges in Scaling and Commercializing Self-Healing Materials	*242*
5.2	Types of Self-Healing Materials for Battery Applications	*243*
5.2.1	Physically Bonded Self-Healing Materials	*243*
5.2.2	Chemically Bonded Self-Healing Materials	*243*
5.2.3	Composite Self-Healing Materials with Multiple Repair Mechanisms	*244*
5.3	Applications of Self-Healing Materials in Batteries	*244*
5.3.1	Gel Polymer Electrolytes	*244*

5.3.2	Solid Polymer Electrolytes	*256*
5.3.3	Composite Electrolytes	*272*
5.3.4	Electrode Binders	*275*
5.4	Conclusions and Outlook	*281*
	References	*282*

6 Application of Low-Dimensional Materials in Batteries *287*
6.1	Introduction	*287*
6.1.1	Lithium-Metal Batteries	*287*
6.1.2	Low-Dimensional Composite Materials	*288*
6.2	Low-Dimensional Composite Cathode Materials	*289*
6.2.1	Composite Methods for Low-Dimensional Cathode Materials	*290*
6.2.2	One-Dimensional Materials in Cathode	*293*
6.2.2.1	Carbon Nanotube (CNT) Materials	*293*
6.2.2.2	Carbon Nanofiber (CNF) Materials	*297*
6.2.3	Two-Dimensional Materials in Cathode	*299*
6.2.3.1	Graphene Materials	*299*
6.2.3.2	MXene Materials	*304*
6.3	Low-Dimensional Composite Materials in Separators	*309*
6.3.1	Zero-Dimensional Materials in Separators	*310*
6.3.2	One-Dimensional Materials in Separators	*312*
6.3.3	Two-Dimensional Materials in Separators	*315*
6.4	Low-Dimensional Composite Current Collectors	*320*
6.4.1	Design of Current Collector	*320*
6.4.2	Nanocomposite Current Collectors	*322*
6.5	Low-Dimensional Composite Anode Materials	*324*
6.5.1	Formation of SEI and Failure Mechanism	*324*
6.5.2	Nanocomposite Lithium Metal Anodes	*325*
6.5.3	Low-Dimensional Materials in 3D-Printing Anodes	*328*
6.6	Conclusion and Outlook	*329*
	References	*330*

7 Applications of Porous Organic Framework Materials in Batteries *339*
7.1	Introduction	*339*
7.1.1	Overview of Energy Demand and Battery Technologies	*339*
7.1.2	Limitations of Traditional Battery Material	*339*
7.1.3	Potential of Porous Organic Framework Materials for Energy Storage	*340*
7.2	Types of Porous Organic Framework Materials	*341*
7.2.1	Metal-Organic Frameworks (MOFs)	*341*
7.2.1.1	Types of MOFs	*342*
7.2.2	Covalent Organic Frameworks (COFs)	*342*
7.2.2.1	Types of COFs	*343*
7.2.3	Hydrogen-Bonded Organic Frameworks (HOFs)	*343*

7.2.3.1	Types of HOF	*343*
7.3	Applications of Porous Organic Framework Materials in Batteries	*345*
7.3.1	Applications in Electrode Materials	*345*
7.3.1.1	MOF as Electrode Materials	*345*
7.3.1.2	COF as Electrode Materials	*352*
7.3.1.3	HOF as Electrode Materials	*356*
7.3.2	Applications in Electrolytes and Electrolyte Additives	*359*
7.3.2.1	MOF as Electrolytes and Electrolyte Additives	*359*
7.3.2.2	COF as Electrolytes and Electrolyte Additives	*364*
7.3.2.3	HOF as Electrolytes and Electrolyte Additives	*367*
7.3.3	Applications in Catalysts and Catalyst Supports	*368*
7.3.3.1	MOF as Catalysts and Catalyst Supports	*368*
7.3.3.2	COF as Catalysts and Catalyst Supports	*371*
7.3.3.3	HOF as Catalysts and Catalyst Supports	*373*
7.3.4	Applications in Battery Separators	*374*
7.3.4.1	MOF as Battery Separator	*374*
7.3.4.2	COF as Battery Separator	*378*
7.3.4.3	HOF as Battery Separator	*380*
7.4	Conclusion and Outlook	*381*
7.4.1	Conclusion	*381*
7.4.2	Outlook	*382*
	References	*383*

Index *389*

Preface

With the growing shortage of global energy and resources, the development of new energy materials has become the focus of worldwide attention. New energy materials through the transformation and utilization of traditional energy (such as solar energy, biomass energy, geothermal energy, wave energy, ocean current energy, and tidal energy, as well as the thermal cycle between the surface and deep layers of the ocean), as well as the development of new energy technologies, effectively solve the energy crisis problem, to achieve sustainable development. They can help to reduce environmental pollution, reduce greenhouse gas emissions, and protect the ecological environment by replacing traditional fossil fuels. For example, the use of renewable energy sources such as solar and wind reduces reliance on fossil fuels, thereby reducing air pollution and carbon emissions. The development of new energy technologies can effectively solve the problem of energy crisis to achieve sustainable development. The research and application of new energy materials require the cross and integration of multiple disciplines, which promotes the development of materials science and other basic disciplines. With the continuous progress of science and technology and the upgrading of industrial structures, the application of new energy materials will be more and more extensive, and the promoting role of economic development will be more and more obvious. The research and development of new materials will continue to promote the development of new energy technologies and provide strong support for solving global energy problems and environmental problems.

As a secondary energy source, electric energy is widely used because it is easy to convert into other forms of energy, such as heat, wind, potential energy, and kinetic energy. Furthermore, it also has the advantages of convenient transmission, easy production and use, clean, safe, and economical. These advantages make electric energy a very important and widely used form of new energy application in modern society. Batteries play a vital role in the supply, storage, conversion, and management of electrical energy; energy storage batteries and power batteries are two important types of batteries. Energy storage batteries are mainly used for the storage of electrical energy, widely used in solar power equipment and wind power equipment energy storage, as well as renewable energy storage; power batteries are mainly used for electric vehicles, electric bicycles, and other mobile equipment to provide power. The development and application of battery materials have greatly promoted the progress of human society, improved people's way of life, and laid a solid foundation for future scientific and technological development and environmental protection.

Batteries consist of positive and negative active materials and other auxiliary materials, which play a key role in the manufacture and performance of batteries. The active material is the core of the electrochemical reaction of the battery, responsible for the storage and release of energy. The auxiliary material provides the necessary structural support and transport pathway for the active material to ensure the smooth progress of the electrochemical reaction. In summary, the active material is the key to the electrochemical performance of the battery, and the auxiliary material provides the necessary structural and transport support for these active materials, which jointly determines the overall performance and application range of the battery.

Auxiliary materials in batteries mainly include separators, conductive agents, binders, current collectors, electrolytes and electrolyte additives, sealing materials, thermal insulation materials, protective layers, and encapsulating materials. The importance of auxiliary materials in batteries is reflected in the following aspects: (1) they can significantly improve the electrical conductivity and mechanical strength of the battery, thereby improving the overall performance and lifespan of the battery; (2) by enhancing the bonding force and stability between the electrode materials, they help to prevent the battery from short circuit and overheating during the charging and discharging processes, thus ensuring the safety of the battery; (3) they can increase the conductivity of the electrolyte, helping to form a stable solid-electrolyte interface film and reduce the burning tendency of the electrolyte, thereby improving the rate performance, low-temperature performance, and the cycle and thermal stability of the battery; (4) they protect the materials in the battery from adverse external factors through various forms, thereby improving the safety, reliability, and service lifespan of the battery. These auxiliary materials play a key role in the production process of batteries, and their performance and selection have an important impact on the performance, lifespan, and safety of batteries. Continuous research and innovation help to improve these auxiliary materials and improve battery performance and reliability. With the development of technology, it is expected that more new materials and auxiliary materials will be used for power batteries in the future.

This book focuses on functional auxiliary materials in batteries; although these components are not a direct source of battery energy, their presence and performance are critical to improving the overall performance of the battery. Organic functional materials and low-dimensional structural materials as the common auxiliary material compositions are widely used in various components of batteries. The first five chapters of this book expand around the application of organic functional materials in batteries, they usually are used as separators, binders, electrolytes, and functional additives. The last two chapters of this book expand around the application of low-dimensional structural materials in batteries, they mainly are used as conductive agents and functional additives. Specifically, the book is divided into the following main chapters:

Preface | xv

Chapter 1, Application of Organic Functional Additives in Batteries: Electrolyte additives, as the "fine-tuning agent" in the electrolyte system, can considerably improve the performance of both the electrolyte and the battery through the introduction of a small number of functional additives. There are many kinds of these additives, including film-forming additives, flame retardant additives, high-voltage additives, overcharge protection additives, and so on, which improve the performance of lithium batteries in different aspects, thus significantly improving the overall performance and safety of the battery.

Chapter 2, Application of Biopolymers in Batteries: Biopolymers and their derivative materials bring new development opportunities in batteries due to their unique structures and properties. These materials are of natural origin, have renewable and degradable properties, and show excellent energy storage and conversion potential. These materials have been successfully applied in various battery systems, such as lithium-ion batteries and supercapacitors, through delicate chemical design and optimization, with remarkable results. Biopolymer materials are rich in designability and tunability, and by adjusting factors such as their molecular structure, composition, and morphology, they can precisely regulate and optimize battery performance. This provides a broad space and unlimited possibilities for the future development of the battery field.

Chapter 3, Application of Synthetic Polymers in Batteries: Synthetic polymer materials are widely used in batteries due to their diversity of structure and function. For example, they can be used as binders, electrolyte materials, separators, functional coatings, flame retardants, and active material carriers. Due to the specificity of the molecular structure, each polymer has its own unique physical and chemical properties. The advancement of polymer technology focuses on developing corresponding functions to serve the performance requirements of batteries through rational use of their properties. Through the functional composite of a variety of different polymers or the introduction of inorganic functional materials to modify polymers to achieve functional enrichment and integration, polymer materials can achieve diversified development in battery applications. These applications demonstrate the diversity and importance of synthetic polymers in battery technology, which can further optimize battery performance and drive the rapid development of battery technology through molecular design and materials engineering.

Chapter 4, Application of Nontraditional Organic Ionic Conductors in Batteries: Energy storage batteries play a crucial role in modern society, and the main challenge in this research area is how to efficiently transport ions. Traditional organic electrolytes provide good ionic conductivity; however, they also face some major challenges, especially security risks. Ionic liquids and single-ion conductors are considered promising new organic ionic conductor candidates with higher security. Ionic liquids have outstanding physical and electrochemical properties, including high safety (nonvolatile, nonflammable), wide operating temperature range, wide electrochemical window, high ionic conductivity, and excellent compatibility with electrode materials, all of which are beneficial to battery performance. In addition to their excellent thermal and chemical stability, single-ion conductors are characterized by their extremely high ion transport numbers, as they can selectively transfer only cations or anions. Their application as battery electrolytes or electrolyte additives has significantly improved the stability and safety of the batteries.

Chapter 5, Application of Self-Healing Materials in Batteries: This chapter explores the use of self-healing materials in batteries, emphasizing their potential to improve battery performance, safety, and lifespan. The chapter outlines the critical need for battery innovation, focusing on energy density, cost reduction, safety, and sustainability. Self-healing materials, capable of autonomously repairing themselves, enhance battery life by addressing issues such as electrode cracking and internal damage during charge cycles. They also improve safety by preventing failures like fires or explosions. Despite their promise, challenges such as technological complexity, cost, and material compatibility must be overcome to facilitate large-scale commercial applications, ensuring sustainable future energy solutions.

Chapter 6, Application of Low-Dimensional Materials in Batteries: The integration of low-dimensional composite materials in high energy density lithium-metal batteries, such as lithium–sulfur batteries, significantly enhances performance across cathodes, separators, current collectors, and anodes. Carbon-based materials like graphene and carbon nanotubes improve active material loading and interfacial stability, addressing volume expansion issues. Additionally, these materials can introduce lithiophilic sites to mitigate polysulfide shuttle effects, thereby extending cycle life. For inactive components, modified separators effectively block polysulfide migration while facilitating lithium-ion transport. Furthermore, innovative designs in current collectors enhance conductivity and mechanical strength. Future research should prioritize optimizing structural design and material stability to boost energy density and overall battery efficiency.

Chapter 7, Applications of Porous Organic Framework Materials in Batteries: This chapter explores the application of porous organic framework materials in battery technology, covering metal-organic frameworks (MOFs), covalent-organic frameworks (COFs), and hydrogen-bonded organic frameworks (HOFs). These materials, with their adjustable pore structures and superior chemical stability, exhibit excellent electrochemical performance, significantly enhancing the energy density and cycle life of lithium-ion batteries, sodium-ion batteries, and zinc-air batteries. By optimizing pore design and functionalization, porous organic framework materials not only improve the performance of electrode materials but also serve as electrolyte additives and catalyst supports, driving the development of next-generation, high-efficiency, and safe energy storage technologies.

In each chapter, the commonly used functional auxiliary materials in batteries are systematically introduced, including their structure, properties, application progress, and roles, as well as corresponding mechanisms in different systems.

In summary, auxiliary materials in batteries play a crucial role in improving battery performance and safety, reducing costs, and promoting technological innovation. I hope this book will be a tool book for applying organic functional materials and low-dimensional structural materials in batteries by providing a comprehensive summary of the progress in these fields. It not only introduces the properties and preparation methods of these materials but also summarizes the application mechanism and conclusions, and puts forward some insights and prospects. It can help new researchers quickly understand the progress of research in related fields, and also can help senior researchers summarize research experience and get inspiration for innovation. With the further development of battery technology and changes in market demand, the research and application of auxiliary materials will continue to deepen.

November 2024

Wei Hu
USTB, Beijing, China

1

Application of Organic Functional Additives in Batteries

1.1 Introduction

Driven by energy transformation and environmental protection, battery technology has received unprecedented attention as key to energy storage and conversion. Lithium-ion batteries (LIBs) are mainly composed of electrolytes, cathodes, and anodes, of which, for liquid electrolytes, separators are often used as supporting materials. As one of the critical materials for battery manufacturing, the electrolyte is mainly used to construct an ion transport channel between cathodes and anodes inside the battery and is a medium for lithium-ion migration and charge transfer (CT) and known as the "blood" of LIBs.

However, with the continuous progress of battery technology, electrolyte performance requirements are also increasing. The traditional electrolyte system has been gradually challenged to meet the needs of modern high-performance batteries, especially in improving energy density, prolonging cycle life, enhancing safety, and other challenges. To address these issues, research on battery electrolyte additives has emerged as a significant area for the advancement of battery science and technology.

Electrolyte additives, as the "fine-tuning agent" in the electrolyte system, can considerably improve the performance of the electrolyte and the battery by introducing a small number of functional additives. Therefore, employing functional additives in the electrolyte is a crucial way to improve the performance of the battery. Additives used in electrolytes need to meet the following primary conditions: (i) soluble in organic electrolytes; (ii) no apparent side effects with other components of the battery; (iii) small dosage and remarkable effect; (iv) no toxicity or negligible toxicity; and (v) low price.

Currently, some widely used additives are fluorinated compounds, nitro compounds, nitrile compounds, phosphate ester compounds, and sulfate ester compounds. This chapter describes the functions of these compounds in batteries in detail and discusses the possible drawbacks of the additives and their solutions.

Functional Auxiliary Materials in Batteries: Synthesis, Properties, and Applications, First Edition. Wei Hu.
© 2025 WILEY-VCH GmbH. Published 2025 by WILEY-VCH GmbH.

1.2 Fluorinated Additives

1.2.1 Functions of Fluorinated Additives

1.2.1.1 Improvement of Safety Performance

Fluorinated electrolytes, such as fluoroethylene carbonate (FEC), difluoroethylene carbonate (DFEC), and methyl(2,2,2-trifluoroethyl) carbonate (FEMC), have a relatively lower heat release and a higher onset and peak temperature compared to nonfluorinated carbonates, which results in improved safety performance. Zhang et al. [1] used differential scanning calorimetry (DSC) to evaluate the thermal stability of various commonly used electrolytes and summarized the solvent exothermic phase diagram. The results of the self-extinguishing time (SET) tests have shown that fluorinated electrolytes such as bis(2,2,2-trifluoroethyl) carbonate (TFEC) have better thermal stability as well as flame retardancy. In addition, Meng et al. [2] used fluorinated electrolytes, such as fluoromethyl 1,1,1,3,3,3-hexafluoroisopropyl ether (HFE), to reduce flammability and improve the safety performance of liquid electrolytes.

1.2.1.2 SEI-Forming Additives

Due to the strong electron absorption characteristics of the F functional group, the fluorinated electrolyte's lowest unoccupied molecular orbital (LUMO) is generally lower than that of the ordinary electrolyte, which means that the fluorinated electrolyte will preferentially react with the anode to generate LiF over other solvents [3]. The high surface energy and low diffusion barrier of LiF are conducive to promoting the rapid distribution of Li$^+$ parallel to the interface to achieve uniform deposition of lithium, which is considered the most favorable inorganic component in the solid electrolyte interface (SEI) [4, 5]. In addition, LiF is a wide-bandgap insulator that prevents electrons from tunneling through the SEI [6]. The formation of dense and uniform inorganic-rich SEI film, on the one hand, can effectively inhibit the further reaction between the anode and the electrolyte [7, 8]. On the other hand, it can improve the reversibility of the anode to form a stable Li$^+$ plating and stripping process and inhibit the growth of lithium dendrites [9].

Liao et al. [10] observed the presence of LiF in the SEI formed by FEC-added electrolytes using X-ray photoelectron spectroscopy (XPS) and confirmed that FEC with low LUMO has a preferential reduction on lithium metal. In addition to lithium metal anodes, Shin et al. [11] and Li et al. [12] found that FEC on the surface of commercial graphite anodes would also reduce to form a dense, uniform, and LiF-rich SEI, thus enabling a stable graphite–electrolyte interface under low-temperature cycling and conferring excellent cycling stability to the battery. Guo et al. [13] combined classical molecular dynamics (CMD) simulation and XPS analysis of the anode to demonstrate that FEC contributes to the formation of a stable and fluorine-rich SEI and improves the reversibility of Na metal. In addition to the widely studied FEC, fluorinated carbonates such as difluoro-substituted

DFEC [14] and hexafluoro-substituted TFEC [1] are also considered effective solvents for the construction of fluorinated SEIs. The decomposition of DFEC and TFEC can similarly produce the inorganic component LiF to form a dense and homogeneous fluorinated SEI.

The dense SEI formed from the fluorinated electrolyte can inhibit further reactions between the electrolyte and the anode. Nitrile compounds, such as succinonitrile (SN), have been widely used in electrolytes due to the benefits of high ionic conductivity and high oxidation stability, but —C≡N is highly reactive and easily reacts with lithium metal, so it is quite essential to impede the contact between —C≡N and lithium metal. The LiF-rich protective layer formed by the degradation of FEC on the surface of lithium metal can effectively inhibit the side reaction between the cyano group and lithium metal, thus improving the cycle performance of the battery [15].

In addition to the decomposition of fluorinated additives to form LiF to change the composition of the SEI, some studies have also shown that the addition of fluorinated electrolytes will change the structure of the SEI. Li et al. [16] observed individual lithium metal atoms and their interfaces with SEI using cryo-electron microscopy and found that the SEI nanostructure formed with FEC was ordered and multilayered: the inner layer was an amorphous polymer matrix, and the outer layer was Li_2O with a large grain size (~15 nm). In contrast, the SEI without FEC is a mosaic structure. The authors concluded that the SEI with multilayer nanostructures has better mechanical stability and is more favorable for a homogeneous lithium plating–stripping process. In contrast, the mosaic-structured SEI has poor mechanical properties due to the random distribution of inorganic substances, resulting in possible fracture during cycling and the formation of dead lithium due to the uneven distribution of its grains. In addition, Aurbach et al. [6] suggested that the addition of fluorinated electrolytes, such as FEC, will form a highly cross-linked polymer network on the lithium metal surface, which accommodates the volumetric changes in the contraction and expansion of the lithium metal for lithium deposition and dissolution during cycling.

Based on the outstanding properties of fluoride, some researchers have also used fluoride to modify the lithium metal to improve the performance of the battery. Yan et al. [17] constructed a double-layer film structure, with an outer layer dominated by the organic constituents ($ROCO_2Li$ and ROLi) and an inner layer dominated by the inorganic constituents (LiF and Li_2CO_3) on the Li metal surface through the spontaneous reaction of lithium metal in FEC, as shown in Figure 1.1a. The double-layer membrane structure achieves columnar deposition of lithium by pre-forming uniform deposition nucleation sites. Through the characterization of the morphology of lithium metal after cycling, it is found that the surface of the pretreated lithium metal was smooth and dense, and the lithium deposition was uniform, which inhibited the formation of lithium dendrites. Kim et al. [3] pretreated metal fluoride (M_xF_y) onto lithium metal at a low annealing temperature, and M_xF_y reacted and decomposed with Li to form metallic M nanoparticles and

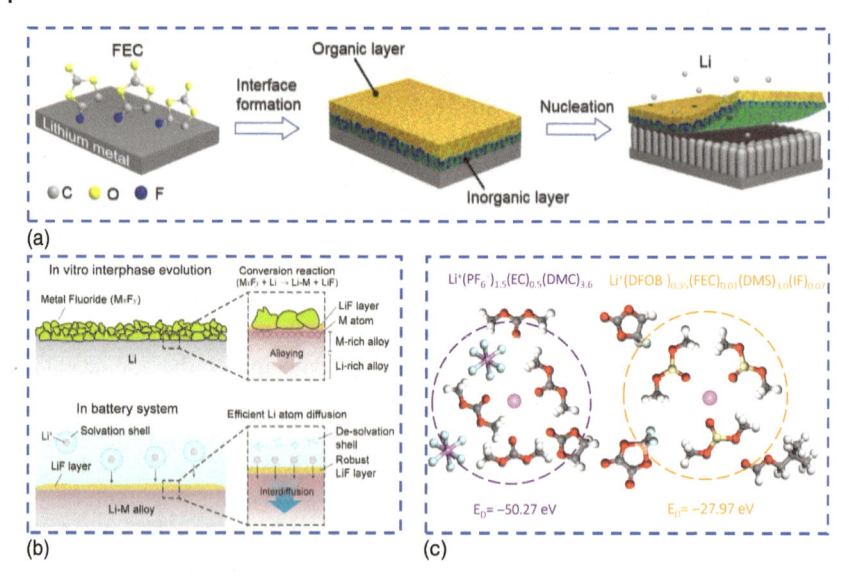

Figure 1.1 Functions of fluorinated additives. (a) The dual-layered film can regulate the uniform deposition of Li ions during repeated charge/discharge cycles and protect the Li metal anode without dendrite formation. Source: Reproduced from Yan et al. [17]/with permission of Wiley-VCH Verlag GmbH & Co. KGaA. (b) Schematic illustration of in vitro interphase evolution employing the conversion reaction of metal fluorides (top) and Li plating/stripping of Li–M alloy with the LiF outermost layer (bottom). Source: Kim et al. [3]/American Chemical Society/CC By NC-ND 4.0. (c) The Li$^+$ solvation structures and corresponding desolvation energies of EC + DMC (left) and 45% IF (right) electrolytes. Source: Reproduced from Liu et al. [18]/with permission from Royal Society of Chemistry.

robust LiF inorganic compounds. The metal M nanoparticles formed a uniform Li–M alloy phase with the Li metal, which can promote the formation of a uniform LiF interphase layer and the uniform diffusion of Li, as shown in Figure 1.1b.

However, at the same time, Kim et al. [3] also proposed that the inorganic-rich SEI formed by organic electrolytes with FEC additives on the surface of Li metal is mainly composed of Li$_2$O rather than LiF and retains an inhomogeneous mosaic-type structure. Tao et al. [19] quantitatively researched the evolution of inactive lithium in lithium-free anode batteries with different electrolytes using mass spectrometry titration and nuclear magnetic resonance (NMR) spectroscopy techniques and found that FEC itself could inhibit the formation of dead lithium metal, but the correlation between LiF formed by the reduction of FEC and dead lithium metal or SEI was weak. Therefore, the assertion that FEC can form a dense and uniform LiF-rich SEI layer needs to be further verified.

1.2.1.3 High Oxidation Stability

Similarly, due to the electron-withdrawing property of the F functional group, the highest occupied molecular orbital (HOMO) of the fluorinated electrolyte is low, indicating high oxidation stability, which can be matched with the high-voltage cathode [20]. Wang et al. [21] proposed a perfluorinated electrolyte FEC/FEMC through potentiostatic testing of graphite|||lithium battery and found that the fluorinated electrolyte showed lower leakage current, which reflects the high-voltage stability of fluorinated electrolyte. The composition and morphology of the cathode electrolyte interface (CEI) layer formed on the recycled graphite revealed that the CEI formed by the fluorinated electrolyte contained less C=O, ROCO$_2$Li, LiF, and Li$_x$PO$_y$F$_z$, suggesting that the reaction between the electrolyte and the graphite cathode was reduced due to the high oxidation stability of the fluorinated electrolyte. It was also observed by transmission electron microscope (TEM) that the fluorinated electrolyte formed a thinner and more uniform CEI.

Although fluorinated electrolytes have relatively low HOMO and are generally challenging to undergo oxidation, some researchers have found that the addition of fluorinated electrolytes may also participate in the formation of CEI. Lu et al. [22] simultaneously introduced 1H,1H,5H-perfluoropentyl 1,1,2,2-tetrafluoroethylether (F-EAE) and FEC into the organic electrolyte of the LiNi$_{0.5}$Mn$_{1.5}$O$_4$ (LNMO)-based battery. The electrochemical floatation tests showed that the addition of fluorinated electrolytes significantly reduced the leakage current and improved the oxidation stability of the electrolyte. TEM also revealed that a uniform and thin passivation layer was formed on the surface of the LNMO cathode with the fluorinated electrolyte. The synergistic effect of the two fluorinated electrolytes formed a modification of the cathode, which could prevent the direct contact between the electrolyte and the active cathode particles and further inhibit the dissolution of the transition metal ions. The XPS results showed that the CEI of the fluorinated electrolyte contained a higher content of F and P substances, and the antioxidant capacity was improved, thus increasing the reversible capacity and the cyclic stability performance of the LNMO cathode. Guo et al. [13] found that the CEI formed by the electrolyte with FEC was thinner and denser. It was also found that this CEI was enriched with NaF by XPS, indicating that FEC was also involved in the formation of CEI, thereby protecting the cathode interface. Nagarajan et al. [23] investigated the composition of CEI at different depths by energy-tunable synchrotron-based hard X-ray photoelectron spectroscopy, and it was also found that the addition of FEC to the carbonate electrolyte was also beneficial in improving the film-forming ability of the cathode.

1.2.1.4 Promotion of the Formation of Anion-Rich Solvation Structure

Compared with the widely used carbonate electrolytes and ether electrolytes, fluorinated electrolytes have a weaker solvation ability. Moreover, the stronger electron-withdrawing ability of fluorine atoms can contribute to the distribution of negative charges, lowering the lattice energy of the salt and facilitating the dissolution of lithium salts in the solvent [24]. Su et al. [25] systematically explored the solvation ability of different electrolytes and found that fluorinated electrolytes have lower solvation ability and are less favorable for coordination with Li$^+$ compared to their nonfluorinated counterparts, but weak solvation can induce more anions to participate in the solvated structure. Zhang et al. [1] compared the solvation ability of ethyl methyl carbonate (EMC), FEC, and TFEC with Li$^+$, and the results showed that the binding energy with Li$^+$ decreased sequentially with the increase of the degree of fluorination, indicating that TFEC is a weakly solvating solvent and rarely participated in the solvation structure of Li⁺. With the introduction of TFEC, the interaction of Li⁺ with the solvent is weakened, but the interaction with the anion is enhanced, again proving that the fluorinated electrolyte can promote the formation of an anion-rich solvation structure. Liu et al. [18] also proposed that the ester-based electrolyte exhibited a weak solvation structure with a low coordination number at low temperatures, and the FEC was free and hardly coordinated with Li⁺, as shown in Figure 1.1c.

In addition, some researchers have suggested that the reason why FEC can decompose to form LiF-rich SEI is also related to the participation in solvation structure. Chen et al. [26] suggested that FEC was selected as the internal solvation complex, thus forming the fluorinated SEI. Su et al. [27] systematically explored the solvation pattern of the SEI-forming agent, FEC, in the electrolyte system and found that the Li⁺ was solvated by at least one FEC molecule on average to ensure the formation of stable SEI. If the solvation number of FEC is <1, other organic electrolyte molecules coordinated with Li⁺ would decompose to form unfavorable SEI. Only when the solvation number of FEC is ≥ 1, almost all lithium complexes can be preferentially reduced during the formation process to construct fluorinated SEI.

1.2.1.5 Reduction of Desolvation Barrier

The desolvation energy of lithium is also reduced due to the weak solvation of the fluorinated electrolyte. Therefore, the addition of fluorinated electrolytes is expected to reduce the desolvation energy and accelerate the transport kinetics of Li⁺ in SEI [28]. Zhang et al. [20] showed that for the electrolyte, from nonfluorinated to perfluorinated electrolyte, the binding energy of solvent molecules to Li⁺ decreased, leading to a lower coordination number. The perfluorinated electrolyte optimized the Li⁺-dipole structure and accelerated the desolvation process of solvated Li⁺, which resulted in the generation of SEI with low transport resistance during the plating/stripping process.

1.2.2 Synergies of Fluoroethylene Carbonate with Other Compounds

As the most widely used fluorinated electrolyte, FEC has many advantages. Furthermore, FEC may also have synergistic effects when used with other electrolytes to improve the electrochemical performance of the battery.

1.2.2.1 Fluoroethylene Carbonate and Other Fluorinated Electrolytes

Both FEC and DFEC are fluorinated electrolytes with excellent lithium anode stability. DFEC further reduces the solvation ability of carbonyl oxygen due to the two strong electron-withdrawing fluorine atoms located on both sides of the carbonate [29], so the solvation energy of DFEC is even lower compared to FEC. As DFEC contains two fluorine atoms, its LUMO is also further reduced, which preferentially undergoes reduction on lithium metal over FEC. Aurbach et al. [6] found that in high-voltage Li∥NMC batteries, in the presence of only FEC, the oxidative decomposition products of the electrolyte diffused from the cathode and ultimately to the lithium metal anode, which produced a thicker and more resistive surface film. However, in the presence of both FEC and DFEC, the DFEC with lower LUMO decomposed and passivated on the Li anode, and then, the FEC acted as a healing agent to continuously "repair" the SEI on the Li anode in the subsequent cycles, which reduced the consumption of FEC.

The bis(trifluoroacetyl)amine (BTFA) molecule has lower LUMO with more fluorine atoms than FEC. Therefore, BTFA will preferentially decompose over FEC to form fluorinated SEI. Wang et al. [30] designed an in situ generation of an atomically rooted SEI (R-SEI) based on the synergistic interaction of BTFA and FEC. The results of ab initio molecular dynamics (AIMD) simulations showed that in the presence of only FEC, the F atoms generated from the C—F bonds in FEC preferred to stay on the surface of Na (110); in the presence of only BTFA, the F atoms within the molecule were released into the Na interior within 200 fs. However, in the presence of both BTFA and FEC, BTFA induced the F atoms of FEC to enter the inner layer of Na, forming a vertical fluorine concentration gradient. The XPS results showed that the content of inorganic species in R-SEI continued to increase with the sputtering depth, while the content of organic species continued to decrease, suggesting that the solution containing both BTFA and FEC realized a large number of F atoms implanted from the outer layer to the inner layer, forming a multilayer SEI, which was conducive to the improvement of the cycling stability performance of batteries.

1.2.2.2 Fluoroethylene Carbonate and Lewis Base

FEC is a Lewis acid that can accept electron pairs from a Lewis base in the electrolyte environment to form a Lewis acid–base complex. The complex not only can retain its respective functions but also have some synergistic effects. Yang et al. [31] introduced Lewis acid, FEC, and Lewis base, tris(trimethylsilyl) phosphite (TMSP), in carbonate electrolyte, where FEC reacted with TMSP by in situ complexation to form a TMSP–FEC complex. TMSP can be used as an impurity scavenger and a CEI-forming additive, while FEC is an SEI-forming additive, and the two synergistically formed an inorganic–organic composite (F/P/Si-rich SEI) and a highly stable CEI. TMSP–FEC complex effectively protected the cathode and the anode and improved the comprehensive performance of the battery, as shown in Figure 1.2.

1.2.2.3 Fluoroethylene Carbonate and Glyme

FEC is thought to reduce to form LiF-rich SEI during the cycling process, but since its structure is similar to that of vinyl carbonate, it is expected to produce some carbon-rich organic substances, such as $ROCO_2Li$. However, too much carbon-rich organic substance may cover up the LiF-rich SEI formed earlier. As a high donor number (DN) solvent, Glyme has a solubilizing effect on substances rich in C—C—O [32]. Therefore, a certain amount of diethylene glycol dimethyl ether (G2) can dissolve undesired carbon-rich substances on SEI, thereby increasing the content of inorganic compounds in SEI and stabilizing the composition of SEI. Biswal et al. [33] introduced both FEC and G2 into a carbonate-based electrolyte and demonstrated using XPS that the addition of G2 reduced the contents of Li_2CO_3 and C—C—O in SEI, decreasing the activation energy of SEI and CT and increasing the SEI diffusivity and exchange current. In summary, the synergistic effect of FEC and G2 can form

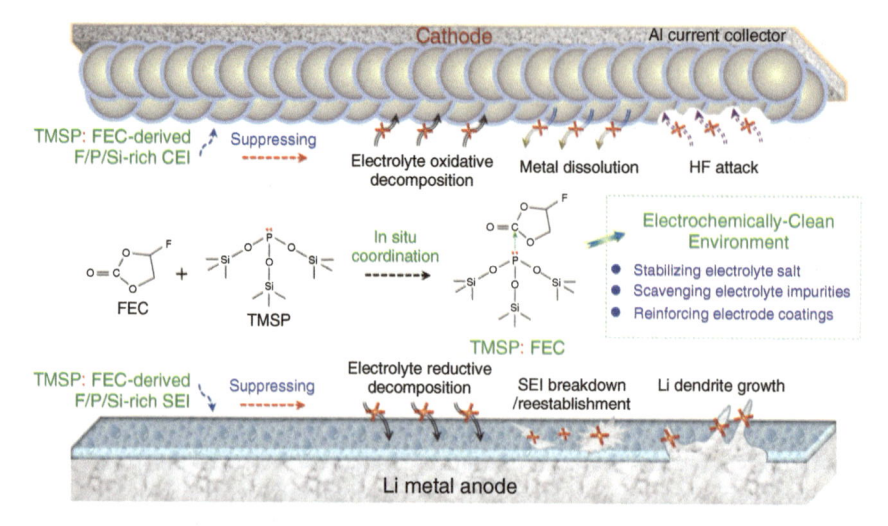

Figure 1.2 Design concept of TMSP–FEC complex as a multifunctional electrolyte additive for lithium metal batteries (LMBs). Source: Reproduced from Yang et al. [31]/with permission of Elsevier.

a stable fluorinated SEI rich in polyene networks, promote the migration of Li$^+$ in SEI, and facilitate the uniform deposition of Li$^+$.

1.2.3 Drawbacks of Fluoroethylene Carbonate

1.2.3.1 Generation of HF Gas

Under the catalysis of Lewis acid PF_6^- or high temperature, FEC is susceptible to dehydrofluorination to produce hydrogen fluoride (HF). On the one hand, the by-product HF will destroy the SEI film, and in the battery with silicon anode, it will corrode silicon particles [34]; on the other hand, HF will lead to the formation of a thicker CEI and catalyze the dissolution of Mn in the case of ternary cathode [11], in addition to corroding the collector aluminum foil [35]. Thus, the generation of HF gas causes irreversible damage to both the anode and cathode, ultimately resulting in a degradation of the cycling performance of the battery.

1.2.3.2 Increase of Impedance and Loss of Impedance

Some researchers have argued that although FEC facilitates the formation of SEI at the anode, it forms a thicker CEI film on the surface of the cathode, which increases the resistance and capacity loss of the battery. Yang et al. [35] found that the initial capacity of Li||LiCoO$_2$ (LCO) batteries was significantly reduced after the addition of the FEC additive, which attributed to the excessive decomposition of FEC and resulted in the formation of a thick interfacial layer on the surface of the cathode that hindered the lithium-ion transport.

1.2.3.3 Incompatibility with Other Electrodes

FEC is widely used on lithium metal and silicon anodes, but FEC cannot form stable SEI on graphite anodes. Shen et al. [36] studied the SEI formed on graphite anodes in different electrolyte systems and found that the SEI formed with the addition of FEC was thicker and denser than that formed with the addition of ethylene carbonate (EC) under the same conditions. Xia et al. [37] investigated the compatibility of FEC-based electrolytes with graphite anodes for the first time. However, the FEC additive was unable to form a protective SEI on the graphite surface because the introduction of F atoms lowered the LUMO of FEC, resulting in a higher reduction potential of FEC than its fluorine-free counterpart EC. However, lithium bis(oxalate)borate (LiBOB) and lithium difluoro(oxalato) borate (LiDFOB), as SEI-forming additives introduced to the electrolyte, could effectively inhibit the reduction of FEC, forming a thin and robust SEI on the graphite anode.

In addition to graphite anodes, for some high-voltage cathodes, FEC may likewise be detrimental to the long-term stable performance of batteries. Aktekin et al. [38] investigated the effect of the FEC additive on LNMO-$Li_4Ti_5O_{12}$ (LTO) cells. The XPS results indicated that with the increase in the FEC content of the electrolyte, the thickness of the formed CEI increased, and the content of organic substances containing C—C—O also increased, contrary to previous results of the formation of LiF-rich SEI. Cycling and rate performance tests also showed that the addition of FEC did not improve the discharge capacity and the cycling stability of the battery but rather negatively affected the long-term stability of the battery system with a high-voltage LNMO cathode.

1.2.3.4 Recycling Issues

Due to their extreme persistence and challenge in biodegrading, halogenated organic pollutants pose a major threat to human health, the ozone layer, and ecological safety when released into the environment. Fluorine is one of the halogen elements with the most environmental impact among them. Therefore, fluorinated additives are detrimental to the environment in the long run. Consequently, it is imperative to both strictly recycle and dispose the used batteries with fluorinated electrolytes and to design and develop new types of fluorine-free electrolytes that will be more conducive to environmental safety and battery recycling [24].

1.3 Nitro Additive

1.3.1 Functions of Nitro (NO_3^-)

1.3.1.1 Participation in Solvation and Desolvation Structures

$LiNO_3$ is a lithium salt with a high DN (22.2 kcal mol^{-1}) [39]. The binding energy of NO_3^- to Li^+ is usually higher than that of conventional lithium salt anions as well as solvent molecules [10]. Therefore, when $LiNO_3$ is dissolved into the electrolyte, NO_3^- is able to expel the solvent molecules from the solvation sheath and preferentially participates in the lithium-ion inner solvation structure, forming an anion-rich solvation environment [40–43]. Zhu et al. [44] confirmed that the addition of NO_3^- altered the solvation structure of Li^+ through Raman spectra, confirming a decrease in the number of both vinylene carbonate (VC) and dimethyl carbonate(DMC) in the solvation structure, as well as a change in the intensity of the peaks in the infrared spectrum. In addition, Wahyudi et al. [45] found that the addition of NO_3^- shifted the peaks of ^7Li NMR toward more positive values, demonstrating the shielding of the electron cloud around Li^+, indicating a larger solvated cluster scale, as well as the entry of the electron-donating anion, NO_3^-, into the solvation sheath of Li^+.

Furthermore, NO_3^- enhances the interaction of $TFSI^-$ with Li^+. In the presence of NO_3^-, the number of free $TFSI^-$ in the electrolyte decreases due to an increase in the number of $TFSI^-$ participating in the Li^+ solvation process. Wahyudi et al. [45] detected the presence of the $TFSI^-$ anion aggregated ion pairs at 747 cm^{-1} in Raman spectroscopy, indicating that NO_3^- enhances the $TFSI-Li^+$ interaction. Fu et al. [46] also demonstrated using Fourier transform infrared (FTIR) and Raman spectra that the addition of NO_3^- caused the peaks to be blueshifted, suggesting that $TFSI^-$ existed more in the contact ion-pair (CIP) or aggregated state.

High-concentration electrolyte (HCE) is considered to preferentially compete with solvent molecules due to the increase in salt concentration, forming an anion-rich solvation structure [47]. However, as the salt concentration in HCE increases, the electrolyte becomes more viscous. To improve the overall performance of the electrolyte, diluents such as hydroflurane need to be added to reduce the viscosity to form localized high-concentration electrolytes (LHCEs). However, the aforementioned measures undoubtedly increase the cost of electrolytes [48, 49]. In contrast, the application of NO_3^- additives not only enables NO_3^- itself to participate in the solvation of Li^+ preferentially but also promotes the interaction between $TFSI^-$ and Li^+, thus achieving an anion-rich solvation environment while ensuring low cost and hardly changing the viscosity of the electrolyte, which is a favorable pathway to realize anion-derived interfacial chemistry.

Additionally, NO_3^- not only participates in the formation of solvation structure but also modulates the distance between the Li^+–solvent–anionic complex and the electrode surface to regulate the properties of the electrolyte and stability of the electrode, thus in turn affecting the thermodynamic and kinetic properties of the Li^+–solvent–anionic complex during desolvation process at the electrode interface [42]. Anion-rich solvation structures can lower the desolvation energy barrier and promote the Li^+ desolvation behavior [41]. Stuckenberg et al. [50] demonstrated that higher oxidation currents with $LiNO_3$ in cyclic voltammetry (CV) tests imply faster kinetics of the lithium electrodeposition/dissolution process, that is, the Li^+ desolvation behavior is effectively enhanced. The additive can change the solvation structure and interface model to promote the desolvation of Li^+. The "distance" between the Li^+–solvent–anionic complex and the electrode surface is also a crucial aspect that affects the stability of the electrolyte and electrode.

1.3.1.2 Formation of Inorganic-Rich SEI

Solvation structures are considered to be precursors of SEI, so anion-rich solvation structures have a greater tendency to form anion-derived SEI. However, it is generally believed that the solvent-rich solvation structures are more likely to generate organic-rich SEI, and such organic substances tend to have low ionic conductivity and mechanical properties, which is detrimental to the stability of SEI.

The anion-rich solvation structures are more likely to generate inorganic-rich SEI with superior ionic conductivity and mechanical properties to form a stable SEI [51].

LiNO$_3$ with low LUMO can preferentially undergo a reduction reaction at the anode over solvent molecules [52]. Reduction reactions tend to go through two processes [53]:

$$\text{LiNO}_3 + 2\text{Li} \rightarrow \text{LiNO}_2 + \text{Li}_2\text{O} \quad \Delta H = -406 \text{ kJ mol}^{-1} \quad (1.1)$$

$$\text{LiNO}_2 + 6\text{Li} \rightarrow \text{Li}_3\text{N} + 2\text{Li}_2\text{O} \quad \Delta H = -857 \text{ kJ mol}^{-1} \quad (1.2)$$

Eight electrons from lithium metal are required for complete decomposition, and the reduction decomposition of NO$_3^-$ anion can lead to more LiN$_x$O$_y$ components in the anode interface [54, 55]. Ma et al. [52] confirmed the occurrence of the aforementioned reactions by XPS characterization of SEI, where Li$_3$N, LiN$_x$O$_y$, and Li$_2$O were observed simultaneously, generating inorganic-rich and robust SEI to achieve better protection of lithium electrodes. Thus, an inorganic-rich SEI layer was formed by the reduction decomposition of LiNO$_3$ with low LUMO. Since NO$_3^-$ enhances the interaction of TFSI$^-$ with Li$^+$, more TFSI$^-$ is involved in the solvation structure and ultimately decomposed on the SEI to produce the favorable inorganic component LiF. Zhang et al. [56] found that the content of Li$_3$N and LiF in the SEI was increased after the introduction of LiNO$_3$ using XPS, indicating that LiNO$_3$ may promote the formation of the interface layer rich in Li$_3$N–LiF. It was further verified that the introduction of NO$_3^-$ also promoted the decomposition of TFSI$^-$. Additionally, the authors simulated the structural configuration changes near the Li anode by AIMD to elucidate the potential mechanism of LiF formation. It was found that the bond lengths of both C—S and C—F bonds were elongated from 0 to 20 ps after the addition of LiNO$_3$ compared to those without LiNO$_3$, which reduced the energy required for bond breaking and accelerated the decomposition of lithium bis(trifluoromethanesulphonyl)imide (LiTFSI). It was demonstrated that LiNO$_3$ led to the breakage of C—S and C—F bonds of LiTFSI molecules, inducing a large amount of LiF generation in the SEI.

Zhu et al. [44] also succeeded in forming a lithium-indium alloy on the lithium metal anode surface with the introduction of In(NO$_3$)$_3$ additive to change the composition of SEI and improve the electrochemical performance of the battery. Kim et al. [57] introduced both AgNO$_3$ and LiNO$_3$ as electrolyte additives to construct SEI with a lithophilic inner layer and a compositionally regulated outer layer on the lithium metal surface sequentially according to their LUMO energy levels. AgNO$_3$ preferentially deposited on the lithium metal surface to form Ag and Ag$_2$O due to the lower LUMO energy level (−3.185 eV) to form an inner layer of Ag-based SEI, which significantly reduced the overpotential of the full battery Li||NCM84.

Due to the high ionic conductivity of the decomposition products of NO$_3^-$, the transport of Li$^+$ in SEI and the diffusion kinetics are improved [58]. Inorganic components with high ionic conductivity can effectively reduce lithium nucleation overpotential, leading to larger grain size, which will grow laterally at low density. This growth mode is conducive to reducing the formation of "dead lithium" during lithium plating/stripping, inducing uniform lithium deposition and inhibiting the growth of lithium dendrites [52, 59, 60]. Liu et al. [61] found that lithium deposition on the Cu surface without LiNO$_3$ was loose dendrite, while lithium deposition with LiNO$_3$ was in the form of dense lumps with Li particles growing along the planar direction, indicating that NO$_3^-$ can induce uniform lithium deposition behavior. The inorganic-rich SEI with outstanding ionic conductivity and mechanical strength can reduce the interface impedance, improve the interface contact performance, ensure the stability of the SEI [10, 62], and inhibit the decomposition of solvent molecules [46, 63].

1.3.1.3 CEI-Forming Additives

LiNO$_3$ can also decompose on the cathode surface to form the CEI to ensure the stability of the cathode and the electrolytes. Fu et al. [46] determined that the addition of LiNO$_3$ resulted in the appearance of small oxidation peaks at approximately 5.2 V using linear sweep voltammetry (LSV), which corresponded to the decomposition of LiNO$_3$. XPS analysis of the CEI showed that the electrolyte with LiNO$_3$ formed a CEI film with a higher F content compared to the electrolyte without LiNO$_3$. In the subsequent static leakage current test, the steady-state oxidative decomposition current of the electrolyte with LiNO$_3$ was lower than that of the electrolyte without LiNO$_3$, which demonstrated that the CEI film with LiNO$_3$ could inhibit the decomposition of the electrolyte on the cathode surface of the NMC811 and reduce the accumulation of high-resistance decomposition products. The CEI membrane impedance (R_{CEI}) and charge transfer impedance (R_{ct}) of the electrolyte with LiNO$_3$ were lower, indicating that LiNO$_3$ accelerated CT kinetics.

Zhu et al. [44] also found that the addition of LiNO$_3$ altered the components of CEI and inhibited the further decomposition of the subsequent electrolyte using XPS. Fang et al. [41] conducted XPS and TEM to clarify the chemical composition and structure of the CEI layer and observed the presence of LiN$_x$O$_y$, a fast lithium-ion conductor, in the CEI layer. Similar to the SEI, the inorganic-rich CEI layer also significantly promoted the interfacial ion diffusion behavior. In addition, only a slight M—O signal was detected in the experimental group with the addition of LiNO$_3$, whereas a stronger M—O signal peak appeared in the control group, indicating that the LiNO$_3$ additive can not only inhibit the decomposition of the electrolyte but also inhibit the dissolution of the transition metal by hindering the vicious reaction between the solvent and the cathode to ensure the structural stability of the cathode.

LiNO$_3$ also contributes to the stability of the cathode interface through the formation of an electric double layer (EDL). Wen et al. [64] found that NO$_3^-$ exhibited a distinct voltage response effect, that is, it would be enriched at the cathode interface once the cathode was charged, thus forming Li$^+$-enriched, thermodynamically favorable EDL with solvent molecules well-coordinated (Figure 1.3). The EDL dramatically accelerated the interfacial reaction kinetics and significantly improved the thermodynamic compatibility between carbonate electrolytes and high-voltage LiTiMnO (LTMO) cathodes.

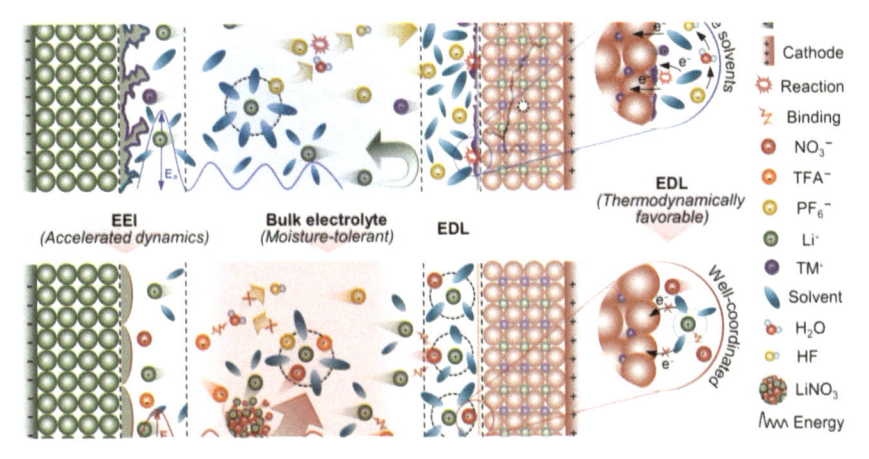

Figure 1.3 Schematic illustration of the reinforced mechanism of the LiTFA–LiNO$_3$ dual-salt additive on conventional carbonate electrolyte. Source: Reproduced from Wen et al. [64]/with permission of Wiley-VCH Verlag GmbH & Co. KGaA.

1.3.1.4 Functions in Lithium–Sulfur Batteries

NO$_3^-$ is also an indispensable additive in lithium-sulfur batteries. It was found that LiNO$_3$ could catalyze the conversion of soluble lithium polysulfide (LiPS) to slightly soluble sulfur as a redox intermediate on the cathode near the end of the charging process (>2.5 V vs. Li/Li$^+$), which inhibited the generation and deposition of polysulfide (PS) and prevented it from dissolving into the electrolyte [39, 45, 65]. Meanwhile, LiNO$_3$ will have a coupling reaction with LiPS to generate a dense passivation layer of LiN$_x$O$_y$ and LiS$_x$O$_y$ on the surface of the lithium metal anode, inhibiting the side reaction of LiPS on lithium metal, reducing the formation of dendrites from PS shuttles, and accelerating the redox kinetics [66].

Kim et al. [67] discovered that NO$_3^-$ can inhibit PS agglomeration through strong coordination with Li$^+$. The authors investigated the low-temperature discharge behavior and found that the second discharge plateau of the electrolyte without LiNO$_3$ disappeared, indicating that the low-order PS formed before the second voltage plateau could not be further reduced to solid Li$_2$S. The researchers regarded PS aggregation as the conversion block from LiPSs to Li$_2$S. In contrast, the electrolyte with LiNO$_3$ showed excellent discharge behavior, and the second voltage plateau of the electrolyte became longer with the increase in LiNO$_3$ content. It suggests that the electrolyte with a high DN value of LiNO$_3$ can promote the conversion of LiPS to Li$_2$S at low temperatures and reduce electrode passivation. Furthermore, the authors demonstrated by density flooding theory (DFT) calculations and molecular dynamics (MD) simulations that anionic NO$_3^-$ with high DN can inhibit PS agglomeration and promote redox kinetics at low temperatures by strongly coordinating with Li$^+$, thus improving the low-temperature performance of the electrolyte.

1.3.1.5 Stabilization of Water Molecules

Zhang et al. [68] found that LiNO$_3$ can stabilize water molecules through strong hydrogen bonding interactions and inhibit the hydrolysis of PF$_6^-$ anions, thus suppressing the formation of highly corrosive HF. Through MD simulations, the authors discovered strong interactions between H$_2$O and LiPF$_6$, including O···Li$^+$ coordination and O—H···F hydrogen bond, which could induce the formation of HF. However, since the solubility of LiNO$_3$ in water is much higher than that in carbonate solvents, H$_2$O molecules are surrounded by NO$_3^-$ anions after the introduction of LiNO$_3$. DFT calculation results showed that the binding energy of H$_2$O–LiNO$_3$ (−0.27 eV) was more negative than that of H$_2$O–LiPF$_6$ (−0.17 eV), which further illustrated that with the introduction of LiNO$_3$, H$_2$O molecules would preferentially coordinate to LiNO$_3$, rather than LiPF$_6$.

All H atoms in H$_2$O can form hydrogen bonds with NO$_3^-$, indicating that NO$_3^-$ anions can effectively stabilize H$_2$O molecules, thereby preventing the hydrolysis of PF$_6^-$ and the formation of HF. Therefore, LiNO$_3$ can effectively stabilize the aqueous electrolyte of the lithium metal battery, avoiding the deterioration of battery performance due to the small amount of water molecules in the electrolyte.

1.3.2 Organic Nitro Additive

1.3.2.1 Complex Nitrate-Based Additives

In addition to the basic metal nitrate salts, many researchers have complexed NO$_3^-$ into organic systems. Wang et al. [69] and Hou et al. [54] proposed to link NO$_3^-$ with structural units containing ether functional groups to form an organic nitrate: isosorbide nitrate (ISDN). The solubility of ISDN in the ester-based electrolyte was dramatically increased to 3.3 M due to the high DN value of the ether functional group. Moreover, the combination of ether functional groups broke the chemical resonance inside NO$_3^-$ and enhanced the anti-reducibility of NO$_3^-$. ISDN can also be decomposed to produce LiN$_x$O$_y$-rich SEI, resulting in uniform lithium deposition.

Wang et al. [70] also complexed NO$_3^-$ with oxygen-containing functional groups to develop a novel nitrate-based additive, triethylene glycol dinitrate (TEGDN), to replace LiNO$_3$. Unlike LiNO$_3$, TEGDN would not interfere with the polymerization of DOL but similarly promoted the formation of a nitrogen-rich SEI layer, which inhibited the parasitic reaction and improved the Coulombic efficiency.

Some other researchers have prepared organic nitrate additives with good solubility in carbonate-based electrolytes by complexing NO$_3^-$ with ionic liquids. Huang et al. [43] combined 1-ethyl-3-methylimidazolium cation [Emim$^+$] with NO$_3^-$ to develop a novel type of ionic liquid, which facilitated the formation of special Li$^+$ coordination dissolved structure of NO$_3^-$, allowing the dissolved NO$_3^-$ to undergo electrochemical reduction and form a stable and conductive SEI. Similarly, Ma et al. [52] complexed 1-methyl-1-decylpyrrolidine cation [Py110$^+$] with NO$_3^-$ to synthesize an ionic liquid electrolyte additive, which could also be dissolved in the electrolyte without adding a solubilizer. Moreover, NO$_3^-$ could be introduced into the Li$^+$ solvation sheath of carbonate electrolyte to form an anion-rich solvation structure. Adiraju et al. [71] combined LiNO$_3$ with 1-trimethylsilyl imidazole [1-TMSI] groups with similar ionic liquid cationic structures to synthesize a soluble carbonate-based electrolyte additive, TMSILN. The solubility of LiNO$_3$ was improved by the bonding interaction between 1-TMSI and LiNO$_3$. The incorporation of TMSILN improved the rate of performance and reduced the overpotential. Ex situ surface analysis by XPS, FE-SEM, and cryo-TEM showed that thin, uniform SEI containing nitrate reduction products could be generated on lithium metal anode by introducing TMSILN additive to the electrolyte.

1.3.2.2 Complex Nitro-Based Additives

Furthermore, nitro, rather than nitrate, can improve the electrochemical performance of batteries, so some researchers have concentrated their studies on nitro-derived additives. Jiang et al. [72] introduced a nitro-C60 derivative (C60(NO$_2$)$_6$) as a bifunctional additive into the electrolyte. Nitro-C60 can gather on electrode protuberances via electrostatic interactions and then be reduced to NO$_2^-$ and insoluble C60. Next, the C60 anchors on the uneven groove of the lithium surface, resulting in a homogeneous distribution of Li ions. Finally, NO$_2^-$ anions can react with metallic Li to build a compact and stable SEI with high ion transport. The nitro and C60 acted synergistically to achieve an inorganic-rich SEI as well as a uniform lithium-deposited surface.

1.3.3 Drawbacks and Solutions of Nitro Additives

1.3.3.1 Low Solubility

The most significant drawback of LiNO$_3$ is its poor solubility ($<10^{-5}$ g ml^{-1}) in ester-based electrolytes. The DN of conventional ester solvents is smaller than that of NO$_3^-$; therefore, the electrostatic interactions between Li$^+$ and NO$_3^-$ are challenging to break in ester-based electrolytes [73]. The macroscopic result is the difficulty of dissolving LiNO$_3$, which dramatically limits its application in the carbonate electrolyte. Many researchers have focused on the solubilization measures of LiNO$_3$ in ester-based electrolytes, mainly in the following ways:

1) Addition of solubilizers with a high DN value

    Introducing solvents with high DN to the electrolyte, such as ethers [74, 75] and sulfolane [46], can effectively improve the solubility of LiNO$_3$ in the ester-based electrolyte. However, solvents with high DN values tend to have strong solvation ability as well, which may compete with NO$_3^-$ to participate in the solvation structure of Li$^+$, thus weakening the solvation effect of NO$_3^-$. In addition, solvents with high DN values may be incompatible with lithium metal, as well as increase the complexity of operation and experimental cost [44]. Wen et al. [75] introduced tetraethylene glycol dimethyl ether (G4) as a solubilizer of LiNO$_3$ and proposed a "high-concentration additive" strategy for lithium nitrate. The DN value of G4 is much higher than that of carbonate solvent, and G4 has abundant solvation sites, which can significantly improve the solubility of LiNO$_3$.

Meanwhile, ethers have good compatibility with lithium metal, which can avoid violent reactions with lithium metal anode. Even in the presence of G4 with a high DN value, the strategy of "high-concentration additive" ensures that NO_3^- participates in the solvation of Li + and forms SEI rich in inorganic substances.

2) Introduction of carrier salts with Lewis acid sites

The electron-deficient Lewis acid can act as an acceptor of electron-rich NO_3^- to decompose $LiNO_3$ clusters, thereby effectively improving the solubility of nitrate. Some metal ions, such as Al^{3+}, Sn^{2+}, Cu^{2+}, In^{3+}, Ag^+, Mg^{2+}, and Zn^{2+}, and some weak Lewis acid groups, such as lithium tetrafluoroborate ($LiBF_4$), have been researched to dissolve $LiNO_3$ in carbonate electrolyte to form nitrogen-derived SEI. Yan et al. [76] achieved the dissolution of 1.0 wt% $LiNO_3$ in carbonate electrolytes by adding trace amounts of CuF_2. $LiNO_3$ and CuF_2 were added to the electrolyte to achieve simultaneous dissolution, and the blue color of the solution was the result of the action of copper ions. However, due to its involvement in SEI formation, the metal cation may be deposited on the lithium metal anode and separator, posing a risk of short-circuiting [77]. Therefore, the addition of Lewis acid should be considered to promote the formation of robust SEI and dense lithium deposition along with the solubilization of $LiNO_3$.

3) Introduction of ions that can coordinate with Li^+ in lithium nitrate

Lewis acid can improve the solubility of $LiNO_3$ through electron-rich NO_3^-, then the same can be done through the Li^+ of $LiNO_3$ by introducing groups that can be strongly coordinated with Li^+ to achieve the decomposition of $LiNO_3$ clusters. Gao et al. [78] and Fang et al. [41] both proposed the introduction of trifluoroacetate anion (TFA^-) to facilitate the dissolution of $LiNO_3$ in carbonate electrolytes. TFA^--containing additives are readily soluble in carbonate electrolytes and have better compatibility with lithium metal compared to solvents with high DN. In particular, the carbonyl group (C=O) of TFA^- can strongly coordinate with Li^+ to promote the dissolution of $LiNO_3$, thus optimizing the solvation structure and transport kinetics of Li^+.

4) Entropy-driven solubilization strategy

Jin et al. [79] correlated the fundamental role of entropy with the limited $LiNO_3$ solubility and proposed a new low-entropy-penalty design that achieves high intrinsic $LiNO_3$ solubility in ester solvents by employing multivalent linear esters (Figure 1.4). The authors concluded that the low-polarity DMC molecules would lose a large amount of conformational entropy during the process of binding with Li^+, forming a transient, unstable solvation structure, which hindered the solubilization of $LiNO_3$. However, multivalent linear ester molecules provided multiple ester-group binding sites, effectively reducing the entropy penalty of the coordination process and thus forming a stable solvation structure. In this way, $LiNO_3$ can directly interact with the primary ester solvents and fundamentally alter the electrolyte properties, resulting in substantial improvements in lithium-metal batteries with high Coulombic efficiency and cycling stability.

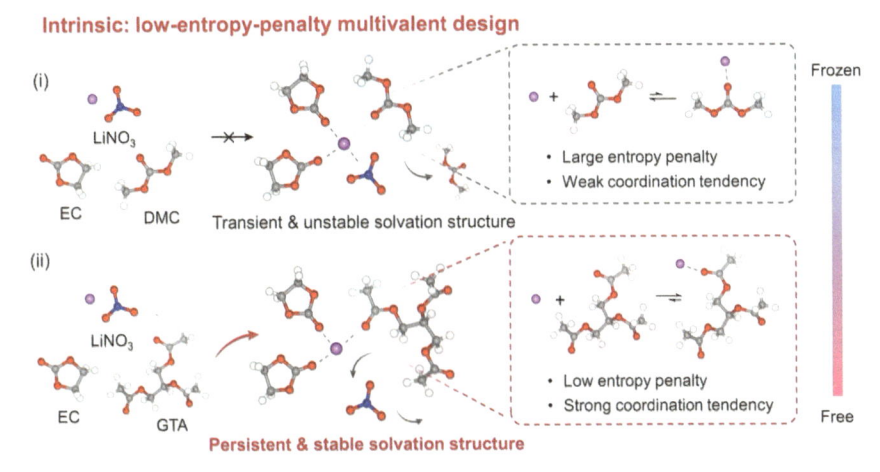

Figure 1.4 Multivalent electrolyte design realizes intrinsic high LiNO$_3$ solubility due to the low entropy penalty without regulating the enthalpy contribution. Source: Reproduced from Jin et al. [79]/with permission of Wiley-VCH Verlag GmbH & Co. KGaA.

1.3.3.2 Sacrificial Additives

Since the decomposition of NO$_3^-$ during the formation of SEI is irreversible, that is, as the cycling process proceeds, NO$_3^-$ continues to react with the deposited lithium metal until it is exhausted, which is considered a "sacrificial" electrolyte additive. Stuckenberg et al. [50] prepared LiNO$_3$-modified separators by dissolving LiNO$_3$ in DME and infiltrating it in the separator to load LiNO$_3$. The authors found that the LiNO$_3$-modified electrolytes showed better cycling stability through cycling tests. Combined with electrochemical tests such as CV, it was concluded that LiNO$_3$-modified separators could continuously release LiNO$_3$ to improve the cycling stability performance of the batteries during the battery cycling process and stabilize its concentration, which effectively overcomes the challenge of depletion of LiNO$_3$ during the cycling process.

Fu et al. [80] incorporated KNO$_3$ uniformly into the metal Li anode to prepare the "salt-in-metal" structure Li/KNO$_3$ (LKNO) composite anode. The concentration change of K$^+$ in the electrolyte proved that the LKNO electrode stably released KNO$_3$ into the electrolyte during the cycle to maintain its concentration, avoiding the situation where NO$_3^-$ was depleted and the electrochemical performance deteriorated. The NO$_3^-$ dissolved in the electrolyte effectively improved the performance of the SEI, thereby effectively improving the electrochemical performance of the LKNO composite anode.

In addition, the researcher embedded LiNO$_3$ powder into metal Li using a simple mechanical kneading method, schematically shown in Figure 1.5a. LiNO$_3$ and Li metal completely reacted to produce inorganic substances Li$_3$N and LiN$_x$O$_y$, thus realizing the protection of lithium metal [53]. Wang et al. [81] preferentially adsorbed and anchored NO$_3^-$ in LiNO$_3$ in the anionic vacancy of MgAl layered bimetallic hydroxides (LDHs) through the memory effect, thus achieving the

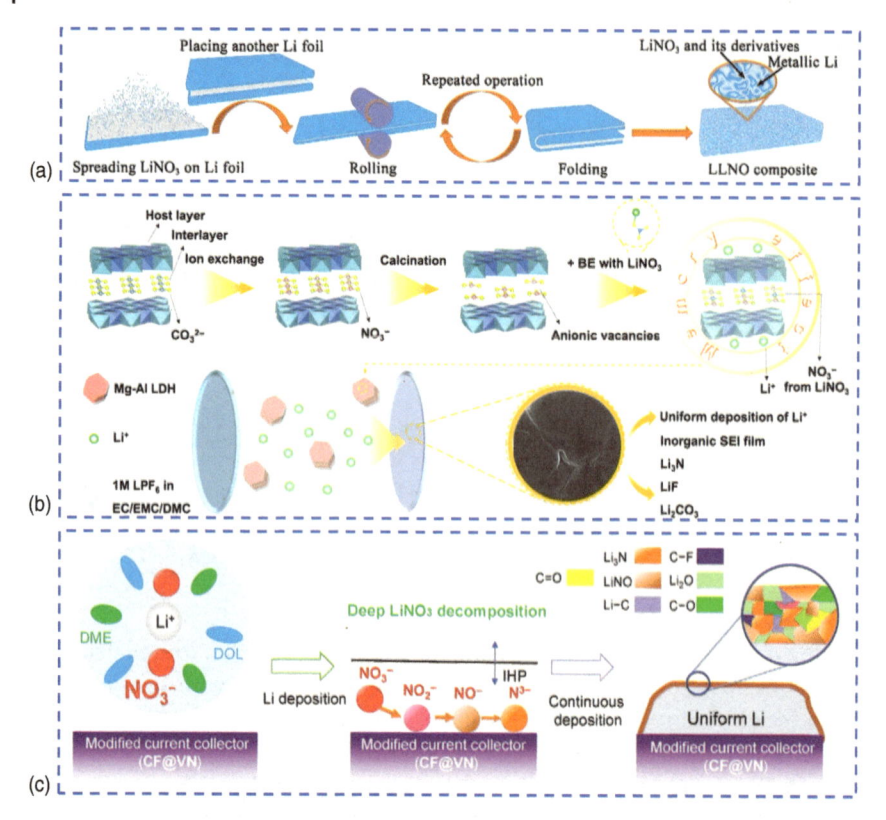

Figure 1.5 Solutions for sacrificial NO$_3^-$ additives. (a) Schematic diagram of the manufacture of Li/LiNO$_3$(LLNO) composites. Source: Reproduced from Fu et al. [53]/with permission of Wiley-VCH Verlag GmbH & Co. KGaA. (b) Schematic illustration of designing LDH additive with anionic vacancies for adsorbing NO$_3^-$ anion to promote high dissolution LiNO$_3$. Source: Reproduced from Wang et al. [81]/with permission of Wiley-VCH Verlag GmbH & Co. KGaA. (c) The deep decomposition of LiNO$_3$ on the IHP interface of the catalytic current collector, CF@VN, and schematic representations of the SEI. Source: Reproduced from Zhang et al. [55]/with permission of Wiley-VCH Verlag GmbH & Co. KGaA.

solubilization of LiNO$_3$ in the carbonate electrolyte and functioning as storing NO$_3^-$ (Figure 1.5b). On the one hand, the reconstituted NO$_3$–MgAl LDHs acted as a sustainable NO$_3^-$ source to avoid the problem of NO$_3^-$ depletion, and on the other hand, the MgAl LDHs could realize the dissolution of LiNO$_3$ in the carbonate electrolyte, which was conducive to the formation of Li$_3$N-rich SEIs with high ionic conductivities. The SEI lowered the energy barriers for Li$^+$ transport, suppressed the growth of lithium dendrites and the side reactions, and formed a uniform and dense Li deposition morphology.

1.3.3.3 High Decomposition Activation Energy of LiNO$_3$

LiNO$_3$ reacts with lithium metal during the formation of inorganic-rich SEI and decomposes into inorganic substances with high ionic conductivity and mechanical strength. One LiNO$_3$ requires eight electrons from the lithium metal to decompose into one Li$_3$N and three Li$_2$O. However, the eight-electron transfer process produces a high energy barrier between LiNO$_3$ and Li$_3$N, decreasing the likelihood of the reaction. Zhang et al. [55] showed that lithium oxynitride (LiNO) was identified as a decomposition intermediate, and experimental and simulation results confirmed that LiNO would hinder the further decomposition of LiNO$_3$, schematically shown in Figure 1.5c. Therefore, the authors modified the collector, where polar V≡N bonds were introduced. The study showed that the dipole–dipole interaction between LiNO and the polar V≡N bond could increase the ionicity of the N=O bond and thus reduce its covalency. The decrease in covalency reduces the separation between the bonding and antibonding orbitals, thereby lowering the cleavage energy barrier and promoting the reduction of LiNO$_3$ to form the inorganic-rich SEI.

1.4 Nitrile Additives

1.4.1 Functions of Nitrile Additives

1.4.1.1 Plasticization

Nitrile compounds are often used as solid polymer electrolytes (SPEs) additives. However, the ionic conductivity of SPEs is generally low. According to the report of Wright et al. [82], the ionic conductivity of fully amorphous polyethylene oxide (PEO) was high, and the activation energy was low, whereas the ionic conductivity of nearly entirely crystalline PEO was rather low at ambient temperature. PEO is a semicrystalline polymer at room temperature, with various forms such as crystal region and amorphous region, and the amorphous region of the activated chain segment is conducive to ion transport. Therefore, lithium ions can only be transported in the amorphous region above the glass transition temperature (T_g). Lithium ions form associations with the active groups on the polymer chain segments, and migration is accomplished by hopping between the chain segments [83].

Therefore, in order to improve the ionic conductivity of polymer electrolytes, some measures will be adopted to destroy the regularity of polymer chain segments to increase the proportion of amorphous regions. Currently, the widely adopted methods are blending, crosslinking, copolymerization, grafting, etc. Among them, adding plasticizer to a polymer matrix is a facile and effective method to improve ionic conductivity. As a micromolecule, nitrile compounds, especially SN, can be introduced into the polymer matrix to reduce the interaction between polymer chain segments to effectively increase the proportion of amorphous region in the polymer matrix, thereby improving the mobility of polymer chain segments and facilitating the migration of Li$^+$. Furthermore, nitrile additives may act as chain transfer agents during polymerization, regulating the molecular weight of the polymer, and lower molecular weight means shorter polymer chains, which are more conducive to providing channels for ion migration [84].

Nguyen et al. [85] introduced SN into the polymer electrolyte system of LIBs and found that the addition of SN decreased the crystallization of PEO, and SN had a synergistic effect with the Nb/Al co-doped $Li_7La_3Zr_2O_{12}$ (NAL), which contributed to form a uniformly dispersed composite electrolyte and improve the overall performance of the electrolyte. Wang et al. [86] also investigated the plasticization effect of SN on sodium-ion batteries. The addition of SN increased the vibrational energy of the coupling between Na^+ and ether-oxygen in PEO, indicating that the presence of SN weakened the interaction between Na^+ and ether-oxygen, which contributed to the migration of Na^+ through the PEO chain segments and improvement of ionic conductivity. In addition to the PEO matrix, SN was also applied to polymer matrices such as polycarbonate (PC) [87], poly(propylene carbonate) (PPC) [88], poly(vinylidene fluoride-co-hexafluoropropylene) (PVDF-HFP) [89, 90], polyacrylonitrile (PAN) [91], and polyacrylates [92] to improve the ionic conductivity.

1.4.1.2 Facilitation of Ion Transport

Although the introduction of a plasticizer into the polymer matrix can effectively improve the mobility of polymer chain segments, the conduction process of lithium ions in the polymer electrolyte is complicated, and the transport of lithium ions may not only be based on the motility of polymer chain segments but also migrate in the plasticizer-rich microphase. Wang et al. [93] systematically investigated the ionic conduction mechanism of the high ionic conductivity in the SPEs plasticized with SN and found that the addition of SN led to the gradual decoupling of Li^+ from the polymer skeleton and Li^+ recoordinating with SN through ^6Li solid-state NMR. The room-temperature ionic conductivity was enhanced significantly by increasing the SN content. When the SN content in the electrolyte reached 50%, the ionic conductivity was comparable to the SN/LiTFSI electrolyte without the polymer, indicating a reduced role of the polymer skeleton in Li^+ conduction. Further experiments revealed that the ionic conduction behavior of SPE shifted from a Vogel–Fulcher–Tammann curve to an Arrhenius curve, suggesting that Li^+ migration occurred mainly through SN rather than the segmental motion of the polymer skeleton. The results indicate that, on the one hand, SN can act as a plasticizer and improve the movement ability of polymer chain segments; on the other hand, when the content of SN is high enough to form a connected phase, the migration mode of Li^+ will change from hopping migration of polymer chain segments to migration in SN-rich microphase.

Liu et al. [84] investigated the transport pathway of Li^+ in SN/polymer matrix and found an interaction between the cyano group (C≡N) and Li^+ in SN and that this coordination may facilitate lithium ion transport by forming the interaction pathway of C≡N⋯Li^+. There were three pathways for lithium-ion transport in the electrolyte (Figure 1.6a): (i) SN → SN → SN; (ii) C=O → C=O → C=O; and (iii) SN → C=O → SN. It suggested that SN could act as a carrier for ionic conduction to improve the ionic conductivity in the SN-rich electrolyte. Hu et al. [94] proposed a Janus quasi-solid-state electrolyte (JSE) design, in which the SN plastic crystal electrolyte was embedded into the Janus host. Through the unique adsorption of SN by LATP, the solvation structures of Li^+ were changed, and a one-dimensional lithium-ion transport channel was constructed inside the solid

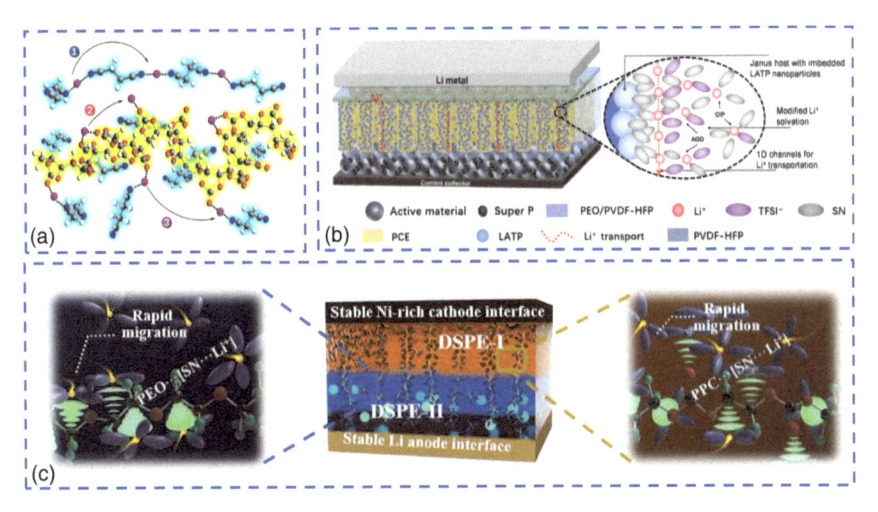

Figure 1.6 Functions of nitrile compounds. (a) Three transference mechanisms of lithium ion in PVCA-SN CPE. The purple balls represent lithium ions. The molecules with blue and yellow light were SN and PVCA, respectively. Source: Liu et al. [84]/John Wiley & Sons/CC BY 4.0. (b) A schematic illustration of the as-designed JSE structure. Source: Reproduced from Hu et al. [94]/with permission of Wiley-VCH Verlag GmbH & Co. KGaA. (c) Structural schematic illustrations of DSPE at room temperature. Source: Reproduced from Wang et al. [88]/with permission of Wiley-VCH Verlag GmbH & Co. KGaA.

electrolyte to enhance the ionic conductivity, as schematically shown in Figure 1.6b. Tong et al. [95] also reported the introduction of SN into Na–CO_2 batteries by utilizing the room-temperature curing property of SN to construct a layer of highly stable ionic conductive network in situ between $Na_3Zr_2Si_2PO_{12}$ and Ru/ carbon nanotubes (CNT) catalysts. The chemical and electrochemical stability of SN contributed to the extent of the cycling life of solid-state batteries.

In addition to the uniform blending of SN with the polymer matrix, some researchers have separated the SN phase from the polymer matrix, and SN acts as a plastic crystal phase for Li-ion conduction. Lee et al. [92] constructed an elastic solid-state electrolyte with a three-dimensional (3D) interconnected plastic crystal phase by the polymerization-induced phase separation method. The polymer matrix and the plastic crystals within the electrolyte achieved a nanoscale phase separation structure, resulting in the formation of a 3D interconnected network structure of SN phases in the polymer matrix, which provided a fast transport path for lithium ions. Jiang et al. [96] synthesized a composite electrolyte with high ceramic content by solvent-free method. $Li_{6.75}La_3Zr_{1.75}Ta_{0.25}O_{12}$ (LLZTO) powder was first bonded with poly(tetrafluoroethylene) (PTFE) and then combined with a nylon mesh framework to form an electrolyte film. The composite electrolyte film was obtained after being immersed in a molten plastic crystal electrolyte (SN/LiTFSI) for 48 hours and by removing excess plastic crystal electrolytes. In this system, the polymer matrix and SN/LiTFSI phase were phase-separated, in which SN no longer acted as a plasticizer but formed a continuous Li^+ transport channel, improving the ion transport efficiency of the electrolyte.

1.4.1.3 Promotion of Lithium Salt Dissolution

Due to the high polarity of the cyano group, compounds containing the C≡N usually have high dielectric constants, such as SN ($\varepsilon = 55$) [97]. Highly polar cyano compounds are able to dissolve various lithium salts up to ≈15 mol% [98], thus increasing the concentration of free Li$^+$ carriers and facilitating rapid ion transport. Wang et al. [88] found that the intermolecular interaction between SN and lithium salt, LiDFOB, can form a specific solvation sheath structure [SN···Li$^+$] through MD simulation. The radial distribution function indicated that the interaction between Li$^+$ and SN mainly occurred on the N atom, and SN preferentially coordinated with Li$^+$, which promoted the dissociation of lithium salt and increased the concentration of Li$^+$ in the electrolyte. DFT combined with FTIR and NMR results further demonstrated the strong interaction between SN and polymer. Combined with the solvated sheath structure of [SN···Li$^+$], the polymer···[SN···Li$^+$] system constructed a fast lithium-ion transport channel, providing a driving force for the transport of lithium ions through the polymer chain segment at room temperature, as schematically shown in Figure 1.6c.

1.4.1.4 Widening of the Electrochemical Window

The cyano groups with low HOMO endow nitrile compounds with high oxidation stability. Therefore, the addition of nitrile compounds to the electrolyte can broaden the electrochemical window of the electrolyte [89], which in turn meets the requirements of matching the high-voltage cathode and expands its application range in LIBs [84]. Zhang et al. [99] found that compared with pure polyethylene glycol methyl ether acrylate (PEGMEA)-based electrolytes, electrolytes with SN showed a wider electrochemical stability window, up to 4.8 V, which met the application requirements of the high-voltage cathode.

1.4.1.5 Inhibiting the Decomposition of the Electrolyte

Researchers have found that the cyano group can also coordinate with transition metal ions on cathodes, thus changing the charge density of the metal ions, lowering their oxidation states, and inhibiting their catalytic decomposition of the electrolyte. Yang et al. [35] explored the mechanism of nitrile compounds on the surface of the LCO cathode by combining Vienna ab initio simulation package calculations and soft X-ray absorption spectroscopy. The results revealed that —CN inhibited the catalytic decomposition of transition metal ions in the LCO cathode to the electrolyte during the charging process. It was observed by TEM that the CEI formed by the electrolyte of 10% FEC without nitrile additives was of high thickness and non-uniform, while the CEI formed with 1% suberonitrile or 1% 1,3,6-hexantrionitrile (HTCN) was dense, thin, and electrically insulated. The thin CEI indicated that the excessive decomposition of FEC on the surface of the cathode was suppressed by nitrile compounds, improving the high voltage stability of the LCO cathode.

1.4.1.6 Low Flammability

SN, as a representative of nitrile additives, is nonflammable and nonvolatile, which has been reported in many articles [96, 100, 101]. The plastic crystals of nonflammable nitrile additives, combined with the polymer matrix with non-leakage properties. The polymer matrix can further enhance the safety of the electrolyte and avoid the risk of thermal runaway at high temperatures [102]. Wang et al. [88] conducted ignition tests on SPE containing SN and found that it was difficult to ignite and only melted gradually even after 90 seconds of flame contact, suggesting that the electrolyte had excellent fire protection potential. Jiang et al. [96] also conducted combustion experiments on a celgard separator and PTFE-LLZTO-SN electrolyte. The separator rapidly shrunk and burned close to the ignition source, whereas the PTFE-LLZTO-SN electrolyte exhibited excellent nonflammability due to the low flammability of SN and the excellent thermal stability of the PTFE-LLZTO framework. Therefore, the nonflammable PTFE-LLZTO-SN electrolyte could eliminate the risk of thermal runaway and effectively improve the safety performance of lithium batteries.

1.4.1.7 Improvement of Polymer Flexibility

SN-based electrolytes turn into a liquid state when heated to their melting point and return to solid state upon cooling, the property that helps to maintain the compactness and flexibility of the composite electrolyte [96]. The addition of an appropriate amount of SN to the electrolyte can improve the elasticity and plasticity of the polymer electrolyte film [86]. Lee et al. [92] proposed that SN, as a plastic crystal, enhanced the mechanical elasticity of the material, enabling the electrolyte to accommodate volume changes during the charge and discharge of lithium metal, thus improving the cycle stability of the battery.

1.4.1.8 Modification of the Cathode Interface

Due to the high oxidative stability of nitrile compounds, nitrile compounds tend to match well with the cathode. Therefore, some researchers have introduced nitrile compounds as functional additives into the electrolyte to stabilize the cathode. Lee et al. [103] introduced adiponitrile (AN) into liquid electrolytes, as a novel bifunctional additive. XPS results showed that the cyano group could affect the oxidation state of the cathode surface, form a strong coordination with Ni^{4+} of the cathode, and inhibit the formation of unfavorable Ni^{2+}. Detrimental Ni^{2+} usually forms an irreversible NiO-type rock salt structure on the cathode surface. Therefore, the addition of AN could effectively inhibit the side reaction on the cathode surface and improve the interfacial stability of the nickel-rich NCM cathode surface.

Moreover, nitrile compounds can also be used as an electrolyte modification layer and compounded into the electrolyte with poor cathode matching to form a multilayer electrolyte film. Chen et al. [104] modified the plastic-crystalline electrolyte (PCE) layer containing SN onto the LLZTO (LLZTO-PCE) surface to reduce the interface impedance between the solid-state oxide electrolyte, LLZTO, and the cathode. PCE film with high ionic conductivity (7.2×10^{-4} S cm^{-1}) could maintain excellent ionic conduction characteristics and maintain the stability of the interface with the cathode. Liu et al. [105] reported a durable $Li_{1.5}Al_{0.5}Ge_{1.5}P_3O_{12}$ (LAGP)-based Li metal battery by employing self-healing polymer electrolytes (SHEs) as Janus interfaces. The SHEs were constructed on both sides of LAGP pellets by in situ polymerization of a functional monomer and a cross-linker in ionic liquid-based (anodic side in contact with Li metal) or AN-based cathodic gel polymer electrolyte (CGPE). The as-developed SHEs show flame-retardant, high ionic conductivity (> 10^{-3} S cm^{-1} at 25 °C), excellent interfacial compatibility with electrodes, and effective inhibition of Li dendrite formation. AN-based CGPE was prepared by polymerizing 1.5 wt% PETEA monomer into 1 M LiTFSI in AN cathode liquid electrolyte (CLE). AN is chosen for cathodic side electrolytes due to its high oxidation stability and ionic conductivity.

1.4.1.9 Involvement in the Solvation Structure of Zn^{2+}

Nitrile compounds also play an essential role in aqueous zinc-ion batteries. During the electrochemical deposition of zinc metal batteries, there is a problem in the decomposition of water molecules into hydrogen due to proton reduction, and current research efforts are devoted to changing the solvation structure of Zn^{2+} to inhibit the dehydrogenation reaction. It has been found that some nitrile compounds, such as SN or acetonitrile, can coordinate with metal zinc ions, thus inhibiting water molecules from participating in the solvation of Zn^{2+} and changing the solvation structure of conventional hydrated zinc ions to restrain the generation of hydrogen [106]. In addition, Yang et al. [107] found that the coordination of SN with Zn^{2+} inhibited the interaction between Zn^{2+} and water molecules and promoted the uniform deposition of metal zinc on the anode, forming a highly uniform Zn deposition with mosaic morphology. The lack of free water in the solvation structure of Zn^{2+} on the cathode inhibited the dissolution of poly(2,3-dithiino-1, 4-benzoquinone) cathode.

1.4.1.10 Weakening of Ionic Association

For the graphite cathode in the dual-ion batteries (DIBs), the severe ion association behavior will cause the anions on the cathode surface to be bound by the multivalent metal cation during the charging process, and the coinsertion of metal cations will occur when the anion is inserted into the cathode. The large size of intercalated anionic complexes leads to a significant decrease in the number of sites practically viable for capacity contribution inside graphite galleries in the energetically favorable single-stacking mode, compromising the intrinsic capacity of graphite cathodes for anion storage. Yang et al. [108] found that the addition of high dielectric constant AN as a cosolvent in the electrolyte could weaken the ion association and inhibit the cation co-insertion behavior, significantly improving the specific capacity of graphite cathode.

1.4.1.11 Contribution to the Formation of SEI

Generally, nitrile compounds with high LUMO will not contribute to the formation of SEI. However, Song et al. [109] found the dipole–dipole interactions between the locally negatively charged N atoms in —C≡N of SN and the locally positively charged C atoms in —O—C=O of PMMA. The interactions induced —C≡N to be reduced to Li$_3$N in the SEI layer, forming a stable SEI layer, which facilitates a uniform lithium deposition and better interfacial stability of lithium metal batteries during cycling.

1.4.2 Compatibility Analysis of Nitrile and Lithium Metal

1.4.2.1 Incompatibility of Nitrile and Lithium Metal

The most critical issue limiting the wide application of nitrile compounds in LMBs is the incompatibility of the cyano group in contact with the lithium metal anode. Lin et al. [9] investigated the mechanism of the degradation of the lithium metal anode with SN-based SPE. Due to the lack of a stable SEI between the lithium metal and the SPE, there was a sustained spontaneous reaction between the lithium metal, SN, and the polyacrylate polymer skeleton, generating large amounts of Li_xNC as decomposition products. Some studies suggest that the SN-based PCE will be catalyzed by lithium metal to undergo nitrile polymerization, which will lead to the interface incompatibility between the anode and the electrolyte and reduce the stability of LMBs [89].

1.4.2.2 Improvement of the Compatibility of Nitrile and Lithium Metal

Currently, the typical solutions for the aforementioned problems mainly include the formation of a stable SEI layer, a decrease in the reduction activity of the cyano group, restriction of the migration of nitrile additives, and in situ formation of a lithium metal protective layer.

1) Formation of a stable SEI layer

    As mentioned earlier, an essential function of fluorinated additives is the SEI-forming agent, so a certain amount of FEC can be added to the electrolyte. FEC can preferentially form an inorganic-rich SEI layer on the surface of lithium metal to effectively inhibit the contact between the cyano group and lithium metal and avoid side reactions [9]. Alternatively, the lithium metal anode is pretreated to form a Li-FEC anode. The Li-FEC anode with a LiF-rich protective layer can prevent the reaction between the lithium anode and the plastic crystal components in the electrolyte, thus improving interfacial compatibility. The surface of the Li-FEC anode was relatively flat after cycling without the presence of by-products and dendrites, which successfully verified that the designed FEC-optimized lithium anode possessed good compatibility with the electrolyte-containing SN [96].

    Dual-salt strategy can also effectively improve the compatibility between SN and lithium metal [15, 110, 111]. Bao et al. [89] introduced a dual salt strategy of LiTFSI and LiBOB to investigate the incompatibility between SN and Li metal. Through SEM and FTIR tests, it was proved that the addition of LiBOB could effectively inhibit the polymerization of SN, catalyzed by Li metal. XPS analysis showed that, compared with the formation of SEI formed by LiTFSI alone, the B—F bond signals were observed on the surface of lithium metal with double salts (Figure 1.7a), indicating the involvement of LiBOB in the formation of SEI. Moreover, it was hypothesized that LiBOB assisted in the formation of the SEI film, thereby inhibiting the harmful side reactions of SN with Li metal.

2) Decrease in reduction activity of cyano group

    The reason why nitrile compounds react adversely with lithium metal is that the reduction activity of the cyano group is too high. The poor reduction stability of the electrolyte will further lead to the decomposition of the electrolyte near the lithium metal anode, resulting in irreversible stripping/plating of lithium and, ultimately, battery failure [113]. Therefore, we can chemically reduce its reduction activity and thus inhibit the reaction between cyano groups and lithium metal. Zhang et al. [101] experimentally and theoretically confirmed the existence of strong electrostatic interactions between α-hydrogen in SN and oxygen-rich ether in 1,3,5-trioxane (TEX) (Figure 1.7b). Intermolecular CT between SN and TEX decreased the reduction activity of SN and increased steric hindrance, both of which together inhibited the side reaction of SN with lithium metal and improved the reduction stability of the electrolyte. Zhang et al. [114] proposed the introduction of the LLZTO, which utilized the La^{3+} in LLZTO to coordinate with the —C≡N of the SN, thereby increasing the electron density of the cyano group, leading to the polymerization of the SN, which in turn reduces the reactive —CN groups or converts them to less reactive —C=N— groups, thus inhibiting the reaction of SN with lithium metal. However, the polymerization of SN will sacrifice the ionic conductivity of the electrolyte. Therefore, better methods can be adopted instead of the polymerization of cyano groups.

3) Restriction of the migration of nitrile additives

Nitrile additives, like SN, migrate in the electrolyte into an anode and thus react with Li metal. However, the contact of nitrile compounds with Li metal can be avoided by constructing a hierarchical solid-state electrolyte (HSE), where nitrile compounds are kept away from the Li metal side, and their migration is limited. Fu et al. [100] designed an HSE structure, in which SN-based quasi-solid electrolyte (slQSE) was matched with NCM622 cathode, and PEO-based polymer solid electrolyte (oPSE) was matched with lithium metal anode. The SN-based antioxidation layer was in contact with the high-voltage cathode, and the PEO-based antireduction layer was in contact with the lithium anode, as schematically shown in Figure 1.7c. However, the layered electrolyte structure was insufficient to restrict the movement of SN, and SN can still diffuse freely through the interface between slQSE and oPSE, and reach the surface of the lithium anode during cycling. Therefore, it was necessary to further introduce substances that could restrict the migration of SN, and nano LLZTO was thus introduced into slQSE. By utilizing the complexation between the La atoms in LLZTO and the N atoms in SN, the free SN molecules were successfully immobilized in the SN-based electrolyte layer, which prevented free SN from diffusing to the lithium anode side and ensured the stability of lithium metal.

Similarly, Wang et al. [88] proposed a heterogeneous bilayer solid-state polymer electrolyte (DSPE) with a structure divided into SN-PPC-LiDFOB (DSPE-I) compatible with the cathode and PEO-$Li_7La_3Zr_2O_{12}$-LiTFSI (DSPE-II) compatible with the lithium anode. Unlike the previous work, LLZO was added into the electrolyte on the anode side, while the limitation of SN relied on the polymer matrix PPC on the cathode side. According to the calculation of electrostatic potential (ESP), the coordination between polymer matrix PPC and SN in DSPE-I was stronger than that of PEO in DSPE-II, that is, SN was more inclined to stay in DSPE-I near the cathode, thus limiting the contact between SN and lithium metal. Zhang et al. [99] also proposed that due to the interaction between

SN and the polymer matrix PEGMEA within the electrolyte, the migration of SN was limited, which enhanced the stability of the interface between SN and the lithium metal anode.

4) In situ formation of a lithium metal protective layer

In addition to the artificial construction of layered solid-state electrolytes, a protective layer can also be in situ constructed on the surface of lithium metal from the composition of the electrolyte. It is worth noting that the protective layer here is different from SEIs, whereas SEIs with excellent performance tend to be membranes dominated by inorganic compounds with certain functions such as rapid Li$^+$ conduction and uniform Li$^+$ deposition, and the protective layer here focuses more on protecting the Li metal from reacting with SN. Wu et al. [112] found through DFT calculation that LiDFOB with low LUMO energy would preferentially reduce to BF3 on lithium metal, and BF$_3$ would further induce TXE in situ polymerization to polyformaldehyde (POM) on the surface of lithium metal (Figure 1.7d). Further calculations revealed that compared to SN, POM would preferentially adhere to the lithium metal surface. POM with a stable LUMO energy level was compatible with Li metal, effectively inhibiting the contact between SN and lithium metal. Liu et al. [84] found that VC would also react on the surface of lithium metal preferentially to SN to form a protective layer in the VC-SN-LiDFOB system, which prevented the direct contact between SN and lithium metal and effectively improved the interface between electrolyte and lithium metal.

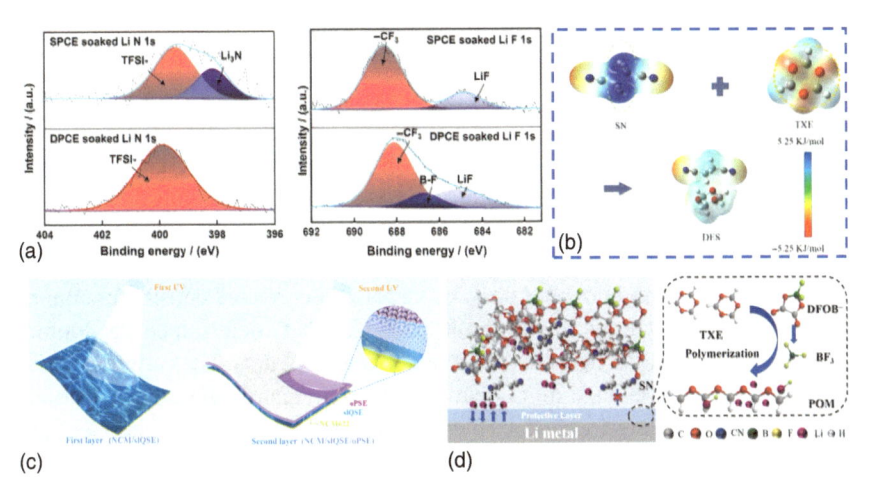

Figure 1.7 Solutions to improve the compatibility of nitrile compounds with lithium metal. (a) XPS profiles of SPCE-soaked Li (top) and DPCE-soaked Li (bottom). Source: Reproduced from Bao et al. [89]/with permission of Wiley-VCH Verlag GmbH & Co. KGaA. (b) ESPs map of total electron density for SN, TXE, and DES. Source: Reproduced from Zhang et al. [101]/with permission of Wiley-VCH Verlag GmbH & Co. KGaA. (c) Schematic representation of the detailed synthesis process and the assembly process of HSE/cathode composite. Source: Reproduced from Fu et al. [100]/with permission of Wiley-VCH Verlag GmbH & Co. KGaA. (d) The formation mechanism of a protective layer on the surface of Li metal. Source: Wu et al. [112]/John Wiley & Sons/CC BY 4.0.

1.4.3 Other Drawbacks of Nitrile Additives

1.4.3.1 Low Mechanical Strength
SN, as a typical nitrile compound, is solid-state at room temperature, but when complexed with lithium salts to prepare PCEs, SN-based PCEs exhibit low melting points and transform into viscous liquid electrolytes [109]. PCEs are insufficient for the formation of self-supporting electrolyte membranes without complexing with a polymer matrix. Self-supporting electrolytes can be achieved by incorporating SN into the polymer network to construct poly(plastic-crystalline electrolytes) (PPCEs) [115]. If nitrile compounds are introduced into polymer-based electrolytes as plasticizers, although the degree of polymerization of the electrolyte will be reduced to improve the ionic conductivity, the mechanical properties of the electrolyte will be decreased as well. Therefore, in the regulation of the ratio of electrolyte to plasticizer, we should pay attention to the balance between ionic conductivity and mechanical properties.

1.4.3.2 Prone to Polymerization
As mentioned earlier, there is a strong coordination between the La^{3+} in LLZTO and the cyano group in SN, which leads to the polymerization of SN. Although the polymerization of SN can inhibit the reaction between the cyano group and lithium metal, the ionic conductivity of the electrolyte system will decrease. Yang et al. [116] induced coordination competition by introducing a strongly polar PAN polymer matrix into the system. There was also a strong interaction between SN and PAN, which weakened the coordination between SN and La^{3+}, thus inhibiting the polymerization of SN. Therefore, the prepared PAN-modified SN electrolyte exhibited a stable and high ionic conductivity (10^{-4} S cm^{-1}).

1.4.3.3 Crystallinity
The crystallinity of SN molecules affects the ionic conductivity, especially at low temperatures, and the increase in crystallinity and viscosity of the molecules will weaken the ionic conductivity of the electrolyte. Kim et al. [115] investigated the effect of the polymer network on the morphology of SN in PPCEs, exploring the binding energies between SN and polymer units by dispersion-corrected density functional theory (DFT-D) simulations. It was found that miscibility determined the conformation and crystallization behavior of SN in polymer networks. The variation of the binding energies between SN and polymer units significantly affected the formation of amorphous phases in PPCEs. Polymer units with high binding energies [e.g. vinyl ethylene carbonate (VEC)] could hinder the aggregation of SN, promote the formation of the amorphous phase, and improve ionic conductivity. These findings highlighted the importance of optimizing the miscibility of components in PPCE to achieve amorphous phases, thus facilitating efficient ion transport. Wang et al. [102] found that through interaction with ethoxylated trimethylpropane triacrylate (ETPTA), crystallization of SN was inhibited, ensuring proper ion migration over a wide temperature range.

1.5 Phosphate Ester Additives

1.5.1 Functions of Phosphate Ester Additives

1.5.1.1 Flame Retardant

Safety is one of the most severe challenges in commercial carbonate electrolytes. Carbonate electrolytes are often flammable. They will cause fires, explosions, and other major safety issues, so the development of electrolytes with flame-retardant properties is an urgent problem to be solved currently. According to the free radical mechanism of combustion, the flame retardant of the electrolyte can be carried out from two aspects: on the one hand, it can reduce the ability of the electrolyte to generate free radicals, such as the use of nonflash point, high flash point, low melting point or nonvolatile solvents; on the other hand, it can introduce flame retardant to improve the ability of the electrolyte to clear free radicals [117].

Phosphate ester additives are one of the most widely used flame retardants in electrolytes. Phosphate ester additives will decompose when heated, forming phosphorus free radicals (PO•), which can remove combustion free radicals (H•, O_2*, HO•), thus blocking the combustion chain reaction [118]. Therefore, the addition of phosphate ester into electrolytes will reduce the exothermic value and self-heating rate of batteries and improve the thermal stability of the electrolytes. Currently, widely researched phosphate flame-retardant additives for LIBs mainly include trimethyl phosphate (TMP) [119, 120], triethyl phosphate (TEP) [121, 122], dimethyl methylphosphonate (DMMP) [123], diethyl ethyl phosphate (DEEP) [124], etc.

TMP is a kind of phosphate ester flame retardant with the simplest structure. Wang et al. [125] described the flame-retardant mechanism of TMP, which could be divided into three parts: (i) gasification of TMP under heat: $TMP_{liquid} \rightarrow TMP_{gas}$; (ii) decomposition of gaseous TMP in flame to produce phosphorus-free radicals: $TMP_{gas} \rightarrow [P]•$; (iii) elimination of hydrogen radicals that sustained the combustion chain reaction by phosphorus-free radicals: $[P]• + H• \rightarrow [P]H$. The combustion chain reaction was interrupted by the lack of hydrogen radicals. The structure of TEP is similar to TMP, and Li et al. [126] used TEP as a flame-retardant electrolyte, which exhibited a low SET of $6.10 \, s \, g^{-1}$, while the SET of 1 M LiPF$_6$ EC/DEC in the conventional carbonate system was as high as $48.92 \, s \, g^{-1}$. SET is the duration for which the electrolyte continues to burn after the external fire source is removed. Generally, if the SET of the electrolyte is $<6 \, s \, g^{-1}$, it is "nonflammable"; $6 \, s \, g^{-1} < SET < 20 \, s \, g^{-1}$ is "flame retardant"; $> 20 \, s \, g^{-1}$, it is "flammable." It could be seen that the electrolyte of the phosphate ester system had excellent flame retardant.

Fluorinated phosphate esters are also widely studied functional additives, such as tris(hexafluoroisopropyl)phosphate (THFP) [127], pentafluorophenyl diethoxyphosphate (FPOP) [118], and tris(2,2,2-trifluoroethoxy) phosphate (TFP) [128]. Gu et al. [128] prepared the electrolyte by mixing TFP with γ-butyrolactone (GBL) solvent, using LiPF$_6$ as lithium salt and adding lithium difluoro(oxalato)borate (LiODFB) as a film-forming additive to improve the interface stability of the electrode. The combustion experiments found that the conventional commercial electrolyte 1 M LiPF$_6$ in EC/DMC was ignited at the instant of contact with the fire source and continued to burn for 110 s. After adding 30 wt% TFP flame retardant, the electrolyte could still be ignited but only lasted for 6 s, indicating that TFP had outstanding flame-retardant performance. However, when the solvent was replaced by GBL and TFP (the mass ratio of the two was 70 : 30), the electrolyte could not be ignited. This indicates that GBL is less flammable than the conventional electrolyte EC/DMC. Combined employment of GBL and TEP provides superior nonflammable performance for the electrolyte, significantly enhancing the safety performance of batteries.

1.5.1.2 Stabilization of Cathodes and Anodes

TMSP, as a typical phosphate ester, is also a widely used electrolyte additive, which is able to passivate the metal oxides of the cathode and form a stable CEI layer [129]. The CEI can reduce the polarization voltage of the charging and discharging process so that the battery can still maintain good cycling and rate performance during the cycling. Zhao et al. [130] proposed an in situ "anchoring + pouring" synergistic CEI construction (Figure 1.8a), realized by using HTCN and TMSP electrolyte additives. HTCN with three nitrile groups could tightly anchor transition metals by coordinative interaction to form the CEI framework, and TMSP would electrochemically decompose to reshape the CEI layer. The uniform and robust in situ constructed CEI layer could suppress the transition metal dissolution, shield the cathode against diverse side reactions, and significantly improve the overall electrochemical performance of the cathode.

However, regarding the stabilizing effect of TMSP on the cathode, some researchers propose that TMSP is not directly involved in the formation of CEI. Liao et al. [133] introduced TMSP into the electrolyte and found that the morphology

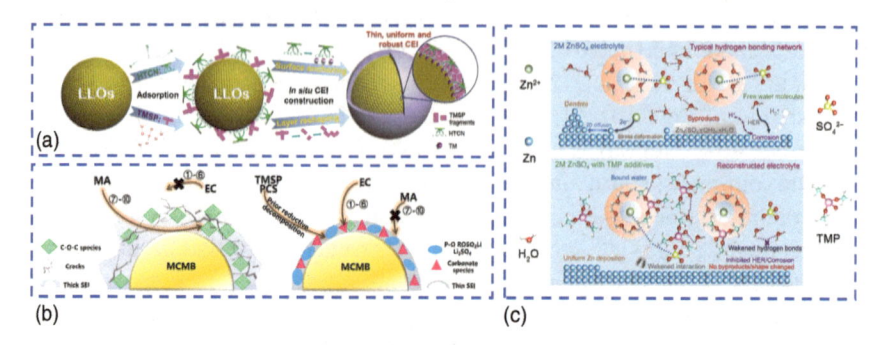

Figure 1.8 Functions of phosphate ester in electrolytes. (a) Schematic illustration of the synergistic effects of HTCN + TMSP on adjusting the CEI structure and cathode electrochemistry. Source: Reproduced from Zhao et al. [130]/with permission of Wiley-VCH Verlag GmbH & Co. KGaA. (b) The possible working mechanism of TMSP and PCS functional additives. Source: Reproduced from Xu et al. [131]/with permission of Elsevier. (c) Schematic illustration of TMP additives to optimize electrolyte/anode interface. Source: Reproduced from Zhang et al. [132]/with permission of Elsevier.

of the NCM111 cathode changed significantly in the electrolyte without TMSP but maintained its morphology with TMSP. XPS results indicated that even with TMSP, the main component of CEI was still the decomposition product of the electrolyte rather than TMSP, indicating that TMSP was not directly involved in the formation of CEI to protect the cathode. Instead, it was regarded as a method of protecting the cathode material through the occurrence of the P—O—M (M = Ni, Co and Mn) complexation on the cathode surface, which resulted in a milder environment at the electrode interface. Therefore, the stabilizing mechanism of TMSP on the cathode still needs to be further explored.

TMSP can also participate in the formation of SEI of graphite anode due to the reactivity of TMSP with lithium alkyl oxides formed during EC degradation [134]. Xu et al. [131] introduced TMSP and 1,3-propanediolcyclic sulfate (PCS) into a carbonate electrolyte with methyl acetate (MA) as a cosolvent. It was found that the surface of the graphitic mesocarbon microbeads (MCMB) anode was smooth after cycling with TMSP and PCS, indicating the formation of a thin and dense SEI layer. The results of XPS on the cycled MCMB electrode revealed that TMSP and PCS would be reduced on the MCMB anode to generate P—O, $ROSO_2Li$, and Li_2SO_4, covering the $CH_3O\bullet$ and C—O—C generated from the decomposition of MA to form a stable SEI (Figure 1.8b).

In addition, some fluorinated phosphate esters are endowed with the ability to participate in the formation of SEI and CEI due to the strong electronegativity of their F atoms. Sun et al. [127] introduced THFP as an additive into a flame-retardant electrolyte and found that THFP would decompose on the surface of NCM622 cathode to form the thin and dense CEI rich in C—F, and the lithium-philic properties of C—F bond would guide the uniform distribution of lithium-ion flux in the CEI layer to improve the cycling stability. Moreover, the strong electron-withdrawing F atoms can reduce the LUMO of phosphate esters, which leads to the preferential reduction of THFP on the lithium surface to form a LiF-rich SEI. In addition to the outstanding film-forming properties, THFP can also dissociate $LiPF_6$ salts through strong bonding with the PF_6^-, facilitating the migration of lithium ions in the electrolyte and modulating the structure and composition of the SEI. Therefore, fluorinated phosphate esters are often introduced into electrolytes as multifunctional additives to improve the thermal and cycling stability of batteries.

1.5.1.3 Involvement in Solvation Structure Regulation

Phosphate ester additives can also regulate the solvation structure of cations in LIBs, potassium ion batteries, and even aqueous zinc-ion batteries. Gao et al. [135] introduced TFP into potassium-ion batteries and found that TFP with weak solvation ability and high oxidative stability could weaken the interaction of solvent molecules with K^+, thereby promoting the retention of FSI^- anions in the solvation sheath of K^+ to form an anion-rich solvation structure to generate anion-derived SEI. In addition, the corrosion of potassium bis(fluoromethanesulfonyl)imide (KFSI)-based electrolytes to the Al collector was suppressed due to the weak solubility of TFP.

In aqueous batteries, phosphate esters are also often used as green and environmentally friendly additives to modulate the solvation structure. Zhang et al. [132] systematically investigated the interaction between phosphate esters and aqueous electrolytes and found that TMP with the highest dielectric constant and the smallest molecular volume among a series of phosphate esters could be easily intercalated into the solvation structure of Zn^{2+} (Figure 1.8c). In addition, abundant theoretical calculations and experimental characterizations proved that TMP molecules could not only modify the solvated structure but also get preferentially adsorbed on the Zn electrode surface to optimize EDL. As a result, a zincophilic–hydrophobic inorganic–organic interface layer was formed, enabling multiple locking of water molecules while thermodynamically and kinetically guarding a continuous and stable zinc plating/stripping process.

1.5.2 Drawbacks of Phosphate Ester

1.5.2.1 Incompatibility with Anodes

Although phosphate ester additives have flame-retardant properties, researchers have revealed phosphate ester additives with low concentrations in the electrolyte are insufficient to act as a flame retardant, and phosphate ester additives with high concentrations tend to decompose on graphite, lithium metal, or sodium metal anodes. The decomposition of phosphate ester participated in the formation of an unstable SEI with poor electronic shielding ability, which had troubles in effectively preventing the coinserting of Li^+-solvent and the continuous decomposition of solvent, further leading to the reduction of battery capacity and poor cycle stability [136]. For graphite anodes, the decomposition of phosphate ester on the anode will lead to delamination and flaking of the graphite electrode, and due to the strong catalytic effect on the surface of the graphite anode, it will further lead to the continuous decomposition of the phosphate ester solvent, which is detrimental to the cycle stability of the battery [137].

1.5.2.2 Improvement of the Compatibility of Phosphate Ester and Lithium Metal

1) Introduction of SEI-forming agent

    LiDFOB, $LiPF_6$, and other lithium salts are commonly used in electrolytes with excellent SEI-forming properties. Li et al. [126] found through characterization by XPS and high-resolution cryo-electron microscopy (cryo-EM) that the thickness of the SEI layer derived from lithium salt without LiDFOB reached 45 nm in thickness. The mosaic SEI layer was randomly distributed with Li_2O, Li_3PO_4, and Li_2CO_3 crystal particles. In contrast, the amorphous layer was mainly composed of low molecular weight organic substances, such as P—O and C—O. The uneven SEI would further lead to uneven Li deposition and result in the growth of lithium dendrites. In contrast, the SEI layer derived with LiDFOB lithium salts was only 13.6 nm, forming a uniform Li_2O/polymer SEI layer, and this uniform SEI prevented the electron tunneling of the Li anode and inhibited the further reaction of the highly active TEP molecules with the lithium metal.

1.5 Phosphate Ester Additives

As mentioned earlier, FEC is one of the most widely used SEI-forming agents. Ma et al. [138] introduced FEC into the TEP-based electrolyte of Na metal batteries. FEC could promote the formation of a robust SEI layer on the anode surface, relieving the interfacial reactivity between the phosphate ester-based electrolytes and the Na metal anode. Through the optimization of the ratio of TEP to FEC, phosphate ester-based electrolytes with high oxidative stability and nonflammability could realize the stable cycle of Na metal batteries. In lithium-sulfur batteries, FEC can still protect the stability of the phosphate ester-based electrolyte through preferential decomposition at the anode. Wang et al. [139] introduced FEC and Sn(OTf)$_2$ additives into a TEP-based electrolyte, and Sn(OTf)$_2$ optimized the solvation structure of Na$^+$ and promoted the TFSI$^-$ to replace a portion of the TEP to participate in the solvation structure (Figure 1.9a), effectively reducing the desolvation energy barrier of Na$^+$. In addition, FEC could preferentially decompose on the anode to form a stable SEI film, thus inhibiting the decomposition of TEP. The synergistic effect of FEC and Sn(OTf)$_2$ effectively enhanced the stability of TEP-based electrolytes and their compatibility with Na metal.

2) Regulation of solvation structures

    Modification of the solvation structure can be achieved by regulating the composition of the electrolyte, thus improving the compatibility of the electrolyte with the anode. Liu et al. [136] explored the ratio of lithium salt, TEP, and EC and found that EC could stabilize TEP in the electrolyte and ensure the compatibility of the electrolyte with the graphite anode when the molar ratio of lithium salt to TEP was 1 : 2. However when the molar ratio of the lithium salt to TEP changed to 1 : 3, it was difficult to regulate the solvation structure of TEP by introducing EC, as shown in Figure 1.9b. Therefore, through the competitive coordination between TEP and EC, a dynamically stable solvation structure could be constructed to weaken the decomposition of TEP and realize a balance between electrolyte flame retardancy and electrochemical stability.

Zeng et al. [121] also found by regulating the molar ratio of salt to solvent that Li$^+$ was more inclined to be complexed with P=O in four TEP molecules to form tetrahedral complexes when the molar ratio of lithium bis(fluorosulfonyl)imide (LiFSI) and TFP was <1 : 4, and Li$^+$ preferred to form an "ion-solvent" complex structure with two TEP molecules and two S=O groups of FSI$^-$ when the molar ratio was >1 : 2. The results showed that when the molar ratio of LiFSI and TFP was high, there were fewer free solvent molecules in the electrolyte, resulting in a negative shift of the solvent reduction potential and inhibiting the decomposition of solvent molecules on the anode surface. However, LiFSI with excessive concentration would cause the increased viscosity of electrolytes and decreased electrochemical performance. Therefore, when the molar ratio of LiFSI to TEP was 1 : 2, the system had lower viscosity and higher ionic conductivity, which could inhibit the irreversible decomposition of solvent and realize the reversible electrochemical cycle of lithium metal anode.

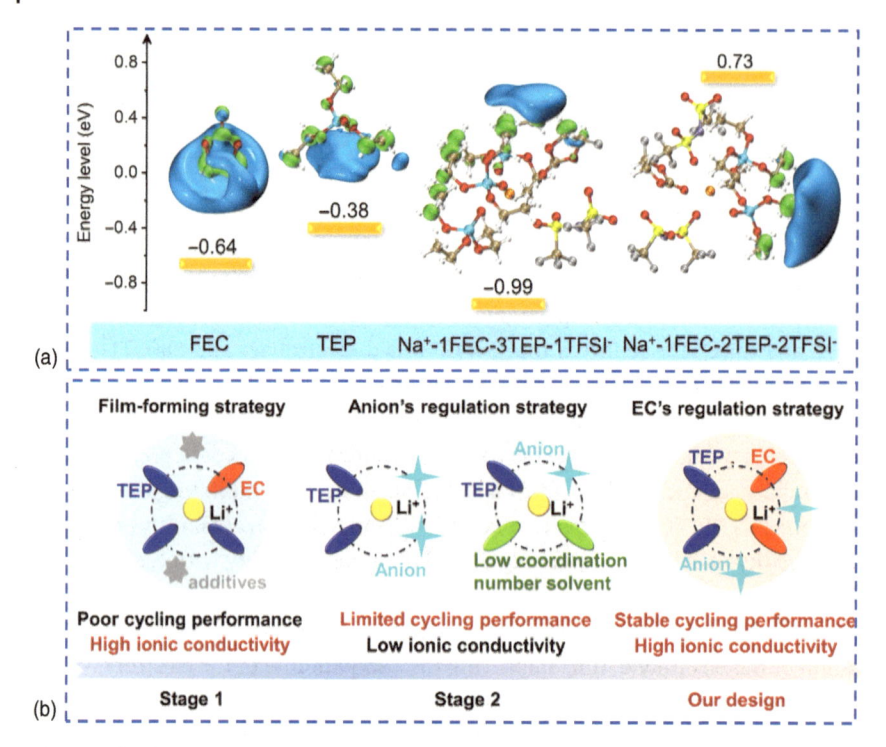

Figure 1.9 Solutions to optimize the compatibility of phosphate ester with anode. (a) LUMO energy level of FEC, TEP, Na$^+$-1FEC-3TEP-1TFSI$^-$ and Na$^+$-1FEC-2TEP-2TFSI$^-$. Source: Reproduced from Wang et al. [139]/with permission of Wiley-VCH Verlag GmbH & Co. KGaA. (b) Features of reported TEP-modified nonflammable electrolytes, their solvation structure models, and the performance of Gr electrodes. Source: Reproduced from Liu et al. [136]/with permission of American Chemical Society.

1.6 Sulfate Ester Additives

1.6.1 Functions of Sulfate Ester Additives

1.6.1.1 SEI-Forming Additives

Sulfate ester is a novel SEI-forming additive with a low LUMO energy level, which can be preferentially reduced and decomposed on the anode surface to form SEI film during cycling. 1,3-Propane sultone (1,3-PS), 1,4-butane sultone (1,4-BS), ethylene sulfate (DTD), and ethylene sulfite (ES) have been investigated for applications in the formation of SEI.

Xu et al. [140] investigated the film-forming performance of 1,4-BS as a novel sulfate ester additive in nonaqueous electrolytes for the first time in 2007. It was found by CV scanning that the electrolyte introduced with 1,4-BS showed a reduction peak at 0.7 V, which might be due to the reduction of 1,4-BS to form SEI film on the surface of the graphite anode. The peak disappeared in the second sweeping, indicating that the 1,4-BS-derived SEI film had been stably formed in the first sweeping and could effectively inhibit the further reduction of the electrolyte on the graphite electrode.

A small amount of 1,4-BS could form a dense and stable SEI film on the surface of the graphite anode, which effectively prevented propylene carbonate and solvation lithium ions from coinsertion into graphite, thereby overcoming the drawbacks of propylene carbonate such as capacity loss and decline in cycle stability. The introduction of 1,4-BS as an additive could change the composition of the SEI film and improve the stability of SEI film and the electrochemical performance of LIBs. Since the structure of 1,3-PS is similar to that of 1,4-BS, Xu et al. also investigated the functions of 1,3-PS in the electrolyte using a similar method in 2009 and found the S element in SEI using XPS characterization, confirming that 1,3-PS indeed involved in the formation of SEI film [141].

Sano et al. [142] systematically investigated the functions of DTD and its derivatives in electrolytes. First, through the CV test, it was found that the reduction current of the electrolyte with DTD was lower than that without DTD, indicating that the DTD decomposition products formed SEI on the surface of the graphite anode, which prevents the decomposition of PC. In addition, the anodic peak with DTD was higher than that without DTD, probably because the decomposition products of DTD prevented the graphite stripping that accompanied the decomposition of PC, which increased the reversible capacity and the anodic current. The charge–discharge curves of the DTD derivatives showed that the initial efficiency of methyl DTD (MDTD) was the same as that of DTD, but the initial efficiency of ethyl DTD (EDTD) was low, indicating that the longer alkyl chain might reduce the performance of SEI. In addition, researchers found that DTD possessed better performance than ES, which had a similar structure to DTD. Therefore, DTD and its alkyl derivatives may be able to replace ES as more efficient SEI-forming additives to effectively inhibit the initial capacity loss of batteries, reduce the expansion of batteries after high-temperature standing, and improve the charging and discharging performance of batteries. Hall et al. [143] found that the reduction onset potential of DTD was related to its addition amount, and the reduction potential of DTD was lower than that of EC based on the differential capacity (dQ/dV) test. It was confirmed that DTD would be preferentially reduced at the anode and participate in the formation of SEI. Li et al. [144] further demonstrated the degradation products of DTD using XPS and found that the SEI formed by the electrolyte with DTD was rich in $LiSO_3$ and $ROSO_2Li$. The inorganic-rich SEI was the key to the reduction of the interface impedance of the battery with DTD.

Previously, researchers proposed the preferential decomposition of DTDs to participate in the formation of SEIs but did not mention the prerequisites required to exert the role of DTDs. Cheng et al. [145] proposed the unique function of DTDs and revealed the prerequisite of solvation for effectively weakening Li^+-solvent interaction in electrolytes for the first time. It was shown that when DTDs with high polar surface area (PSA) and strong coordination of Li^+ were introduced into the flame-retardant TMP-based electrolytes, the DTDs would not weaken the Li^+-TMP interaction. TMP with a low dielectric constant ($\varepsilon = 21.3$) possessed weak dissociation ability for $LiPF_6$, unable to dissociate Li^+ and PF_6^- ions in the electrolyte completely. Therefore, DTDs were challenging to coordinate with Li^+

and existed near the second solvation sheath, failing to inhibit Li$^+$–TMP interactions. In contrast, when EC with high dielectric constant and strong solvation ability was introduced into the electrolyte, DTD was able to enter the first solvation sheath and coordinate with Li$^+$ to inhibit Li$^+$–TMP interactions due to the strong dissociation of LiPF$_6$ by EC (Figure 1.10a). With the gradual addition of DTD, EMC, and EC, the content of TMP molecules in the solvation structure of lithium ions gradually decreased while the content of DTD increased, indicating the strengthened Li$^+$–DTD interactions and weakened Li$^+$–TMP interactions were achieved through the regulation of solvents and additives. Therefore, when using DTD as an electrolyte additive, we need to pay attention to the dielectric constant of the solvent in the electrolyte, ensuring the prerequisite conditions for the dissociation of lithium salts to enable DTD to perform its function.

ES is structurally similar to VC and has been used in LMBs to form stable SEI. It can effectively inhibit the coinsertion behavior of propylene carbonate when used as a solvent or co-solvent [148]. Pham et al. [149] used ES and DES as electrolytes, and ES possesses a high boiling point (173 °C), low melting point (−11 °C), and high dielectric constant (41.0). However, ES with a high dielectric constant has strong solvation ability, thus forming a solvation structure dominated by solvent, which is not conducive to forming rich inorganic SEI. Therefore, DES with a low dielectric constant (16.2) was introduced to balance the solvation ability of the mixed solvents. In the electrolyte with sulfate ester, an anion-rich solvation structure was formed due to the low dielectric constant of DES, resulting in an anion-derived SEI rich in inorganic components LiF as well as LiPS, which were mainly produced by the decomposition of phosphate esters. These inorganic components played a crucial role in passivating lithium deposition, inhibiting dendrite formation, and ultimately improving the electrochemical performance of batteries. In contrast, the SEI formed by the carbonate electrolyte was mainly composed of organic components with poor ionic conductivity and mechanical strength. It could be concluded that the phosphate ester-based electrolytes are able to produce stable and robust SEIs and improve the coulomb efficiency and cycle stability of batteries. Liu et al. [146] revealed that electrolytes with dimethyl sulfite (DMS) and diethyl sulfite (DES) exhibited excellent compatibility with composite SiO/C anodes. The sulfide-rich SEI with low impedance was formed on the surface of the anode through pre-cycling, which could inhibit the oxidative decomposition of sulfur-containing organic solvents at higher potentials during the subsequent cycling process (Figure 1.10b). In addition, this sulfide-rich, SEI-modified layer endowed the battery with cycling stability performance at low temperatures, which was of great significance in promoting the development of low-temperature lithium batteries.

1.6.1.2 CEI-Forming Additives

Sulfur-containing compounds can also be used as CEI-forming agents due to their natural insolubility and the high oxidative stability of sulfates, like Li$_2$SO$_4$ and ROSO$_3$Li. Kim et al. [150] showed through theoretical calculation that the —S(=O)$_2$—, —OS(=O)$_2$— and —OS(=O)$_2$O— groups in sulfur-containing compounds and their decomposition products could be complexed with the

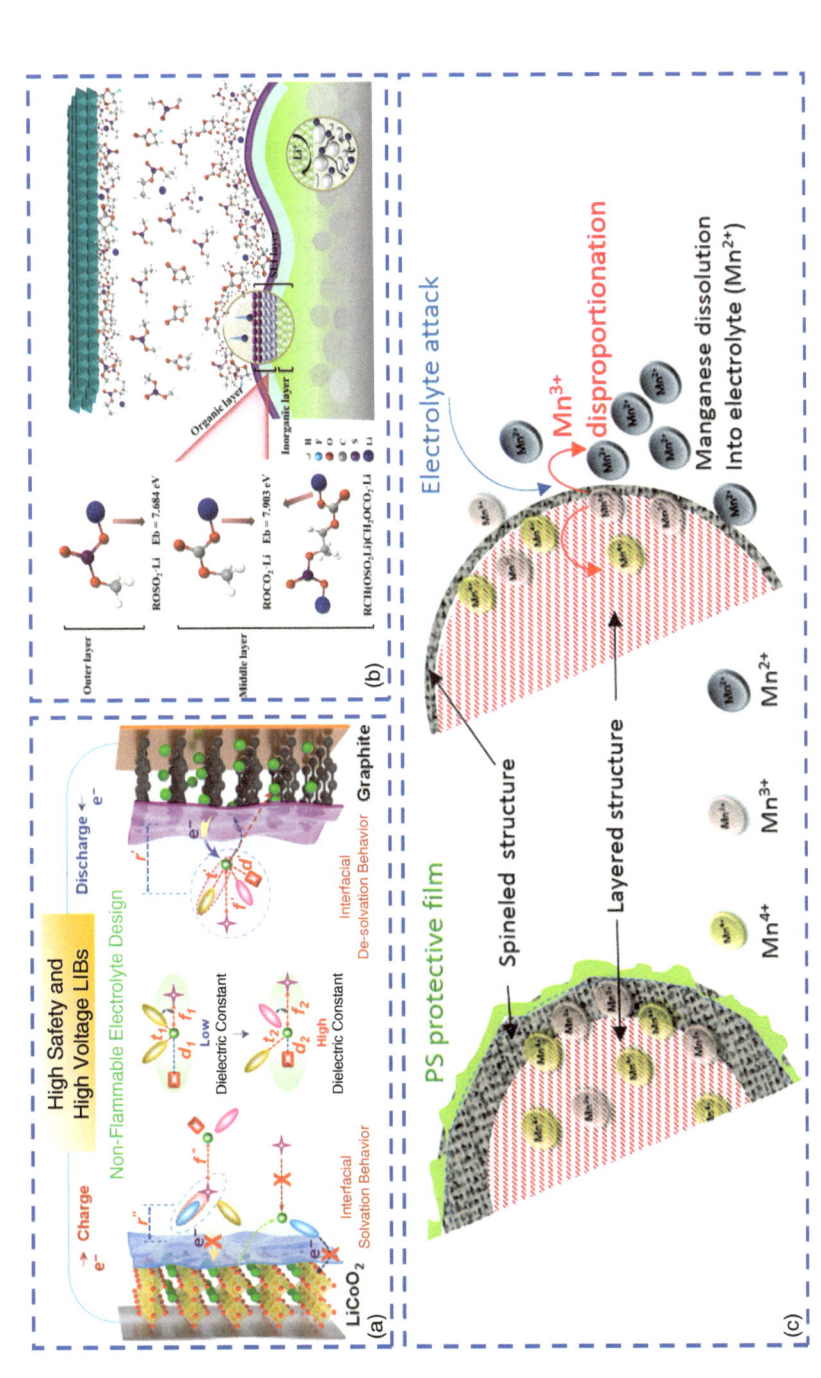

Figure 1.10 SEI and CEI forming properties of sulfate. (a) Schematics illustrating solvation structure and interfacial model in designed electrolytes. Source: Reproduced from Cheng et al. [145]/with permission of American Chemical Society. (b) The binding energy for Li$^+$ with organic components in a modified layer by DFT and simulation model of MD in the battery. Source: Liu et al. [146]/John Wiley & Sons/CC BY 4.0. (c) A schematic of the PS protection mechanism on the Li-rich NMC cathode during cycling and progressive transformation from the layered to the spinel structure. Source: Reproduced from Pires et al. [147]/with permission of Royal Society of Chemistry.

transition metal (Ni^{2+}) in the cathode, thus enhancing the structural stability of the Ni-containing cathodes, improving the interfacial stability of the cathode and electrolytes, and inhibiting the dissolution of the transition metal atoms.

DTD has been proven to be applied to lithium metal and graphite anodes to form stabilized SEIs, while it can also be decomposed on the cathode to generate CEIs. Ren et al. [151] added DTD to Na—S batteries, where a ring-opened intermediate was created between the electrophilic DTD additive and Na_2S_x ($1 \leq x \leq 8$). The intermediate was an organic sodium salt, consisting of a PS and a sulfate group, which were linked by a —CH_2—CH_2— group. When the sodium salt intermediate aggregated in a solvent with low DN, such as DME, PS and sulfate reacted to form insoluble polymeric products, explaining the formation of an interphase layer. Instead, in the solvent with high DN, the anion would be separated by a solvent and would not be in direct contact with other anions. Therefore, there was no aggregation-induced polymerization, explaining why the organic sodium salt could be dissolved in the solvent with high DN. As a result, the DTD additives spontaneously reacted with sodium sulfide to form a robust sulfate-based CEI on the sulfated polyacrylonitrile cathode, which protected the cathode and inhibited the dissolution of sodium polysulfide.

Due to the high oxidation state of the sulfur atom in 1,3-PS, its decomposition products can be more stable at high potentials and match with high-voltage cathodes [152]. Pires et al. [147] added 1,3-PS to the electrolyte of high-voltage LIB and found that 1,3-PS could form a protective film with a wide electrochemical stable window (5.0 V) on the surface of Li-rich-NMC cathode, which effectively suppressed the occurrence of side reactions on the electrode surface as well as the dissolution of the metal ions (Figure 1.10c). Similar to 1,3-PS, prop-1-ene-1,3-sultone (PES) can also be used as a CEI-forming agent in high-voltage LIBs. Li et al. [153] introduced PSE into the electrolyte and demonstrated that PSE formed a protective CEI film on the LNMO cathode, which inhibited the decomposition of the electrolyte on the cathode surface, improved the interfacial properties between the cathode and electrolyte, and significantly enhanced the cycling stability of LNMO. With the addition of 1 wt% PES, the capacity retention rate increased from 49% to 90% after 400 cycles at the rate of 1C.

Methylene methanedisulfonate (MMDS) is also capable of improving the cycling stability of batteries at high voltage, which not only forms a stable SEI film on the graphite anode but also facilitates the formation of a thin CEI to improve the performance of batteries. Zuo et al. [154] showed that the electrolyte without MMDS was decomposed at 5.3 V, while the electrolyte with 0.5 wt% MMDS was oxidized at 5.1 V through LSV tests. A decrease in the oxidation potential indicated that MMDS might decompose on the LCO cathode before the solvent, so as to inhibit the decomposition of the electrolyte. A further CV test showed that two reduction peaks appeared at 3.6 and 3.8 V in the electrolyte with MMDS, which were attributed to the insertion of lithium ions into LCO and the decomposition of MMDS on the LCO cathode to form a CEI, respectively. It was observed by TEM that the thickness of CEI obtained with MMDS was only 3–5 nm, compared with that formed without MMDS (15–20 nm), and the thinner CEI could reduce the surface impedance and thus improve the cycling performance of the battery.

1.7 Conclusion and Outlook

After the discussion of the functions of widely used electrolyte additives, we can find that electrolyte additives, as a core component of electrolytes, directly affect the overall performance and safety of batteries. With the rapid expansion of markets for electric vehicles, smartphones, and other electronic products, there is a growing demand for batteries with high energy density, long cycle life, and stable safety performances, so it is essential to carry out continuous research and optimization of the functions and performances of electrolytes additives.

Currently, the functions of electrolyte additives are mainly concentrated in the following aspects:

i) SEI and CEI film-forming agents.

Additives with low LUMO and high HOMO can be preferentially decomposed on the electrode surface, which prevents the excessive decomposition of the electrolyte on the electrodes. Some additives can regulate the solvation sheath to form anion-rich solvation structures, which also essentially facilitates the formation of inorganic-rich SEIs.

ii) Improvement of ion transport.

Nitrile compounds can act as plasticizers in polymer-based electrolytes and ionic transport carriers in nitrile compounds-rich electrolytes, thereby facilitating the transport of ions.

iii) Flame retardant.

For flammable and explosive electrolytes, the addition of flame retardants, like phosphate esters, is crucial to improve the thermal stability of batteries effectively.

However, although the application of additives has achieved certain results, the employment of electrolyte additives is still limited, and the stability of electrolytes in extreme environments, like high temperature and high pressure, needs to be improved. Moreover, the action mechanism of some additives is still unclear and still needs further exploration and research.

Therefore, we consider that future research on electrolyte additives should be mainly focused on the following aspects:

i) the development of novel additives to meet the requirements of extreme conditions of battery systems;
ii) in-depth research on the action mechanism of additives to provide theoretical support for the optimization of the design of additives;
iii) to strengthen the study on synergistic effects of additives with electrolytes, electrodes, and other components of batteries to achieve the optimization of the overall performance of batteries.

In summary, research on electrolyte additives is of great significance for improving the performance and safety of batteries. Although the current research has achieved certain results, there are still many challenges and problems to be solved to promote the application of battery electrolyte additives in industrial production.

References

1 Zhang, S., Li, S., Wang, X. et al. (2023). Nonflammable electrolyte with low exothermic design for safer lithium-based batteries. *Nano Energy* 114: 108639.
2 Meng, Y., Zhou, D., Liu, R. et al. (2023). Designing phosphazene-derivative electrolyte matrices to enable high-voltage lithium metal batteries for extreme working conditions. *Nature Energy* 8: 1023–1033.
3 Kim, M.-H., Wi, T.-U., Seo, J. et al. (2023). Design principles for fluorinated interphase evolution via conversion-type alloying processes for anticorrosive lithium metal anodes. *Nano Letters* 23: 3582–3591.
4 Yu, Y., Wang, S., Zhang, J. et al. (2023). Long-life lithium batteries enabled by a pseudo-oversaturated electrolyte. *Carbon Energy* 6: e383.
5 Li, G.-X., Jiang, H., Kou, R. et al. (2022). A superior carbonate electrolyte for stable cycling Li metal batteries using high Ni cathode. *ACS Energy Letters* 7: 2282–2288.
6 Aurbach, D., Markevich, E., and Salitra, G. (2021). High energy density rechargeable batteries based on Li metal anodes. The role of unique surface chemistry developed in solutions containing fluorinated organic co-solvents. *Journal of the American Chemical Society* 143: 21161–21176.
7 Cui, Z., Zou, F., Celio, H. et al. (2022). Paving pathways toward long-life graphite/LiNi0.5Mn1.5O4 full cells: electrochemical and Interphasial points of view. *Advanced Functional Materials* 32: 2203779.
8 Li, Z., Tang, W., Deng, Y. et al. (2022). Enabling highly stable lithium metal batteries by using dual-function additive catalyzed in-built quasi-solid-state polymer electrolytes. *Journal of Materials Chemistry A* 10: 23047–23057.
9 Lin, R., He, Y., Wang, C. et al. (2022). Characterization of the structure and chemistry of the solid–electrolyte interface by cryo-EM leads to high-performance solid-state Li-metal batteries. *Nature Nanotechnology* 17: 768–776.
10 Liao, C., Han, L., Wang, W. et al. (2023). Non-flammable electrolyte with lithium nitrate as the only lithium salt for boosting ultra-stable cycling and fire-safety lithium metal batteries. *Advanced Functional Materials* 33: 2212605.
11 Shin, H., Park, J., Sastry, A.M. et al. (2015). Effects of fluoroethylene carbonate (FEC) on anode and cathode interfaces at elevated temperatures. *Journal of the Electrochemical Society* 162: A1683–A1692.
12 Li, Z., Yao, N., Yu, L. et al. (2023). Inhibiting gas generation to achieve ultralong-lifespan lithium-ion batteries at low temperatures. *Matter* 6: 2274–2292.
13 Guo, X.F., Yang, Z., Zhu, Y.F. et al. (2022). High-voltage, highly reversible sodium batteries enabled by fluorine-rich electrode/electrolyte interphases. *Small Methods* 6: 2200209.
14 Su, C.-C., He, M., Amine, R. et al. (2019). Cyclic carbonate for highly stable cycling of high voltage lithium metal batteries. *Energy Storage Materials* 17: 284–292.

15 Fu, C., Ma, Y., Lou, S. et al. (2020). A dual-salt coupled fluoroethylene carbonate succinonitrile-based electrolyte enables Li-metal batteries. *Journal of Materials Chemistry A* 8: 2066–2073.

16 Li, Y., Li, Y., Pei, A. et al. (2017). Atomic structure of sensitive battery materials and interfaces revealed by cryo-electron microscopy. *Science* 358: 506–510.

17 Yan, C., Cheng, X.B., Tian, Y. et al. (2018). Dual-layered film protected lithium metal anode to enable dendrite-free lithium deposition. *Advanced Materials* 30: 1707629.

18 Liu, J., Yuan, B., He, N. et al. (2023). Reconstruction of LiF-rich interphases through an anti-freezing electrolyte for ultralow-temperature $LiCoO_2$ batteries. *Energy & Environmental Science* 16: 1024–1034.

19 Tao, M., Xiang, Y., Zhao, D. et al. (2022). Quantifying the evolution of inactive Li/lithium hydride and their correlations in rechargeable anode-free Li batteries. *Nano Letters* 22: 6775–6781.

20 Zhang, W., Yang, T., Liao, X. et al. (2023). All-fluorinated electrolyte directly tuned Li+ solvation sheath enabling high-quality passivated interfaces for robust Li metal battery under high voltage operation. *Energy Storage Materials* 57: 249–259.

21 Wang, Y., Zhang, Y., Dong, S. et al. (2022). An all-fluorinated electrolyte toward high voltage and long cycle performance dual-ion batteries. *Advanced Energy Materials* 12: 2103360.

22 Lu, H., He, L., Yuan, Y. et al. (2020). Synergistic effect of fluorinated solvents for improving high voltage performance of $LiNi_{0.5}Mn_{1.5}O_4$ cathode. *Journal of the Electrochemical Society* 167: 120534.

23 Nagarajan, S., Weiland, C., Hwang, S. et al. (2022). Depth-dependent understanding of cathode electrolyte interphase (CEI) on the layered Li-ion cathodes operated at extreme high temperature. *Chemistry of Materials* 34: 4587–4601.

24 Hernández, G., Mogensen, R., Younesi, R. et al. (2022). Fluorine-free electrolytes for lithium and sodium batteries. *Batteries & Supercaps* 5: e202100373.

25 Su, C.-C., He, M., Amine, R. et al. (2019). Solvating power series of electrolyte solvents for lithium batteries. *Energy & Environmental Science* 12: 1249–1254.

26 Chen, L., Nian, Q., Ruan, D. et al. (2023). High-safety and high-efficiency electrolyte design for 4.6 V-class lithium-ion batteries with a non-solvating flame-retardant. *Chemical Science* 14: 1184–1193.

27 Su, C.C., He, M., Shi, J. et al. (2020). Solvation rule for solid-electrolyte interphase enabler in lithium-metal batteries. *Angewandte Chemie International Edition* 59: 18229–18233.

28 Weng, S., Yang, G., Zhang, S. et al. (2023). Kinetic limits of graphite anode for fast-charging lithium-ion batteries. *Nano-Micro Letters* 15: 215.

29 Su, C.-C., He, M., Cai, M. et al. (2022). Solvation-protection-enabled high-voltage electrolyte for lithium metal batteries. *Nano Energy* 92: 106720.

30 Wang, C., Sun, Z., Liu, L. et al. (2023). A rooted interphase on sodium via in situ pre-implantation of fluorine atoms for high-performance sodium metal batteries. *Energy & Environmental Science* 16: 3098–3109.

31 Yang, X., Cheng, F., Yang, Z. et al. (2023). Multifunctionalizing electrolytes in situ for lithium metal batteries. *Nano Energy* 116: 108825.

32 Zhang, W., Koverga, V., Liu, S. et al. (2024). Single-phase local-high-concentration solid polymer electrolytes for lithium-metal batteries. *Nature Energy* 9: 386–400.

33 Biswal, P., Rodrigues, J., Kludze, A. et al. (2022). A reaction-dissolution strategy for designing solid electrolyte interphases with stable energetics for lithium metal anodes. *Cell Reports Physical Science* 3: 100948.

34 Huang, W., Wang, Y., Lv, L. et al. (2021). 1-Hydroxyethylidene-1,1-diphosphonic acid: a multifunctional interface modifier for eliminating HF in silicon anode. *Energy Storage Materials* 42: 493–501.

35 Yang, X., Lin, M., Zheng, G. et al. (2020). Enabling stable high-voltage LiCoO$_2$ operation by using synergetic interfacial modification strategy. *Advanced Functional Materials* 30: 2004664.

36 Shen, C., Wang, S., Jin, Y. et al. (2015). In situ AFM imaging of solid electrolyte interfaces on HOPG with ethylene carbonate and fluoroethylene carbonate-based electrolytes. *ACS Applied Materials & Interfaces* 7: 25441–25447.

37 Xia, L., Lee, S., Jiang, Y. et al. (2017). Fluorinated electrolytes for Li-ion batteries: the lithium difluoro(oxalato)borate additive for stabilizing the solid electrolyte interphase. *ACS Omega* 2: 8741–8750.

38 Aktekin, B., Younesi, R., Zipprich, W. et al. (2017). The effect of the fluoroethylene carbonate additive in LiNi$_{0.5}$Mn$_{1.5}$O$_4$-Li$_4$Ti$_5$O$_{12}$ lithium-ion cells. *Journal of the Electrochemical Society* 164: A942–A948.

39 Tan, J., Ye, M., and Shen, J. (2022). Deciphering the role of LiNO$_3$ additives in Li-S batteries. *Materials Horizons* 9: 2325–2334.

40 Zhou, P., Hou, W., Xia, Y. et al. (2023). Tuning and balancing the donor number of lithium salts and solvents for high-performance Li metal anode. *ACS Nano* 17: 17169–17179.

41 Fang, W., Wen, Z., Chen, L. et al. (2022). Constructing inorganic-rich solid electrolyte interphase via abundant anionic solvation sheath in commercial carbonate electrolytes. *Nano Energy* 104: 107881.

42 Cai, T., Sun, Q., Cao, Z. et al. (2022). Electrolyte additive-controlled interfacial models enabling stable antimony anodes for lithium-ion batteries. *The Journal of Physical Chemistry C* 126: 20302–20313.

43 Huang, G., Liao, Y., Zhao, X. et al. (2022). Tuning a solvation structure of lithium ions coordinated with nitrate anions through ionic liquid-based solvent for highly stable lithium metal batteries. *Advanced Functional Materials* 33: 2211364.

44 Zhu, Y., Li, X., Si, Y. et al. (2022). Regulating dissolution chemistry of nitrates in carbonate electrolyte for high-stable lithium metal batteries. *Journal of Energy Chemistry* 73: 422–428.

45 Wahyudi, W., Ladelta, V., Tsetseris, L. et al. (2021). Lithium-ion desolvation induced by nitrate additives reveals new insights into high performance lithium batteries. *Advanced Functional Materials* 31: 2101593.

46 Fu, J., Ji, X., Chen, J. et al. (2020). Lithium nitrate regulated sulfone electrolytes for lithium metal batteries. *Angewandte Chemie International Edition* 59: 22194–22201.

47 Fan, X., Chen, L., Ji, X. et al. (2018). Highly fluorinated interphases enable high-voltage Li-metal batteries. *Chem* 4: 174–185.

48 Yamada, Y., Wang, J., Ko, S. et al. (2019). Advances and issues in developing salt-concentrated battery electrolytes. *Nature Energy* 4: 269–280.

49 Ren, X., Zou, L., Cao, X. et al. (2019). Enabling high-voltage lithium-metal batteries under practical conditions. *Joule* 3: 1662–1676.

50 Stuckenberg, S., Bela, M.M., Lechtenfeld, C.T. et al. (2023). Influence of LiNO$_3$ on the lithium metal deposition behavior in carbonate-based liquid electrolytes and on the electrochemical performance in zero-excess lithium metal batteries. *Small* 20: 2305203.

51 Zheng, T., Zhu, B., Xiong, J. et al. (2023). When audience takes stage: pseudo-localized-high-concentration electrolyte with lithium nitrate as the only salt enables lithium metal batteries with excellent temperature and cathode adaptability. *Energy Storage Materials* 59: 102782.

52 Ma, X., Yu, J., Zou, X. et al. (2023). Single additive to regulate lithium-ion solvation structure in carbonate electrolytes for high-performance lithium-metal batteries. *Cell Reports Physical Science* 4: 101379.

53 Fu, L., Wang, X., Wang, L. et al. (2021). A salt-in-metal anode: stabilizing the solid electrolyte interphase to enable prolonged battery cycling. *Advanced Functional Materials* 31: 2010602.

54 Hou, L.P., Yao, N., Xie, J. et al. (2022). Modification of nitrate ion enables stable solid electrolyte interphase in lithium metal batteries. *Angewandte Chemie International Edition* 61: e202201406.

55 Zhang, Q., Xu, L., Yue, X. et al. (2023). Catalytic current collector design to accelerate LiNO$_3$ decomposition for high-performing lithium metal batteries. *Advanced Energy Materials* 13: 2302620.

56 Zhang, Z., Wang, J., Zhang, S. et al. (2021). Stable all-solid-state lithium metal batteries with Li$_3$N-LiF-enriched interface induced by lithium nitrate addition. *Energy Storage Materials* 43: 229–237.

57 Kim, S., Lee, T.K., Kwak, S.K. et al. (2021). Solid electrolyte interphase layers by using lithiophilic and electrochemically active ionic additives for lithium metal anodes. *ACS Energy Letters* 7: 67–69.

58 Cheng, X.B., Yang, S.J., Liu, Z. et al. (2023). Electrochemically and thermally stable inorganics–rich solid electrolyte interphase for robust lithium metal batteries. *Advanced Materials* 36: 2307370.

59 Jung, T.-J., Lee, H., Park, S.H. et al. (2022). Statistical and computational analysis for state-of-health and heat generation behavior of long-term cycled LiNi$_{0.8}$Co$_{0.15}$Al$_{0.05}$O$_2$/graphite cylindrical lithium-ion cells for energy storage applications. *Journal of Power Sources* 529: 231240.

60 Wang, K., Ni, W., Wang, L. et al. (2023). Lithium nitrate regulated carbonate electrolytes for practical Li-metal batteries: mechanisms, principles and strategies. *Journal of Energy Chemistry* 77: 581–600.

61 Liu, S., Xia, J., Zhang, W. et al. (2022). Salt-in-salt reinforced carbonate electrolyte for Li metal batteries. *Angewandte Chemie International Edition* 61: e202210522.

62 Liang, Y., Wu, W., Li, D. et al. (2022). Highly stable lithium metal batteries by regulating the lithium nitrate chemistry with a modified eutectic electrolyte. *Advanced Energy Materials* 12: 2202493.

63 Jiang, Z., Yang, T., Li, C. et al. (2023). Synergistic additives enabling stable cycling of ether electrolyte in 4.4 V Ni-rich/Li metal batteries. *Advanced Functional Materials* 33: 2306868.

64 Wen, Z., Fang, W., Wang, F. et al. (2024). Dual-salt electrolyte additive enables high moisture tolerance and favorable electric double layer for lithium metal battery. *Angewandte Chemie International Edition* 63: e202314876.

65 Li, X., Zhao, R., Fu, Y. et al. (2021). Nitrate additives for lithium batteries: mechanisms, applications, and prospects. *eScience* 1: 108–123.

66 Coke, K., Johnson, M.J., Robinson, J.B. et al. (2024). Illuminating polysulfide distribution in lithium sulfur batteries; tracking polysulfide shuttle using operando optical fluorescence microscopy. *ACS Applied Materials & Interfaces* 16: 20329–20340.

67 Kim, S., Jung, J., Kim, I. et al. (2023). Tuning of electrolyte solvation structure for low-temperature operation of lithium–sulfur batteries. *Energy Storage Materials* 59: 102763.

68 Zhang, X., Wang, Y., Ouyang, Z. et al. (2023). Dual-functional lithium nitrate mediator eliminating water hazard for practical lithium metal batteries. *Advanced Energy Materials* 14: 2303048.

69 Wang, Z., Hou, L.P., Li, Z. et al. (2022). Highly soluble organic nitrate additives for practical lithium metal batteries. *Carbon Energy* 5: e283.

70 Wang, Z., Wang, Y., Shen, L. et al. (2023). Towards durable practical lithium–metal batteries: advancing the feasibility of poly-DOL-based quasi-solid-state electrolytes via a novel nitrate-based additive. *Energy & Environmental Science* 16: 4084–4092.

71 Adiraju, V.A.K., Chae, O.B., Robinson, J.R. et al. (2023). Highly soluble lithium nitrate-containing additive for carbonate-based electrolyte in lithium metal batteries. *ACS Energy Letters* 8: 2440–2446.

72 Jiang, Z., Zeng, Z., Yang, C. et al. (2019). Nitrofullerene, a C60-based bifunctional additive with smoothing and protecting effects for stable lithium metal anode. *Nano Letters* 19: 8780–8786.

73 Linert, W., Jameson, R.F., and Taha, A. (1993). Donor numbers of anions in solution: the use of solvatochromic Lewis acid-base indicators. *Journal of the Chemical Society, Dalton Transactions* 3181–3186.

74 Xu, X., Yue, X., Chen, Y. et al. (2023). Li plating regulation on fast-charging graphite anodes by a triglyme-LiNO$_3$ synergistic electrolyte additive. *Angewandte Chemie International Edition* 62: e202306963.

75 Wen, Z., Fang, W., Wu, X. et al. (2022). High-concentration additive and triiodide/iodide redox couple stabilize lithium metal anode and rejuvenate the inactive Lithium in carbonate-based electrolyte. *Advanced Functional Materials* 32: 2204768.

76 Yan, C., Yao, Y.X., Chen, X. et al. (2018). Lithium nitrate solvation chemistry in carbonate electrolyte sustains high-voltage lithium metal batteries. *Angewandte Chemie International Edition* 57: 14055–14059.

77 Chen, W., Hu, Y., Lv, W. et al. (2019). Lithiophilic montmorillonite serves as lithium ion reservoir to facilitate uniform lithium deposition. *Nature Communications* 10: 4973.

78 Gao, Y., Wu, G., Fang, W. et al. (2024). Transesterification induced multifunctional additives enable high-performance lithium metal batteries. *Angewandte Chemie International Edition* 63: e202403668.

79 Jin, Z., Liu, Y., Xu, H. et al. (2024). Intrinsic solubilization of lithium nitrate in ester electrolyte by multivalent low-entropy-penalty design for stable lithium-metal batteries. *Angewandte Chemie International Edition* 63: e202318197.

80 Fu, L., Wang, X., Chen, Z. et al. (2022). Insights on "nitrate salt" in lithium anode for stabilized solid electrolyte interphase. *Carbon Energy* 4: 12–20.

81 Wang, F., Wen, Z., Zheng, Z. et al. (2023). Memory effect of MgAl layered double hydroxides promotes $LiNO_3$ dissolution for stable lithium metal anode. *Advanced Energy Materials* 13: 2203830.

82 Wright, P.V. (1998). Polymer electrolytes – the early days. *Electrochimica Acta* 43: 1137–1143.

83 Liu, Q., Peng, B., Shen, M. et al. (2014). Polymer chain diffusion and Li+ hopping of poly(ethylene oxide)/$LiAsF_6$ crystalline polymer electrolytes as studied by solid state NMR and ac impedance. *Solid State Ionics* 255: 74–79.

84 Liu, Z., Zhang, S., Zhou, Q. et al. (2023). Insights into quasi solid-state polymer electrolyte: the influence of succinonitrile on polyvinylene carbonate electrolyte in view of electrochemical applications. *Battery Energy* 2: 20220049.

85 Nguyen, H.L., Luu, V.T., Nguyen, M.C. et al. (2022). Nb/Al co-doped $Li_7La_3Zr_2O_{12}$ composite solid electrolyte for high-performance all-solid-state batteries. *Advanced Functional Materials* 32: 2207874.

86 Wang, H., Sun, Y., Liu, Q. et al. (2022). An asymmetric bilayer polymer-ceramic solid electrolyte for high-performance sodium metal batteries. *Journal of Energy Chemistry* 74: 18–25.

87 He, Y., Liu, N., and Kohl, P.A. (2020). High conductivity, lithium ion conducting polymer electrolyte based on hydrocarbon backbone with pendent carbonate. *Journal of the Electrochemical Society* 167: 100517.

88 Wang, S., Sun, Q., Zhang, Q. et al. (2023). Li-ion transfer mechanism of ambient-temperature solid polymer electrolyte toward lithium metal battery. *Advanced Energy Materials* 13: 2204036.

89 Bao, D., Tao, Y., Zhong, Y. et al. (2023). High-performance dual-salt plastic crystal electrolyte enabled by succinonitrile-regulated porous polymer host. *Advanced Functional Materials* 33: 2213211.

90 Qiu, G., Shi, Y., and Huang, B. (2022). A highly ionic conductive succinonitrile-based composite solid electrolyte for lithium metal batteries. *Nano Research* 15: 5153–5160.

91 Ren, Z., Li, J., Gong, Y. et al. (2022). Insight into the integration way of ceramic solid-state electrolyte fillers in the composite electrolyte for high performance solid-state lithium metal battery. *Energy Storage Materials* 51: 130–138.

92 Lee, M.J., Han, J., Lee, K. et al. (2022). Elastomeric electrolytes for high-energy solid-state lithium batteries. *Nature* 601: 217–222.

93 Wang, L., He, Y., and Xin, H.L. (2023). Transition from Vogel-Fulcher-Tammann to Arrhenius ion-conducting behavior in poly(ethyl acrylate)-based solid polymer electrolytes via succinonitrile plasticizer addition. *Journal of the Electrochemical Society* 170: 090525.

94 Hu, Y., Li, L., Tu, H. et al. (2022). Janus electrolyte with modified Li+ solvation for high-performance solid-state lithium batteries. *Advanced Functional Materials* 32: 2203336.

95 Tong, Z., Wang, S.-B., Fang, M.-H. et al. (2021). Na–CO_2 battery with NASICON-structured solid-state electrolyte. *Nano Energy* 85: 105972.

96 Jiang, T., He, P., Wang, G. et al. (2020). Solvent-free synthesis of thin, flexible, nonflammable garnet-based composite solid electrolyte for all-solid-state lithium batteries. *Advanced Energy Materials* 10: 1903376.

97 Fan, L.Z., Hu, Y.S., Bhattacharyya, A.J. et al. (2007). Succinonitrile as a versatile additive for polymer electrolytes. *Advanced Functional Materials* 17: 2800–2807.

98 Reber, D., Borodin, O., Becker, M. et al. (2022). Water/ionic liquid/succinonitrile hybrid electrolytes for aqueous batteries. *Advanced Functional Materials* 32: 2112138.

99 Zhang, D., Liu, Y., Sun, Z. et al. (2023). Eutectic-based polymer electrolyte with the enhanced lithium salt dissociation for high-performance lithium metal batteries. *Angewandte Chemie International Edition* 62: e202310006.

100 Fu, F., Liu, Y., Sun, C. et al. (2022). Unveiling and alleviating chemical "crosstalk" of succinonitrile molecules in hierarchical electrolyte for high-voltage solid-state lithium metal batteries. *Energy & Environmental Materials* 6: e12367.

101 Zhang, J., Wu, H., Du, X. et al. (2022). Smart deep eutectic electrolyte enabling thermally induced shutdown toward high-safety lithium metal batteries. *Advanced Energy Materials* 13: 2202529.

102 Wang, A., Geng, S., Zhao, Z. et al. (2022). In situ cross-linked plastic crystal electrolytes for wide-temperature and high-energy-density lithium metal batteries. *Advanced Functional Materials* 32: 2201861.

103 Lee, S.H., Hwang, J.Y., Park, S.J. et al. (2019). Adiponitrile ($C_6H_8N_2$): a new bi-functional additive for high-performance Li-metal batteries. *Advanced Functional Materials* 29: 1902496.

104 Chen, S., Zhang, J., Nie, L. et al. (2020). All-solid-state batteries with a limited lithium metal anode at room temperature using a garnet-based electrolyte. *Advanced Materials* 33: 2002325.

105 Liu, Q., Zhou, D., Shanmukaraj, D. et al. (2020). Self-healing Janus interfaces for high-performance LAGP-based lithium metal batteries. *ACS Energy Letters* 5: 1456–1464.

106 Liu, H., Xin, Z., Cao, B. et al. (2023). Polyhydroxylated organic molecular additives for durable aqueous zinc battery. *Advanced Functional Materials* 34: 2309840.

107 Yang, W., Du, X., Zhao, J. et al. (2020). Hydrated eutectic electrolytes with ligand-oriented solvation shells for long-cycling zinc-organic batteries. *Joule* 4: 1557–1574.

108 Yang, Y., Wang, J., Du, X. et al. (2023). Cation co-intercalation with anions: the origin of low capacities of graphite cathodes in multivalent electrolytes. *Journal of the American Chemical Society* 145: 12093–12104.

109 Song, H., Xue, S., Chen, S. et al. (2022). Polymeric wetting matrix for a stable interface between solid-state electrolytes and Li metal anode. *Chinese Journal of Structural Chemistry* 41: 48–69.

110 Li, S., Chen, Y.-M., Liang, W. et al. (2018). A superionic conductive, electrochemically stable dual-salt polymer electrolyte. *Joule* 2: 1838–1856.

111 Hu, Z., Xian, F., Guo, Z. et al. (2020). Nonflammable nitrile deep eutectic electrolyte enables high-voltage lithium metal batteries. *Chemistry of Materials* 32: 3405–3413.

112 Wu, H., Tang, B., Du, X. et al. (2020). LiDFOB initiated in situ polymerization of novel eutectic solution enables room-temperature solid lithium metal batteries. *Advanced Science* 7: 2003370.

113 Moon, H., Jung, G.Y., Lee, J.E. et al. (2023). Starving free solvents: toward immiscible binary liquid electrolytes for Li-metal full cells. *Advanced Functional Materials* 33: 2302543.

114 Zhang, X., Fu, C., Cheng, S. et al. (2023). Novel PEO-based composite electrolyte for low-temperature all-solid-state lithium metal batteries enabled by interfacial cation-assistance. *Energy Storage Materials* 56: 121–131.

115 Kim, B., Yang, S.H., Seo, J.H. et al. (2023). Inducing an amorphous phase in polymer plastic crystal electrolyte for effective ion transportation in lithium metal batteries. *Advanced Functional Materials* 34: 2310957.

116 Yang, Y.N., Jiang, F.L., Li, Y.Q. et al. (2021). A surface coordination interphase stabilizes a solid-state battery. *Angewandte Chemie International Edition* 60: 24162–24170.

117 Zhang, S.S. (2006). A review on electrolyte additives for lithium-ion batteries. *Journal of Power Sources* 162: 1379–1394.

118 Li, L., Liu, J., Li, L. et al. (2024). Pentafluorophenyl diethoxy phosphate: an electrolyte additive for high-voltage cathodes of lithium-ion batteries. *Journal of Energy Storage* 87: 111364.

119 Shi, P., Zheng, H., Liang, X. et al. (2018). A highly concentrated phosphate-based electrolyte for high-safety rechargeable lithium batteries. *Chemical Communications* 54: 4453–4456.

120 Wang, J., Yamada, Y., Sodeyama, K. et al. (2017). Fire-extinguishing organic electrolytes for safe batteries. *Nature Energy* 3: 22–29.

121 Zeng, Z., Murugesan, V., Han, K.S. et al. (2018). Non-flammable electrolytes with high salt-to-solvent ratios for Li-ion and Li-metal batteries. *Nature Energy* 3: 674–681.

122 Chen, S., Zheng, J., Yu, L. et al. (2018). High-efficiency lithium metal batteries with fire-retardant electrolytes. *Joule* 2: 1548–1558.

123 Wang, X.F., He, W.J., Xue, H.L. et al. (2022). A nonflammable phosphate-based localized high-concentration electrolyte for safe and high-voltage lithium metal batteries. *Sustainable Energy & Fuels* 6: 1281–1288.

124 Jiang, L., Cheng, Y., Wang, S. et al. (2023). A nonflammable diethyl ethylphosphonate-based electrolyte improved by synergistic effect of lithium difluoro(oxalato)borate and fluoroethylene carbonate. *Journal of Power Sources* 570: 233051.

125 Wang, X.M., Yasukawa, E., and Kasuya, S. (2001). Nonflammable trimethyl phosphate solvent-containing electrolytes for lithium-ion batteries – I. Fundamental properties. *Journal of the Electrochemical Society* 148: A1058–A1065.

126 Li, S., Zhang, S., Chai, S. et al. (2021). Structured solid electrolyte interphase enable reversible Li electrodeposition in flame-retardant phosphate-based electrolyte. *Energy Storage Materials* 42: 628–635.

127 Sun, H., Liu, J., He, J. et al. (2022). Stabilizing the cycling stability of rechargeable lithium metal batteries with tris(hexafluoroisopropyl)phosphate additive. *Science Bulletin* 67: 725–732.

128 Gu, Y., Fang, S., Yang, L. et al. (2021). Tris(2,2,2-trifluoroethyl) phosphate as a cosolvent for a nonflammable electrolyte in lithium-ion batteries. *ACS Applied Energy Materials* 4: 4919–4927.

129 Han, Y.-K., Yoo, J., and Yim, T. (2015). Why is tris(trimethylsilyl) phosphite effective as an additive for high-voltage lithium-ion batteries? *Journal of Materials Chemistry A* 3: 10900–10909.

130 Zhao, J., Liang, Y., Zhang, X. et al. (2020). In situ construction of uniform and robust cathode–electrolyte interphase for Li-rich layered oxides. *Advanced Functional Materials* 31: 2009192.

131 Xu, G., Huang, S., Cui, Z. et al. (2019). Functional additives assisted ester-carbonate electrolyte enables wide temperature operation of a high-voltage (5 V-class) Li-ion battery. *Journal of Power Sources* 416: 29–36.

132 Zhang, T., Yang, J., Wang, H. et al. (2024). A solubility-limited, non-protonic polar small molecule co-solvent reveals additive selection in inorganic zinc salts. *Energy Storage Materials* 65: 103085.

133 Liao, X., Zheng, X., Chen, J. et al. (2016). Tris(trimethylsilyl)phosphate as electrolyte additive for self-discharge suppression of layered nickel cobalt manganese oxide. *Electrochimica Acta* 212: 352–359.

134 Yim, T. and Han, Y.-K. (2017). Tris(trimethylsilyl)phosphite as an efficient electrolyte additive to improve the surface stability of graphite anodes. *ACS Applied Materials & Interfaces* 9: 32851–32858.

135 Gao, Y., Li, W., Ou, B. et al. (2023). A dilute fluorinated phosphate electrolyte enables 4.9 V-class potassium ion full batteries. *Advanced Functional Materials* 33: 2305829.

136 Liu, M., Zeng, Z., Gu, C. et al. (2023). Ethylene carbonate regulated solvation of triethyl phosphate to enable high-conductivity, nonflammable, and graphite compatible electrolyte. *ACS Energy Letters* 9: 136–144.

137 Yayathi, S., Walker, W., Doughty, D. et al. (2016). Energy distributions exhibited during thermal runaway of commercial lithium ion batteries used for human spaceflight applications. *Journal of Power Sources* 329: 197–206.

138 Ma, Y., Qin, B., Du, X. et al. (2022). Delicately tailored ternary phosphate electrolyte promotes ultrastable cycling of $Na_3V_2(PO_4)_2F_3$-based sodium metal batteries. *ACS Applied Materials & Interfaces* 14: 17444–17453.

139 Wang, L., Ren, N., Jiang, W. et al. (2024). Tailoring Na+ solvation environment and electrode-electrolyte interphases with $Sn(OTf)_2$ additive in non-flammable phosphate electrolytes towards safe and efficient Na-S batteries. *Angewandte Chemie International Edition* 63: e202320060.

140 Xu, M.Q., Li, W.S., Zuo, X.X. et al. (2007). Performance improvement of lithium ion battery using PC as a solvent component and BS as an SEI forming additive. *Journal of Power Sources* 174: 705–710.

141 Xu, M., Li, W., and Lucht, B.L. (2009). Effect of propane sultone on elevated temperature performance of anode and cathode materials in lithium-ion batteries. *Journal of Power Sources* 193: 804–809.

142 Sano, A. and Maruyama, S. (2009). Decreasing the initial irreversible capacity loss by addition of cyclic sulfate as electrolyte additives. *Journal of Power Sources* 192: 714–718.

143 Hall, D.S., Allen, J.P., Glazier, S.L. et al. (2017). The solid-electrolyte interphase formation reactions of ethylene sulfate and its synergistic chemistry with prop-1-ene-1,3-sultone in lithium-ion cells. *Journal of the Electrochemical Society* 164: A3445–A3453.

144 Li, X., Yin, Z., Li, X. et al. (2013). Ethylene sulfate as film formation additive to improve the compatibility of graphite electrode for lithium-ion battery. *Ionics* 20: 795–801.

145 Cheng, H., Ma, Z., Kumar, P. et al. (2024). Non-flammable electrolyte mediated by solvation chemistry toward high-voltage lithium-ion batteries. *ACS Energy Letters* 9: 1604–1616.

146 Liu, X., Zhang, T., Shi, X. et al. (2022). Hierarchical sulfide-rich modification layer on SiO/C anode for low-temperature Li-ion batteries. *Advanced Science* 9: 2104531.

147 Pires, J., Timperman, L., Castets, A. et al. (2015). Role of propane sultone as an additive to improve the performance of a lithium-rich cathode material at a high potential. *RSC Advances* 5: 42088–42094.

148 Xia, J., Aiken, C.P., Ma, L. et al. (2014). Combinations of ethylene sulfite (ES) and vinylene carbonate (VC) as electrolyte additives in $Li(Ni_{1/3}Mn_{1/3}Co_{1/3})O_2$/graphite pouch cells. *Journal of the Electrochemical Society* 161: A1149–A1157.

149 Pham, T.D., Bin Faheem, A., Kim, J. et al. (2023). Unlocking the potential of lithium metal batteries with a sulfite-based electrolyte. *Advanced Functional Materials* 33: 2305284.

150 Kim, D.Y., Park, I., Shin, Y. et al. (2019). Ni-stabilizing additives for completion of Ni-rich layered cathode systems in lithium-ion batteries: an ab initio study. *Journal of Power Sources* 418: 74–83.

151 Ren, Y., Lai, T., and Manthiram, A. (2023). Reversible sodium–sulfur batteries enabled by a synergistic dual-additive design. *ACS Energy Letters* 8: 2746–2752.

152 Yu, B.T., Qiu, W.H., Li, F.S. et al. (2006). A study on sulfites for lithium-ion battery electrolytes. *Journal of Power Sources* 158: 1373–1378.

153 Li, B., Wang, Y., Tu, W. et al. (2014). Improving cyclic stability of lithium nickel manganese oxide cathode for high voltage lithium ion battery by modifying electrode/electrolyte interface with electrolyte additive. *Electrochimica Acta* 147: 636–642.

154 Zuo, X., Fan, C., Xiao, X. et al. (2012). High-voltage performance of $LiCoO_2$/graphite batteries with methylene methanedisulfonate as electrolyte additive. *Journal of Power Sources* 219: 94–99.

2

Application of Biopolymers in Batteries

2.1 Introduction

Biopolymers and their derivatives have shown tremendous potential in enhancing battery technology, offering both environmental and performance benefits that address many of the limitations associated with traditional materials. Their key advantages biocompatibility, biodegradability, and renewability make them an attractive, environmentally friendly alternative to conventional materials used in batteries. These attributes are particularly valuable as the world shifts toward greener energy-storage solutions to meet the increasing demand for cleaner technologies.

In binder applications, biopolymers play a crucial role by providing strong mechanical integrity to battery electrodes, ensuring that active materials adhere securely to current collectors even during prolonged cycling. This prevents the detachment of active materials and improves the overall durability and reliability of the battery. The mechanical properties of biopolymer binders, such as flexibility and strength, help to stabilize the electrode structure, which is critical in maintaining performance under mechanical stress during charge and discharge cycles. Additionally, the use of biopolymer-based binders reduces the reliance on toxic and environmentally harmful chemicals, further contributing to safer and greener battery production.

When used in electrolytes, biopolymers significantly enhance ionic conductivity, which is vital for efficient ion transport between the electrodes. By optimizing the discharge performance and reducing internal resistance, these materials help improve battery efficiency, energy density, and cycle life. For example, cellulose-based electrolytes offer not only enhanced electrochemical stability but also improved thermal and chemical resilience, making them a safer alternative to conventional liquid organic electrolytes. Electrolytes derived from biopolymers also open the door to ecofriendly, biodegradable electrolyte solutions that replace hazardous organic solvents traditionally used in batteries.

In separators, biopolymers provide improved thermal stability, mechanical strength, and ionic conductivity, which are essential for the safety and efficiency of modern batteries. Conventional polyolefin-based separators can fail under high temperatures, leading to short circuits and potential safety hazards. However,

Functional Auxiliary Materials in Batteries: Synthesis, Properties, and Applications, First Edition. Wei Hu.
© 2025 WILEY-VCH GmbH. Published 2025 by WILEY-VCH GmbH.

biopolymer-based separators, such as those made from cellulose nanofibers, offer superior performance by maintaining structural integrity even under extreme conditions, which is critical for high energy density applications like electric vehicles and grid storage. Furthermore, these biopolymer separators are biodegradable, contributing to a reduction in battery waste and environmental impact.

Biopolymers also serve as protective layers for battery anodes, addressing the major challenge of dendrite formation in metal-ion batteries such as lithium-ion and zinc-ion systems. Dendrites are needle-like structures that form on the surface of the anode during repeated charging, eventually penetrating the separator and causing short circuits. By suppressing dendrite growth and providing a stable interface between the anode and electrolyte, biopolymer-based protective layers can extend the lifespan of the battery and improve its overall safety. Chitosan (CS)-based protective layers, for instance, help to regulate the deposition of zinc and lithium ions, ensuring uniform plating and preventing dendrite formation, which is crucial for the long-term stability and efficiency of metal-ion batteries.

In addition to these roles, biopolymers are increasingly being used as additives in electrolytes to enhance ionic transport and stabilize the electrode–electrolyte interface. Biopolymer electrolyte additives, such as cellulose nanocrystals (CNCs), can significantly improve ion mobility, reduce interfacial resistance, and prevent the formation of harmful side products that degrade the electrolyte over time. These additives help to extend the cycle life of batteries, improve energy efficiency, and enhance overall performance. Moreover, the use of nontoxic and biodegradable additives aligns with the industry's push toward more sustainable and environmentally friendly battery technologies.

Despite the promising advancements in the application of biopolymers in battery technology, several challenges remain. One major hurdle is the need for further optimization of biopolymer materials to ensure consistently high performance across a range of operating conditions, including high current densities, extreme temperatures, and prolonged cycling. To compete with traditional materials, biopolymers must demonstrate enhanced ionic conductivity, mechanical durability, and chemical stability without compromising their biodegradability or renewability.

Another challenge lies in scaling up the production of biopolymers in a cost-effective manner. While biopolymers are renewable and biodegradable, their large-scale manufacturing must be economically viable to support widespread adoption in commercial batteries. Advances in biopolymer synthesis and processing technologies are needed to ensure that these materials can be produced at a competitive cost without sacrificing performance. Additionally, securing a reliable supply of raw materials and minimizing the environmental impact of biopolymer production are key considerations for the sustainable development of these materials.

Furthermore, the compatibility of biopolymers with different battery chemistries, such as sodium-ion and magnesium-ion systems, needs to be explored. While biopolymers have shown significant promise in lithium-ion batteries (LIBs) and zinc-ion batteries (ZIBs), further research is needed to understand their interactions with other electrochemical systems and to optimize their performance in a broader range of energy-storage technologies.

Finally, long-term durability is a critical challenge. For biopolymers to be viable in commercial battery applications, they must withstand mechanical stress, chemical degradation, and temperature fluctuations over extended periods. Future research should focus on enhancing the mechanical strength and chemical stability of biopolymers to ensure their longevity in real-world applications.

In conclusion, biopolymers and their derivatives offer a promising path forward for battery technology, providing sustainable and environmentally friendly solutions to many of the challenges associated with conventional materials. By integrating biopolymers into battery components such as binders, electrolytes, separators, and protective layers, the industry can move toward more efficient, safer, and greener energy-storage solutions. As research continues to optimize their performance and scalability, biopolymers are expected to play an increasingly important role in the development of next-generation batteries that meet the growing demand for clean energy and sustainability.

2.2 Overview of Biopolymers

Biopolymers are polymer compounds derived from living organisms, such as cellulose, CS, and sodium alginate. They have the advantages of being renewable, degradable, and biocompatible. Biopolymer derivatives are materials obtained by modifying or functionalizing biopolymers through chemical or physical methods and have broader application prospects. Compared with synthetic polymers in the traditional chemical industry, biopolymers have many attractive properties and beautiful functions. For example, biopolymers have high strength and are usually environmentally friendly. Biopolymers are derived from living organisms, and their industrial application can achieve sustainability. As shown in Figure 2.1, biopolymer materials are rich in polar groups such as carboxyl, hydroxyl, or amine groups, so biopolymers can specifically interact with many substances and materials, showing extreme affinity [1]. The molecular structures of biopolymers mentioned in this chapter are shown in Figure 2.2.

2.2.1 Carboxymethyl Cellulose (CMC)

Carboxymethyl cellulose (CMC) is a surface-active colloid polymer compound. It is an odorless, tasteless, nontoxic water-soluble cellulose derivative. It is made of absorbent cotton through physical–chemical treatment. The obtained organic cellulose binding agent is a kind of cellulose ether, and its sodium salt is generally used, so its full name should be carboxymethyl cellulose sodium (CMC-Na) [2]. CMC has many carboxyl groups and hydroxyl groups in its molecular chain, so it has excellent water solubility and film-forming properties [3].

The presence of carboxyl and hydroxyl functional groups in CMC enables it to interact with a wide range of materials, especially through hydrogen bonding. These polar groups promote strong adhesion to active materials, which is crucial in battery applications. CMC's hydrophilic nature also allows it to retain moisture,

Figure 2.1 Application diagram of biopolymers in battery. Source: Ding et al. [1]/American Chemical Society/CC BY 4.0.

improving its ability to form stable, flexible films. Additionally, its biodegradability and nontoxicity make CMC an environmentally friendly choice for applications where sustainability is critical, such as in the development of green batteries. Its high surface activity, ion-exchange capacity, and ability to regulate pH further expand its utility in diverse chemical systems.

CMC is widely used in battery applications, particularly as a binder in lithium-ion and lithium–sulfur (Li–S) batteries [4]. Its ability to form strong hydrogen bonds with active materials enhances the adhesion of electrode components, contributing to improved structural stability during charge-discharge cycles. CMC's water solubility and film-forming properties make it an ideal binder for the aqueous processing of electrodes, reducing the need for organic solvents. In Li–S batteries, CMC effectively regulates polysulfide migration, helping to minimize the shuttle effect and improve cycling stability.

2.2.2 Chitosan (CS)

CS is the product of the deacetylation of chitin, which is widely distributed in nature and is the second largest natural polymer in terms of reserves after cellulose. The molecular formula of CS can be expressed as $(C_6H_{11}NO_4)_n$, where n represents the degree of polymerization, that is, the number of repeating units in the molecule. CS comprises β-(1 → 4)-linked 2-amino-2-deoxy-D-glucose unit [5–8].

Physical and chemical properties of CS: (i) Density: The density of CS is usually about $1.75\,\mathrm{g\,cm^{-3}}$, but the specific value may vary due to factors such as preparation method, purity, and degree of polymerization. (ii) Melting point: The melting point

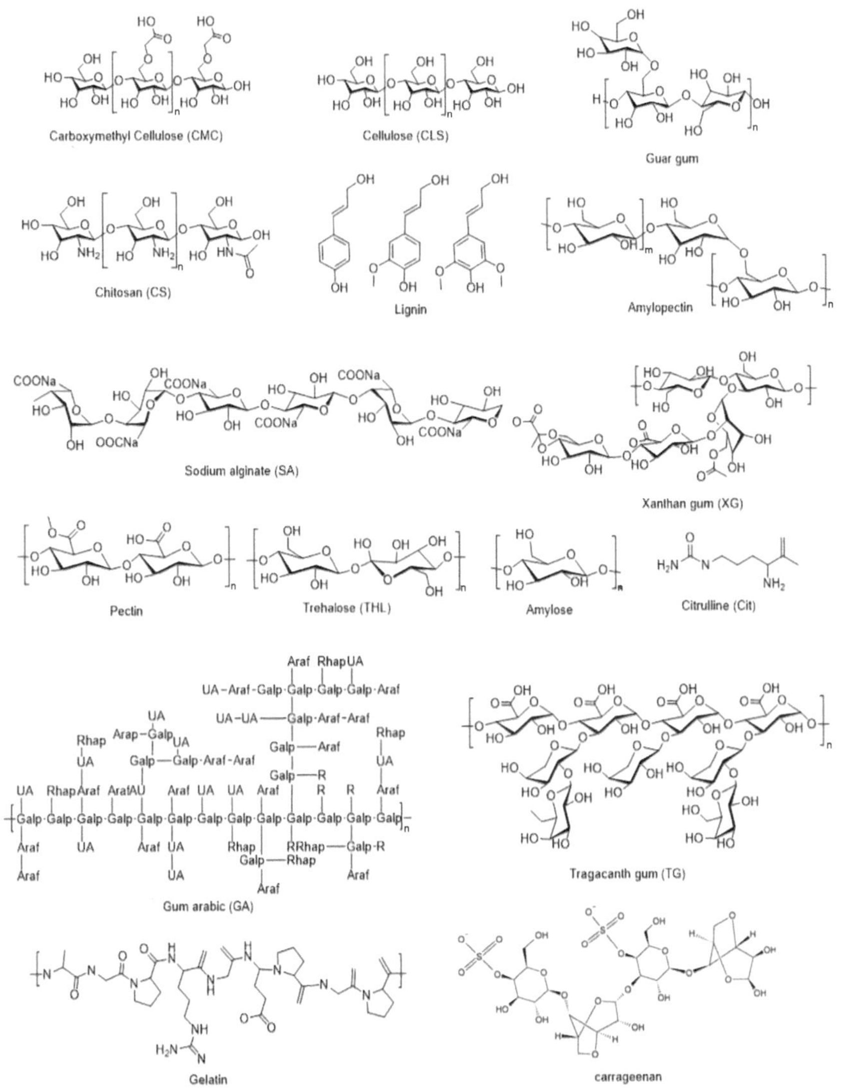

Figure 2.2 Molecular structure of biopolymers mentioned in this chapter.

of CS is usually higher, but due to its significant molecular weight and easy formation of hydrogen bonds, it is not easy to melt under normal conditions. Some data indicate that its melting point may be between 88 and 102.5 °C. (iii) Solubility: CS has limited solubility in water but has good solubility in dilute acid solutions (such as acetic acid solutions). This is because the amino groups of CS will be protonated under acidic conditions, thereby increasing its hydrophilicity. (iv) Stability: CS is stable at room temperature but may degrade under high temperature, strong acid, or alkali conditions. In addition, CS also has good biocompatibility and biodegradability. (v) Film-forming properties: CS has good film-forming properties and can form transparent or translucent films in solution. These films have good air permeability and biocompatibility and broad application prospects in medicine, food packaging, and other fields. (vi) Hygroscopicity: CS has strong hygroscopicity, second only to glycerin and higher than polyethylene glycol (PEG) and sorbitol. This hygroscopicity gives CS good potential for moisturizing and hygroscopic applications. To summarize, CS has unique physicochemical properties, which make it widely used in medicine, food, agriculture, environmental protection, and other fields.

CS has found significant applications in battery technologies, particularly as a binder and electrolyte material in LIBs and ZIBs. Its excellent film-forming capabilities, mechanical strength, and biocompatibility make it an effective binder that enhances the structural integrity of electrodes. Additionally, CS's ability to chelate metal ions improves ionic conductivity and facilitates uniform deposition of active materials, such as zinc or lithium. In ZIBs, CS-based electrolytes help suppress dendrite formation and mitigate side reactions, extending battery life and improving performance.

2.2.3 Sodium Alginate (SA)

SA is a carbohydrate-based polymer material derived from brown algae cell membranes. It is widely used in food and tissue engineering because of its biodegradability and sustainability. The SA molecule is an anionic polysaccharide with a rigid polymer chain composed of two aldehydic acids. One is α-D-mannuronic acid (M-block), and the other is β-l-gluconic acid (G-block), which are linked together by glycosides [7]. Physical and chemical properties of SA: (i) Appearance and properties: SA is an odorless, tasteless, and hygroscopic white or light yellow powder. Its relative density is about 1.59, and its bulk density is 87.39 kg m^{-3}. SA can form a viscous colloidal solution in a 1% aqueous solution with a pH value between 6 and 8. (ii) Solubility: SA is easily soluble in hot and cold water, but almost insoluble in organic solvents such as ethanol, ether, or chloroform. (iii) Stability: SA has good stability, but it may cause chemical reactions under the action of strong acids, strong bases, or strong oxidants. In summary, SA has broad application prospects in the battery field with its unique physical and chemical properties and advantages. With the continuous development of science and technology, the application of SA in the battery field will become more and more extensive, making more significant contributions to the advancement of battery technology.

SA has gained significant attention for its use as a sustainable and high-performance binder in battery technologies, especially in lithium-ion, zinc-ion, and sodium-ion batteries. Its unique gel-forming properties, biocompatibility, and non-toxic nature make it an environmentally friendly alternative to synthetic binders like polyvinylidene fluoride (PVDF). In batteries, SA enhances the mechanical strength of electrodes, allowing them to accommodate the volume expansion and contraction that occurs during repeated charge-discharge cycles, particularly in high-capacity materials such as silicon and sulfur. Its abundant hydroxyl and carboxyl groups enable strong hydrogen bonding with active electrode materials, improving electrode adhesion and minimizing the risk of active material detachment. Furthermore, SA helps optimize electrolyte interaction by promoting better ionic conductivity and maintaining stable electrochemical interfaces. These properties result in improved cycling performance, reduced degradation, and extended battery lifespan. SA's versatility, along with its eco-friendly and cost-effective nature, has led to its growing adoption in the development of next-generation energy-storage systems, offering a pathway toward more sustainable and high-performance battery technologies.

2.2.4 Lignin

Lignin is a polydisperse amorphous natural polymer, mainly derived from natural plant raw materials, composed of highly substituted phenylpropane basic structural units (including guaiacyl (G type), syringyl (S type) and *p*-hydroxyphenyl (H type)) randomly polymerized which is the main component of the plant skeleton. In lignin molecules, common functional groups connected to the benzene ring include alcoholic hydroxyl groups, methoxy groups, phenolic hydroxyl groups, carbonyl groups, etc. Different lignins have phenylpropanoid structures connected with various functional groups. These functional groups give lignin unique chemical properties and reactivity. At the same time, there are many types of connecting bonds in lignin, such as β-O-4 bonds, β-5 bonds, β-1 bonds, etc. [9, 10]. The connecting bonds are related to multiple functional groups that jointly determine the diversity and complexity of lignin molecules, making lignin have extensive research and application value in fields such as biology, chemistry, and materials science [11].

Lignin has emerged as a promising material for battery applications, particularly in lithium-ion, Li–S, and sodium-ion batteries. Its complex structure, composed of aromatic rings and functional groups such as hydroxyl and methoxyl, makes it an effective binder that can significantly enhance electrode stability and performance. In Li–S batteries, for example, lignin-based binders have been shown to mitigate the polysulfide shuttle effect, which is a major cause of capacity loss, by effectively trapping sulfur species within the electrode matrix. This not only improves sulfur utilization but also extends the battery's cycle life.

Furthermore, lignin's ability to form strong chemical bonds with active materials, combined with its natural mechanical strength, provides a robust framework that can accommodate the expansion and contraction of materials like silicon during charge-discharge cycles, reducing electrode degradation. In addition to its role as a binder, lignin can also be chemically modified to improve its ionic conductivity and interaction with metal ions, making it a multifunctional component in advanced battery systems. For instance, by introducing conductive additives or functionalizing its structure, lignin can serve as a pathway for ion transport, further enhancing the overall electrochemical performance of the battery. Lignin's versatility allows it to support not only silicon anodes, which undergo large volume changes, but also sulfur- and sodium-based cathodes, where its binder and structural properties help maintain electrode integrity under demanding cycling conditions [12, 13].

2.2.5 Gum Arabic (GA)

Gum arabic (GA) is a nontoxic, water-soluble natural polymer extracted from Senegalese soap locust. Its main components are polymer polysaccharides and their calcium, magnesium, and potassium salts, including gum aldose, galactose, and glucural acids, etc. GA has a more extended D-galactose amino sugar backbone and side D-glucuronic acid, l-arabinose amino sugar, and hydroxyproline groups, which can be easily modified to introduce functional groups [14, 15].

This natural gum has the following physical and chemical properties: (i) Viscosity and water solubility: GA gradually dissolves in water into an acidic, viscous liquid, making it an ideal binder in manufacturing processes. (ii) Stability: odorless, tasteless, flammable, has good stability, and can be stored for a long time without deterioration. (iii) Hydrophilicity and lipophilicity: The structure contains some protein substances and rhamnose, which have excellent hydrophilicity and lipophilicity. It is a perfect natural oil-in-water emulsion stabilizer. (iv) Electrical conductivity and chemical stability: The physical and chemical properties of GA itself make it have high electrical conductivity and chemical stability, which is conducive to efficient energy conversion and storage. Because GA has high viscosity and stickiness, it can effectively increase the energy density inside the energy-storage device. Using GA as a binder or filler can significantly increase the energy-storage capacity of energy-storage devices compared to traditional materials. At the same time, because GA is a very lightweight material, it can reduce additional loads during the manufacturing process of energy-storage devices, allowing the energy-storage devices to store more energy under the same volume or mass. However, despite the many advantages of GA in the manufacture of batteries and energy-storage devices, there are some challenges and limitations, such as viscosity and adhesion, that may lead to some operational difficulties and may even affect productivity. In addition, its sustainability and renewability are of concern.

GA, a natural polysaccharide extracted from the sap of acacia trees, plays a crucial role in the development of advanced battery technologies, particularly in Li–S batteries and LIBs, where it is used as an effective binder. One of the most critical challenges in these batteries is managing the mechanical stress caused by the expansion and contraction of active materials during charge and discharge cycles. GA addresses this issue through its high mechanical flexibility and excellent binder properties, which enable it to maintain electrode integrity and prevent the detachment of active materials. In Li–S batteries, GA is particularly beneficial in mitigating the polysulfide shuttle effect, a phenomenon where sulfur species migrate between the cathode and anode, leading to significant capacity loss and reduced cycle life. By forming a strong network that traps these polysulfides within the cathode, GA enhances sulfur utilization and reduces material loss, resulting in more stable and efficient battery performance.

Furthermore, GA's rich composition of hydroxyl and carboxyl groups allows it to establish strong hydrogen bonds with both the sulfur and lithium ions, thereby improving ionic conductivity and facilitating smoother charge transport within the battery. This ability to enhance ion transport is crucial for maintaining high energy efficiency and power density in batteries, especially under high-stress conditions. Its excellent water solubility and biocompatibility also make it easy to process into a binder solution, offering a more sustainable and environmentally friendly alternative to traditional synthetic binders like PVDF, which require toxic solvents during processing.

2.2.6 Guar Gum (GG)

Guar gum (GG) is a natural polymer widely used as a stabilizer and thickener in various foods. GG has many advantages, such as abundant reserves, water-soluble, nontoxic, renewable, biocompatible, and cost-effective [10]. The main component of GG is galactomannan. The molecular weight varies depending on the source. It is about 1 million to 2 million. Its structure is composed of D-mannose connected through β-1,4 glycoside bonds to form a main chain. On some mannose, D-galactose forms side chains through α-1,6 glycoside bonds to form a multi-branched polysaccharide [16]. GG has the following physical and chemical properties: (i) Solubility: GG can be dispersed in hot or cold water to form a viscous liquid and is a natural gum with high viscosity. In addition, GG can be dissolved in water-miscible solvents (such as ethanol, dimethylformamide, and dimethyl sulfoxide) at a limited concentration but is insoluble in organic reagents. (ii) Physical form: Commercially available GG is usually a white to light yellow-brown free-flowing powder that is almost odorless and tasteless. As a natural polymer polysaccharide, GG has broad application prospects in the battery field. With the continuous advancement of science and technology and the deepening of research, the application of GG in the battery field will become more extensive and in-depth.

GG has gained significant attention for its role in advanced battery systems, particularly as a binder in Li–S batteries and LIBs. One of the critical functions of a binder in batteries is to maintain the structural integrity of the electrodes, ensuring that the active materials remain bound to the current collector during charge and discharge cycles. GG, with its high viscosity and excellent binder properties, forms a robust, three-dimensional (3D) network that binds the active materials firmly together, reducing the likelihood of material detachment and preventing the disintegration of the electrode structure during repeated cycling. This helps to enhance the long-term stability of the battery, leading to extended cycle life and improved durability.

In Li–S batteries, GG plays a pivotal role in mitigating the polysulfide shuttle effect, a common challenge that leads to capacity fading and performance degradation. The hydroxyl groups in GG can form strong hydrogen bonds with polysulfide species, effectively trapping them within the cathode and preventing their migration to the anode. By doing so, GG enhances sulfur utilization and helps maintain higher energy efficiency throughout the battery's life cycle. This property is especially important in high-sulfur-loading batteries, where the risk of polysulfide dissolution is greater. Additionally, GG's hydrophilic nature improves electrolyte absorption and ionic conductivity, promoting faster lithium-ion transport and contributing to overall improved electrochemical performance.

2.2.7 Xanthan Gum (XG)

Xanthan gum (XG) is a natural nontoxic polysaccharide that can be used as a food additive, rheology modifier, and polymer stabilizer. XG has a unique double helix superstructure, which can achieve a strong mechanical interlocking effect [17–19]. The physical and chemical properties of XG: (i) Unique rheological properties: XG has special rheological properties, allowing it to exhibit diverse characteristics under different conditions. (ii) Good water solubility: XG can dissolve quickly in water, especially in cold water, making it easy to use. However, it should be noted that if water is added directly without sufficient stirring, the outer layer will absorb water and swell into micelles, which will prevent water from entering the inner layer and affect the performance of the function. (iii) Stability to heat, acid, and alkali: The aqueous solution of XG has almost no change in viscosity between 10 and 80 °C, showing stability to heat, acid, and alkali. (iv) It has good compatibility with a variety of salts: XG is compatible with a variety of salts, but it should be noted that it is intolerant to salts and acidic substances. (v) Thickening: XG solution has the characteristics of low concentration and high viscosity. The viscosity of 1% aqueous solution is equivalent to 100 times that of gelatin. It is an efficient thickener. (vi) Pseudoplasticity: XG aqueous solution has high viscosity under static or low shear and shows a sharp decrease in viscosity under high shear, but the molecular structure remains unchanged. When the shear force is eliminated, the original viscosity is restored immediately. With its unique physical and chemical properties and significant advantages, XG has broad application prospects in many fields and shows its unique value in the battery field.

XG has emerged as a versatile and sustainable binder material in advanced battery technologies, particularly in Li–S batteries and LIBs. Its primary role in batteries is to maintain the structural integrity of the electrodes by acting as a strong binder that holds the active materials together during charge and discharge cycles. Due to its high viscosity, excellent binder properties, and unique molecular structure, XG can form a robust, 3D network within the electrode, effectively preventing the detachment of active materials from the current collector. This leads to enhanced cycling stability and prolonged battery life, which is critical for ensuring consistent performance in long-term applications. Additionally, XG has a high affinity for polar functional groups and can interact with polysulfides in Li–S batteries, helping to mitigate the polysulfide shuttle effect, a common issue that causes capacity fading and reduces battery efficiency. Its hydrophilic nature further enhances electrolyte wettability, improving ionic conductivity and promoting faster lithium-ion transport within the electrode.

2.2.8 Starch

Starch is the most common high-molecular carbohydrate, which is a polysaccharide polymerized from glucose molecules. Its basic structural unit is α-D-glucopyranose, which is divided into two categories: amylose and amylopectin (AP) [20]. The former is an unbranched helical structure; the latter is composed of 24–30 glucose residues connected end to end by α-1,4-glycosidic bonds, with α-1 at the branch, 6-glycosidic bond.

Starch has the following characteristics: (i) Chemical stability: Starch molecules have excellent chemical stability and can maintain their structure and properties in a variety of environments. (ii) Biodegradability: Starch is a biodegradable material that can be quickly decomposed in the natural environment and will not cause long-term pollution to the environment. (iii) Low cost: As a biological material that widely exists in nature, starch has a relatively low acquisition cost, making it highly cost-effective in industrial production. (iv) Gelability: Starch solution can form a gel-like substance. This characteristic allows starch to act as an electrolyte in batteries while providing high conductivity. Since starch is renewable and biodegradable, using starch as a battery material is beneficial to environmental protection and sustainable development, and the oxidation reaction of starch in the battery will not produce harmful substances and is harmless to the environment. With its unique physical and chemical properties and advantages, starch shows broad application prospects in the battery field. As a renewable energy material, starch helps promote the green and sustainable development of battery technology. With the continuous deepening of scientific research work, the application of starch in the battery field will be more widely expanded.

Starch has gained attention as an eco-friendly and cost-effective material in battery applications, particularly as a binder in lithium-ion and Li–S batteries. The use of starch in batteries primarily focuses on its role as a natural binder that helps maintain the structural integrity of the electrodes during cycling. Starch, with its abundant hydroxyl groups, can form strong hydrogen bonds with active materials like silicon and sulfur, ensuring good adhesion between the active particles and the current collector. This results in enhanced mechanical stability, preventing the delamination and pulverization of the electrode, which is critical for the long-term performance of batteries. Furthermore, starch-based binders improve electrolyte wettability and ionic conductivity, promoting faster ion transport within the electrode structure. In Li–S batteries, starch's ability to trap polysulfides through chemical interactions helps to mitigate the polysulfide shuttle effect, which is a significant cause of capacity loss in these systems. Researchers have also explored modified starch derivatives, such as carboxymethyl starch or cross-linked starch, to further enhance their electrochemical performance and mechanical properties. Starch's natural abundance, low cost, and biodegradable nature make it a promising alternative to traditional synthetic binders, contributing to the development of more sustainable and environmentally friendly energy-storage systems.

2.2.9 Gelatin

Gelatin is a natural polymer protein material extracted from animal bones, skin, fish scales, cartilage, and other parts of the body, and is one of the most versatile natural biopolymers. The relative molecular mass of gelatin ranges from 15,000 to 250,000, with an average of 50,000 to 70,000, and is a heterogeneous mixture of single- or multi-chain polypeptides composed of glycine and proline residues, which can be obtained by hydrolysis of collagen, and can form a very stable 3D spatial structure [21, 22]. Gelatin has the following physical and chemical properties: (i) Gelling property: After gelatin is heated and dissolved in an aqueous solution, it can gel into a solid colloid after cooling. This gelling property can be used for adhesion, coagulation, thickening, and other functions in food, medicine, cosmetics, and other industries. (ii) Stability: Gelatin has good stability in aqueous solutions at room temperature and does not quickly deteriorate. It can be stored at low temperatures and can withstand certain pH and temperature changes. (iii) Solubility: Gelatin has good solubility in water, but it quickly reacts with acid and precipitates under acidic conditions. (iv) Conductivity: Since the chemical composition of gelatin contains many conductive elements, such as carbon, hydrogen, and nitrogen, gelatin has good conductive properties. Under appropriate cross-linking conditions, gelatin materials can exhibit ideal conductive properties. In summary, gelatin has broad application prospects in the battery field with its unique physical and chemical properties and advantages. Through further research and development, it is believed that the application of gelatin in the battery field will be more extensive and in-depth.

Gelati has found promising applications in battery technology, particularly as a binder and electrolyte material in LIBs and ZIBs. Gelatin's unique properties, such as its ability to form flexible, biocompatible gels and films, make it an attractive option for enhancing battery performance. As a binder, gelatin helps maintain the mechanical stability of electrodes by forming strong binder bonds with active materials like silicon, sulfur, or zinc, ensuring better cohesion within the electrode structure. This improves the cycling stability and prevents electrode degradation caused by volume expansion and contraction during charge and discharge cycles. Additionally, gelatin's functional groups, such as amine and carboxyl groups, can interact with metal ions, enhancing ion transport and conductivity within the battery. In ZIBs, gelatin-based hydrogels have been used as electrolyte materials due to their excellent ionic conductivity and ability to suppress dendrite formation, which can improve the lifespan and safety of the battery. Gelatin's natural biodegradability and low cost make it a sustainable alternative to synthetic materials, aligning with the growing demand for environmentally friendly battery technologies. Gelatin-based electrolytes have been explored for use in flexible and wearable energy-storage devices, leveraging their flexibility and mechanical strength to withstand the mechanical stress encountered in these applications.

2.2.10 Tragacanth Gum (TG)

Tragacanth gum (TG) is easily soluble in water, environmentally friendly, renewable, and has good thermal stability. When dissolved in water at a low concentration, it can form a highly viscous hydrogel. TG is a water-soluble anionic polysaccharide with galactose residues as the main chain and arabinose residues as the branch chain. The molecular structure of its main components has more branch chains and contains a large number of polar functional groups such as carboxyl and hydroxyl groups, so it is easy to design its molecular structure [23, 24]. The physical and chemical properties of TG are: (i) Solubility: TG is difficult to dissolve in water, but part of TG is easily absorbed by water and swells into a gel-like substance. This characteristic enables tragacanth to form a stable gel system to a certain extent, providing the possibility for its application in various fields. (ii) Stability: TG is relatively stable to changes in pH value, and the viscosity is maximum when the pH value is 5. This characteristic allows TG to maintain good stability in different pH environments, providing a guarantee for its application in food, cosmetics, and other fields. (iii) Viscosity:

The viscosity of TG is relatively high. A 1% gum solution will become a smooth, thick, milky white, nonbinder gel-like liquid after being fully hydrated. This high viscosity characteristic gives TG significant advantages in terms of thickening and stabilization. (iv) Emulsifying properties: When TG is used in an oil-in-water emulsion stabilization system, there is no need to add other surfactants because TG also has the function of reducing surface tension. This characteristic gives gum tragacanth broad application prospects in emulsification, suspension, and other aspects. It should be noted that there are currently relatively few studies on the application of tragacanth in the battery field, and further research and exploration are needed in the future. At the same time, due to the particularity and complexity of the battery field, the application of tragacanth in batteries needs to take into account a variety of factors, such as cost, safety, and performance. Therefore, in practical applications, various factors need to be considered comprehensively to select the most suitable additives and processes.

TG has shown potential applications in battery technology, particularly as a binder in lithium Li–S and other advanced battery systems. Its unique binder properties, combined with its ability to form viscous gels in aqueous solutions, make TG an effective binder for maintaining the structural integrity of electrodes. In Li–S batteries, for instance, TG helps to stabilize the sulfur cathode by mitigating the dissolution and shuttle effect of polysulfides, which is a common issue that leads to capacity fade during cycling. TG's strong binding ability enhances the cohesion between active materials and conductive additives within the electrode, improving the overall cycling stability of the battery.

2.2.11 Cellulose (CLS)

Cellulose (CLS) is a natural polymer and the most precious natural renewable resource for humanity. It is a linear polymer compound composed of β-D-glucopyranose repeating units, with the chemical formula $(C_6H_{10}O_5)_n$, where n represents the degree of polymerization, with values ranging from hundreds to tens of thousands. The macromolecule of CLS is in a chair conformation [25], and each glucose ring has free hydroxyl groups at carbon positions 2, 3, and 6 [26, 27]. Due to its many polar chemical groups (such as —OH, —O—), it has ionic conductivity, which makes it a candidate polymer matrix for quasi-solid polymer electrolyte (QPE). In addition, flexible energy-storage devices assembled with CLS as the matrix have excellent mechanical properties and stability in use and can still work usually after withstanding multiple cycles of external force. Therefore, CLS has extensive application prospects in the battery field.

CLS one of the most abundant natural polymers, has gained significant attention in battery applications, particularly as a sustainable and eco-friendly material for separators, binders, and electrolytes. In LIBs and ZIBs, CLS-based materials are used to fabricate separators with excellent mechanical strength, thermal stability, and ionic conductivity. Due to its high porosity and hydrophilicity, CLS allows for efficient electrolyte uptake and ion transport, which enhances the overall performance of the battery. Moreover, its chemical modification, such as grafting or cross-linking with other polymers, can further improve its electrochemical properties, such as reducing dendrite formation on metal anodes, a critical issue in lithium and zinc batteries. As a binder, CLS derivatives like CMC offer strong adhesion and flexibility, helping to maintain the structural integrity of electrodes during cycling. Furthermore, CLS-based materials are biodegradable and non-toxic, making them an ideal choice for developing green and sustainable battery technologies. These attributes underscore CLS's potential in advancing high-performance and environmentally friendly energy-storage solutions.

2.2.12 Trehalose (THL)

Trehalose (THL) is a disaccharide widely present in organisms, including bacteria, insects, and plants [28]. Excellent chemical stability and high glass transition temperature make THL an ideal barrier for microorganisms. THL has fewer internal hydrogen bonds and can establish strong interactions with polar groups in biomolecules due to the high flexibility of the two glucose units. The amorphous THL shell formed can even exclude water from the biomolecule's hydration shell to solid hydrogen bonding sites around the THL molecule. In addition, THL can deconstruct the tetrahedral hydrogen bond network of water and inhibit the crystallization of ice in the THL/water system [29].

THL has found promising applications in battery technology, particularly in enhancing the performance and longevity of zinc-ion and lithium-based batteries. THL is utilized primarily as an electrolyte additive, where it helps regulate the ionic environment, stabilize the electrode interface, and inhibit detrimental reactions such as hydrogen evolution and dendrite formation. Its unique ability to form a stable coordination environment around metal ions, like zinc and lithium, promotes uniform deposition and reduces the formation of irregular structures that can cause battery failure. THL's ability to control crystallization and prevent degradation mechanisms extends the cycling stability of the battery and enhances its overall efficiency. Additionally, its biocompatibility and water solubility make it an environmentally friendly option for electrolyte modifications, aligning with the growing demand for sustainable energy-storage solutions. The multifunctional role of THL in improving battery interface chemistry and electrochemical performance highlights its potential as a novel and green additive in next-generation battery technologies.

2.2.13 Citrulline (Cit)

Citrulline (Cit) is an α-amino acid with the chemical formula $C_6H_{13}N_3O_3$. It contains the typical —NH_2 and —COOH groups in general amino acids [30]. It is worth noting that the side chain of Cit is composed of polar groups, which makes it hydrophilic. Amino acids have an apparent affinity for water molecules, resulting in high water solubility [31]. Citrulline has the following advantages: (i) Antioxidant properties: Citrulline has antioxidant properties, which makes it worthwhile in many. The field has broad application potential. (ii) Diverse synthesis methods: The synthesis methods of citrulline include the reduction reaction method, synthetic enzyme method, amination reaction method, substrate exchange method, etc.

Various synthesis methods help to meet the needs of different fields. Citrulline has a variety of physical and chemical properties and advantages, and has broad application prospects in many fields. However, its application in the battery field still requires further research and exploration.

Cit has gained attention as an electrolyte additive in battery systems, particularly in ZIBs. Citrulline's unique chemical structure, featuring highly polar functional groups such as amine (—NH$_2$) and carboxyl (—COOH), allows it to interact strongly with both Zn^{2+} and the zinc metal anode. This dual interaction helps regulate the solvation shell of zinc ions and stabilizes the double electric layer at the anode-electrolyte interface. By effectively controlling these interactions, citrulline inhibits dendrite formation, a major issue in zinc-based batteries, and minimizes side reactions such as hydrogen evolution, which can lead to battery degradation. The result is enhanced cycling stability and improved performance under high current densities and capacities. Citrulline's dual role as a zinc ion modulator and an interface stabilizer makes it a promising additive for improving the longevity, safety, and efficiency of zinc-ion and other metal-based batteries, while also offering the advantages of being biocompatible and environmentally benign.

2.2.14 Pectin

Pectin is a water-soluble polysaccharide that is abundant in nature and has low toxicity. It is the polysaccharide with the most complex structure and function. It is rich in galacturonic acid, including homogalacturonic acid, rhamnogalacturonic acid I, and Substituting galacturonic acid rhamnogalacturonic acid II and xylosegalacturonic acid [32–34]. Pectin content by ester is divided into high methoxyl pectin and low methoxyl pectin. High methoxy pectin has a polysaccharide backbone with a carboxyl content of 50%, while low methoxy pectin has a methyl ester content of <50% [35]. The anionic structure of pectin plays a crucial role in forming coordination sites with metal cations in polymer electrolyte systems. The increase in the number of coordination sites in the polymer structure will increase ion dissociation, which is beneficial for improving ionic conductivity [36].

Pectin, a natural polyanionic polysaccharide commonly found in plant cell walls, has been explored as a versatile electrolyte additive and binder in battery applications, particularly in zinc-ion and Li–S batteries. Its abundance of oxygen-containing functional groups, such as carboxyl and hydroxyl, enables pectin to form strong hydrogen bonds with metal ions and facilitate uniform ion transport. In ZIBs, pectin helps create a stable interfacial adsorption layer on the zinc anode, reducing the desolvation activation energy of hydrated zinc ions and thereby improving zinc ion transport. This adsorption layer also inhibits unwanted side reactions, such as hydrogen evolution, and prevents dendrite formation, significantly enhancing the cycling stability and longevity of the battery. Furthermore, the zincophilic nature of pectin's functional groups promotes uniform zinc deposition, preventing performance degradation. Pectin's ability to act as both a binder and an electrolyte additive, combined with its eco-friendly and biodegradable nature, makes it an attractive material for developing sustainable, high-performance battery systems.

2.2.15 Carrageenan

Carrageenan is a natural polysaccharide derived from red seaweed, known for its distinctive gel-forming and thickening properties. It consists of repeating galactose units and anhydrogalactose, often containing sulfate groups, which give it unique functional attributes. The physical properties of carrageenan are largely determined by its structure and type (kappa, iota, or lambda). Kappa carrageenan forms strong, rigid gels in the presence of potassium ions, while iota forms softer, elastic gels with calcium ions. Lambda carrageenan does not gel but contributes to high viscosity in aqueous solutions. Carrageenan is highly soluble in hot water and can form viscous solutions, making it useful as a thickening and stabilizing agent. Its gelation ability and film-forming capacity are exploited in various industries, including food, and pharmaceuticals, and increasingly in bio-material applications like energy storage, where it can create stable, flexible films, or hydrogels.

Chemically, carrageenan is characterized by the presence of sulfate ester groups, which make it anionic and allow it to interact with various cations. These ionic interactions, especially with metal ions, play a crucial role in its gel-forming behavior and potential applications like ion transport in batteries. The degree of sulfation affects the mechanical and chemical properties of carrageenan, such as its solubility, viscosity, and thermal stability. Carrageenan is thermally stable under moderate conditions but can degrade under extreme heat or highly acidic or basic environments. Its hydrophilic nature allows it to absorb water, forming hydrogels, which are useful in electrolyte applications. Additionally, carrageenan is biodegradable and environmentally friendly, making it an appealing material for sustainable technologies, including its use in bio-based battery systems where it can enhance ionic conductivity and mechanical stability.

Carrageenan has emerged as a promising bio-based material for battery applications due to its excellent gel-forming ability, ionic conductivity, and mechanical properties. In particular, kappa carrageenan, known for its strong gelation in the presence of metal ions like potassium, has been explored as a component in solid or quasi-solid electrolytes for batteries, especially in lithium-ion and zinc-ion systems. Its ability to form hydrogels and its high water-absorption capacity make carrageenan a suitable candidate for creating electrolyte separators, which facilitate ion transport while maintaining structural integrity. Carrageenan-based electrolytes can enhance battery performance by improving ionic conductivity and preventing electrolyte leakage. Additionally, the sulfate groups in carrageenan interact with metal ions, which helps stabilize the electrode/electrolyte interface and reduces issues such as dendrite formation in metal anodes, thereby extending the battery life. Due to its biodegradability and sustainability, carrageenan is also viewed as an environmentally friendly alternative to synthetic polymers in green energy-storage solutions.

2.3 Application of Biopolymers in Binders

The role of binders in the development of high-performance batteries is increasingly gaining attention as researchers strive to enhance battery capacity, stability,

2.3 Application of Biopolymers in Binders

and overall efficiency. In particular, biopolymers and their derivatives have emerged as promising alternatives to traditional polymeric binders in energy-storage systems such as lithium-ion, sodium-ion, and Li–S batteries. This section critically examines the application of biopolymers and their derivatives in battery binders, with a focus on their structural advantages, functional performance, and the mechanisms underlying their improvements in battery efficiency.

Binders are critical components in battery systems, especially in cathodes, where they ensure adhesion between the active materials and the current collector. Despite being used in small amounts, binders significantly affect key battery parameters such as capacity, life cycle, and safety. Traditional binders, while effective in providing adhesion, often hinder ionic conductivity and exhibit structural limitations under prolonged cycling. This has spurred the search for multifunctional binders that offer both strong adhesion and improved electrochemical performance, including ionic conductivity and interface stability.

Three major design criteria have been identified for advanced binders: (i) Strong adhesion that facilitates better contact between the binder and active materials. (ii) A stable 3D cross-linked structure that ensures the integrity of the electrode during repeated charge-discharge cycles. (iii) Interface stability that maintains efficient electron and ion transport throughout the battery's life. Biopolymers, with their inherent functional groups and adaptable molecular structures, have the potential to meet these criteria more effectively than conventional binders. Biopolymers and their derivatives have shown significant promise in battery applications due to their unique chemical and physical properties.

2.3.1 Carboxymethyl Cellulose

CMC plays a crucial role in regulating polysulfides in Li–S batteries, improving cycling stability. Researchers [37] developed a co-polymer binder combining CMC and glucose (G), which regulates polysulfides and forms viscoelastic fibers, optimizing the sulfur cathode's microstructure. The cathode electrode with a CMC/G binder system shows a more favorable separation structure. Sulfur and carbon are exposed to a large extent because of dispersed particle-level connections rather than aggregated networks. This network structure enables the sulfur-positive electrode to have a short internal path for maximum exposure to the active substance, enhanced electrolyte accessibility and low resistance, and lithium-ion transfer. This leads to a pouch cell prototype with 206 Wh kg^{-1} energy, showing strong potential for practical applications.

Yu et al. [38] introduced small molecule epichlorohydrin (ECH) into CMC through a simple ring-opening reaction and developed a novel water-based novel sodium CMC-ECH binder with a 3D network crosslinking structure. The preparation process is simple, low cost, and environmentally friendly, which can be used for Si anodes of lithium-ion half batteries and full batteries. The excessive degradation of electrolytes caused by silicon anode grinding can be effectively prevented by the abundant covalent bond and hydrogen bond in the CMC-ECH binder.

Sun et al. [39] reported a water-soluble CW-20 binder consisting of sodium CMC-Na and waterborne polyurethane (WPU). CW-20 has an abundant hydrogen bond site 3D structure. This new binder can not only establish a cross-linked 3D network through hydrogen bonding, effectively maintaining the integrity of the electrode, but also form a stable solid-electrolyte interphase (SEI) layer, thereby improving cycle stability and durability. In addition, the long PEG chain in the binder is conducive to promoting the diffusion of Li$^+$ due to the abundance of O atoms.

CMC-based binders have strong rigidity and brittleness. After vacuum drying, cracks are obviously visible on the electrode surface with CMC as the binder, which may even lead to gaps between the electrode material coating and the fluid collector, resulting in electrode "dropping". This defect can be improved by blending CMC with a high-elastic polymer such as styrene-butadiene rubber.

2.3.2 Chitosan

Various innovative binder systems on CS developed by researchers have made significant progress in improving battery performance with Li–S, zinc–iodine, and silicon anode technologies. These binders, including CS-based networks, polyaniline (PANI)-grafted CS, natural rubber cross-linked CS, and 3D polymer networks, provide mechanical strength, elasticity, and improved polysulfide or ion adsorption. By addressing key challenges like electrode expansion, shuttle effects, and charge transfer resistance, these binders enhance cycling stability, capacity retention, and overall battery longevity, paving the way for next-generation, sustainable energy-storage solutions.

The elastic CS-based network developed by Kim et al. [40] combining CS with carboxyl nitrile butadiene rubber in water, exhibits excellent mechanical strength, elasticity, and polysulfide adsorption capabilities. This binder system plays a critical role in stabilizing Li–S batteries by providing structural integrity and managing polysulfide behavior. Its strong mechanical properties and elasticity help accommodate the expansion and contraction of the sulfur cathode during charge-discharge cycles, preventing degradation and maintaining the cathode's integrity. Additionally, the polysulfide adsorption capability of the CS component effectively mitigates the polysulfide shuttle effect, minimizing capacity loss and enhancing long-term cycling performance. Despite using only 3 wt% of the binder, the system achieves remarkable cycling stability, with capacity retention decay as low as 0.026% and 0.029% after 500 cycles at 5C and 10C, respectively. Furthermore, the cell demonstrates a high specific energy of 228 Wh kg^{-1} at an ultrahigh charge-discharge rate of 20C, showcasing its potential for next-generation, commercially viable energy-storage devices. The combined principles of polysulfide adsorption, mechanical reinforcement, and optimized material usage make this binder a promising solution for enhancing the performance and longevity of Li–S batteries.

The Janus functional binder developed by Yang et al. [41] is based on CS, a material with a unique double-helix structure and strong ion coordination abilities. CS, a natural, abundant, and low-cost biopolymer, provides multiple advantageous properties that make it ideal for battery applications. Structurally, CS has a distinctive double-helix formation and is rich in hydrogen-bonded groups, which support the formation of a robust hydrogen-bonded cross-linked network. This network acts

Figure 2.3 CS acts as a binder to limit electrode material expansion and iodine shuttling. Source: Reproduced from Yang et al. [41]/with permission of American Chemical Society.

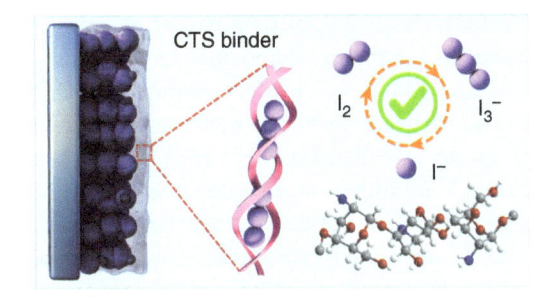

as a stable framework within the battery, effectively restricting electrolyte swelling and countering the expansion of host materials, essential for maintaining electrode integrity over time.

Functionally, CS contains polar functional groups such as —NH$_2$, —OH, and —O—, which exhibit a high affinity for iodine species. As shown in Figure 2.3. These groups not only physically and chemically adsorb iodine but also play a key role in facilitating quasi-iodine catalytic conversion, thereby improving iodine retention and significantly reducing the undesirable shuttle effect common in iodine-based systems. Unlike traditional binders like PVDF, which lacks these capabilities, the CS binder ensures consistent iodine distribution and stability by forming an interaction that supports iodine's redox behavior within the electrode. This mechanism results in a uniform, dense iodine layer that adheres well to the active material, improving the overall electrode functionality and mitigating capacity loss over time. Furthermore, CS is inherently degradable and recyclable, supporting sustainability by enabling reused or recovered battery components. The binder's natural biocompatibility and renewability align with growing demands for eco-friendly materials in energy storage. In the context of Zn–I$_2$ systems, CS offers a dual benefit: high performance through structural and chemical stability and reduced environmental impact, as it allows for material reuse without significant degradation. This approach underscores CS's potential as a multifunctional binder, providing new perspectives on binder design for sustainable, high-performance zinc-based batteries.

The PANI-grafted CS binder, developed by Kim et al. [42] plays a crucial role in improving the performance of silicon anodes. By grafting PANI onto CS, the binder becomes water-soluble and demonstrates enhanced electrical conductivity. The degree of grafting and cross-linking directly influences the reduction in charge-transfer resistance of the silicon electrodes. The optimal binder composition, consisting of 50% PANI and 50% CS, results in a specific capacity of 1091 mAh g^{-1} after 200 cycles, with a high Coulombic efficiency of 99.4%. The balance of electrical conductivity and mechanical properties ensure that this binder is highly effective for silicon anodes, supporting efficient charge transport and stability.

The natural rubber cross-linked with CS binder, developed by Lee et al. [43] is designed to address the significant volume expansion in silicon anodes during cycling. This study employs a CS/natural rubber network as a binder material for silicon-based anodes in LIBs. The binder network combines CS, known for its binder properties through hydrogen bonding, and natural rubber, chosen for its elasticity, toughness, abundance, and low cost. The CS component of the network contains amine and hydroxyl groups, which form strong hydrogen bonds with silicon particles, anchoring them securely. The natural rubber, which is epoxidized to create epoxidized natural rubber (ENR), contributes elasticity, allowing the network to stretch and contract reversibly in response to the silicon's volume changes during charging and discharging cycles.

This combination of properties provides a stable, flexible structure that can accommodate the significant volume variations of silicon, reducing mechanical failure and maintaining electrode integrity. The cross-linking between CS and ENR, particularly between the amine groups in CS and the epoxy groups in ENR, ensures a robust and flexible network. This adaptability and resilience of the CS/ENR binder enable the silicon electrode to withstand repeated expansion and contraction without degradation, providing enhanced longevity and stability in battery applications. The study illustrates that balancing binder strength with elasticity is key to managing the stress associated with silicon's volume expansion in high-performance LIBs.

Cao et al. [44] introduced a 3D polymer network by grafting 3,4-dihydroxy benzaldehyde onto CS and cross-linking with glutaraldehyde. The catechol group in this binder enhances adhesion between the silicon electrode and the binder, while the 3D polymer network significantly improves mechanical strength. This combination results in a silicon nanoparticle-based (SiNP) anode with 91.5% capacity retention after 100 cycles and a specific capacity of 2144 mAh g^{-1}. The strong adhesion and mechanical reinforcement provided by this binder make it an excellent candidate for high-performance polymer binders in silicon and other alloy-based electrodes, contributing to the stability and long-term efficiency of LIBs.

As an environmentally friendly material with excellent properties, CS-based binders have broad application prospects in the field of battery manufacturing. The future research direction of CS-based binders for batteries includes: developing a new type of CS-based conductive binders to improve the performance of batteries by doping highly conductive materials. Explore composite binders that combine conductive materials and polymers to improve electrical conductivity and mechanical bonding. Research on self-healing binders that utilize reversible and non-covalent bonds to mitigate battery damage during long-term use. In the future, through continuous research and technological innovation, it is expected to further optimize its performance to meet the needs of batteries with higher energy density and longer cycle life.

2.3.3 Sodium Alginate

SA serves as an effective binder in silicon-based anodes, offering enhanced mechanical strength and structural stability through strong hydrogen bonds formed between its polar groups and the silicon surface. This bonding creates a stable interface that helps mitigate the volume expansion of silicon during battery cycling. SA's ability to form a cross-linked structure, particularly when enhanced with additives like calcium ions or combined with other polymers, further improves its

2.3 Application of Biopolymers in Binders

capacity to maintain electrode integrity under mechanical stress. Additionally, SA enhances ionic conductivity and provides flexibility, making it an ideal material for managing mechanical stresses and maintaining the performance of silicon anodes in LIBs. The research by Hu et al. [45] highlights the strong hydrogen bonds formed between SA's polar groups, such as carboxyl and hydroxyl, and the hydroxyl groups on the silicon surface. This bonding creates a stable interface that significantly improves the mechanical properties of nano-silicon electrodes, surpassing those of traditional PVDF-based electrodes. SA enhances the peeling strength of the electrode, which contributes to its overall superior electrochemical performance. The principle behind this is the strong hydrogen bonding that provides a more robust and stable electrode structure, making SA an ideal binder for silicon anodes in improving battery efficiency and longevity.

Zhang et al. [46] developed an SA-based binder enhanced with calcium ions to form a 3D crosslinked network, which significantly improved the material's properties. This network provided excellent mechanical strength and resistance to the volume expansion of silicon particles, which is a common issue in silicon-based anodes. The role of this material in the system is to act as a binder that maintains electrode integrity while allowing for stable, high-capacity performance. The principle behind its effectiveness is the formation of a robust 3D cross-linked structure that mitigates the mechanical stresses caused by silicon's volume expansion, thus enhancing the durability and performance of silicon-based anodes in LIBs.

Gendensuren and Oh [47] developed a composite binder by grafting polyacrylamide (PAM) onto SA and utilizing both ionic and covalent cross-linking. This binder significantly enhances the structural stability of the silicon anode while also improving its ionic conductivity. The dual-crosslinked structure is particularly effective in controlling the volume expansion of the active materials during cycling, as confirmed by in situ electrochemical dilatometry. As a result, the Si/C electrode maintains a high capacity of nearly 840 mAh g^{-1} even after 100 cycles, demonstrating excellent cycle stability. The principle behind its effectiveness lies in the combined cross-linking, which strengthens the binder's mechanical properties and provides better ionic transport, leading to improved electrochemical performance and durability.

Guo et al. [48] developed a blend of SA and sodium carboxymethylcellulose (CMC) to create a flexible yet mechanically strong binder for silicon anodes. This blend formed a cross-linked hydrogen bond structure that allowed the silicon anode to endure significant volume changes during cycling, a common challenge in silicon-based anodes. This showcases the potential of SA blends to enhance the durability and performance of silicon anodes in LIBs.

Li et al. [49] investigated the partial carbonization of SA at 250 °C. This process improved the conductivity of the binder while retaining its bonding properties, which are critical for maintaining the electrode's structural integrity. As a result, the cycling stability of silicon electrodes bound with SA was significantly enhanced. The carbonization technique offered a cost-effective method for boosting the performance of SA as a binder, making it a promising approach for large-scale applications in energy storage.

In 2011, Kovalenko et al. [50] demonstrated the effectiveness of marine-derived SA as a binder for silicon anodes. They showed that SA could form a uniform, elastic passivation layer on the silicon surface, which buffered the volume expansion of nano-silicon particles during cycling. This elastic property preserved both the electronic and ionic conductivity of the silicon anode, resulting in a reversible capacity that was eight times higher than that of traditional graphite anodes. Their work established SA as a viable alternative binder material with superior performance, particularly for managing the mechanical stresses associated with silicon's large volume changes during battery operation.

SA-based binders play an important role in battery manufacturing because of their good chemical and electrochemical stability, dispersion performance, and certain environmental adaptability. However, it is more sensitive to environmental conditions, such as temperature. Under high-temperature conditions, additional treatment or modification may be required to enhance its performance. Therefore, its application may need to be appropriately adjusted and optimized according to the specific battery type and application needs.

2.3.4 Lignin

Lignin, when used as a binder material in lithium-based batteries, plays a crucial role in providing strong adhesion between active materials, conductors, and current collectors. Its natural binder properties, enhanced through chemical modification, enable it to create stable and conductive interfaces, essential for maintaining electrode integrity during cycling. Additionally, lignin-based binders offer flexibility and mechanical strength, helping to buffer volume changes in active materials, particularly in silicon and sulfur electrodes. These properties make lignin a promising material for improving the durability, stability, and overall performance of lithium-ion and Li–S batteries.

Inspired by the natural binding properties of lignin in plants and the adhesion structures of mussels, Chen et al. [51] synthesized a novel Li–S battery cathode material using lysine-modified catecholin lignin (AL-Lys-D). As shown in Figure 2.4, this material leverages alkali lignin (AL), a byproduct of pulping, with modifications that enhance its adhesion properties and dispersibility. The amino acid moieties improve the adsorption between lignin and active materials, while the catechol groups provide strong adhesion to the current collector, even after electrolyte wetting. AL-Lys-D's amphiphilic nature ensures good dispersion within the active and conductive materials, allowing it to function effectively at an ultra-low dosage (2 wt%), which is unprecedented in Li–S batteries.

The material's semi-rigid 3D network structure and elasticity help it manage the volume variations of sulfur during cycling, while its functional groups interact with sulfur species to suppress the polysulfide shuttle effect. This combination of properties enables AL-Lys-D to act as a robust, water-dispersible binder that enhances the stability, adhesion, and performance of sulfur cathodes in Li–S batteries. The unique structure and chemical composition of AL-Lys-D allows it to serve as an effective and sustainable alternative to traditional synthetic binders, underscoring its potential for high-performance, eco-friendly battery applications.

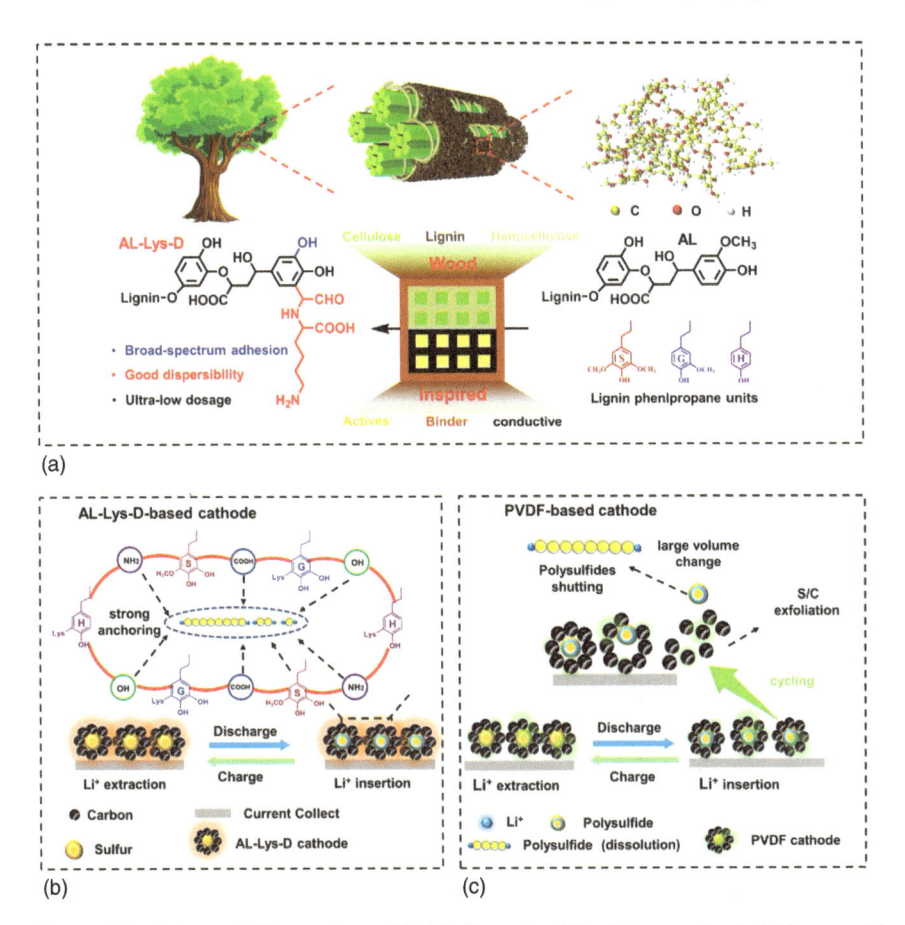

Figure 2.4 Schematic illustrations of (a) binding role of lignin in wood, mimicking wood to fabricating lignin bound sulfur cathode in Li–S battery, (b) mussel-like lignin binder can effectively maintain the electrode stable and anchor polysulfides during the electrochemical process, and (c) traditional PVDF bound cathode would exhibit large volume change, active material exfoliation, and polysulfides dissolution during cycling. Source: Reproduced from Chen et al. [51]/with permission of Wiley-VCH Verlag GmbH & Co. KGaA.

In another study, Yuan et al. [52] developed a 3D lignin polyacrylic acid (PAA) copolymer binder (L-co-PAA) with a lignin-to-PAA mass ratio of 1 : 3. This binder demonstrated ultra-high elasticity with a tensile strain of 630%, thanks to the copolymerization of lignin and PAA, which weakens the hydrogen bonding between PAA chains and enhances flexibility. This method provides a cost-effective way to utilize lignin in polymer binders with superior mechanical properties, paving the way for its application in high-performance LIBs.

With the rapid development of electric vehicles and portable electronic devices, the requirements for battery performance are constantly increasing. Lignin-based binders are considered one of the promising materials for improving battery performance due to their advantages of environmental protection, renewable and low cost. Although lignin-based binders have good development prospects, there are still some challenges, such as how to improve their weather resistance and how to further optimize their performance. At the same time, with the advancement of technology and the reduction of cost, lignin binders are expected to be applied in more fields, bringing new growth points to the binder market.

2.3.5 Gum Arabic

GA serves as an effective binder in Li–S batteries due to its excellent mechanical properties, such as high bonding strength and ductility. These characteristics allow GA to accommodate the volume changes that occur during battery operation, helping to maintain the integrity of the sulfur cathode structure. Additionally, GA's strong bonding capability helps mitigate the shuttle effect of sulfur polysulfides, which is a common challenge in Li–S batteries. As a natural biopolymer, GA not only enhances the structural stability and cycling performance of the cathode but also offers an environmentally friendly alternative for battery material development.

Li et al. [53] explored the use of GA as a binder for Li–S batteries, leveraging its excellent mechanical properties, including high bonding strength and ductility. These characteristics enable the GA binder to effectively buffer the volume changes that occur during the charge and discharge cycles, helping to suppress the shuttle effect of sulfur polysulfides, a significant issue in Li–S batteries. As a result, the GA binder supports the formation of stable, high-capacity Li–S batteries with improved cycling stability. The use of this natural biopolymer not only enhances the electrochemical performance of sulfur-based cathodes but also offers an environmentally friendly solution for battery materials. The principle behind GA's role lies in its mechanical flexibility and ability to form strong bonds, which stabilize the cathode structure during operation.

Zhong et al. [54] prepared acacia grafted polyacrylic acid (GA-g-PAA) by free radical grafting polymerization of acrylic acid on acacia gum. This material is used as a cross-linked water-based binder for the silicon anode of LIBs, forming a cross-linked network by binding with branched polyols such as pentaerythritol (PER) and triethanolamine (TEOA). This cross-linking reaction takes place at approximately 110 °C, matching the processing temperature of industrial electrode sheets. GA-g-PAA/PER binder has higher bonding strength than GA-g-PAA and GA-g-PAA/TEOA. In addition, DSC and FTIR measurements confirmed esterification crosslinking of GA-g-PAA with PER/TEOA at about 110 °C. Silicon anode using GA-g-PAA/P shows slightly better cycling performance and better magnification capability than GA-g-PAA/T and GA-g-PAA. SEM observation shows that GA-g-PAA/P and GA-g-PAA/T binders can inhibit the formation of silicon-negative electrode cracks during the cycle, compared with GA-g-PAA. This environmentally friendly polymer cross-linked GA-g-PAA water-based binder has great application potential in the silicon negative electrode of next-generation LIBs.

The main challenge with GA binders is that their performance may not be suitable for all types of materials, especially in application scenarios that require high strength and durability. In addition, its production may be limited by the availability of raw materials, especially in the context of global resource constraints. With the advancement of technology and the increasing requirements for environmental protection, the market demand for GA binders as a natural and sustainable choice is expected to continue to grow. At the same time, through innovation research and development, existing performance limitations can be overcome and new application areas can be opened up.

2.3.6 Guar Gum and Xanthan Gum

GG and XG act as an effective and sustainable binder for Li–S batteries due to their rich oxygen-containing functional groups, which form a strong 3D biopolymer network through intermolecular binding. They have their characteristics and play an indispensable role in battery manufacturing. For specific battery systems and working conditions, the most suitable adhesive type and dosage need to be selected through experiments.

Liu et al. [55] synthesized an efficient and environmentally friendly binder for Li–S batteries using non-toxic, low-cost GG and XG. These materials, rich in oxygen-containing functional groups, formed a strong 3D biopolymer network (N-GG-XG) through intermolecular binding, which effectively trapped polysulfides. This robust binder enabled the battery to achieve an exceptionally high sulfur loading of 19.8 mg cm^{-2} and an ultrahigh areal capacity of 26.4 mAh cm^{-2}. The binder's mechanical strength and ability to bind polysulfides contributed to its success, offering a promising approach for developing high-energy-density Li–S batteries with low-cost and sustainable materials. The principle behind its effectiveness lies in the strong intermolecular network that maintains structural integrity while enhancing sulfur utilization and capacity.

GG and XG as important binders, its development prospects are broad. As battery technology continues to advance, the two may be used in combination or developed with other novel adhesives to meet the need for higher performance. With the expansion of application areas and the growth of market demand, both binders are expected to achieve wider applications in the coming years.

2.3.7 Starch

Starch-based binders play a crucial role in improving the adhesion and structural integrity of silicon anodes in LIBs. The natural polymers in starch, such as AP and modified starches, form strong hydrogen bonds with silicon particles, enhancing the binder's ability to maintain electrode stability during cycling. Additionally, starch-based binders can be easily modified, such as through cross-linking or incorporating other components, to improve their mechanical strength and ionic conductivity. These properties help the silicon anodes better withstand volume changes, ensuring more reliable performance and long-term durability in energy-storage applications.

Inspired by the traditional use of glutinous rice as a binder in ancient construction, Ling et al. [56] extracted AP from glutinous rice and applied it as a water-based binder for silicon anodes. Their research demonstrated that AP-bonded silicon electrodes possess better adhesion to silicon particles compared to conventional binders like CMC and PVDF. Nano-scratch tests confirmed that AP increases the friction coefficient, while nanoindentation tests showed enhanced mechanical strength, allowing the electrodes to better withstand silicon's volume changes during cycling. The Si-AP electrodes, containing 60% silicon, achieved a high discharge capacity of 1517.9 mAh g^{-1} after 100 cycles, outperforming traditional binders.

In a similar effort, Hapuarachchi et al. [57] used PEG-modified cassava starch as a water-soluble binder for silicon anodes. The hydroxyl groups in cassava starch formed strong hydrogen bonds with silicon particles, ensuring excellent adhesion, while the PEG component improved lithium-ion conductivity. This combination resulted in better cycling stability than conventional binders like PVDF and CMC. The modified starch binder enhanced the overall electrochemical performance of silicon anodes, highlighting its potential for use in advanced LIBs.

Further advancing the use of natural starches, Rohan et al. [58] cross-linked corn starch with maleic anhydride in situ to create a 3D network structure for silicon anode binders. Their tests showed that the cross-linked corn starch binder exhibited an adhesion force 4.9 times stronger than uncrosslinked corn starch, providing superior structural integrity. Additionally, the Si composite cathode paired with this binder delivered a high specific capacity of 3720 mAh g^{-1}, significantly improving cycle life. This approach underscores the cost-effectiveness and environmental sustainability of corn starch as a promising binder for silicon-based electrodes.

With the increasing requirements for sustainable development and environmental protection, starch-based binders are expected to be widely used in large-scale battery production because of their cost-effectiveness and environmental friendliness. Future research will focus on improving the electrochemical properties of the binders, enhancing their environmental stability, and expanding their application range.

2.3.8 Gelatin

Gelatin-based battery binders play an important role in the manufacture of LIBs, mainly for the bonding of positive electrode materials, whose bonding is mainly dependent on intermolecular van der Waals forces. In the positive electrode material, chemical bonds and chelates are formed between the gelatin and the active material, thereby improving the bond strength and overall stability of the electrode. This kind of binder usually has good adhesion and electrochemical stability, but there are some limitations, such as brittleness, functional groups, easy-to-cause side reactions, and poor water resistance. In order to improve these properties, researchers usually adopt the technical route of blending with other water-based binders.

Sun et al. [59] developed a water-soluble functional binder by cross-linking gelatin (GN) with boric acid (BA) for use in Li–S batteries. This binder effectively maintains the stability of the electrode structure during cycling, buffers the volume changes, prevents active materials from detaching from the current collector, and anchors polysulfides through the formation of chemical bonds (B-O-Li, C-O-Li, and

C-N-Li), thereby suppressing the "shuttle effect." The loose porous structure of the sulfur electrode prepared with this binder facilitates better electrolyte penetration and rapid ion transmission, leading to significantly improved electrochemical performance compared to traditional PVDF binders. The principle behind this material's role lies in its ability to form strong chemical bonds with polysulfides, thus enhancing electrode stability and performance, making it a cost-effective and environmentally friendly solution for high-energy Li–S batteries.

Liu et al. [60] prepared a polymer binder with a highly stretchable and elastic network structure, called CPAU, by in situ cross-linking of PAA with isocyanate end-based polyurethane oligomers (PUOs). This binder adheres firmly to the silicon particles by forming hydrogen bonds with the hydroxyl group on the surface of the Si particles, while the PEG chain introduces the flexibility of the polymer network, and the 2-ureido-4-pyrimidinone (UPy) group gives the polymer network the required mechanical strength by forming a reversible strong quadruple hydrogen bond crosslinker. The role of this binder is to provide adequate adaptability when the silicon particles undergo volume changes and to effectively maintain the integrity of the silicon anode through strong mechanical support, thereby enhancing cycle stability and rate performance. The working principle is that through the covalent bond crosslinking network formed during electrode preparation, the stress is allowed to be dispersed by the stretching of the PEG chain when the silicon particles are lithium, while the slip of the molecular chain is limited by the cross-linking structure, maintaining good mechanical properties. After the delimitation of silicon particles, the network can be restored to the initial state under the contraction of the PEG chain, and the strong quadruple hydrogen bond formed by the UPy group ensures that the polymer network has satisfactory mechanical properties. When the volume of lithium silicon particles expands, the PAA chain firmly adheres to the silicon particles through hydrogen bonding, the soft PUO chain can adapt to the volume change, and the cross-linked structure can keep the electrode material from disintegrating, which together improve the cycle and rate performance of the battery.

Ye et al. [61] synthesized a crosslinked recombination binder called PGC based on PAA, gelatin (GN), and β-cyclodextrin (β-CD). PAA as the main chain provides a large number of carboxyl groups (—COOH), and GN provides a wealth of carboxyl and amide groups, which can form a cross-linked network with PAA. β-CD provides a rich hydroxyl group and conical hollow ring structure that can alleviate stress accumulation by forming new dynamic cross-linked coordination configurations during stretching. This binder can fully maintain the stability of the silicon negative electrode structure during the battery cycle and has a simple preparation method and good capacity retention ability. Therefore, it plays an important role in silicon anode materials, helping to stabilize the electrode, reduce the increase of SEI layers, and achieve stable cycling performance.

Zhou et al. [62] designed GB-Y, a new water-soluble binder based on supramolecular chemistry. The polymer chains of the binder are low-cost gelatin and BA, and the rare earth yttrium-pyridine complex (3,5-pyridine dicarboxylate y) are cross-linked by hydrogen bonds. This binder plays a crucial role in Li–S battery systems. It has three main properties: First, it maintains the integrity of the sulfur cathode through the synergistic effect of hydrogen bonding and strong chemical bonding; Second, it anchors lithium polysulfide (LiPSs) through the lithium affinity effect. Finally, it accelerates the conversion of LiPSs to Li_2S_2/Li_2S by electrocatalysis of the rare earth catalytic site (Y^{3+}). The principle of this binder lies in its unique supramolecular chemical structure, which can effectively adsorb LiPSs and accelerate its transformation kinetics, thereby fundamentally solving the shuttle effect problem. In addition, this binder also has good mechanical toughness, which can effectively alleviate the volume expansion of the sulfur cathode during the battery cycle, helping to maintain the integrity of the electrode structure.

These studies show that the selection of suitable water-soluble polymer with good compatibility to blend gelatin as a binder can significantly improve the charge-discharge cycle performance of the battery. With the continuous progress of LIB technology, the performance requirements of battery binders are becoming higher and higher. Future research directions include the development of environmentally friendly, low-cost, high-performance water-based binders, as well as multi-functional binders with self-healing capabilities and improved ionic conductivity of electrode sheets. This will help improve the electrochemical performance and cycle stability of LIBs, and promote the further development of LIB technology.

2.3.9 Tragacanth Gum (TG)

TG-based binder is a widely used binder in battery manufacturing, mainly used to enhance the stability and electrochemical performance of electrode structures. This binder closely combines solid powder such as active material and conductive agent with the fluid collector to form a stable electrode structure, thus ensuring the battery has good electrochemical performance and cycle stability.

Senthil et al. [63] developed a composite polymer biopolymer binder based on TG for sulfur cathodes in Li–S batteries. This binder allows for high sulfur loading over $12\,mg\,cm^{-2}$ while maintaining excellent sulfur utility, reversibility, and improved fire safety. The saccharide conformal fraction in the binder contains multifunctional polar units that serve as active channels, enhancing lithium-ion access to sulfur particles by 80%. This results in 46% better polysulfide entrapment and limits volume changes to within 16%, even at a high charge-discharge rate of 4C. The flexibility of the binder contributes to a Li–S battery with a gravimetric energy density of $243\,Wh\,kg^{-1}$, demonstrating both high reactivity and shape conformality. The material's role lies in its ability to enhance sulfur reactivity while preventing polysulfide loss and managing volume changes, thus enabling the development of compact and flexible high-power Li–S batteries.

Although TG-based battery binders have many advantages, their high cost and sensitivity to the environment are a challenge. In addition, the preparation method of the binder needs to be simple in order to facilitate industrial promotion. However, with continued research and innovation, these problems are expected to be solved, so that TG-based battery binders are more widely used in the battery industry.

Biopolymers and their derivatives offer a promising pathway toward the development of multifunctional binders that enhance both the mechanical and electrochemical performance of battery electrodes. Their unique chemical structures, coupled with the ability to form strong networks through hydrogen bonding and other interactions, provide significant advantages over traditional binders. As research in this field progresses, biopolymers are likely to play an increasingly vital role in the sustainable development of high-performance energy-storage systems.

2.4 Application of Biopolymers in Electrolytes

Electrolytes play an essential role in the operation and performance of electrochemical energy-storage devices by facilitating the transport of ionic species between electrodes. The performance and stability of electrolytes significantly influence battery energy density, power output, and longevity. In recent years, biopolymers and their derivatives have garnered increasing interest as sustainable and high-performance materials for electrolytes in advanced energy-storage systems such as LIBs, ZIBs, and solid-state batteries. This section evaluates the potential of biopolymers in enhancing electrolyte performance, focusing on their ionic conductivity, mechanical properties, and environmental benefits.

The electrolyte is a critical component of electrochemical devices, determining both the ionic conductivity and the electrochemical stability window (ESW), which ultimately governs the battery's performance. The choice of electrolyte impacts not only the power and energy density of the device but also its safety and lifespan. Poor electrolyte performance, such as degradation or instability, can lead to decreased capacity, shortened cycle life, and potential safety hazards like thermal runaway [64]. Therefore, developing high-performance electrolytes is vital for achieving safe, efficient, and high-energy-density batteries.

An ideal electrolyte should possess several key properties, including: (i). High ionic conductivity and electronic insulation. (ii) A wide ESW to support both high-voltage cathodes and lithium metal anodes. (iii) Thermal and chemical stability. (iv) Low cost, environmental friendliness, and the ability to be mass produced [65]. However, no current electrolyte meets all these criteria, which has prompted the exploration of new materials, such as biopolymers, for innovative electrolyte designs.

In recent years, natural polymers, particularly CLS, CS, GA, and SA, have been widely used to prepare gel polymer electrolytes (GPEs). These biopolymers can form a polymer network or serve as additives in various gels, including hydrogels, ion gels, and organic gels, making them highly practical for different energy-storage applications.

2.4.1 Cellulose

Cellulose plays a crucial role in the development of advanced battery materials by providing structural support, enhancing ionic conductivity, and improving electrolyte stability. When modified or combined with other materials, cellulose creates one-dimensional channels or cross-linked networks that facilitate rapid ion transport and maintain mechanical integrity during cycling. Its oxygen-containing functional groups allow for effective ion binding and transport, while its inherent flexibility and adaptability make it ideal for use in both solid and quasi-solid electrolytes. These properties contribute to better electrode-electrolyte interfaces, reduced side reactions, and improved durability, making cellulose a valuable component for high-performance, sustainable batteries.

In the research conducted by Yang et al. [66] the cellulose separator was modified with Cu^{2+} to create a highly conductive polymer solid electrolyte for LIBs. The Cu^{2+}-modified cellulose creates a one-dimensional channel for rapid lithium ion diffusion. The material's unique structure, featuring expanded segment spacing and a large number of oxygen-containing functional groups, allows for independent lithium ion transport. This results in high ionic conductivity (1.5 mS cm^{-1} at room temperature), a high transference number (0.78), and a broad ESW (0–4.5 V). This innovation enables efficient ion transport in thick LiFePO$_4$ cathodes and demonstrates the potential for high-energy-density, solid-state batteries.

Wang et al. [67] focused on creating a low-cost, esterified cellulose-based QPE with excellent Li$^+$ migration performance. The cellulose acetate (CLA) matrix demonstrated high stability and a rapid Li$^+$ transmission channel. With a high Li$^+$ transference number (0.85), the material showed long-term cycling stability, with a capacity retention of 97.7% after 1200 cycles at 1 and 25 °C. This approach highlights the potential for cost-effective, stable electrolytes in solid-state lithium batteries, providing a sustainable alternative to traditional materials.

Ma et al. [68] synthesized an alkali-resistant, super-stretchable dual-network hydrogel electrolyte by introducing cellulose into a polyacrylonitrile network. This hydrogel could stretch up to 1200% while maintaining its ionic conductivity and mechanical strength. The zinc–air batteries using this hydrogel electrolyte demonstrated high power densities (108.6 mW cm^{-2}, rising to 210.5 mW cm^{-2} when stretched by 800%). The flexible nature of the electrolyte also supports stable performance under extreme deformation, making it ideal for powering flexible electronics and wearable devices.

Cao et al. [69] developed a viscoplastic gel electrolyte based on biodegradable regenerated cellulose for ZIBs. This gel exhibited viscoplasticity above 2% strain, improving electrode contact and preventing side reactions, such as hydrogen evolution and zinc dendrite formation. Its unique micro-elastic properties adapt to volume changes during cycling, maintaining the electrode's interlocking structure. The gel's ability to chelate zinc ions promoted uniform zinc deposition along the (002) plane, resulting in stable cycling performance even under high areal capacities.

Yang et al. [70] designed a polymer electrolyte composite using CNCs grafted with acrylonitrile. The CNCs act as a hard skeletal structure within the electrolyte, lending high mechanical strength, while PAN provides a soft, flexible component to improve interfacial compatibility and boost ionic mobility. A key feature of CNCs is their surface functionalization with sulfonate radicals and oxygen-containing polar groups, which enable effective Li$^+$ solvation, thereby facilitating easier ion transport. This combination of hard CNC and soft PAN results in a durable, porous structure that can be compressed into thin membranes with uniform filler distribution, achieved by an innovative dry-film process that minimizes solvent use. Furthermore, the nitrile groups in PAN interact with ethylene carbonate (EC) molecules to form ion-conductive channels, enhancing overall conductivity. This approach highlights the potential of biomass-derived materials, with the CNC-PAN SPE providing an environmentally friendly, scalable, and cost-effective solution for developing high-performance, safe lithium battery technology.

Cellulose-based binders are not only showing potential in traditional battery applications, but are also being explored for applications in emerging technologies such as solid-state batteries. These batteries use solid electrolytes and may not require traditional liquid binders, thus providing a new market opportunity for cellulose-based binders. Cellulose-based battery binders are becoming an important development direction in the battery industry due to their environmental protection, economic, and performance improvement potential. In the future, with the continuous progress of technology and the gradual development of the market, the application prospects of cellulose-based battery binders will be broader.

2.4.2 Chitosan

CS plays a pivotal role in the development of advanced hydrogel electrolytes for ZIBs due to its high ionic conductivity, mechanical strength, and biocompatibility. As a flexible and sustainable biopolymer, CS enhances the structural stability of electrolytes by forming strong, uniform networks that facilitate zinc ion deposition while reducing the formation of zinc dendrites. Its ability to form cross-linked structures with other polymers allows it to adapt to volume changes and mitigate side reactions, extending the lifespan of batteries. Additionally, CS's environmentally friendly and biodegradable properties make it an ideal material for developing safe and sustainable energy-storage systems.

A key material in this research is the C-PAMCS hydrogel electrolyte, which combines PAM and CS to create a flexible, high-performance binder. The electrolyte allows for uniform Zn deposition and provides high fatigue resistance, enabling Zn∥Zn cells to achieve a lifespan of 1500 hours at a high current density of $10\,\text{mA}\,\text{cm}^{-2}$ and an areal capacity of $10\,\text{mAh}\,\text{cm}^{-2}$. This material's flexibility and fatigue resistance make it ideal for use in advanced, flexible ZIBs. The uniform deposition of zinc reduces the likelihood of dendrite formation, a common issue in metal anodes, and enhances the overall stability of the battery.

Wu et al. [71] demonstrated the effectiveness of a CS-zinc gel electrolyte for high-rate and long-life zinc metal batteries. This bio-based polymer electrolyte provides high ionic conductivity, mechanical strength, and sustainability. It also facilitates the ideal deposition of zinc in the form of parallel hexagonal plates, leading to excellent cycling stability with a coulombic efficiency of 99.7%. The Zn anode showed superior performance across 1000 cycles at high current densities of $50\,\text{mA}\,\text{cm}^{-2}$. The non-flammable and biodegradable nature of this electrolyte also underscores its potential for safe and sustainable energy-storage solutions, marking a significant step toward green, high-efficiency energy-storage systems.

Lu et al. [72] further expanded the scope of hydrogel electrolytes by developing recyclable and biodegradable options using CS and polyaspartic acid. This combination created a dual coupling network that effectively inhibited zinc dendrite formation and controlled side reactions at the zinc anode. The Zn||CPZ-H symmetric cells exhibited remarkable cycle stability, maintaining excellent reversibility for 2200 hours at a current density of 10 mA cm^{-2}. The hydrogel electrolyte's ability to redistribute H$^+$ and regulate the two-electron redox reaction in Zn|MnO$_2$ batteries enabled these cells to achieve a capacity retention of 92.5% after 5000 cycles. The hydrogel's sustainable and recyclable properties present a promising avenue for the future development of environmentally friendly, high-performance aqueous energy-storage devices.

Wang et al. [73] proposed a unique "open" soft-pack battery design that uses a gel electrolyte containing cross-linked kappa-carrageenan and CS. This design allows for hydrogen release and the refilling of consumed electrolyte components during the battery cycle. The gel electrolyte binds water molecules, reducing side reactions and preventing electrolyte leakage or rapid evaporation. In tests, the Zn||Zn$_x$V$_2$O$_5$·nH$_2$O multi-layer pouch cell with the carrageenan/CS electrolyte demonstrated a discharge capacity of 0.9 Ah and a retention rate of 84% after 200 cycles. This innovative design offers a new method for mass-producing water-based zinc batteries with greater stability and longer cycle lives.

The research conducted by Liu et al. [74] examined the antifatigue properties of hydrogel electrolytes. This study employs a dual-network hydrogel electrolyte system based on PAM integrated with carboxymethyl CS and Zn(ClO$_4$)$_2$, termed C-PAMCS. As shown in Figure 2.5. The materials used are PAM, which provides high elasticity, CMC, which enhances tensile elongation and low-temperature performance, and Zn(ClO$_4$)$_2$, which acts as a chaotropic agent that reinforces antifatigue properties. The C-PAMCS electrolyte leverages the structural resilience and elasticity from the PAM backbone, while CMC improves mechanical stability and electrolyte integrity under cyclic compression. Zn(ClO$_4$)$_2$ further stabilizes the electrolyte, preventing failure under high-current conditions. The role of C-PAMCS in the Zn-ion battery system is to deliver improved stability and durability through its dual-network structure, ensuring a homogeneous Zn deposition while preventing dendrite formation and maintaining flexibility. This combination of properties allows the hydrogel electrolyte to support reliable and stable performance in flexible and wearable Zn-ion batteries, making it a promising solution for advanced energy-storage systems.

Research on CS as a binder for batteries is ongoing. Scientists are exploring ways to improve the bonding strength and battery performance of CS by chemically modifying it or combining it with other materials. With the deepening of scientific research, the technology and performance of CS-based binders are expected to be further improved. In the future, this binder may be applied to more types of batteries, providing new impetus for the innovation and development of battery technology.

2.4.3 Lignin

Lignin-based electrolyte is a kind of high-performance battery electrolyte based on biomass material, which has good electrochemical stability and environmental protection characteristics. Recent research has focused on developing ultra-thin,

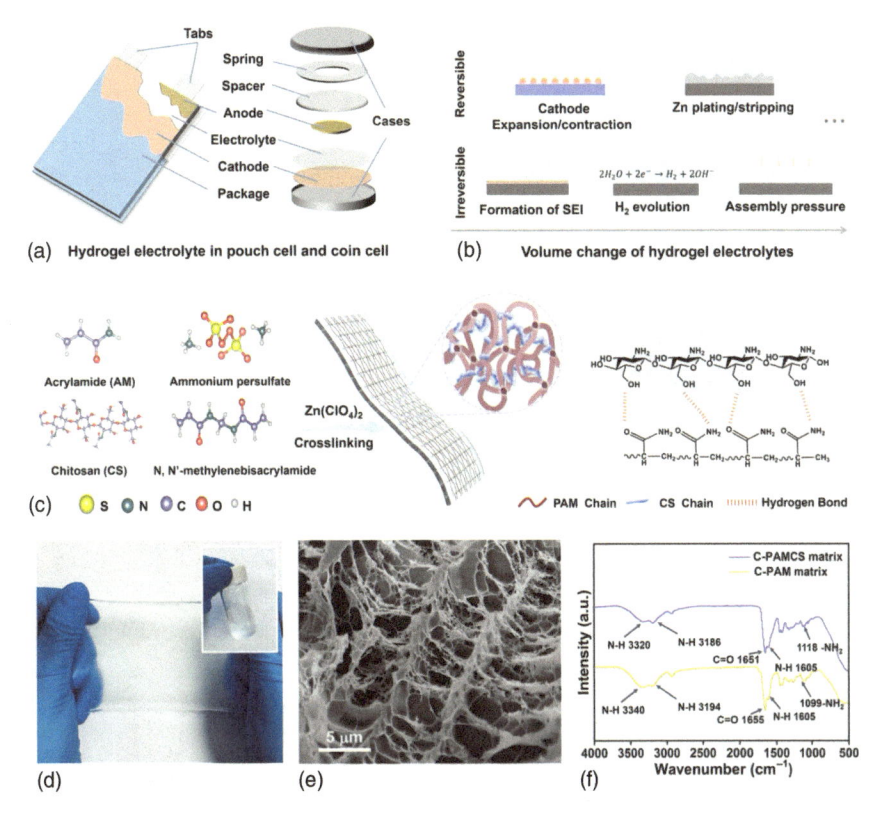

Figure 2.5 (a) Structure of traditional zinc button batteries and pouch batteries. (b) Demonstration of hydrogel electrolyte volume changes as affected by elements. (c) Schematic diagram of the preparation process of C-PAMCS. (d) Photos of precursor and C-PAMCS; (e) Cross-sectional SEM image of C-PAMCS; (f) FT-IR spectra of C-PAM and C-PAMCS matrix. Source: Reproduced with permission from Liu et al. [74]/John Wiley & Sons.

all-solid lignin-based polymer electrolytes that not only have high ionic conductivity, but also maintain good electrochemical performance at low temperatures, which is of great significance for the development of next-generation lithium batteries.

The ultra-thin all-solid composite polymer electrolyte (CPE) developed by Liu et al. [75] is based on lignin and is only 13.2 μm thick. As shown in Figure 2.6. This article presents the development of a lignin-based CPE for LIBs that incorporate lithium lignosulfonate and lithium trifluoromethanesulfonyl imide groups, along with polyvinylidene fluoride-hexafluoropropylene (PVDF-HFP) and polyethylene oxide (PEO) as fillers. The unique properties of these materials contribute to the CPE's performance. Lithium lignosulfonate provides abundant lithium ions and creates a 3D single-ion nanofiber ionic bridge network, which facilitates uniform Li$^+$ distribution and forms robust Li$^+$ channels. The PVDF-HFP framework enhances mechanical strength, while PEO assists in ionic conduction by creating additional pathways. This nanofiber structure, created through electrospinning and hot pressing, offers high flexibility and ultra-thin thickness (13.2 μm), minimizing ion

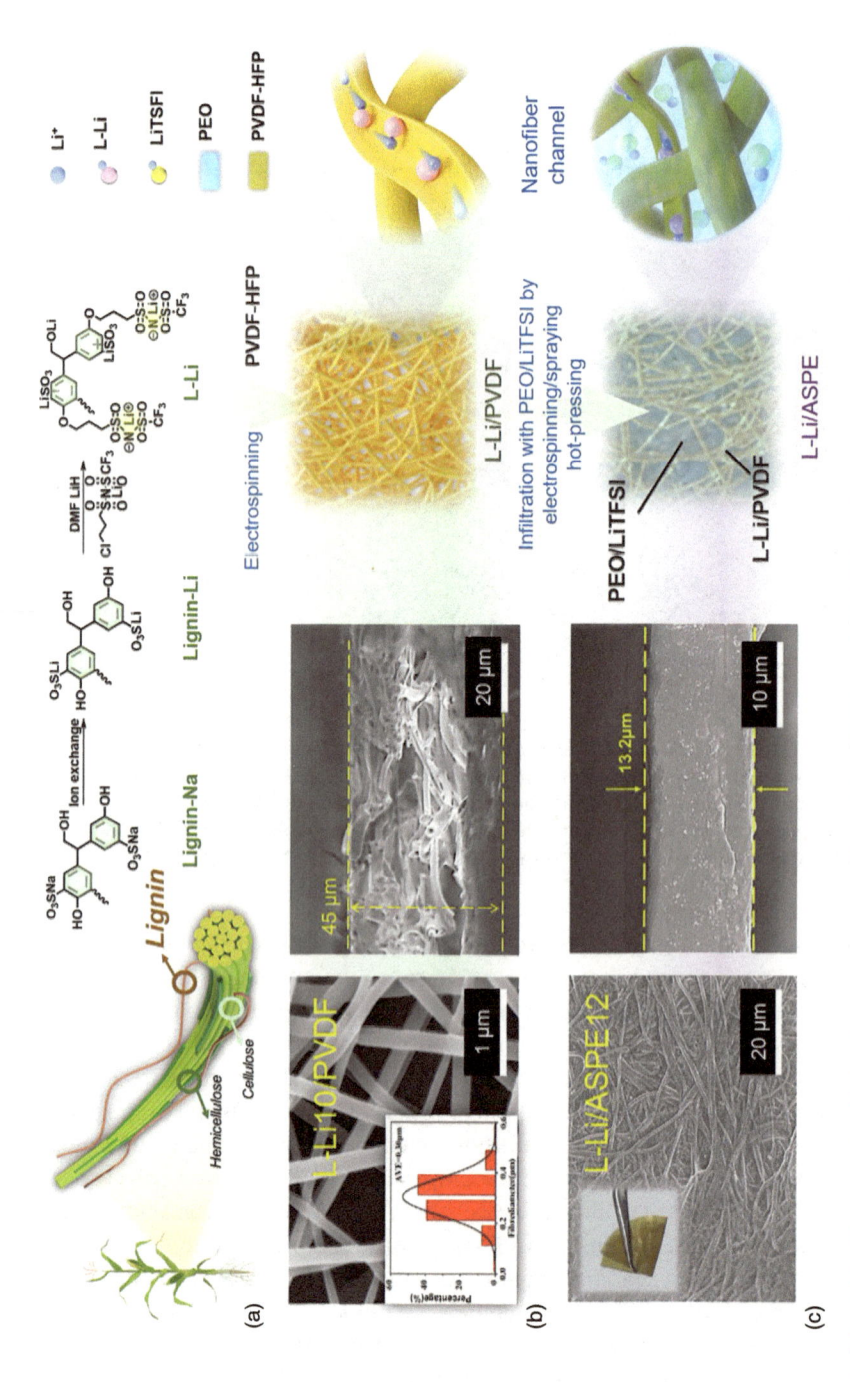

Figure 2.6 (a) Synthesis route of L-Li, the preparation method of L-Li/PVDF and L-Li/ASPE, and internal sketch map of nanofiber networks. SEM images of the surface and cross section of (b) L-Li/PVDF and (c) L-Li/ASPE. Source: Reproduced with the permission from Liu et al. [75]/John Wiley & Sons.

transmission distance and improving Li⁺ diffusion. The interconnected nanofiber network aids in preventing lithium dendrite formation, ensuring interface compatibility and stability between the CPE and lithium electrodes. This innovative, lignin-based CPE design provides a sustainable, high-performance solution for solid-state lithium-ion batteries (SLIBs) by leveraging its structural integrity, favorable ionic conductivity, and interfacial compatibility to enhance safety and efficiency.

The pure lignin was directly isolated from poplar and the gel electrolyte was prepared by using deep eutectic solvent (DES) as solvent. Lignin-derived carbon electrodes with high porosity were obtained by activating lignin with KOH. These electrode materials have a large specific surface area (1125 m² g⁻¹), layered pore size distribution, and a high degree of graphitization. In addition, a double-network interpenetrating ionic gel was synthesized from lignin and acrylamide in DES as a biodegradable electrolyte by catalyzing ring-opening polymerization and crosslinking. This gel electrolyte has an interpenetrating cellular network structure and excellent electrical conductivity (13.5 S m⁻¹). As a result, this material in supercapacitors is able to provide high specific capacitance (181.5 F g⁻¹) and good capacitance retention (80.2%) for more than 2000 cycles. In addition, the supercapacitor can maintain an energy density of 29.0 Wh kg⁻¹ at an extremely high power density (3.2 KW kg⁻¹) [76].

Su et al. [77] proposed a simple, versatile, and fast gel reaction based on dealkaline lignin-alkali metal ions. This material has anti-freeze properties and good mechanical properties, and can gel quickly (within 4 min) at room temperature without additional energy input. In this system, the catechol group of lignin forms a complex with alkali metal ions, which accelerates the establishment of a reversible REDOX reaction. The resulting free radicals ($SO_4^-\cdot$, $OH\cdot$, and singlet oxygen $^1O_2$) promote the rapid polymerization of vinyl monomers. Alkali metal ions play a dual role in the rapid polymerization and freeze resistance of hydrogel electrolytes. By adjusting the amount of DL and the concentration of metal ions, a preferred hydrogel electrolyte with excellent freeze resistance (0.51 mS cm⁻¹ at −40 °C) and strong mechanical properties (0.4 MPa tensile stress, 1125% strain) can be prepared in an alkaline aqueous solution.

These research results provide an effective strategy for solid-state lithium batteries to achieve ultra-thin bio-based all-solid-state polymer electrolytes with high safety, long life, and high electrochemical performance. This not only provides a new strategy for the development of lithium batteries, but also opens up new possibilities for the application of lignin-based materials in the field of high-performance batteries.

2.4.4 Gelatin

Gelatin plays a crucial role in the development of advanced hydrogel electrolytes for ZIBs due to its ability to form robust, flexible, and repairable networks. As a biocompatible and biodegradable material, gelatin enhances the mechanical strength and ionic conductivity of the electrolyte while maintaining adaptability to battery interfaces. Its unique structural properties, such as the formation of cross-linked networks and helical chains, facilitate uniform zinc ion transport and help prevent dendrite formation. Additionally, gelatin's dynamic sol–gel transition enables the continuous repair of the electrode–electrolyte interface, improving the stability and longevity of the battery during cycling. This makes gelatin an ideal material for sustainable and high-performance energy-storage applications.

The research conducted by Zhou et al. [78] involved grafting a gelatin-based electrolyte into a 3D porous cellulose aerogel (CA) framework, resulting in the creation of an all-solid electrolyte membrane known as CAG. The CAG film features a highly porous structure with exceptional liquid storage capacity, which enhances its ionic conductivity while maintaining mechanical strength and biodegradability. This innovation was applied to ZIBs, where α-MnO$_2$ nanorod/rGO composite cathodes and conductive zinc ink anodes were developed using in-situ synthesis and hydrothermal methods. The transient battery design demonstrated excellent electrochemical performance, including a specific capacity of 211.5 mAh g^{-1} and wide voltage stability, showing great potential for transient electronic technologies and clinical applications.

Kumankuma-Sarpong et al. [79] developed a new multi-network hydrogel electrolyte (ODGelMA) based on gelatin, oxidized dextran (OD), and methacrylic anhydride (MA). As shown in Figure 2.7. The hydrogel is synthesized through Schiff

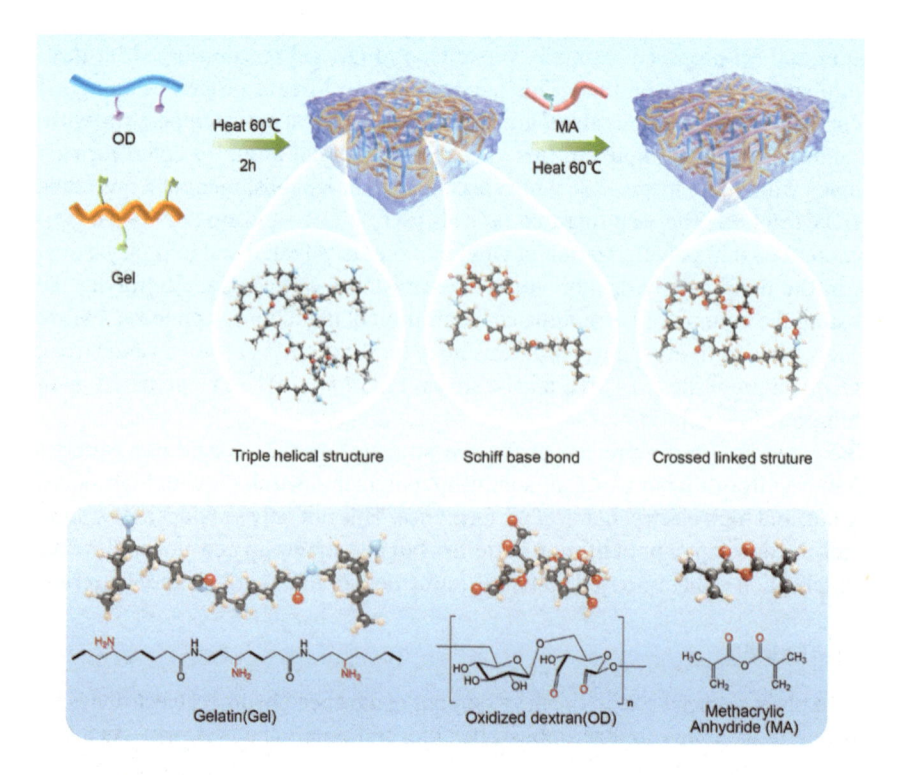

Figure 2.7 Schematic representation for the synthesis of ODGelMA hydrogel electrolyte. Source: Reproduced from Kumankuma-Sarpong et al. [79]/with permission of Wiley-VCH Verlag GmbH & Co. KGaA.

base reactions, involving multiple crosslinking mechanisms where aldehyde groups in OD form chemical bonds with the amino groups of gelatin, further strengthened by MA-induced crosslinking. This structure creates a robust, entangled network that maintains its toughness and structural stability. The resulting hydrogel's interconnected pore channels facilitate Zn^{2+} desolvation and enhance ionic conductivity, reducing electrostatic forces. This property contributes to uniform, reversible Zn stripping and plating, leading to dendrite-free operation. Additionally, in large-format pouch cells, the ODGelMA hydrogel supports high cycling stability by preventing side reactions and improving reversibility, thus making it suitable for scalable energy-storage applications.

Yang et al. [80] designed a dynamically repairable gelatin-based hydrogel electrolyte with a reversible sol–gel transition, enabling the construction of a conformal electrode–electrolyte interface. This dynamic interface helps prevent dendrite formation in zinc anodes and maintains a stable solid–solid interface during battery operation. The gelatin's unique helical chain structure, combined with sulfate groups, creates a uniform zinc ion transport channel. The reversible sol–gel transition ensures the repeated repair of the anode-electrolyte interface, significantly improving the battery's longevity. ZIBs utilizing this design exhibited 86% capacity retention after 2100 cycles, demonstrating the effectiveness of the repairable interface and offering potential advancements for other metal-ion batteries.

Due the its unique biocompatibility and versatility, gelatin-based electrolytes have shown great application potential in many high-tech fields. Current challenges include improving the conductivity, stability, and processability of electrolytes. The future development direction focuses on the development of multi-functional, intelligent, and environmentally friendly gelatin-based electrolyte systems. Combined with new synthesis techniques and advanced characterization methods, the application potential in emerging fields is further explored. The gelatin-based electrolyte has become one of the hot areas of scientific research and technological innovation because of its excellent properties and wide application prospects.

Biopolymers and their derivatives offer a compelling pathway toward the development of next-generation electrolytes for energy-storage devices. Their inherent biodegradability, mechanical flexibility, and ability to form stable ion-conducting networks make them ideal candidates for use in batteries and other energy-storage applications. Continued research and innovation in this field will be crucial in addressing the remaining challenges and unlocking the full potential of biopolymer electrolytes in creating safe, efficient, and sustainable energy-storage systems.

2.5 Application of Biopolymers in Electrolyte Additives

For liquid LIBs, organic electrolytes play an essential role in the battery and can improve the cycle stability and life of the battery. While searching for the best organic solvents, additives have attracted significant attention due to their small dosage, almost no increase in cost, and the ability to improve battery performance significantly.

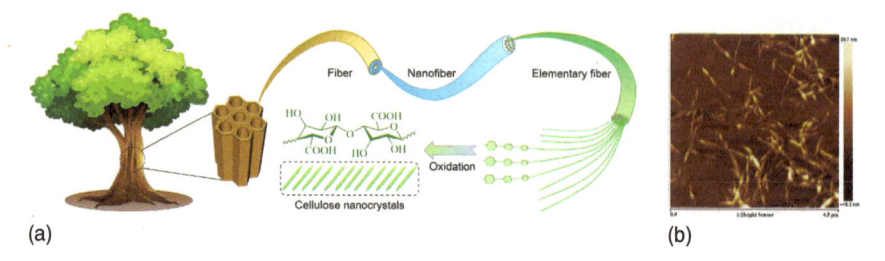

Figure 2.8 Structural characterizations of the CNCs-ZnSO$_4$ electrolyte. (a) Schematic illustration of fabricating the CNCs from wood. (b) The AFM image showing the whisker-like shape of CNCs. Source: Reproduced with the permission from Wu et al. [81]/John Wiley & Sons.

With continuous research, the types of additives are increasing, and their scope of use is also expanding. The number of additives generally does not exceed 5 vol%. There are several requirements: (i) Although the amount is small, it can significantly improve the battery performance. (ii) Does not have a side reaction with the battery material, and organic solvents have good compatibility. (iii) Relatively low price. (iv) Nontoxic or less toxic. (v) To reduce the flammability of the organic electrolyte. (vi) To provide overcharging protection or increase the overcharge tolerance. (vii) To terminate the battery operation under abusive conditions. This chapter describes the use of CLS, THL, Cit, and pectin in electrolyte additives.

2.5.1 Cellulose

CNCs are bio-based, nanoscale materials with high aspect ratio, thermal stability, abundant surface functional groups, and self-assembly properties. It is widely used in electrolyte additives to optimize the zinc ion coordination environment and enhance interfacial chemistry to achieve ultra-high rate performance and ultra-stability in ZIBs.

Wu et al. [81] found that CNC electrolyte additives significantly improve the zinc ion coordination environment and enhance interface chemistry, leading to ultra-high rate performance and stable zinc anodes. Schematic illustration of fabricating the CNCs from wood as Figure 2.8. Thanks to their unique physical and chemical properties, CNCs act as fast ion carriers, optimizing the diffusion dynamics of zinc ions while forming a dynamic, self-healing protective layer at the interface. These findings suggest that CNCs offer a sustainable, scalable solution for enhancing the performance and longevity of aqueous ZIBs.

As a multifunctional electrolyte additive, cellulose can effectively optimize the electrochemical behavior of zinc ions through their unique physical and chemical properties, providing a new way to develop efficient electrochemical energy-storage technology.

2.5.2 Trehalose

Trehalose as an electrolyte additive has recently attracted a lot of attention in the field of electrochemistry. It is found that trehalose can effectively improve the electrochemical performance of zinc metal anode, especially in improving its reversibility and cycle stability.

It was applied by researchers [82] in electrolyte modification for zinc anodes. Trehalose plays a key role in the stability of the zinc anode as an electrolyte additive in this system. Trehalose promotes the reversibility of zinc anodes through its many functions, including as a hydrogen bond-breaking agent, solvent regulator, and interfacial reaction accelerator. From a thermodynamic point of view, Trehalose's abundant hydrogen bond sites and strong trapping ability of Zn^{2+} help to reduce the reducing activity of water molecules and weaken the solvation environment of Zn^{2+}, thus promoting the desolvation process of Zn^{2+}. In terms of kinetics, the interfacial adsorption layer of Trehalose molecules acts as an aid to Zn^{2+} desolvation, providing a uniform and efficient supply of zinc ions, which helps to improve the limited zinc plating/stripping kinetics. Finally, with the addition of Trehalose, a zinc anode with a high average coulomb efficiency (99.8%) was achieved and a stable cycle of more than 1500 hours (9.0% discharge depth) was achieved in a zinc-symmetric cell.

In order to solve the instability of the zinc anode (including surface corrosion, hydrogen precipitation, and irregular zinc deposition), a polyhydroxylated organic molecular additive trehalose is added to the electrolyte. In this system, trehalose inhibits the decomposition of water molecules by optimizing the solvation structure of zinc ions, thereby improving the stability of zinc anodes. The polyhydroxyl structure of trehalose is involved in the reconstruction of the hydrogen bond network, which increases the overpotential of the water molecular decomposition reaction, thereby inhibiting the production of hydrogen. In addition, the chemisorption of trehalose at the zinc metal-molecular interface enhances corrosion resistance and promotes the deposition of zinc in the plane. This additive significantly improves the cycle stability of the battery by improving the reversibility of zinc stripping/plating [83].

Trehalose, as an additive of aqueous zinc ion electrolyte, can significantly improve the electrochemical performance of zinc metal cathode through interfacial dynamic regulation. These studies not only develop a new type of electrolyte additive, but also provide new ideas and methods for the research in the field of electrochemistry.

2.5.3 Citrulline

Citrulline as an electrolyte additive in aqueous ZIBs has demonstrated excellent stability and cycle life. Citrulline can not only regulate the solvated shell of zinc ions, but also improve the double electric layer structure of the negative electrode/electrolyte interface, so as to effectively slow down dendrite growth and inhibit side reactions.

The research by Chen et al. [31] explored citrulline (Cit), a biocompatible compound, as an electrolyte additive to enhance the stability of zinc anodes. As a bifunctional additive, Cit has an abundance of highly polar groups (—NH_2 and —COOH) that can strongly interact with Zn^{2+} and Zn metals. This interaction helps to regulate the solvated shell of Zn^{2+} and the electric double layer at the electrode/electrolyte interface, thus effectively inhibiting dendrite growth and interface side reactions, giving the Zn anode excellent stability. Thus, Zn||Zn symmetric batteries with citric acid additive can operate stably for more than 1600 hours at $1.0\,\text{mA cm}^{-2}$ and maintain a stable cycle for 650 hours at $10.0\,\text{mA cm}^{-2}$. In addition, the Zn||Cu asymmetric battery exhibits a coulomb efficiency of up to 99.6% at $1.0\,\text{mA cm}^{-2}$ and $0.5\,\text{mAh cm}^{-2}$. The principle of this action of citric acid is to reshape the solvated structure of Zn^{2+} by partially substituting the coordination water molecules and selectively adsorbing on the Zn anode to alleviate the undesirable "tip effect," thereby inhibiting parasitic reactions and dendrite growth.

The use of citrulline not only improves the cycle stability and coulomb efficiency of batteries, but also enhances the overall performance of batteries. This is of great significance for promoting the application of water ZIBs in the field of large-scale energy storage. Citrulline, as a highly efficient electrolyte additive, improves the performance of ZIBs through its dual-functional properties and provides a new solution for the development of high-performance water ZIBs.

2.5.4 Pectin

As a kind of electrolyte additive, pectin has shown excellent performance in the application of zinc metal cathode. By forming a protective layer on the surface of the zinc negative electrode, pectin effectively inhibits the growth of dendrites and improves the stability of the zinc negative electrode.

Hong et al. [84] introduced the concept of using desolvation activation energy as a criterion for selecting electrolyte additives. They conducted a systematic study on pectin, a polyanionic polysaccharide, as an electrolyte additive. DFT calculations revealed that pectin molecules adsorb parallel to the surface of the Zn anode, forming an interfacial adsorption layer. Hydrogen bonding between pectin and water molecules lowers the desolvation activation energy, aiding in the faster removal of H_2O from hydrated zinc ions and reducing interface voltage polarization. Pectin also raises the energy barrier for water decomposition, effectively inhibiting the hydrogen evolution reaction. Furthermore, the oxygen-containing groups in pectin act as zincophilic sites, promoting uniform zinc deposition. Therefore, the anionic polysaccharide additive can significantly improve the cycle stability and reversibility of ZIBs, while inhibiting dendrite growth and reducing overpotential.

Pectin can significantly improve the electrochemical performance of zinc metal cathode through its unique physical and chemical properties, which provides strong support for the development of efficient and stable zinc ion batteries.

2.6 Application of Biopolymers in Separators

Battery separators are critical components in electrochemical energy-storage systems, playing a pivotal role in ensuring the safe and efficient operation of batteries by physically separating the anode and cathode while allowing the transport of ions between them [85]. As the demand for higher energy densities, improved safety, and longer battery life spans continues to grow, the design of advanced separators with enhanced properties has become increasingly important. Among the

2.6 Application of Biopolymers in Separators

materials explored, biopolymers and their derivatives have emerged as promising candidates for high-performance battery separators due to their excellent mechanical properties, chemical and thermal stability, and environmental sustainability. This review explores the application of various biopolymers, such as cellulose, CS, and others, in battery separators and discusses their advantages, challenges, and future directions.

Battery separators serve multiple functions in a battery, primarily to prevent physical contact between the anode and cathode while permitting the free movement of ions during charge and discharge cycles. The performance of a battery separator directly impacts the efficiency, safety, and lifespan of the battery. Separators must meet several key requirements, including: (i) Mechanical Strength and Thermal Stability: The separator must maintain its structural integrity under mechanical stress and high temperatures encountered during battery operation, particularly in high-energy-density batteries where elevated temperatures are common. (ii) Ion Conductivity and Electrolyte Wettability: Efficient ion transport through the separator is critical to the overall performance of the battery. The separator should have good wettability with the electrolyte, ensuring that the ions can migrate easily across the separator. (iii) Chemical and Thermal Stability: To prevent degradation and failure, separators must withstand exposure to harsh chemical environments and maintain stability under high temperatures. (iv) Thinness and Porosity: A thin separator with high porosity facilitates faster ion transport and reduces internal resistance, contributing to improved battery performance. (v) Environmental Friendliness and Sustainability: With increasing concerns over the environmental impact of batteries, the development of eco-friendly separators from renewable and biodegradable materials is becoming a key focus.

Conventional separators, typically made from synthetic polymers such as PE or polypropylene (PP), have limitations in terms of thermal stability, electrolyte wettability, and sustainability. Biopolymers offer an alternative, with their natural origin, biodegradability, and potential for functionalization making them ideal candidates for next-generation battery separators. Biopolymers, particularly CLS, CS, and other natural polymers, have shown significant promise in battery separator applications due to their desirable mechanical, chemical, and environmental properties. Below, we examine several biopolymer-based separators and their specific applications in LIBs and ZIBs.

2.6.1 Cellulose

Cellulose serves as a versatile and highly effective material for battery separators, playing a crucial role in enhancing the performance and stability of both LIBs and ZIBs Its natural abundance, excellent mechanical strength, hydrophilicity, and ionic conductivity make it an ideal candidate for improving electrolyte wettability and facilitating ion transport. Cellulose's ability to form strong, flexible membranes aids in preventing dendrite formation, reducing harmful side reactions, and ensuring uniform ion deposition, particularly in zinc-metal batteries. Through various modifications, such as carbonylation or cyanoethyl grafting, cellulose can be tailored to optimize ion conduction and thermal stability, making it suitable for high-energy, sustainable energy-storage systems.

Zhang et al. [86] developed a new natural polymer nanofiber separator based on cyanoethyl-grafted chitin nanofibers (CCN). This material demonstrates strong mechanical strength, thermal stability, electrolyte wettability, and ionic conductivity. The separator outperforms conventional PP separators by providing a better lithium-ion transport mechanism, as evidenced by the superior rate capability and enhanced capacity retention in $LiFePO_4||Li_4Ti_5O_{12}$ full cells. Additionally, the CCN separator remains effective even at elevated temperatures of 120 °C, showcasing its robustness for use in high-performance LIBs. The development of this chemically modified natural polymer nanofiber separator marks a significant step toward sustainable, high-performance battery separators.

Zhang et al. [87] introduced a strong renewable membrane using carbonylation-modified biosynthetic cellulose nanofibers for rechargeable ZIBs. The carbonylated side groups in the cellulose enable weak interactions between zinc ions and the polymer chains, allowing for improved zinc ion transport through the separator. This innovation ensures uniform zinc plating on the anode and prevents dendritic damage. The separator's effectiveness is demonstrated by the ultralong cycle life in symmetric Zn cells, which achieved over 2800 hours of cycling at 15 mA cm^{-2}. The membrane's high intrinsic zinc ion transport capability promises enhanced performance in ZIBs, making it a valuable addition to battery separator technology.

Ge et al. [88] demonstrated the potential of a single-ion Zn^{2+} conductive nanocellulose membrane as a separator for zinc anodes. The structure and mechanical properties of CNF films are shown in Figure 2.9. This CNF-based separator incorporates physical properties such as high mechanical toughness, strong hydrophilicity, and uniform pore distribution while promoting selective Zn^{2+} transport with a Zn^{2+} transference number of 0.70 ± 0.12. The CNF membrane, functioning with minimal water content and no additional zinc salts, significantly enhances Zn anode stability and Coulombic efficiency by suppressing common issues like hydrogen evolution, corrosion, and dendrite formation. Additionally, the CNF membrane doubles as a macromolecular zinc salt or GPE, offering flexible applications even in moisture-absorbing conditions. This innovation leverages cellulose's sustainability and adaptability in the papermaking process, making it scalable and cost-effective, paving the way for practical, reliable zinc battery technology.

Zhou et al. [89] explored the use of cotton-based cellulose separators to suppress zinc dendrites and harmful side reactions in zinc-metal batteries. Cellulose, the most abundant natural polymer, exhibits excellent hydrophilicity, mechanical strength, and ionic conductivity. Compared to traditional glass fiber separators, cellulose-based separators promote zinc ion migration, reduce nucleation overpotential, and accelerate zinc deposition at the electrode interface. The Zn||Zn symmetric cells using cellulose separators demonstrated impressive stability, with the ability to handle ultra-large areal capacities of 20 mAh cm^{-2}. This research provides valuable insights into designing efficient, cost-effective separators for electrochemical energy-storage devices, especially for improving the performance and safety of zinc-metal batteries.

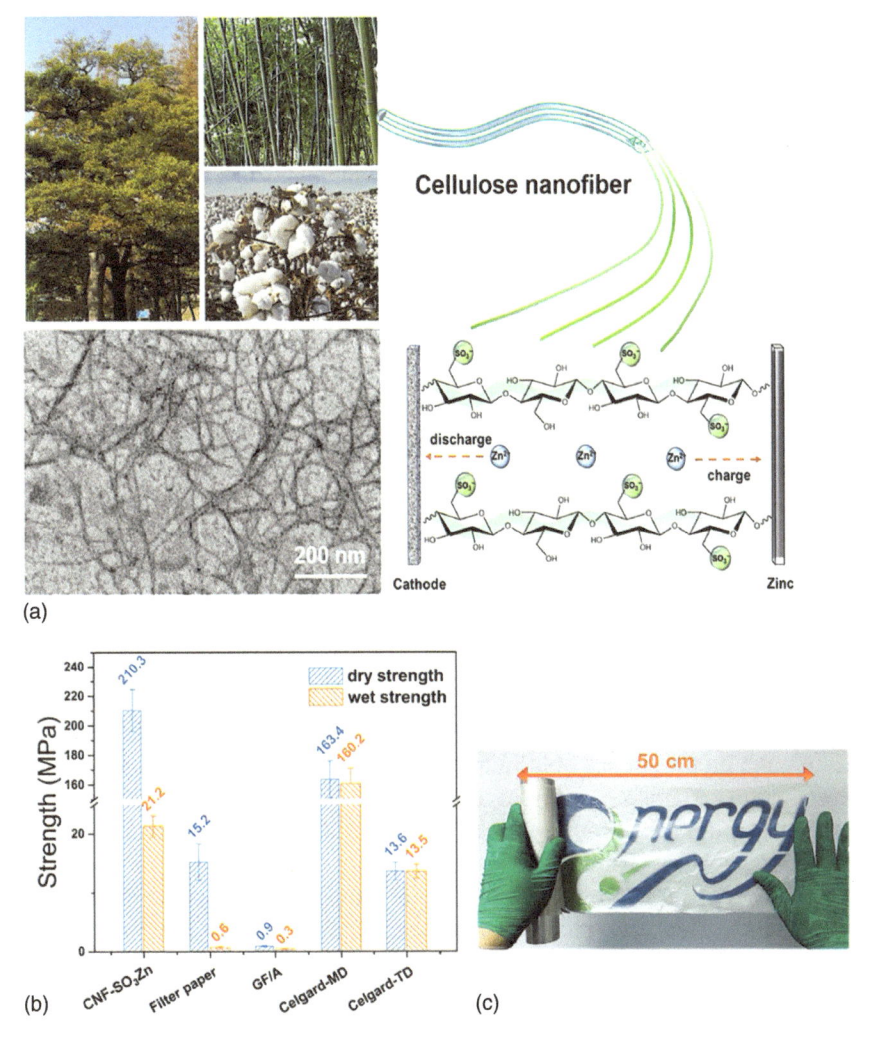

Figure 2.9 The structure and mechanical properties of the CNF membranes. (a) Schematic illustration of the CNF-SO$_3$Zn structure. The raw materials can be derived from plants, such as wood, bamboo, cotton, etc. Sulfonate anions are covalently grafted on the cellulose chains, which makes Zn^{2+} the dominant mobile ions under the electric field. The transmission electron microscopy image shows the micromorphology of CNFs. (b) The dry strengths and wet strengths of different membranes. (c) A digital photo image of an as-synthesized CNF-SO$_3$Zn film with a size of 50 × 20 cm. Source: Reproduced with the permission from Ge et al. [88]/John Wiley & Sons.

Although cellulose-based separators show great potential and advantages, they still face some challenges in practical applications, such as insufficient mechanical strength and pore structure regulation. Future studies need to further optimize these properties to achieve commercial applications and large-scale development of cellulose-based separators.

2.6.2 Starch

The extracted material is amyloid fibril biomolecules, which exhibit strong sodiophilic properties and are highly effective in controlling sodium-metal deposition. Their key properties include the ability to homogenize electric field distribution and sodium ion concentration, thus preventing the formation of loosely packed surface crystals and mitigating dendrite growth. The amino acids in the amyloid fibrils are particularly sodiophilic, meaning they interact favorably with sodium ions, aiding in the formation of stable SEI films. This results in uniform sodium ion transport and dendrite-free sodium deposition during cycling. In this system, the amyloid fibrils play the critical role of modifying the separator, enabling the sodium-metal battery to achieve more compact and reversible deposition, leading to increased cycling stability and prolonged battery life. The principle behind this role is the fibrils' ability to maintain a uniform ion transport environment, which enhances battery performance and longevity by preventing harmful dendritic growth [90].

The β-lamellar lysozyme membrane is used as a modification layer in lithium metal batteries (LMBs) to improve the performance of lithium metal anodes. Lysozyme membrane can regulate the deposition behavior of lithium ions through its complete β-lamellar structure, high lithium affinity mercaptan groups, and columnar nanopores, thereby improving the stability and cycling performance of lithium metal anodes. Its working principle is to resist the dendrite impact under overcurrent through the high mechanical strength and lithium-ion binding energy of lysozyme membrane and promote the uniform distribution of lithium-ion concentration on the surface of lithium metal through the uniform distribution of lithium-ion flux. Therefore, the lysozyme membrane plays a key role in improving the stability of lithium-lithium symmetric batteries and the cycle performance of lithium-iron phosphate batteries [91].

The starch-based separator is a new type of battery separator material, which has good biodegradability, relatively low cost, certain mechanical strength, and chemical stability. Compared with the traditional polymer membrane, the tensile strength of the starch-based membrane needs to be improved. Performance may decline at high temperatures, requiring further formulation and process optimization. At present, the production cost is still higher than that of traditional materials, and the realization of large-scale commercial applications is facing economic pressure. The future direction of improvement is the development of multifunctional composite separators, combining the advantages of different materials to improve performance stability, reduce cost, and expand production scale. With the rapid development of the new energy industry and the growing demand for green environmental protection materials, starch-based battery separator is expected to become one of the important development directions in the future battery field. Further research will focus on improving performance stability, reducing cost, and expanding production scale. With the rapid development of the new energy industry and the growing demand for green environmental protection materials, starch-based battery separator is expected to become one of the important development directions in the future battery field.

2.6.3 Carrageenan

Carrageenan-based battery separator has significant characteristics and potential in the field of battery technology, including good biocompatibility and degradation, not easy to burn or explosion, strong tensile strength and puncture resistance, good absorption and retention of electrolytes, high ionic conductivity, a wide range of sources and renewable, and high-cost effectiveness.

The iota-carrageenan biopolymer exhibits properties such as high porosity (over 90%) with interconnected pore sizes of 50–70 μm, and electrochemical parameters that include an ionic conductivity of 1.34 mS cm^{-1}, a tortuosity of 3, a MacMullin number of 7, and a lithium transference number of 0.48 [92]. These properties make it an ideal candidate for use as a separator in LIBs. The freeze-dried separators enhance ion transport and provide mechanical stability within the battery system, crucial for maintaining the structure during charge-discharge cycles. The principle behind this functionality lies in the material's porosity, which allows efficient electrolyte penetration and ion transfer, while its mechanical strength ensures long-term battery stability. This makes iota-carrageenan an effective, sustainable solution for improving the performance and durability of next-generation LIBs.

With the continuous improvement of safety performance and environmental protection requirements in the battery industry, carrageenan-based membranes are expected to become one of the key components of the next generation of batteries. However, the production process needs to be further optimized to improve yield and quality, and composite applications with other materials need to be explored to broaden their performance boundaries.

2.7 Application of Biopolymers in Anode Functional Layers

Introducing a functional protective layer into the battery system means constructing a dense protective layer on the surface of the electrode by using inorganic or organic substances through electrodeposition or direct coating, which is an effective strategy to improve the stability of the negative electrode. Compared with electrolyte modification, the artificial protective layer can avoid consumption during long-term circulation. Compared with separator modification, the construction of an artificial protective layer is also more straightforward, so the negative electrode protective layer has received excellent research interest from scientific researchers. The ideal artificial protective layer has the following characteristics: (i) High mechanical strength and good flexibility to adapt to volume changes and maintain structural integrity. (ii) Uniform ionic conductivity. (iii) Sizeable electrochemical window to avoid decomposition during the cycle. (iv) Low solubility in the electrolyte to ensure its structural stability. (v) High corrosion resistance and good stability [93, 94]. This chapter will introduce in detail several typical biopolymers and their derivative materials, such as the application of CLS, CS, and SA in the negative electrode protective layer, and explore their modification methods and mechanism of action.

2.7.1 Cellulose

Cellulose anode functional coating has shown great application potential in the field of environmentally friendly energy storage, especially in water ZIBs. The formation of dendrites can be effectively inhibited by constructing the cellulose coating with ion affinity on the zinc anode, thus improving the stability and cycle life of the battery.

Liu et al. [95] developed a new lignocellulose-containing nanofiber-MXene protective layer (LM) to stabilize zinc metal anodes. This LM layer, with excellent mechanical properties (43.7 MPa), serves two critical functions: it acts as a desolvent layer to prevent the release of free water molecules, thereby reducing water-induced corrosion, and as a zinc gating layer, guiding lateral zinc deposition to suppress dendrite formation. By promoting zinc crystal growth along the (002) Zn plane, the LM layer enhances the structural stability of zinc anodes during cycling. This unique protective layer significantly improves the performance of zinc symmetric and zinc-copper asymmetric batteries, delivering high stability and long cycle life. This work offers new insights into designing dendrite-free zinc anodes for long-lasting battery applications.

Chen et al. [96] prepared a bio-based multifunctional cellulose L-glucate (CLE) that has the ability to regulate the chemical properties of the zinc anode/electrolyte interface. CLE additive exhibits strong adsorption capacity through its acetylacetone group in the levoxylic acid portion, and it provides both acceptors and donors of hydrogen bonds due to keto-enol tautomerism. This material constructs an in situ protective layer on the surface of the zinc anode, effectively inhibits the growth of zinc dendrites, reduces corrosion, and improves cycle stability. The principle is to realize the application of high-performance zinc metal anode in water-based zinc ion batteries by constructing a protective layer on the zinc anode in situ.

These research results not only provide an effective method to inhibit zinc dendrites and side reactions, but also provide an important theoretical and practical basis for developing advanced rechargeable zinc-based battery systems. The application of cellulose anode coating is expected to promote the commercial application of water ZIBs in the field of environmentally friendly energy storage.

2.7.2 Chitosan and Sodium Alginate

CS and SA as natural polysaccharides, because of their unique physical and chemical properties, have been widely used in the field of battery anode coating. Their application not only improves battery performance, but also brings environmental and economic benefits.

Cai et al. [97] developed a biodegradable electrode/electrolyte interface protective layer (LbL-(CS/SA)$_4$) using layer-by-layer (LbL) self-assembly technology, aimed at improving the zinc ion solvation structure and cycle performance of aqueous ZIBs. Construction of LbL-(CS/SA)$_4$ film on Zn anodes as shown in Figure 2.10. This article describes a protective layer for zinc anodes in ZIBs using CS and SA as the primary materials. Both CS and SA are natural, biodegradable polyelectrolytes with

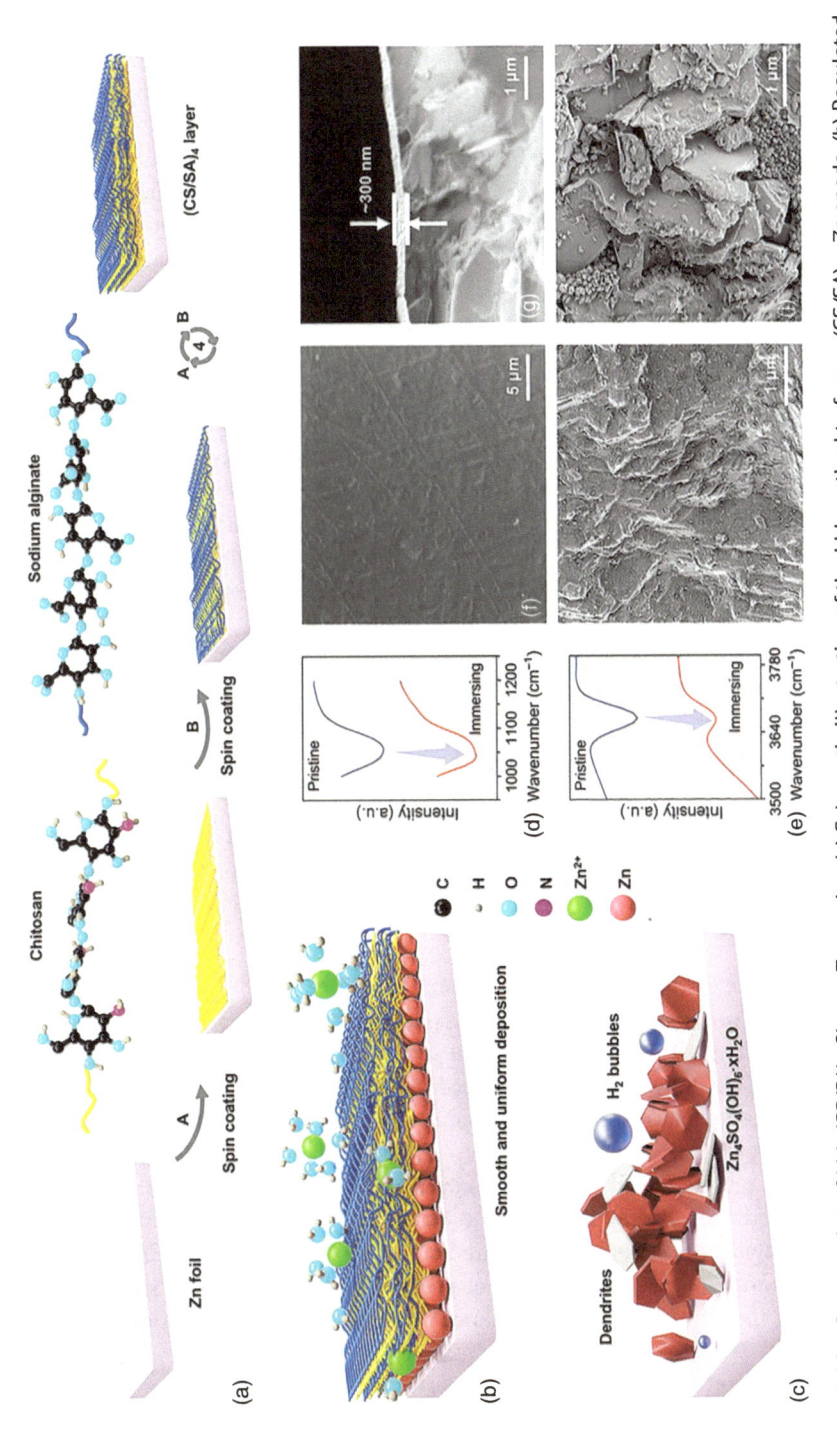

Figure 2.10 Construction of LbL-(CS/SA)$_4$ film on Zn anodes. (a) Schematic illustration of the LbL method to form a (CS/SA)$_4$-Zn anode. (b) Regulated interface using the (CS/SA)$_4$ film. (c) Side reaction and dendrite growth on the bare-Zn anode. (d,e) Fourier transform infrared spectroscopy (FT-IR) spectra of the LbL Layer with/without immersing in the electrolyte for 24 h. SEM images of the pristine LbL film: (f) top view and (g) side view. Deposition morphology of the Zn anode after 50 cycles: (h) (CS/SA)$_4$-Zn anode and (i) bare-Zn anode. The applied current density was 1 mA cm^{-2}, and the areal capacity was 1 mAh cm^{-2}. Source: Reproduced with the permission from Cai et al. [97]/John Wiley & Sons.

opposite charges, which enables them to form a stable gel-like layer through the LbL assembly process driven by electrostatic interactions. The CS/SA layer possesses key properties including mechanical flexibility, abundant hydroxyl groups, and strong adherence to the zinc surface. These properties collectively create a robust barrier that isolates the zinc anode from the bulk electrolyte, while also facilitating desolvation of Zn^{2+}. Additionally, the hydroxyl groups on the CS/SA layer form hydrogen bonds with Zn^{2+}, reducing desolvation energy and enhancing Zn^{2+} transport, leading to more uniform zinc deposition. The flexible nature of this layer allows it to accommodate volume changes during Zn plating/stripping cycles, effectively suppressing dendrite growth and corrosion reactions. This work demonstrates how the LbL method can effectively regulate the electrode/electrolyte interface in aqueous metal batteries.

Although the application of CS in battery anode coating shows great potential, its mechanical strength and electrical conductivity still need to be further improved to meet the needs of high-performance batteries. The application of SA in battery anode coatings also faces challenges, such as improving its stability and cycle life at high current densities. The application of CS and SA as natural polysaccharides in battery anode coating shows its advantages of environmental protection and low cost. However, in order to achieve its widespread application in high-performance batteries, challenges in mechanical strength, conductivity, and stability still need to be overcome.

2.8 Conclusion and Outlook

The future of biopolymers in battery technology holds significant promise as the demand for sustainable energy solutions continues to grow. As research progresses, several key areas offer exciting opportunities for further development and optimization of biopolymers in battery systems.

1) Performance Optimization Under Extreme Conditions
   One of the primary challenges in the application of biopolymers in commercial battery systems is ensuring consistent performance under extreme operating conditions. Batteries are often required to function at high current densities, extreme temperatures, and under prolonged cycling. In these demanding environments, biopolymers must maintain their structural integrity, ionic conductivity, and chemical stability without degradation. Future research should focus on enhancing the thermal, mechanical, and electrochemical properties of biopolymers to meet these requirements. Innovations in molecular design, such as cross-linking and functional group modifications, could significantly improve the durability and performance of biopolymer-based materials in harsh conditions.
2) Scalability and Economic Viability
   To achieve widespread commercial adoption, biopolymers must be cost-effectively mass produced. While biopolymers are renewable and environmentally

friendly, their large-scale manufacturing processes must be further refined to ensure they are economically viable. This will require advancements in both the synthesis of biopolymers and the development of more efficient processing technologies. Additionally, securing a reliable and sustainable supply of raw materials is crucial for minimizing production costs. The development of low-cost, high-performance biopolymers could accelerate their integration into mainstream battery manufacturing, making sustainable energy-storage solutions more accessible.

3) Expanding Applications Across Different Battery Chemistries

While biopolymers have demonstrated significant potential in lithium-ion and ZIBs, their application across a wider range of battery chemistries, such as sodium-ion, magnesium-ion, and other emerging systems, remains underexplored. These alternative chemistries are gaining attention due to their abundance, cost-effectiveness, and potential for large-scale energy storage. Future research should investigate how biopolymers can be adapted to work within these systems, focusing on their compatibility with different electrolytes, electrodes, and operating environments. Successful integration of biopolymers into a broader spectrum of battery chemistries could further expand their role in various energy-storage applications, from consumer electronics to grid-scale storage.

4) Development of Multifunctional Biopolymers

As battery technology evolves, there is a growing demand for multifunctional materials that can enhance multiple aspects of battery performance simultaneously. Biopolymers have the potential to fulfill this role by combining properties such as high ionic conductivity, mechanical flexibility, and thermal stability. Future innovations could involve the development of biopolymers with self-healing properties, improved fire resistance, and the ability to suppress side reactions that degrade battery performance over time. These multifunctional materials could lead to batteries that are not only more efficient and longer-lasting but also safer and more reliable.

5) Environmental Sustainability and Circular Economy

One of the most compelling advantages of biopolymers is their environmental sustainability. As the world shifts toward greener technologies, the use of biodegradable and renewable materials in batteries aligns with the principles of a circular economy. Biopolymers can be sourced from renewable resources, reducing the environmental impact of battery production. Moreover, their biodegradability allows for easier recycling and disposal at the end of a battery's life cycle, reducing waste and minimizing the environmental footprint of battery systems. Future research should continue to explore ways to further reduce the carbon footprint of biopolymer production and ensure that these materials are fully integrated into sustainable supply chains.

6) Collaboration Between Academia and Industry

The successful commercialization of biopolymer-based batteries will require close collaboration between academic researchers and industry stakeholders. Academia can continue to push the boundaries of innovation, exploring new materials and applications, while industry can focus on scaling these technologies and bringing them to market. By working together, both sectors can accelerate the development of biopolymer-based batteries and ensure that they meet the performance, safety, and cost requirements necessary for widespread adoption.

In summary, biopolymers represent a promising and environmentally friendly alternative to traditional battery materials, with the potential to transform energy-storage technologies. As research continues to optimize their performance, scalability, and compatibility with different battery chemistries, biopolymers are poised to play an increasingly important role in the development of sustainable, high-performance batteries. The next phase of innovation will likely focus on creating multifunctional biopolymer materials, expanding their application across various battery systems, and ensuring that these solutions are both economically viable and environmentally responsible. By addressing these challenges, biopolymers could revolutionize the energy-storage industry, providing the foundation for safer, more efficient, and sustainable batteries that meet the growing global demand for clean energy.

References

1 Ding, J., Yang, Y., Poisson, J. et al. (2024). Recent advances in biopolymer-based hydrogel electrolytes for flexible supercapacitors. *ACS Energy Letters* 9: 1803–1825.
2 Chen, Z., Tang, Y., Zhu, P. et al. (2022). Progress in preparation and applications of carboxymethyl cellulose. *Transactions of China Pulp and Paper* 37: 144–154.
3 Mohamed, M.A. (2012). Swelling characteristics and application of gamma-radiation on irradiated SBR-carboxymethylcellulose (CMC) blends. *Arabian Journal of Chemistry* 5: 207–211.
4 Ye, Y., Oguzlu, H., Zhu, J. et al. (2022). Ultrastretchable ionogel with extreme environmental resilience through controlled hydration interactions. *Advanced Functional Materials* 33: 2209787.
5 Liang, T.-W., Huang, C.-T., Dzung, N. et al. (2015). Squid pen chitin chitooligomers as food colorants absorbers. *Marine Drugs* 13: 681–696.
6 Fan, Z., Qin, Y., Liu, S. et al. (2018). Synthesis, characterization, and antifungal evaluation of diethoxyphosphoryl polyaminoethyl chitosan derivatives. *Carbohydrate Polymers* 190: 1–11.
7 Diana, M.I., Selvin, P.C., Selvasekarapandian, S. et al. (2021). Investigations on Na-ion conducting electrolyte based on sodium alginate biopolymer for all-solid-state sodium-ion batteries. *Journal of Solid State Electrochemistry* 25: 2009–2020.
8 Joseph, S.M., Krishnamoorthy, S., Paranthaman, R. et al. (2021). A review on source-specific chemistry, functionality, and applications of chitin and chitosan. *Carbohydrate Polymer Technologies and Applications* 2: 100036.

References

9 dos Santos, P.S.B., Erdocia, X., Gatto, D.A. et al. (2014). Characterisation of Kraft lignin separated by gradient acid precipitation. *Industrial Crops and Products* 55: 149–154.

10 Messali, M., Lgaz, H., Dassanayake, R. et al. (2017). Guar gum as efficient non-toxic inhibitor of carbon steel corrosion in phosphoric acid medium: electrochemical, surface, DFT and MD simulations studies. *Journal of Molecular Structure* 1145: 43–54.

11 Xu, C., Arancon, R.A.D., Labidi, J. et al. (2014). Lignin depolymerisation strategies: towards valuable chemicals and fuels. *Chemical Society Reviews* 43: 7485–7500.

12 Thakur, V.K., Thakur, M.K., Raghavan, P. et al. (2014). Progress in green polymer composites from lignin for multifunctional applications: a review. *ACS Sustainable Chemistry & Engineering* 2: 1072–1092.

13 Lu, X. and Gu, X. (2023). A review on lignin-based epoxy resins: lignin effects on their synthesis and properties. *International Journal of Biological Macromolecules* 229: 778–790.

14 Randall, R.C., Phillips, G.O., and Williams, P.A. (1989). Fractionation and characterization of gum from Acacia senegal. *Food Hydrocolloids* 3: 65–75.

15 Prasad, N., Thombare, N., Sharma, S.C. et al. (2022). Gum arabic – a versatile natural gum: a review on production, processing, properties and applications. *Industrial Crops and Products* 187: 115304–115322.

16 Ji, Y., Li, Z., and Qiao, W. (2005). Chemical modification of guar gum. *China Surfactant Detergent and Cosmetics* 35: 111–114.

17 Zhang, G., Qiu, B., Xia, Y. et al. (2019). Double-helix-superstructure aqueous binder to boost excellent electrochemical performance in Li-rich layered oxide cathode. *Journal of Power Sources* 420: 29–37.

18 Parija, S., Misra, M., and Mohanty, A.K. (2001). Studies of natural gum adhesive extracts: an overview. *Journal of Macromolecular Science, Polymer Reviews* 41: 175–197.

19 Chen, H., Ling, M., Hencz, L. et al. (2018). Exploring chemical, mechanical, and electrical functionalities of binders for advanced energy-storage devices. *Chemical Reviews* 118: 8936–8982.

20 Tester, R.F., Karkalas, J., and Qi, X. (2004). Starch – composition, fine structure and architecture. *Journal of Cereal Science* 39: 151–165.

21 Thakur, S., Govender, P.P., Mamo, M.A. et al. (2017). Recent progress in gelatin hydrogel nanocomposites for water purification and beyond. *Vacuum* 146: 396–408.

22 Scott, S., Terreblanche, J., Thompson, D.L. et al. (2022). Gelatin and alginate binders for simplified battery recycling. *The Journal of Physical Chemistry C* 126: 8489–8498.

23 He, S., Liu, J., Gong, Z. et al. (2021). Application of tragacanth gum in silicon anode lithium-ion battery. *Journal of Xiamen University. Natural Science* 60: 849–854.

24 Nazarzadeh Zare, E., Makvandi, P., and Tay, F.R. (2019). Recent progress in the industrial and biomedical applications of tragacanth gum: a review. *Carbohydrate Polymers* 212: 450–467.

25 Suo, L., Hu, Y., Li, H. et al. (2013). A new class of solvent-in-salt electrolyte for high-energy rechargeable metallic lithium batteries. *Nature Communications* 4: 1481–1490.

26 Ummartyotin, S. and Manuspiya, H. (2015). A critical review on cellulose: from fundamental to an approach on sensor technology. *Renewable and Sustainable Energy Reviews* 41: 402–412.

27 Xin, S., Gu, L., Zhao, N.-H. et al. (2012). Smaller sulfur molecules promise better lithium-sulfur batteries. *Journal of the American Chemical Society* 134: 18510–18513.

28 Wu, C. and Wang, C. (2014). Strategies for desymmetrising trehalose to synthesise trehalose glycolipids. *Organic & Biomolecular Chemistry* 12: 5558–5562.

29 Olgenblum, G.I., Sapir, L., and Harries, D. (2020). Properties of aqueous trehalose mixtures: glass transition and hydrogen bonding. *Journal of Chemical Theory and Computation* 16: 1249–1262.

30 Wang, Y., Wu, X., Jiang, X. et al. (2022). Citrulline-induced mesoporous CoS/CoO heterojunction nanorods triggering high-efficiency oxygen electrocatalysis in solid-state Zn-air batteries. *Chemical Engineering Journal* 434: 134744–134755.

31 Chen, J., Liu, N., Dong, W. et al. (2024). Simultaneous regulation of coordination environment and electrode interface for highly stable zinc anode using a bifunctional citrulline additive. *Advanced Functional Materials* 34: 2313925–2313936.

32 Perumal, P., Christopher Selvin, P., and Selvasekarapandian, S. (2018). Characterization of biopolymer pectin with lithium chloride and its applications to electrochemical devices. *Ionics* 24: 3259–3270.

33 Mohnen, D. (2008). Pectin structure and biosynthesis. *Current Opinion in Plant Biology* 11: 266–277.

34 Burapapadh, K., Takeuchi, H., and Sriamornsak, P. (2012). Novel pectin-based nanoparticles prepared from nanoemulsion templates for improving in vitro dissolution and in vivo absorption of poorly water-soluble drug. *European Journal of Pharmaceutics and Biopharmaceutics* 82: 250–261.

35 Naqash, F., Masoodi, F.A., Rather, S.A. et al. (2017). Emerging concepts in the nutraceutical and functional properties of pectin – a review. *Carbohydrate Polymers* 168: 227–239.

36 Mobarak, N.N., Jumaah, F.N., Ghani, M.A. et al. (2015). Carboxymethyl carrageenan based biopolymer electrolytes. *Electrochimica Acta* 175: 224–231.

37 Huang, Y., Shaibani, M., Gamot, T.D. et al. (2021). A saccharide-based binder for efficient polysulfide regulations in Li-S batteries. *Nature Communications* 12: 5375–5390.

38 Yu, L., Tao, B., Ma, L. et al. (2024). A robust network sodium carboxymethyl cellulose-epichlorohydrin binder for silicon anodes in lithium-ion batteries. *Langmuir* 17930–17940.

39 Sun, X., Lin, X., Wen, Y. et al. (2024). A water-soluble binder in high-performance silicon-based anodes for lithium-ion batteries based on sodium carboxymethyl cellulose and waterborne polyurethane. *Green Chemistry* 26: 9874–9887.

40 Kim, S., Kim, D.H., Cho, M. et al. (2020). Fast-charging lithium-sulfur batteries enabled via lean binder content. *Small* 16: 2004372–2004379.

41 Yang, J., Liu, H., Zhao, X. et al. (2024). Janus binder chemistry for synchronous enhancement of iodine species adsorption and redox kinetics toward sustainable aqueous Zn-I2 batteries. *Journal of the American Chemical Society* 146: 6628–6637.

42 Rajeev, K.K., Kim, E., Nam, J. et al. (2020). Chitosan-grafted-polyaniline copolymer as an electrically conductive and mechanically stable binder for high-performance Si anodes in Li-ion batteries. *Electrochimica Acta* 333: 135532.

43 Lee, S.H., Lee, J.H., Nam, D.H. et al. (2018). Epoxidized natural rubber/chitosan network binder for silicon anode in lithium-ion battery. *ACS Applied Materials & Interfaces* 10: 16449–16457.

44 Cao, P., Yang, G., Li, B. et al. (2019). Rational design of a multifunctional binder for high-capacity silicon-based anodes. *ACS Energy Letters* 4: 1171–1180.

45 Hu, J., Wang, Y., Li, D. et al. (2018). Effects of adhesion and cohesion on the electrochemical performance and durability of silicon composite electrodes. *Journal of Power Sources* 397: 223–230.

46 Zhang, L., Zhang, L., Chai, L. et al. (2014). A coordinatively cross-linked polymeric network as a functional binder for high-performance silicon submicro-particle anodes in lithium-ion batteries. *Journal of Materials Chemistry A* 2: 19036–19045.

47 Gendensuren, B. and Oh, E. (2018). Dual-crosslinked network binder of alginate with polyacrylamide for silicon/graphite anodes of lithium ion battery. *Journal of Power Sources* 384: 379–386.

48 Guo, R., Zhang, S., Ying, H. et al. (2019). Preparation of an amorphous cross-linked binder for silicon anodes. *ChemSusChem* 12: 4838–4845.

49 Li, Z., Ji, J., Wu, Q. et al. (2020). A new battery process technology inspired by partially carbonized polymer binders. *Nano Energy* 67: 104234.

50 Kovalenko, I., Zdyrko, B., Magasinski, A. et al. (2011). A major constituent of brown algae for use in high-capacity Li-ion batteries. *Science* 334: 75–79.

51 Chen, Z., Lu, M., Qian, Y. et al. (2023). Ultra-low dosage lignin binder for practical lithium-sulfur batteries. *Advanced Energy Materials* 13: 2300092–2300103.

52 Yuan, J., Ren, W., Wang, K. et al. (2021). Ultrahighly elastic lignin-based copolymers as an effective binder for silicon anodes of lithium-ion batteries. *ACS Sustainable Chemistry & Engineering* 10: 166–176.

53 Li, G., Ling, M., Ye, Y. et al. (2015). Acacia senegal-inspired bifunctional binder for longevity of lithium-sulfur batteries. *Advanced Energy Materials* 5: 1500878–1500886.

54 Zhong, H., He, J., and Zhang, L. (2021). Crosslinkable aqueous binders containing Arabic gum-grafted-poly (acrylic acid) and branched polyols for Si anode of lithium-ion batteries. *Polymer* 215: 123377–123384.

55 Liu, J., Galpaya, D.G.D., Yan, L. et al. (2017). Exploiting a robust biopolymer network binder for an ultrahigh-areal-capacity Li–S battery. *Energy & Environmental Science* 10: 750–755.

56 Ling, H.Y., Wang, C., Su, Z. et al. (2020). Amylopectin from glutinous rice as a sustainable binder for high-performance silicon anodes. *Energy & Environmental Materials* 4: 263–268.

57 Hapuarachchi, S.N.S., Wasalathilake, K.C., Nerkar, J.Y. et al. (2020). Mechanically robust tapioca starch composite binder with improved ionic conductivity for sustainable lithium-ion batteries. *ACS Sustainable Chemistry & Engineering* 8: 9857–9865.

58 Rohan, R., Kuo, T., Chiou, C. et al. (2018). Low-cost and sustainable corn starch as a high-performance aqueous binder in silicon anodes via in situ cross-linking. *Journal of Power Sources* 396: 459–466.

59 Sun, R., Hu, J., Shi, X. et al. (2021). Water-soluble cross-linking functional binder for low-cost and high-performance lithium-sulfur batteries. *Advanced Functional Materials* 31: 2104858–2104867.

60 Liu, Z., Fang, C., He, X. et al. (2021). In situ-formed novel elastic network binder for a silicon anode in lithium-ion batteries. *ACS Applied Materials & Interfaces* 13: 46518–46525.

61 Ye, R., Liu, J., Tian, J. et al. (2024). Novel binder with cross-linking reconfiguration functionality for silicon anodes of lithium-ion batteries. *ACS Applied Materials & Interfaces* 16: 16820–16829.

62 Zhou, F., Mei, Y., Wu, Q. et al. (2024). Sulfur electrode tolerance and polysulfide conversion promoted by the supramolecular binder with rare-earth catalysis in lithium-sulfur batteries. *Energy Storage Materials* 67: 103315.

63 Senthil, C., Kim, S., and Jung, H.Y. (2022). Flame retardant high-power Li-S flexible batteries enabled by bio-macromolecular binder integrating conformal fractions. *Nature Communications* 13: 145–159.

64 Famprikis, T., Dawson, J.A., Fauth, F. et al. (2019). A new superionic plastic polymorph of the Na+ conductor Na3PS4. *ACS Materials Letters* 1: 641–646.

65 Chen, S., Zhang, M., Zou, P. et al. (2022). Historical development and novel concepts on electrolytes for aqueous rechargeable batteries. *Energy & Environmental Science* 15: 1805–1839.

66 Yang, C., Wu, Q., Xie, W. et al. (2021). Copper-coordinated cellulose ion conductors for solid-state batteries. *Nature* 598: 590–596.

67 Wang, D., Xie, H., Liu, Q. et al. (2023). Low-cost, high-strength cellulose-based quasi-solid polymer electrolyte for solid-state lithium-metal batteries. *Angewandte Chemie International Edition* 62: e202302767.

68 Ma, L., Chen, S., Wang, D. et al. (2019). Super-stretchable zinc-air batteries based on an alkaline-tolerant dual-network hydrogel electrolyte. *Advanced Energy Materials* 9: 1803046–1803054.

69 Cao, F., Wu, B., Li, T. et al. (2021). Mechanoadaptive morphing gel electrolyte enables flexible and fast-charging Zn-ion batteries with outstanding dendrite suppression performance. *Nano Research* 15: 2030–2039.

70 Yang, J., Cao, Z., Chen, Y. et al. (2023). Dry-processable polymer electrolytes for solid manufactured batteries. *ACS Nano* 17: 19903–19913.

71 Wu, M., Zhang, Y., Xu, L. et al. (2022). A sustainable chitosan-zinc electrolyte for high-rate zinc-metal batteries. *Matter* 5: 3402–3416.

72 Lu, H., Hu, J., Wei, X. et al. (2023). A recyclable biomass electrolyte towards green zinc-ion batteries. *Nature Communications* 14: 4435–4449.

73 Wang, F., Zhang, J., Lu, H. et al. (2023). Production of gas-releasing electrolyte-replenishing Ah-scale zinc metal pouch cells with aqueous gel electrolyte. *Nature Communications* 14: 4211–4221.

74 Liu, Q., Yu, Z., Zhuang, Q. et al. (2023). Anti-fatigue hydrogel electrolyte for all-flexible Zn-ion batteries. *Advanced Materials* 35: 2300498–2300509.

75 Liu, Y., Wang, P., Yang, Z. et al. (2024). Lignin derived ultrathin all-solid polymer electrolytes with 3D single-ion nanofiber ionic bridge framework for high performance lithium batteries. *Advanced Materials* 36: 2400970–2400981.

76 Huang, J., Hu, Y., Wang, H. et al. (2022). Lignin isolated from poplar wood for porous carbons as electrode for high-energy renewable supercapacitor driven by lignin/deep eutectic solvent composite gel polymer electrolyte. *ACS Applied Energy Materials* 5: 6393–6400.

77 Su, H., Guo, Q., Qiao, C. et al. (2024). Lignin-alkali metal ion self-catalytic system initiated rapid polymerization of hydrogel electrolyte with high strength and anti-freezing ability. *Advanced Functional Materials* 34: 2316274.

78 Zhou, J., Zhang, R., Xu, R. et al. (2022). Super-assembled hierarchical cellulose aerogel-gelatin solid electrolyte for implantable and biodegradable zinc ion battery. *Advanced Functional Materials* 32: 2111406–2111418.

79 Kumankuma-Sarpong, J., Chang, C., Hao, J. et al. (2024). Entanglement added to cross-linked chains enables tough gelatin-based hydrogel for Zn metal batteries. *Advanced Materials* 36: 2403214.

80 Yang, Z., Zhang, Q., Wu, T. et al. (2024). Thermally healable electrolyte-electrode interface for sustainable quasi-solid zinc-ion batteries. *Angewandte Chemie International Edition* 63.

81 Wu, Q., Huang, J., Zhang, J. et al. (2024). Multifunctional cellulose nanocrystals electrolyte additive enable ultrahigh-rate and dendrite-free Zn anodes for rechargeable aqueous zinc batteries. *Angewandte Chemie International Edition* 63: e202319051.

82 Li, H., Ren, Y., Zhu, Y. et al. (2023). A bio-inspired trehalose additive for reversible zinc anodes with improved stability and kinetics. *Angewandte Chemie International Edition* 62: e202310143.

83 Liu, H., Xin, Z., Cao, B. et al. (2023). Polyhydroxylated organic molecular additives for durable aqueous zinc battery. *Advanced Functional Materials* 34: 2309840.

84 Hong, L., Guan, J., Tan, Y. et al. (2024). An effective descriptor for the screening of electrolyte additives toward the stabilization of Zn metal anodes. *Energy & Environmental Science* 17: 3157–3167.

85 Arora, P. and Zhang, Z.M. (2004). Battery separators. *Chemical Reviews* 104: 4419–4462.

86 Zhang, T.W., Chen, J.L., Tian, T. et al. (2019). Sustainable separators for high-performance lithium ion batteries enabled by chemical modifications. *Advanced Functional Materials* 29: 1902023–1902032.

87 Zhang, Y., Liu, Z., Li, X. et al. (2023). Loosening zinc ions from separator boosts stable Zn plating/striping behavior for aqueous zinc ion batteries. *Advanced Energy Materials* 13: 2302126–2302135.

88 Ge, X., Zhang, W., Song, F. et al. (2022). Single-ion-functionalized nanocellulose membranes enable lean-electrolyte and deeply cycled aqueous zinc-metal batteries. *Advanced Functional Materials* 32: 2200429–2200440.

89 Zhou, W., Chen, M., Tian, Q. et al. (2022). Cotton-derived cellulose film as a dendrite-inhibiting separator to stabilize the zinc metal anode of aqueous zinc ion batteries. *Energy Storage Materials* 44: 57–65.

90 Wang, J., Gao, Y., Liu, D. et al. (2023). A sodiophilic amyloid fibril modified separator for dendrite-free sodium-metal batteries. *Advanced Materials* 36: 2304942.

91 Liang, S., Miao, J., Shi, H. et al. (2022). Tuning interface mechanics via β-configuration dominant amyloid aggregates for lithium metal batteries. *ACS Nano* 16: 19584–19593.

92 Serra, J.P., Fidalgo-Marijuan, A., Teixeira, J. et al. (2022). Sustainable lithium-ion battery separator membranes based on carrageenan biopolymer. *Advanced Sustainable Systems* 6: 2200279.

93 Xu, R., Cheng, X., Yan, C. et al. (2019). Artificial interphases for highly stable lithium metal anode. *Matter* 1: 317–344.

94 Zhou, H., Yu, S., Liu, H. et al. (2020). Protective coatings for lithium metal anodes: recent progress and future perspectives. *Journal of Power Sources* 450: 227632.

95 Liu, C., Li, Z., Zhang, X. et al. (2022). Synergic effect of dendrite-free and zinc gating in lignin-containing cellulose nanofibers-MXene layer enabling long-cycle-life zinc metal batteries. *Advanced Science* 9: 2202380–2202390.

96 Chen, K., Chen, Y., Xu, Y. et al. (2024). Regulating interfacial chemistry with biobased multifunctional cellulose levulinate ester for highly reversible zinc ion batteries. *Energy Storage Materials* 71: 103597.

97 Cai, X., Wang, X., Bie, Z. et al. (2023). A layer-by-layer self-assembled bio-macromolecule film for stable zinc anode. *Advanced Materials* 36: 2306734–2306746.

3A

Application of Synthetic Polymers in Batteries: Carbon-chain Polymers

3A.1 Introduction

Synthetic polymers are long chains of large molecules made from many small molecules connected by chemical reactions. They have unique properties and a wide range of uses and are an indispensable part of modern industry and scientific research. A polymer material is a material consisting of a large number of repeating units connected by covalent bonds. These macromolecular compounds usually have high molecular weights and are capable of forming complex structures. These materials are highly malleable and functional, and their physical and chemical properties can be regulated by adjusting the type, number, and connection of monomers. In 1920, German scientist Staudinger proposed the long-chain structure of polymers, forming the concept of polymers, which began the era of synthetic polymers prepared by chemical methods. In 1929, Wallace Hume Carothers studied a series of condensation reactions, validated and developed the macromolecular theory, and led to the creation of polyamide 66. After entering the 1960s, polymer science entered a turning point, from basic regularity research to design and creation, synthesis of a large number of high-performance polymers, such as ultra-high modulus, ultra-high strength, flame resistance, high-temperature resistance, oil resistance, and other materials, as well as biomedical materials, semiconductor or superconductor materials, and low-temperature flexible materials.

3A.2 Overview of Synthetic Polymers Materials

Synthetic polymer materials are widely used in batteries due to their diversity of structure and function, for example, they can be used as binders, electrolyte materials, separators, functional coatings, flame retardants, and active material carriers. These applications demonstrate the diversity and importance of synthetic polymers in battery technology, which can further optimize battery performance and drive further development of battery technology through molecular design and materials engineering.

Functional Auxiliary Materials in Batteries: Synthesis, Properties, and Applications, First Edition. Wei Hu.
© 2025 WILEY-VCH GmbH. Published 2025 by WILEY-VCH GmbH.

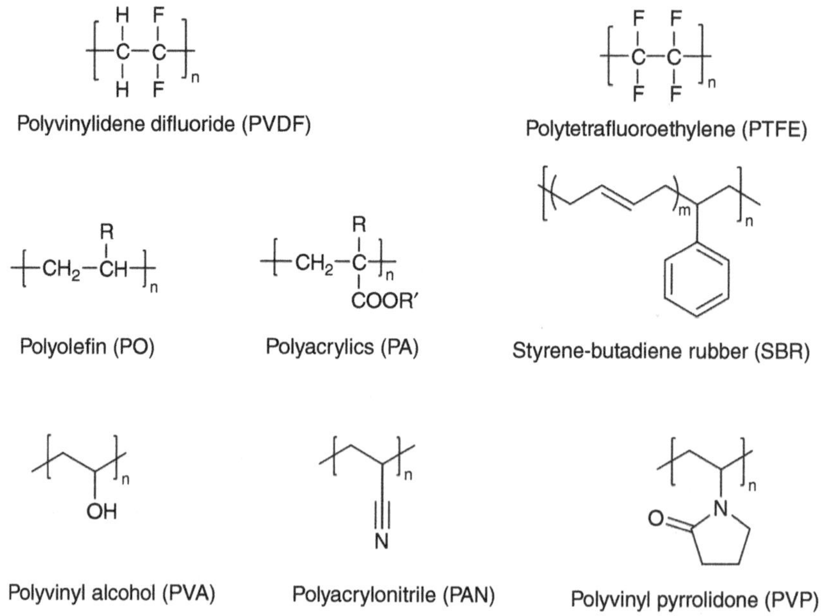

Figure 3A.1 Typical chemical structure of carbon-chain polymers commonly used in batteries.

The structure, physicochemical properties, processing methods, and application fields of polymers (their typical chemical structures are shown in Figure 3A.1) commonly used in batteries are summarized in the subsequent text.

3A.2.1 Polyvinylidene Difluoride (PVDF)

PVDF mainly refers to the copolymer of vinylidene fluoride homopolymer or vinylidene fluoride and a small amount of other fluorine-containing vinyl monomers, which can be synthesized by suspension polymerization or emulsion polymerization of 1,1-fluoroethylene. It is a highly nonreactive thermoplastic fluoropolymer with the characteristics of a fluorine resin and general resin, in addition to good chemical corrosion resistance, high-temperature resistance, oxidation resistance, weather resistance, and radiation resistance, but also has special properties such as piezoelectric, dielectric, thermoelectricity, etc. It is the second-largest fluoroplastic product with a global annual production capacity of >53,000 tons.

Its chemical structure is bound by F—C bonds, which have the short bond property and can form the most stable hydrogen bond with H. Therefore, fluorocarbon coatings have specific physicochemical properties, not only have strong wear resistance and impact resistance, but also have high fading resistance and ultraviolet resistance in extremely harsh environments. Compared to other fluoropolymers, PVDF has a lower melting point (about 177 °C), making it easier to melt. Therefore, it can be processed by general thermoplastic processing methods, such as extrusion, injection molding, pouring, molding, and transfer molding.

The applications of PVDF are mainly concentrated in three major fields of petrochemical, electrical and electronic, and fluorocarbon coatings, due to its excellent flexibility, light weight, low thermal conductivity, high chemical corrosion resistance, and heat resistance as well as other multiple excellent properties. PVDF can be used as the insulating skin of electric wires and also can be used to make professional monofilament fishing lines as an alternative to traditional nylon monofilament.

It can be used in the production of tactile sensor arrays, inexpensive strain gauges, and light weight audio transducers due to its piezoelectric properties. PVDF sensors are more suitable for dynamic modal testing than semiconductor piezoresistive sensors and have some advantages over piezoelectric ceramic transducers in terms of structural integration. Due to the lower cost and greater compatibility, active sensors using PVDF are important for the future development of structural health monitoring. In the biomedical field, PVDF films are often used in western blotting, on which proteins are electrophoresed. Because PVDF is resistant to solvent corrosion, the film used in the detection can be easily stripped and reused to detect other proteins. They can also be used to make syringes or wheeled membrane filtration devices. The material's heat and chemical resistance and low protein binding properties allow it to be used as a disinfection filter in the preparation of drugs, and as a filter in the preparation of samples for analysis, such as HPLC, to prevent expensive equipment from being damaged by small amounts of particulate matter in these samples.

In the battery field, PVDF is also the typical binder for lithium battery composite electrodes: Mixing PVDF dissolved in 1–2% N-methyl-2-pyrrolidone (NMP) with active lithium storage materials such as graphite, silicon, tin, $LiCoO_2$, $LiMn_2O_4$, or $LiFePO_4$, and conductive additives such as carbon black or carbon nanofibers. The slurry is then poured onto the metal current collector, and the NMP is evaporated to form a composite electrode or paste electrode. PVDF can be used in this situation because it is chemically inert within the potential range of the battery charge and discharge and does not react with the electrolyte or lithium.

3A.2.2 Polytetrafluoroethylene (PTFE)

PTFE is a polyethylene (PE)-type synthetic polymer material in which all hydrogen atoms in the molecular structure are replaced by fluorine atoms, which can be produced by suspension polymerization or emulsion polymerization of tetrafluoroethylene monomer in industry. The molecular structure of PTFE is similar to that of PE, which is a linear polymer with symmetry and no branched chain, so the molecule does not show polarity. Due to the large repulsive force between the non-bonded fluorine atoms, the carbon chain of PTFE is twisted and becomes a helical structure, and the fluorine atoms just cover the carbon main chain completely, which prevents the carbon atoms from reacting with other substances. In addition, the fluorocarbon bond has a high bond energy of 466 kJ mol^{-1}, which is the highest bond energy of all chemical single bonds, so the stability of PTFE is very high.

PTFE is currently the most widely used fluoropolymer, plastic is the most resistant to chemical corrosion, has the best dielectric properties, and has a wide operating temperature range of the variety, known as "plastic king." It has high-temperature resistance (can be used for a long time in the temperature range of 200–260 °C), low-temperature resistance (can be used for a long time below −100 °C), corrosion resistance (has a very high chemical inertness), climate resistance (very high atmospheric aging resistance), high lubrication (has a very small coefficient of friction), nonadhesion (it is hard to adhere because of its very low surface tension), good human compatibility (has good physiological inertia, nontoxic), good electrical insulation (high insulation, and is not affected by temperature and humidity), and a series of use advantages, these characteristics of PTFE are closely related to its molecular configuration. However, PTFE also has its own shortcomings, mainly manifested as cold flow, linear expansion coefficient, thermal conductivity, wear resistance, creep resistance, and outstanding surface nonviscosity, PTFE defects make its application to a certain extent limited.

In the battery field, the application of PTFE is mainly concentrated as a binder, film material, etc., and its application in solvent-free processes, lithium battery dry electrode materials, hydrogen fuel cells, and other fields shows its important position and role in new energy technology.

3A.2.3 Styrene-Butadiene Rubber (SBR)

SBR mainly refers to a class of polystyrene butadiene copolymers, which are formed by emulsion polymerization or solution polymerization of butadiene and styrene monomers and other functional monomers. By adjusting the ratio of butadiene and styrene monomers, a series of SBRS with different cross-linking degrees and different glass transition temperatures can be obtained.

SBR, as the backbone product of the rubber industry, is the largest and earliest general synthetic rubber variety to achieve industrial production. According to the polymerization process, it is divided into emulsion styrene-butadiene rubber (ESBR) and soluble styrene-butadiene rubber (SSBR). Compared with the soluble styrene butadiene process, the emulsion styrene butadiene process is more dominant in terms of cost savings, and about 75% of the global styrene butadiene plant capacity is based on the emulsion styrene butadiene process. ESBR has good comprehensive properties, mature technology and wide application, the production capacity, output, and consumption are the first in SBR.

In the battery field, SBR is mainly used as an aqueous binder in lithium-ion batteries. The groups on the surface of SBR react with the groups on the surface of copper foil to form chemical bonds. SBR emulsion itself is a product of hydrophilic and hydrophobic balance, on the one hand through hydrophobic and graphite organic bonding, on the other hand through hydrophilic groups and copper foil surface groups condensation reaction.

3A.2.4 Polyvinyl Alcohol (PVA)

PVA is a kind of stable nontoxic water-soluble polymer with white powder, flake or flocculent solid appearance, which can be prepared from ethylene acetate by first polymerization and then alcoholysis. It can be regarded as a linear polymer with a secondary hydroxyl group. The hydroxyl group in the molecule has high activity and can carry out typical chemical reactions of low alcohols, such as esterification, etherification, and acetalization, and can also react with many inorganic compounds or organic compounds. It contains many alcohol groups, has a polarity, and can form hydrogen bonds with water, so it can dissolve in polar water; It is also soluble in hot hydroxy solvents such as glycerol and phenol, and insoluble in methanol, benzene, acetone, gasoline, and other general organic solvents. PVA with an alcoholysis degree of <95% can be dissolved in room temperature water, and the PVA resin with an alcoholysis degree of >99.5% can only be dissolved in hot water above 95 °C.

PVA will soften when heated, there is no significant change below 40 °C, glass transition temperature: 75–85 °C, heating to >100 °C slowly discoloration, embrittlement. Above 160 °C, long-term heating will gradually color, dehydration etherification, and loss of solubility. Above 220 °C, decomposition occurs and water, acetic acid, acetaldehyde, and butenal are formed. Above 250 °C becomes a polymer containing conjugated double bonds. It is almost unaffected by weak acids, alkalis or organic solvents and has high oil resistance. It is a low viscosity polymer, and its aqueous solution is stable at room temperature. The aqueous solution will not deteriorate during storage. Due to the high adhesion between PVA molecules, PVA is easy to form a film, the formed film is colorless and transparent, has good mechanical strength, the surface is smooth and not sticky, and the solubility is good. The molecular film has good light transmittance, high moisture transmittance, no electricity, no dust suction, and good printability.

PVA can be used as a protective colloidal aqueous solution during polymer emulsification and suspension polymerization because its aqueous solution has good film formation and emulsification. It can be used in textile slurry, Vinnie fiber, paper coating agent, building material, binder, PVA film, Polyvinyl Butyral (PVB) raw material, food, and medicine. It can also be used as a soil improver, polymerization suspension agent, emulsifier, quenching agent, and so on. In addition, with the continuous development and improvement of PVA performance, its use is constantly expanding.

In the battery field, the application of PVA is mainly reflected in the electrolyte, binder, and separator material to improve the safety, electrochemical performance, and cycling stability of batteries. By combining PVA with other materials, it can further improve the overall performance of batteries and provide new strategies for efficient and safe applications of batteries.

3A.2.5 Polyacrylics (PA)

PA is a kind of polymer obtained by polymerization of acrylic acid, methacrylic acid, and its esters, collectively referred to as acrylic resin. The polymerization method contains emulsion polymerization, suspension polymerization, bulk polymerization, and solution polymerization. According to the structure and film-forming characteristics of acrylic resin coating, it can be divided into two categories: thermoplastic and thermosetting. Thermoplastic PA resin is a special type of solvent-based acrylic resin, which has the characteristics of melting and dissolving in appropriate solvents. Thermosetting PA resin is a special type of thermosetting resin, that is formed by polymerization of acrylic monomers and their derivatives. This resin can be cross-linked when heated to form a hard, heat-resistant material. According to the form and properties of acrylic resin polymers are divided into three types: solvent-based, water-based, and solvent-free.

PA has the following characteristics: Due to the particularity of its molecular structure, It has high transparency and brightness, which is suitable for the preparation of highly transparent products; It can withstand ultraviolet radiation, high temperature, low temperature, and other environmental conditions, not easy to occur oxidation decomposition and aging; It has good tolerance to chemicals such as water, hydrochloric acid, alkali, and organic acid, and can maintain stability in complex chemical environments. It has high tensile strength, elastic modulus, and wear resistance, and can meet various mechanical properties requirements.

PA is a high-quality raw material for the preparation of a variety of water-based, solvent-based, and powder coatings with high gloss and excellent pollution resistance. It can be used to prepare a variety of glue and tape, with excellent adhesion and adhesion, widely used in various industries. It can be used to manufacture a variety of plastic products, such as injection parts, extruded parts, films, plates and optical fiber, and cable outer cover, these products have high transparency, good toughness, and other characteristics. It can be used as a fiber optic cable sheath material, with high transparency, good processing performance, and mechanical strength. It can be used to prepare electronic and electrical components, such as electronic packaging, wire and cable insulation materials, and liquid crystal displays, which have good insulation properties and mechanical strength. These properties make PA resins an important raw material for industries such as coatings, binders, plastics, optical fiber and cable, and electronic appliances.

In the battery field, PA can be used to prepare solid polymer electrolytes (SPEs) to improve their ionic conductivity and electrochemical stability. It can also be used as a lithium battery binder and separator coating modified materials, thereby improving the overall performance and safety of the battery.

3A.2.6 Polyacrylonitrile (PAN)

PAN is a kind of polymer material that is formed by free radical polymerization of acrylonitrile monomer, the polymerization method contains solution and suspension polymerizations. Its appearance is white or slightly yellow opaque powder, and it is difficult to crystallize. Unlike other polar olefin polymers, even random PANs also can form two-dimensional ordered quasicrystals or 3D ordered crystal structures. The difficulty of crystallization of PAN is mainly reflected in its molecular structure and properties. Its molecular chain has high flexibility, which makes it difficult to form an ordered crystal structure during crystallization. In addition, its glass transition temperature is about 95 °C and the melting temperature is 322 °C. These thermodynamic properties also make it difficult to crystallize under conventional conditions. Although PAN is difficult to crystallize, under certain conditions, such as through stretching and heat treatment, it can form a certain crystalline structure. The crystal structure includes two dimensional ordered quasicrystal structure and 3D ordered crystal structure, belonging to the hexagonal crystal system and orthorhombic crystal system respectively. It is soluble in a variety of polar solvents, such as dimethylformamide, dimethylacetamide, vinyl carbonate, and so on. Its dielectric constant is 5.5 (1 kHz, 25 °C), 4.2 (1 MHz, 25 °C), and it has good electrical insulation properties.

PAN fiber has excellent strength and wear resistance, and is widely used in the production of clothing, carpets, curtains, and other textiles. It can be used to make air filters, water treatment filters, etc., due to its excellent filtration performance. It can be used to make cable insulation materials, electronic components, etc. because of its excellent electrical insulation properties and high-temperature resistance. It can be used to make medical bandages, surgical gowns, and other medical materials, due to its good biocompatibility and antibacterial properties. PAN is also the raw material for the preparation of carbon fiber, which is first formed by thermal oxidation in the air at 230 °C to oxidize PAN fiber, and then carbonized in an inert atmosphere above 1000 °C to eventually form carbon fiber. Carbon fiber has become a key material in the modern aerospace industry due to its superior strength and lightweight properties.

In the battery field, especially lithium-metal batteries and lithium-sulfur batteries, PAN is mainly used as an electrolyte and separator material. Its electrochemical stability and wide electrochemical window make it suitable as a polymer electrolyte for lithium-metal batteries, matching with high-voltage cathode materials to achieve high-energy-density batteries. It shows the potential to improve battery stability, safety, and energy density.

3A.2.7 Polyvinyl Pyrrolidone (PVP)

PVP is a nonionic polymer polymerized from *N*-vinyl pyrrolidone (NVP) by bulk polymerization or solution polymerization. It is a readily flowing white or nearly white powder with definite hygroscopicity, and soluble in water, ethanol, amines, nitroalkanes, and low molecular fatty acids, insoluble in acetone and ether. PVP is classified mainly based on its average molecular weight, including PVP-K90, PVP-K60, PVP-k30, PVP-K25, PVP-K17, and so on. Generally, the greater the *K* value, the greater the viscosity and the stronger the adhesion. These PVP with different polymerization degrees are different in chemical structure and physical properties and are suitable for different application fields.

Solution polymerization is generally used to synthesize PVP in industry, one method is solution polymerization in an organic solvent followed by steam stripping; another method is solution polymerization of NVP monomers with water-soluble cationic, anionic, or nonionic monomers. The structure and properties of PVP obtained by different polymerization methods are different, and the composition and structure of PVP obtained by free radical solution polymerization are more uniform. The performance is also relatively stable and is the most common method of NVP homopolymerization. By adjusting the monomer concentration, polymerization temperature, initiator amount, and other reaction conditions can obtain different molecular weights and different water-soluble PVP homopolymerization. The unique advantages of PVP are its excellent water solubility, nontoxic, nonirritating, and strong chemical stability, which makes it an ideal choice in many situations where water solubility, film formation, or binders are required.

In the battery field, PVP can enhance the adhesion and dispersion of the electrode material, ensuring that the active material and the conductive agent are evenly distributed on the electrode, which is essential to improve the performance and efficiency of the battery. It also can effectively disperse CNT in lithium iron phosphate battery positive slurry as a dispersing agent, so as to produce positive slurry with excellent performance.

3A.2.8 Polyolefin (PO)

PO is a general term for a class of thermoplastic resins derived from the single or co-polymerization of ethylene, propylene, 1-butene, 1-pentene, 1-hexene, 1-octene, 4-methyl-1-pentene, and certain cycloolefins. Due to the abundant raw materials, low price, easy processing and forming, and excellent comprehensive performance, it is a kind of polymer material with the largest output and a wide range of applications. Among them, PE and polypropylene (PP) are the most important. The main varieties are PE and some ethylene-based copolymers, such as ethylene-vinyl acetate copolymers, ethylene-acrylic, or acrylate copolymers, as well as PP and some propylene copolymers, poly1-butene, poly4-methyl-1-pentene, cycloolefin polymers. Here, we focus on PE and PP.

PE is the simplest polymer structure and the most widely used polymer material, which is formed by the addition polymerization of ethylene. It can be divided into high-density PE, low-density PE, and linear low-density PE according to the polymerization method, molecular weight, and chain structure. It is a typical thermoplastic, odorless, tasteless, nontoxic flammable white powder. PE has excellent chemical resistance and can resist the attack of most acids, bases, and organic solvents. Therefore, it is often used in chemical storage containers, pipes, and anticorrosion coatings to ensure the long-term durability of the material. As a plastic with high toughness, PE can maintain its flexibility at low temperatures and has strong impact resistance. Its elasticity makes it an ideal packaging material and protective equipment; PE material has excellent waterproof performance, often used in the production of waterproof film, rain gear, and so on. In addition, it is also a material with good electrical insulation performance and is widely used in the insulation layer and protective layer of wire and cable. With the improvement of environmental awareness, the recycling of PE materials has become a major advantage. Recycled PE not only reduces waste, but also extends the life of the material, in line with sustainable design trends.

PP is a thermoplastic resin made from the polymerization of propylene. It is a non-toxic, odorless, tasteless, milky white highly crystalline polymer with a density of only about 0.90 g cm^{-3}, which is one of the lightest varieties of all plastics. It is divided into isotactic PP, atactic PP, and syntactic PP according to the methyl arrangement position. PP has excellent mechanical properties because of its high crystallinity and regular structure. The absolute value of the mechanical properties of PP is higher than that of PE, but it is still a low variety in plastic materials, and its tensile strength can only reach 30 MPa or a slightly higher level. The specific application areas of PP are similar to PE.

In the battery field, PO is mainly used as a PO separator, they are mainly made of PE and PP materials, which have excellent mechanical properties and good chemical stability. Due to its low cost and excellent performance, PO separators are widely used in various lithium-ion batteries, especially in the field of power batteries, where the improvement of safety and energy density is particularly important.

3A.3 Application of Synthetic Polymers in Binders

The main functions of the binder inside the battery include: improving the uniformity of the electrode components as a dispersant or thickener; Bonding active material, conductive agent, and fluid collector to maintain electrode structural integrity. Providing the required electron conduction within the electrode. Improving the wettability of electrolytes at the interface to enhance the transmission performance of Li$^+$ at the battery interface [1–5]. Although the proportion of binders in the batteries is small, the internal binder is an important part and the most easily ignored component of the battery. Protecting the electrode structure from damage is a key factor to ensure the long-term operation of the battery. With the further development of thick electrode technology, solid-state battery, and silicon anode material technology, the importance of the binder is further highlighted. At present, the binders used for battery products are mainly synthetic.

3A.3.1 Polyvinylidene Difluoride

PVDF is a non-polar thermoplastic fluoropolymer known for its excellent chemical stability, thermal stability, mechanical strength, and excellent processability. In the manufacture of lithium batteries, PVDF acts as a binder, with the active material (such as positive electrode material), conductive agent, and fluid collector (such as copper foil or aluminum foil) firmly combined together to form an electrode structure, which is essential to improve the energy density, cycle life, and assurance of the batteries. The PVDF molecular chain contains strongly polar fluorine atoms, which can form a strong intermolecular effect to ensure close contact between the active substance particles; It can effectively prevent the active substances from falling off during the charging and discharging process, so as to extend the battery cycle life.

It has very high chemical stability, which enables it to resist the erosion and frustration of various components in the electrolyte. It is easily dissolved in a specific solvent to form a uniform and viscosity-controllable binder solution, which is conducive to the operation of the electrode coating process and enhances production efficiency.

The mechanism of PVDF binder mainly includes two aspects of physical adsorption and chemical bonding: PVDF molecules contain polar fluorine atoms and nonpolar hydrogen atoms, so it has two polar and nonpolar regions. When the PVDF binder contacts the surface of the bonded object, its polar and polar regions of the surface of the object will attract each other, this physical adsorption can increase the contact area between the binder and the surface of the object, improving the adhesion effect. Fluorine atoms in PVDF molecules have a high electronegativity, making them have a strong electrophilicity. When the PVDF binder contact with the surface of the object, a chemical bond will form between the fluorine atoms in the PVDF molecule and the electrophilic groups (such as hydroxyl and carboxyl) on the surface of the object. This chemical bonding can increase the strength and stability of the bonding at the molecular level, thus improving the bonding performance of the binder [6, 7]. Therefore, PVDF exhibits outstanding comprehensive properties in the practical application of batteries. Although PVDF is the most widely used commercial battery binder, all the relevant research literature points out that it has weak mechanical properties, poor adhesion effect, and relatively low electronic and ionic conductivity, which hinder its wide application in high power capacity battery materials with a thick electrode [1–5].

In order to improve the performance of PVDF binder-based batteries, many PVDF modification and composite strategies have been proposed. He et al. [8] presented a high-performance intelligent nanosol binder (sol-binder) designed with polyvinylidene fluoride (PVDF) and propylene carbonate as feedstock. The processing of electrode paste based on sol-binder and the evolution of electrode microenvironment structure were studied deeply. They first prepared a uniform electrode slurry similar to neutral ink based on a nanosol binder. In addition, the sol–gel transformation of the sol binder occurred spontaneously during the high-temperature drying process, which greatly inhibited the uncontrollable component aggregation/separation behavior in the conventional solution-type binder slurry. Thus a "local concentration" binder distribution mode on demand was realized, which ultimately helped the electrode to establish a more solid and healthy electrode microenvironment structure, and significantly improved its electrochemical performance.

Jing et al. [9] reported the application of PVDF and isobutyl isobutyrate as a combination of binder and solvent to fabricate sheet electrodes for sulfide-based all-solid-state batteries (ASSBs). $LiClO_4$ was used to modify PVDF to promote the interfacial transmission of Li^+ and further improve the performance of ASSB. Although PVDF bonding in the composite cathode electrode reduced the direct contact between the components, the reduced contact area led to a lower degree of side reactions, which made the battery composed of sheet electrodes exhibit better cycling performance than the battery composed of pellet electrodes. By facilitating interfacial Li⁺ transport with LiClO₄-modified PVDF, the battery exhibited excellent capacity retention rates at low manufacturing pressures and during long-term cycling.

Gu et al. [10] successfully modified a commercial PVDF binder by introducing polyacrylamide (PAA) as an auxiliary agent by hydrogen bonding between PVDF and PAA. Density functional theory calculations showed that the designed hydrogen bond interconnect binder had higher electron conductivity and faster Na⁺ migration kinetics than PVDF and PAA alone. This design realized the structural revolution of PVDF binders, further improved the electrochemical performance of Na₃V₂(PO₄)₂O₂F (NVPOF) cathode materials, and provided a new perspective for directly designing the molecular structure of polymer-based binders. The energy density of lithium batteries based on nickel-rich layered oxide cathode materials can effectively be further promoted by elevating operating voltage. However, high-voltage cycling usually induces severe decomposition of commercial liquid electrolytes and degradation of active material, resulting in the rapid decline of the battery capacity. Especially, reactive oxidative species including ¹O₂ induced by high-voltage cycling are the main cause of severe electrolyte decomposition. In order to solve this problem, Liu et al. [11] prepared a vinylphenol-grafted PVDF binder for nickel-rich layered oxide cathode materials, which can eliminate ¹O₂ like dopamine. It was demonstrated that the eliminating behavior of the as-prepared binder can help the formation of a better cathode/electrolyte interphase (CEI) layer. This work offers an easy and effective strategy to improve the CEI compatibility and electrochemical performances of high-voltage batteries based on nickel-rich layered oxide cathode materials.

As the battery industry moves toward greater energy efficiency, longer life, and higher security, PVDF binders as the most widely used commercial binders will face more opportunities and challenges.

3A.3.2 Polytetrafluoroethylene

Compared to traditional battery binder PVDF, PTFE has a similar chemical structure, but has superior chemical resistance and thermal stability [12]. PVDF as a binder usually requires to use of solvent NMP, which is expensive, toxic, and harmful to the environment, and this high boiling point solvent requires more energy to volatilize during battery manufacturing. In order to meet the requirements of green battery production, new functional binders are needed to complete battery manufacturing. PTFE has the potential to achieve green manufacturing, so it has become a hot spot of scientific research. According to the processing procedure, PTFE-based binder can be classified into two processes: slurry-casting process and solvent-free process. Many researchers attempted to develop a water-based PTFE binder, due to PTFE can form an aqueous solution containing a stabilizer. The specific process is to grind PTFE particles and water into emulsion, and then add the emulsion to lithium iron phosphate cathode material slurry for homogenization. In addition, PTFE has flexibility, low resilience, moderate tensile strength, and high elongation. It also has excellent nonviscous properties, low coefficient of friction, and chemical and thermal stability. Based on these properties, when a shear load is applied to PTFE particles, the flat spherical particles exhibit flexibility, allowing them to stretch and form fibers. The solvent-free process uses a solid binder to form a 3D "fiber-mesh" structure, so that the electrode powder is cross-linked by this 3D fiber-mesh structure, so as to achieve the bonding function [13]. In the solvent-free process, PTFE is suitable for both battery anode and cathode.

Wei et al. [14] present a solvent-free process of PTFE binder for high-loading graphite anodes. The reduction of PFTE at low potential results in a large capacity loss, so the use of graphite anodes is hard to achieve. They developed a PEO coating to prevent electrical contact between graphite and the PTFE binder, successfully inhibiting the reduction of PTFE. The coating promotes the efficient use of PTFE in high-load lithium pouch batteries. Specifically, to prevent this electrochemical reaction, they interrupted the connection between the PTFE and the graphite particle by coating the graphite particles with a polymer substance, PEO, which is both ionic conductive and insulating and is known to be reduction stable. The presence of PEO effectively inhibits the reduction of PTFE, thereby improving the initial performance. The coulomb efficiency of the graphite/lithium battery and ultimately the high-load graphite anode is achieved by using a solvent-free coating of PTFE.

Dong et al. [15] prepared a water-soluble binder based on PTFE and alginate polysaccharide and used it in water zinc-ion batteries. This water-soluble binder facilitates the in situ formation of the CEI protective layer and adjusts the interface morphology. In addition to the manipulation of the surface morphology, the composite binder also contributes to the formation of the CEI layer, and its in situ formation helps to improve battery stability. This general strategy offers a promising way to modify the CEI interface of water-based batteries through a cost-effective combined bonder.

Therefore, to achieve a lithium battery process without solvent (NMP) addition, regardless of the use of slurry-casting or solvent-free electrode processes, PTFE is the key material.

3A.3.3 Styrene-Butadiene Rubber

SBR as a traditional water-based binder has been widely used in the anode materials of lithium batteries because of its good bonding strength, flexibility, and mechanical stability. SBR is mainly used in the homogenization and coating process of graphite anode materials in lithium-ion batteries, and its dosage is small but very important. Graphite and carbon black particles are difficult to disperse in water due to surface hydrophobicity, while SBR and CMC cooperate as binders, which can significantly improve the stability and coating performance of the slurry. SBR not only enhances the bonding performance of the graphite negative electrode, but also improves the bonding force between the negative electrode and the copper foil, which effectively prevents the polarization and falling off of the electrode sheet during the charging and discharging process, thus prolonging the service life of the battery [2–4]. SBR works synergistically with CMC, mainly acting as a thickening agent to achieve the viscosity range of the paste suitable for processing, while SBR is mainly responsible for bonding, and both work together to optimize the performance of the paste.

Chang et al. [16] systematic studies of the physicochemical behavior of CMC and SBR binders and directly observed their distribution in real graphite electrodes. The study shows that the migration of CMC to the surface during the drying process depends on the degree of formation of the cross-linked binder-graphite network, which is determined by the surface properties of the graphite and CMC materials. Because the CMC migrates to the electrode surface during the drying process, the CMC is usually concentrated at the top of the electrode. The properties of the graphite surface and CMC material are key to controlling the CMC migration. The vertical distribution of SBR is different from CMC, SBR is more evenly distributed or more evenly concentrated at the bottom of the electrode, contributing to the adhesion with the collector.

Isozumi et al. [17] proposed that SBR can be modified by changing the degree of crosslinking or partially replacing styrene with acrylonitrile. They prepared four SBR-based latex binders (SBRx) and applied them to composite electrodes for comparison with conventional PVDF. The results showed that the electrode performance of SBRx is better than that of PVDF binder. The crosslinking degree is obviously dependent on its cycle performance, and the modified SBR_{low} with a low crosslinking degree has the best cycle performance and rate ability. The acrylonitrile-modified SBR_{CN} has higher solvent absorptivity, better rate performance, and self-discharge inhibition performance.

With the increasing demand for green solvent-free battery manufacturing technology, aqueous binders represented by SBR will receive more attention.

3A.3.4 Polyvinyl Alcohol

PVA as a water-based binder is used in the preparation of lithium battery electrodes due to its good bonding ability and environmentally friendly properties. Using water as a solvent is more environmentally friendly and less costly than other binders. It is rich in —OH functional groups to help form strong hydrogen bonds, and high bonding properties to help maintain electrode integrity. It has a tendency to generate mechanically flexible films, showing strong adhesion and elasticity, which helps to achieve uniform and thin SEI film formation in solid-state water batteries. The formed homogeneous and thin KF-enriched SEI film can greatly inhibit electrolyte reduction decomposition and improve the cycling performance and capacity retention of the electrode.

Binder plays an important role in maintaining the structural integrity and electrochemical performance of electrodes in batteries. In addition, the properties of the binder affect CEI chemistry and the composition of the SEI layer. In potassium batteries, due to the high activity of potassium ions and the high catalytic surface of graphite, it is easy to induce electrolyte decomposition and side reactions. Therefore, the binder is more critical in potassium batteries than in lithium batteries. Appropriate binders can inhibit electrolyte decomposition and form stable and uniform SEI, which helps to improve the cycling stability and rate performance of graphite anode. Mao et al. [18] use environmentally friendly, low-cost, and water-soluble binder PVA to achieve long-term cycling stability of commercial graphite in potassium batteries. As shown in Figure 3A.2., compared to traditional PVDF binders, PVA binders offer several advantages: (i) Oxygen-rich contains functional groups (e.g. —OH) that can form strong hydrogen bonds, which is beneficial to maintain electrode integrity better than the weak mechanical nested forces in PVDF. (ii) It has a tendency to produce mechanically flexible films with strong adhesion and elasticity. (iii) It helps to form a uniform and thin KF-enriched SEI film, greatly inhibits electrolyte reduction decomposition, and improves the cycling performance and capacity retention of the electrode. (iv) It uses environmentally friendly water as a solvent, unlike PVDF which requires harmful and expensive NMP as a solvent. Therefore, the PVA binder can promote the formation of a stable and uniform KF-rich SEI layer on the graphite surface, thus inhibiting the decomposition of the electrolyte. This work demonstrates a simple, effective, and environmentally friendly method to improve the cycling stability of graphite anode in potassium batteries using a water-soluble binder.

ASSBs based on lithium metal anodes have a higher energy density, much higher than traditional graphite-based batteries, which can meet the increasing demand for high-performance batteries. However, it is prone to uneven lithium deposition during the cycle, resulting in internal short circuits and obstruction of charge transfer, which compromises battery performance. Lee et al. [19] proposed a "bottom deposition" strategy based on the use of a lipophilic collector and a protective layer composed of polymer binders and carbon black to optimize lithium metal deposition. Bottom deposition, in which lithium plating is performed between the protective layer and the collector, avoids internal short circuits and promotes uniform volume change of lithium. The prepared protective layer functional binder has excellent mechanical strength and bonding properties and can withstand the volume expansion caused by metal growth. The bottom deposition strategy specifically refers to the plating of lithium between the protective layer and the collector, thereby avoiding internal short circuits and promoting uniform volume changes of lithium. The prepared protective functional binder layer has excellent mechanical strength and bonding properties and can withstand the volume expansion caused by metal growth. Silver incorporated on the collector provides a lipophilic surface, alleviating dendrite growth and promoting uniform lithium deposition.

The silicon-based material has a high electrochemical performance with a specific capacity of 3590 mAh g^{-1} and a low operating voltage of 0.5 V vs. Li/Li$^+$, making it one of the most promising anode materials. However, due to multiple problems such as large volume expansion, serious pulverization of silicon particles, irreversible damage of electrode structure, and incompatibility of SEI, the practical application of silicon-based anode is greatly limited. Li et al. [20] developed a lithium polymer binder with the potential for mass production and application of silicon-based anodes with high cycling stability. The binder is a PVA-grafted PAA, which is a neutral partially lithium binder with rich —COOLi and —OH groups. The abundant neutral -OH and -COOLi groups showed strong interaction with Si particles during

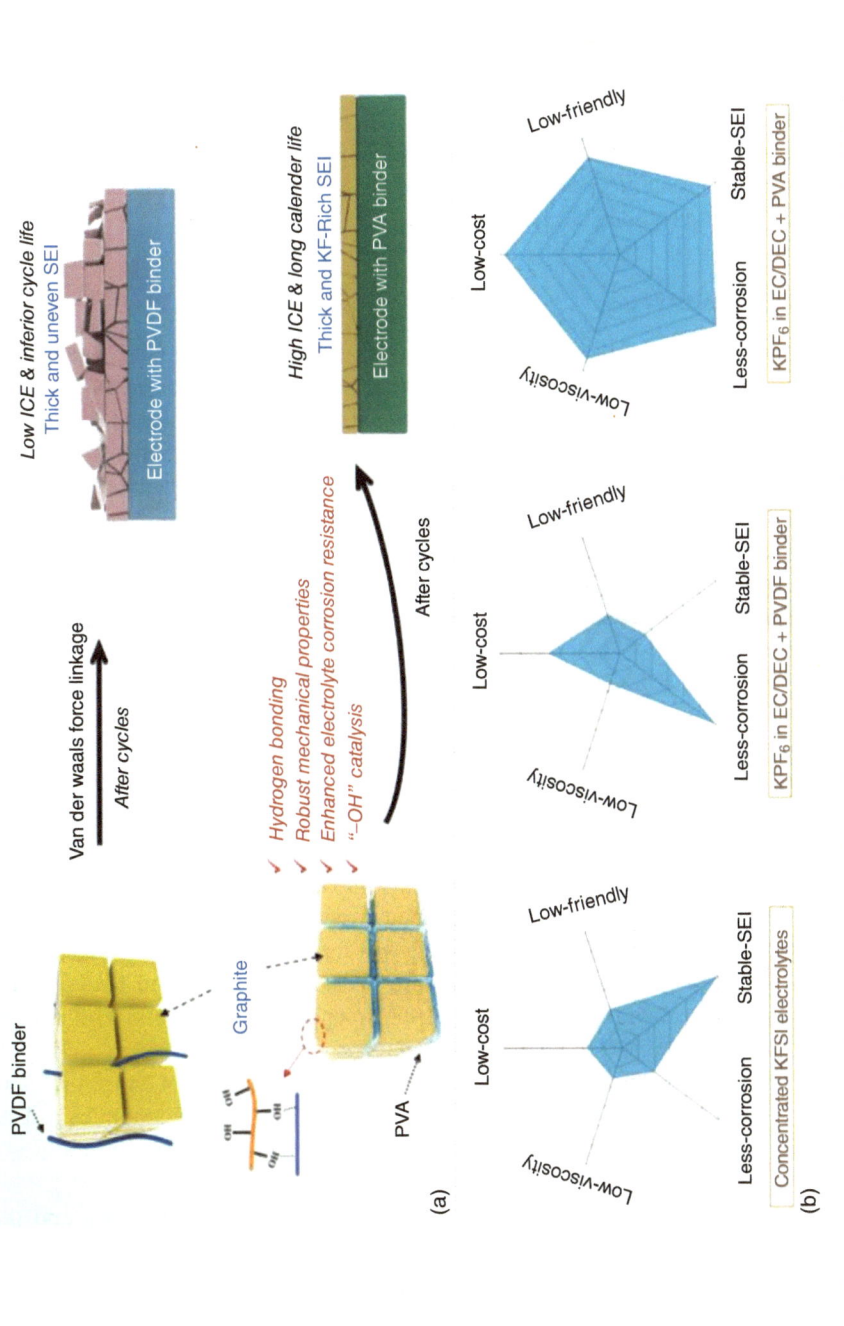

Figure 3A.2 (a) The advantages of PVA compared to PVDF. (b) The properties radar plots of different systems in potassium-ion batteries. Source: Reproduced from Mao et al. [18]/with permission of Wiley-VCH Verlag GmbH & Co. KGaA.

the long-term cycle. The neutral PVA-g-LiPAA binder can induce the formation of a uniform and stable "core-shell" SEI film on the SiO$_x$ anode, preventing structural evolution and volume expansion. This —CF$_3$-rich SEI film can also increase ionic conductivity and accelerate electrochemical kinetic reactions. This work can promote the large-scale commercial development of silicon-based nodes with high energy density and long cycle life, and provide a new idea for the design of SiO$_x$ anodes in advanced lithium batteries.

The application and development of PVA battery binder in the lithium battery industry has broad prospects, its improved versions will play an important role in both anode and cathode binder markets.

3A.3.5 Polyacrylics

Compared with traditional oil-based binders such as PVDF, PA binders have the advantages of solvent-free release, green production environment, low cost, nonflammable, and safe use. Due to its rich polar functional groups, it is insoluble in organic electrolytes, so they have high chemical stability and better lithium-ion transport capacity. In Li batteries, the PA binder can effectively promote the contact between lithium-ions and active substances, reduce side reactions, and improve the overall cycle efficiency. In silicon-based batteries, PA binders can enhance the bonding force between active particles and fluid collectors by forming hydrogen bonds with silicon-based materials, thus improving the cycle performance and life of the batteries.

Silicon-based batteries have a large volume change during the process of lithiation/delithiation, requiring the binder to have mechanical robustness and electrochemical stability. Huang et al. [21] propose to use of PAA and PVA to form a cross-linked and partially neutralized composite with a mass fraction of 60/40 wt% as a binder, which can accommodate the volume change of silicon active material electrode during the battery charging and discharging process to the greatest extent. The combination of PAA and PVA takes advantage of the strong adhesion of PAA and the mechanical robustness of PVA, and has the potential to overcome the current technical challenges faced by traditional PVDF binder systems. Compared with unmodified pure PAA, PAA/PVA mixed binders have higher stiffness, bonding strength, and electrochemical properties. The current density can maintain 100 cycles at 240 mA g^{-1} and 40 cycles at 400 mA g^{-1}. The pure PAA bonder can only maintain 38 cycles at the current density of 240 mA g^{-1} and 25 cycles at 400 mA g^{-1}. PAA/PVA blends partially neutralized by sodium hydroxide exhibit higher electrochemical performance, sustaining >140 cycles at a current density of 240 mA g^{-1} and >60 cycles at 400 mA g^{-1}.

In Silicon anode batteries, the choice of binder, carbon, electrolytes, and the morphology of the silicon itself plays a key role in improving capacity retention. It has been proved that the carboxyl group in PA binder is beneficial to improve the cycle performance of the battery. As a kind of soft matrix skeleton, binder has been widely studied for its role in resisting the volume expansion of anode and maintaining its morphological integrity. However, the effect of binders on

SEI formation rates in different cycles and their distribution around silicon nanoparticles has not been comprehensively studied. Parikh et al. [22] assembled and characterized batteries based on two different binders (PAA and CMC) to understand the effects of binders on SEI. They characterized and analyzed in detail the evolution of SEI on Si electrodes for two binders by scanning transmission electron microscopy-energy dispersive X-ray spectroscopy, electron energy loss spectroscopy, and X-ray photoelectron spectroscopy. The results showed that the faster decomposition of FSI- with PA binder resulted in the formation of LiF, preventing F- from participating in subsequent SEI formation cycles. This promoted the further decomposition of LiFSI salts into sulfates and sulfides, which became the key components of SEI coated with silicon nanoparticles after cycling 100 times in the PAA binder system. The combination effect of the rapid consumption of F- to form LiF and the passivation of the sulfide distributed in SEI makes the PAA binder system perform better than CMC.

Similarly, to solve the submitted expansion problem of silicon-based battery, Preman et al. [23] synthesized a series of acrylic copolymer binders with randomly distributed acrylic acid and n-butyl carbamate groups. The acrylic group ensures good adhesion to the silicon surface, while the n-butyl carbamate group provides self-healing as well as thermal and mechanical properties. By changing the composition of these functional groups, the mechanical, adhesion, and self-healing properties of the binder and the absorption capacity of the electrolyte can be adjusted, and the electrode binder with the best electrochemical performance can be obtained. Figure 3A.3. shows a functional diagram of the newly designed copolymer binder: there are covalent bonds and various types of hydrogen bonds, which may strongly interact with the Si surface and the conductive agent. These dynamic interactions can regenerate the connection after it is broken by gross expansion, which helps maintain the structural integrity of the Si anode during (i) lithiation and (ii) decay. In addition, they systematically investigated the relationship between structure-properties in this copolymer, such as thermal, mechanical, self-healing, and adhesion properties, and correlated them to the electrochemical properties of the Si electrode. This kind of binder with balanced performance in all aspects is conducive to realizing the practical application of silicon anode batteries with high electrochemical performance.

In order to obtain high energy density, it is necessary to manufacture high-load electrodes, but the usual mechanical degradation of high-load electrodes seriously hinders their practical application. In addition, the low cost, ecological friendliness, and easy operation of the battery manufacturing process are also important factors to be considered. Wang et al. [24] proposed a new 3D network binder with an efficient damping capacity, obtained by thermal condensation of polyacrylic acid and xanthan gum (c-PAA-XG), whose covalent cross-linking provides strong force property to confront mechanical degradation. The designed binder with strong mechanical properties enabled to achievement of thick electrodes with high loading: For example, the loading of a micro-graphite anode can reach 27.4 mg cm^{-2}; the loading of nano/micro-Si/C anode can reach 18.3 mg cm^{-2}; the loading of nano/micro-LiMn$_2$O$_4$ cathode can reach 47.8 mg cm^{-2}; and the loading

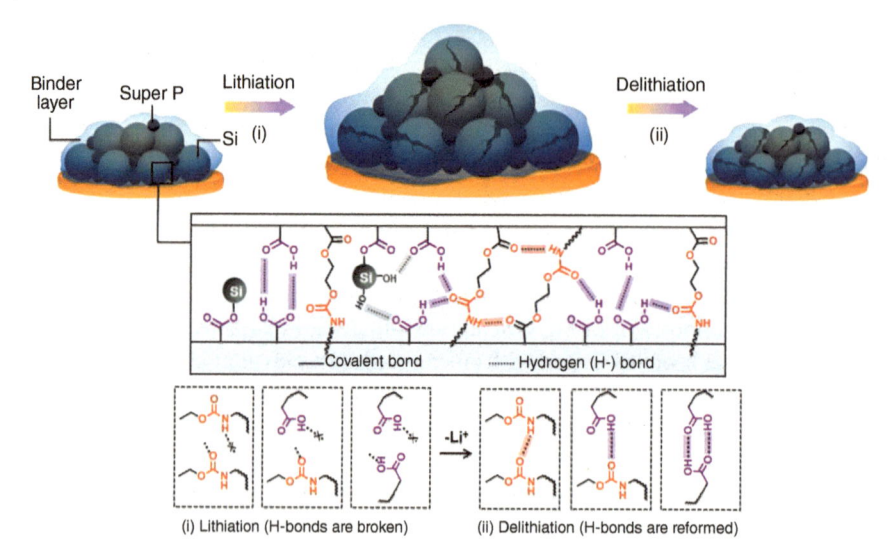

Figure 3A.3 Expansion resistance and self-healing mechanism diagram of PA composite binder. Source: Reproduced from Preman et al. [23]/with permission of American Chemical Society.

of a nano-LiFePO$_4$ cathode can reach 56.3 mg cm^{-2}. These results clearly show that the binder can be widely used in anode and cathode materials, and has great practical application potential in the manufacture of high-energy-density battery materials.

Due to the unique structure and properties of PA, it is a new development direction to form cross-linked binders with other materials.

3A.4 Application of Synthetic Polymers in Electrolytes

Polymer electrolyte is a material composed of a polymer matrix and electrolyte components that can conduct ions in batteries or other electrochemical devices. They are typically made from high molecular weight polymers that can combine with lithium salts to form a conductive network. Polymer electrolytes offer significant advantages for battery applications [25–35], including improved interface stability due to their good compatibility with electrode materials, which enhances cycling stability and efficiency while reducing interfacial resistance. Their ionic conductivity and mechanical strength can be increased by doping with various salts and incorporating liquid plasticizers or inorganic fillers. Compared to traditional liquid electrolytes, polymer electrolytes are generally more environmentally friendly and less toxic. They also have deformable properties that can be prepared into flexible batteries, making them suitable for flexible and wearable devices. Their mechanical toughness helps to withstand volume changes in electrodes during cycling, benefiting interface stability and charge transfer resistance. Additionally, the polymer matrix and functional filler can inhibit the growth of lithium dendrites, thereby improving the safety and cycling stability of the battery. Collectively, these attributes position polymer electrolytes as a promising solution for enhancing battery performance, safety, and environmental sustainability.

3A.4.1 Polyvinylidene Difluoride

PVDF is a kind of fluorinated polymer with good chemical stability and thermal stability. PVDF-based electrolytes have been widely used in solid-state batteries [36]. The ionic conductivity of PVDF-based solid electrolytes usually depends on their interaction with lithium salts. By adjusting the concentration and type of lithium salt, its conductive properties can be optimized. It has good mechanical strength and flexibility, making it excellent in battery assembly and application, and can withstand a certain deformation without breaking. The interface between the PVDF-based electrolyte and the electrode material is relatively stable, which helps to reduce the interface reaction generated by the battery during the charging and discharging process, thereby improving the cycle life of the battery. In addition, due to the high chemical stability of PVDF, PVDF-based solid electrolytes show high safety at high temperatures or extreme conditions, reducing the risk of thermal runaway in the battery. PVDF-based solid electrolyte has a wide range of potential applications in lithium-ion batteries, solid-state batteries, and other electrochemical devices, especially in the development of high-energy density and high-safety batteries. In recent years, researchers have been exploring ways to further improve the ionic conductivity and mechanical properties of PVDF-based solid electrolytes by adding nano-fillers or other polymers to meet the needs of modern battery technology [37–39].

Combining PVDF with other organics to prepare solid electrolytes can increase ionic conductivity, improve electrochemical stability, enhance mechanical properties, and optimize interface compatibility. PVDF has a strong electron absorbing ability of F atom and a high dielectric coefficient (~8), but due to its high crystallinity and poor dissociation ability of lithium salt, resulting in low ionic conductivity at room temperature. The PVDF-based polymer electrolyte exhibits high room temperature ionic conductivity (~10^{-4} S cm^{-1}) and high oxidation potential (>4.5 V) when containing a certain amount of N,N-dimethylformamide (DMF) as a plasticizer. However, in the preparation process of DMF/PVDF-based polymer electrolyte, the gradual volatilization of DMF usually leads to the phase separation of PVDF-based polymer and creates voids, and it is difficult to accurately control the content of DMF in the electrolyte. In order to solve the above problems, Xu et al. [40] adopted the bottom-to-up design strategy and prepared a quasi-ionic liquid [Li(DMF)$_3$][TFSI] by using the strong complexation between DMF and Li$^+$. Then the quasi-ionic liquid [Li(DMF)$_3$][TFSI] was mixed with PVDF-HFP by using easily volatilized THF, to avoid the phase separation of the polymer. In addition, the content of DMF in the electrolyte was accurately controlled. The prepared single-phase dense polymer electrolyte has a high ionic conductivity of 1.55×10^{-3} S cm^{-1}, a tensile strength of 3.4 MPa, elongation at break of ~1550%, toughness of 43 MJ cm^{-3}, good thermal stability, and flame-retardant properties.

The use of solvents as plasticizers in PVDF-based polymer electrolytes is beneficial for obtaining high ionic conductivity, but the side reaction between solvents and Li metal can lead to poor stability. Zhang et al. [41] proposed a balanced strategy for the preparation of PVDF-based polymer electrolytes with high ionic conductivity and electrochemical stability, which was achieved by using 2,2,2-trifluoro-N,N-dimethylacetamide (FDMA) as a nonside reaction solvent for polyvinylidene fluoro-hexafluoropropylene (PVHF)-based polymer electrolytes. The synthesized FDMA solvent can promote the formation of stable SEI through the interfacial reaction with lithium metal, so as to effectively reduce the side reaction and dendrite growth on the lithium metal electrode. Therefore, even under limited lithium supply and high load cathode, Li||Li symmetrical batteries and Li||LiFePO$_4$ batteries have excellent cyclic performance. The Li||LiCoO$_2$ battery can steadily run at a cut-off voltage of 4.48 V, which indicates that the prepared polymer electrolyte has high voltage stability. This research provides valuable insights into the development of advanced PVDF-based polymer electrolytes to improve the performance and lifetime of Li metal batteries.

Introducing ionic liquids (ILs) into the polymer electrolyte can improve its ionic conductivity at room temperature, and endow it with the advantages of flame retardancy, chemical stability, and high electrochemical window. However, the transfer rate of Li$^+$ at the ILs phase and polymer-ILS phase interface is much higher than that in the polymer phase, which will lead to the uneven transfer rate of Li$^+$ and induce the growth of lithium dendrites. Liu et al. [42] used polyvinylidene-trifluoroethylene trifluorovinyl chloride [P-(VDF-TrFE-CTFE), PTC] as the skeleton to prepare the electrolyte, and the transport rate of Li$^+$ in the polymer phase was greatly improved. Unlike PVDF, PTC with moderate local polarity has weaker adsorption capacity for IL ions, which reduces the possibility of them occupying Li$^+$ hops. The dielectric constant of PTC is significantly higher than that of PVDF, which is conducive to the dissociation of lithium-ion clusters. These two factors promote the transport rate of Li$^+$ along the PTC polymer chain, narrow the difference between the transport of Li$^+$ in different phases, and facilitate the uniform deposition of lithium. Their proposed strategy of promoting the transfer of Li$^+$ and the dissociation of ion clusters through the molecular design of the polymer matrix provides a new idea for the preparation of high-safety IL polymer electrolytes.

It is also a good choice to combine PVDF and polymer materials with different structures and functional groups, their advantages can work together to improve the comprehensive performance of polymer electrolytes. Wang et al. [43] reported a tangle-associated polymer electrolyte (PVFH-PVCA) based on PVFH and copolymer stabilizer (PVCA) prepared from acrylonitrile, maleic anhydride and vinyl carbonate. The tangle structure of the electrolyte endowed it with excellent mechanical properties eliminated the stress caused by dendrite growth during the cycle, and formed a stable interfacial layer. PVCA can promote the formation of electrochemically stable CEI, so that the battery has a high specific capacity and excellent cycling stability. The composite design of polymer electrolyte can combine the performance advantages of different materials, and this strategy can be adapted

3A.4 Application of Synthetic Polymers in Electrolytes

to different cathode materials, so that the assembled battery can achieve good cycling stability.

In addition to organic materials, inorganic materials also are often used to improve the performance of PVDF-based electrolytes. In order to solve the danger of the existence of highly active residual solvents, such as DMF, to the stability of the long-term cycle, Yang et al. [44] prepared a new kind of gel polymer electrolyte (GPE) using micro-nano structure PVDF (vinylidene fluoride) as matrix and zeolite ZSM-5 and SiO_2 nanoparticles as inorganic filler, which can accelerate the transport of lithium-ions and inhibit the growth of Li dendrites. The coenhancement effect of SiO_2 and ZSM-5 can greatly improve the ion transfer rate and ion transfer number of Li^+. This combination design of multiscale materials provides an effective strategy for the exploration of electrolytes applied in the lithium metal battery (LMB) with high performances.

Yang et al. [38] proposed to limit the residual solvent molecules by introducing a low-cost 3A zeolite molecular sieve as a filler. The strong interaction between molecular sieve and DMF can weaken the ability of DMF to participate in Li^+ solvation and make more anions participate in solvation. The interfacial side reactions between the lithium anode and high voltage NCM811 cathode are effectively suppressed by the tailored anion-rich coordination environment. Barbosa et al. [39] also proposed a way to use a zeolite molecular sieve to improve the cycling performance of the batteries based on polymer electrolytes. The polymer electrolyte consists of three parts, poly(vinylidene fluoride-co-hexafluoropropylene) and poly(vinylidene fluoride-trifluoroethylene-chlorofluoroethylene) as the host polymer; clinoptilolite zeolite as cycle performance stabilizer; 1-butyl-3-methylimidazolium thiocyanate, 1-methyl-1-propylpyrrolidinium bis(trifluoromethylsulfonyl) imide or lithium bis-(trifluoromethanesulfonyl) imide as ionic conductive additives. Their experimental results show that the polymer solid state battery with the best performance can be obtained by optimizing the selection of polymer matrix, IL, and lithium salt in the ternary electrolyte system.

Metal organic framework (MOF) materials are also common fillers used to improve the overall performance of polymer electrolytes due to their mechanical properties and electrochemical catalytic capabilities. Li et al. [45] reported a strategy for a solid-state composite polymer electrolyte (CPE) consisting of in situ bridging a soft polymer and a robust MOF UiO-66 for potassium-organic batteries, achieving long cycling stability for the batteries. UiO-66 was anchored in a 3D polyvinylidene fluoro-co-hexafluoropropylene (PVDF-HFP) network framework by in situ polymerization of photo polymeric monomers, obtaining a soft 3D framework containing the polymer matrix and UiO-66. The composite electrolyte had high mechanical strength and chemical stability, a stable ion migration channel, and can inhibit the growth of potassium dendrites. The abundant channels within UiO-66 allow cation transport while limiting anion transport, improving cation mobility. In addition, UiO-66 nanoparticles can disturb the order degree of the polymer and reduce the glass transition temperature, thereby increasing the movement capacity of the polymer chain segment and achieving the purpose of improving the ionic conductivity.

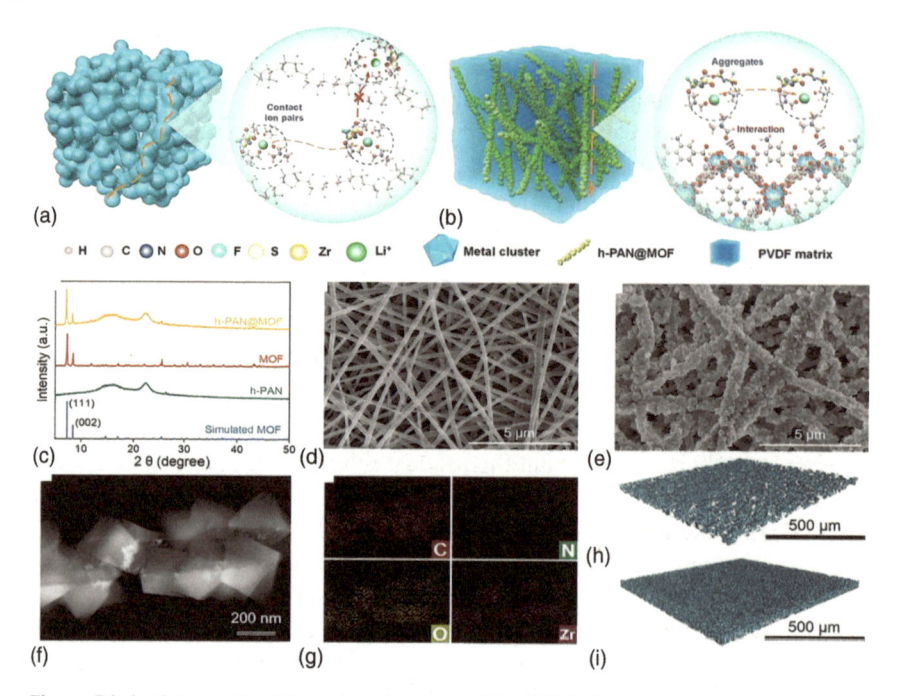

Figure 3A.4 Schematic of the microstructure of the PVDF electrolyte (a) and the composite electrolyte (b); characterization data of the electrolyte composition. XRD patterns (c); SEM images of PAN nanofibers (d) and PAN@MOF networks (e); TEM image (f) and EDS element mappings (g) of PAN@MOF fibers; computed tomography structure of PVDF (h) and composite electrolyte (i). Source: Reproduced with permission from Ma et al. [46]/Royal Society of Chemistry.

Ma et al. [46] reported a dual modification strategy using polymer composite and MOF material doping to improve the overall performance of PVDF-based polymer electrolytes (Figure 3A.4.). They loaded the interconnected MOF UIO-66 coatings onto PAN fibers by heat treatment, which were then poured together with PVDF to produce an electrolyte with a 3D continuous Li$^+$ transport. The strong interaction between the surface of MOF crystal and the C=O in DMF effectively weakened the binding strength of Li$^+$ and DMF in the solvated structure. Highly efficient Li$^+$ transmission channels and networks are formed, and the ionic conductivity is as high as 1.03×10^{-3} S cm^{-1}. The combination of MOF, PAN, and PVDF promotes the formation of stable phase interfaces, and the compact electrolyte structure also provides high tensile strength (20.84 MPa). With a high Li$^+$ transmission rate and excellent mechanical properties, the assembled solid-state LMB has achieved stable cycling at a high rate of 5C. This study provides a simple and effective strategy for the development of solid electrolyte-based lithium metal batteries with fast charging capability.

In summary, PVDF and other polymer composite electrolytes achieve comprehensive performance improvement by combining the flexibility of PVDF and the electrochemical stability of polymer electrolytes, as well as the high ionic conductivity and mechanical strength of inorganic fillers, which provides an important material basis for the development of solid-state batteries.

3A.4.2 Polyacrylonitrile

PAN-based electrolyte has excellent electrochemical and thermal stability, high mechanical strength, and good compatibility with lithium metal. Nevertheless, the limited chain motion of PAN results in poor ionic conductivity, limiting commercial applications. In addition, PAN is a flammable polymer with low thermal stability, which poses a safety hazard to batteries. Therefore, improving the ionic conductivity and safety performance of PAN can meet the standards of high-performance lithium batteries.

Combining PAN with other organic compounds is a common strategy to prepare composite SPE, Ma et al. [47] combined PAN matrix and PEO/Li salt ion conductors to prepare a new scalable, ultra-thin, and high-temperature resistant SPE for LMBs. The SPE provides a stable SPE/Li interface containing LiF and Li_3N, and its unique interface and good mechanical strength inhibit lithium dendrites and prevent short circuits. As a result, the solid electrolyte exhibits better rate performance and cycling stability in the battery, as well as excellent cycling stability at high temperatures of up to 150 °C. Shen et al. [48] used an in situ free radical polymerization process to prepare GPE-containing electrolyte with (PAN spinning film as a supporting matrix and pentaerythritol tetraacrylate (PETEA) as polymerization precursor. The system constructs an efficient ion transport channel through the oxygen-containing group of PETEA and the nitrile group of PAN, which effectively promotes Li^+ migration. The ionic conductivity and lithium-ion transport number of the GPE at 30 °C are 1.48×10^{-3} S cm^{-1} and 0.77, respectively, which can be applied to high voltage, high load positive, and high-energy-density quasi-solid state batteries.

Yang et al. [49] proposed an in situ graft polymerization strategy to prepare SPE with a 3D network by combining hard and soft polymer composite systems. The composite SPE was composed of a hard cellulose nanocrystal (CNC) matrix and a soft PAN filler. The effective Li^+ transportation pathway formed between CNC-PAN powders, in which rich sulfonate radicals and hydroxyl groups on the CNC surface acted as the bridge of Li^+ transport. In addition, the mechanical strength and ionic conductivity of the composite SPE were improved by the dipole–dipole interaction of the —CN groups in the PAN molecules and the —OH groups in the CNC molecules. Therefore, it exhibited good mechanical strength (tensile strength of 9.5 MPa), high ionic conductivity (3.9×10^{-4} S cm^{-1}, 18 °C), and lithium-ion transport number ($t_{Li+} = 0.8$). Zhang et al. [50] used 1,4-dichlorobutane modified tetra-bromo-bisphenol A (TBBA) to synthesize a new organic polymer flame retardant, TBBA-dichlorobutane, and mixed it into PAN matrix to prepare SPE, realizing its application in LMBs. This composite SPE will pass bromine ions to achieve a flame-retardant effect at high temperatures. In the normal cycle process, the Lewis base action of bromine ions is conducive to the dissociation of lithium salts, which can promote the conduction of lithium-ions. This strategy of preparing flame-retardant polymer electrolytes by the organic modification method significantly improves the safety of the batteries while ensuring the performances of the batteries.

The combination of PAN and inorganic materials is also a common preparation strategy for composite SPE, Chen et al. [51] prepared LLZTO@PAN composite electrolyte by uniformly coating the surface of LLZTO particles with PAN layer. The tight chemical interaction at the ceramic/polymer interface contributes to the induction of partial dehydrogenation of PAN and the formation of locally conjugated structures, which interact more easily with lithium-ions and trigger fast Li$^+$ exchange at the LLZTO/PAN interface. This thin film electrolyte prepared by the casting of polymer-coated ceramic particles shows good electrochemical properties and can achieve stable cycling of symmetric Li/Li batteries and all-solid-state LMBs. Yao et al. [52] constructed a SiO$_2$-doped composite SPE by in situ hydrolyzing tetraethoxysilane (TEOS) within a PAN matrix. The interlinked inorganic network formed in situ not only acts as a sturdy support for the entire SPE, but also provides enough continuous surface Lewis acid sites to facilitate the dissociation of lithium salts. As a result, the prepared SPE has high ionic conductivity (0.35 mS cm^{-1}), excellent mechanical properties (Young's modulus, 8.627 GPa), and a satisfactory lithium-ion transport number (0.52). Li et al. [53] proposed a novel dielectric-bridged ultra-thin polymer electrolyte that utilizes a dense PAN-SiO$_2$ nanofiber film as a unique multifunctional medium. The SPE had superior mechanical properties and efficient thermal conductivity, thus achieving high-temperature structural stability. The PAN skeleton can effectively lock anions to achieve rapid ion transport, thus achieving a Zn^{2+} transport number of up to 0.71. The prepared SPE had wide temperature adaptability and excellent cycling performance, and can be used to construct a wide temperature, high rate, and durable solid Zn battery.

Wang et al. [54] prepared a CPE with a heterogeneous structure using PAN and PEO, and modified the electrolyte matrix and interface using Na$_3$Zr$_2$Si$_2$PO$_{12}$ (NZSP) and nanocellulose (NC), respectively. The prepared electrolyte can be matched to the high-voltage positive electrode without being oxidized and also exhibits outstanding high-temperature performance and mechanical strength. In this heterogeneous SPE, combined with the different characteristics of PAN and PEO, the PAN layer is oriented toward the positive side, and the PEO side is oriented toward the negative side, and the introduction of NZSP in the polymer array increases the PEO chain motion and the diffusion of Na$^+$ ions. Nanocellulose with abundant hydroxyl and carboxylic acid groups is introduced into the composite SPE, which exhibits high surface density, thin thickness, and excellent high-temperature stability. Chai et al. [55] proposed a novel asymmetric composite SPE that was prepared by electrospinning and in situ polymerization to simultaneously inhibit dendrite growth and resist high pressure. In this double-layered composite solid electrolyte, the PAN layer with its high voltage tolerance is in contact with the cathode, while the PVDF-HFP layer with its excellent mechanical strength and good compatibility with the Li anode is oriented toward the Li anode. MOFs with high specific surface area were pre-embedded in the double-layer structure, which not only enhanced the conductivity of the SPE, but also improved the uniformity of lithium deposition. The contact of the PVDF-HFP layer with the Li anode induced the formation of a self-regulating gradient SEI with rich LiF and Li$_3$N, which significantly improved the cycling stability and safety of the batteries.

Liu et al. [56] used alkaline Li$_{6.4}$La$_3$Zr$_{1.4}$Ta$_{0.6}$O$_{12}$ nanoparticles to promote the rate-opening polymerization of ethylene carbonate (EC) and formed dipole–dipole interaction with PAN chain to prepare PAN-based composite SPE. In this polymerization process, the PAN chain segment is reconstructed to form a fast Li$^+$ transport channel, which improves the ionic conductivity and enhances the stability of the interface with Li metal. The prepared PAN-based SPE had a high ionic conductivity of 2.96×10^{-4} S cm^{-1} and a lithium transport number of 0.56 at 25 °C. Shi et al. [57] prepared a novel polymer-in-salt composite SPE using PAN, ZnCl$_2$, and niobium pentoxide (Nb$_2$O$_{5-x}$) with oxygen vacancy. In this SPE, the high concentration of ZnCl$_2$ caused Zn^{2+} to decouple from the polymer segment and offered a more amorphous region with a high ionic transport rate. Functional inorganic fillers containing oxygen vacancies (Nb$_2$O$_{5-x}$) enhanced the mechanical strength of the SPE. The anchoring effect of oxygen vacancies of Nb$_2$O$_{5-x}$ on Cl$^-$ ions significantly increased the cation transport number, reduced concentration polarization, and enhanced ion transport dynamics, resulting in uniform zinc-ions deposition and excellent wide temperature performance. The optimized SPE offered excellent performance in terms of long cycle times, wide temperature ranges, flexibility, and compatibility with electrode materials.

Although PAN applications in LMBs show positive prospects, there are still some challenges, such as increasing electrical conductivity, improving compatibility with cathode materials, and reducing production costs. Future research will aim to address these challenges and further optimize the performance of PAN electrolytes to promote their widespread application in LMBs.

3A.4.3 Polyacrylics

PA-based electrolyte has good electrochemical stability and can ensure the integrity of the electrode cycle and structural stability due to its good bonding properties. It can form a stable CEI layer, which can effectively reduce the occurrence of side reactions with the cathode and inhibit the migration of transition metals. PA-based SPE exhibited great development potential due to the advantages of low cost and obvious environmental protection.

In situ polymerization of acrylate monomer mixed with plasticizer and lithium salt is a common preparation method for PA-based SPEs. Lee et al. [58] used the polymerization-induced phase separation (PIPS) strategy to prepare elastomer electrolytes embedded in plastic crystals with different phase separation structures. In this system, the acrylates include butyl acrylate (BA) and PEGDA, butanedionitrile (SN) as a plasticizer, and lithium salt as lithium bis(trifluoromethanesulfonyl)imide (LiTFSI), which are mixed and induced in situ polymerization with azodiisobutyronitrile (AIBN) to form phase separation structures. The elastomer electrolyte has the characteristics of strong mechanical strength, high ionic conductivity, low interfacial resistance, and high lithium-ion transport number. It is formed in situ on copper foil and can adapt to the volume change during the long-time lithium plating and stripping process, and the coulomb efficiency is 100.0%. They found that when the elastomer phase and plastic-crystal phase are at an optimal volume ratio (1 : 1), the bicontinually structured elastomer electrolyte consists of highly efficient ionic conductivity, a cross-linked elastomer matrix, and a plastic-crystal with a remote connectivity channel [59]. It exhibits extreme lithium-ion transport performance (ionic conductivity of 10^{-3} S cm^{-1}, transport number of 0.7, at 25 °C) and excellent mechanical resilience (~300% elongation at break).

Wang et al. [60] prepared a PEG-based solid electrolyte by adding deep eutectic solvent (DES) into a double-cross-linked network. Specifically, they did so through the copolymerization of PEGDA and 2-(3-[6-methyl-4-oxy-1,4-dihydropyrimidine-2-yl] ureyl) methacrylate (UpyMA), the DES molecules between N-methylurea (NML) and LiTFSI were captured within a double cross-linked network. In this system, the soft polymer PEGDA framework served as the amorphous ion transport zone. The dynamically bonded UPyMA dimer with a rigid flat aromatic structure self-assembled in the condensed phase, strengthening the soft polymer network and easing its deformation during the cycle; The non-combustible, asymmetric bifunctional NML molecules and the carbonyl groups in the polymer chain participated in the solvation of Li$^+$, promoting the dissociation of lithium salt and improving the ionic conductivity. The prepared SPE also had a wide electrochemical stability window, high electrical conductivity, excellent mechanical strength, and toughness. Gong et al. [61] developed a novel biphasic SPE with shape memory effects through in situ thermal crosslinking of 2-ethyl cyanoacrylate (CA), polyethylene glycol methyl ether acrylate (PEGMEA), SN, and vinyl fluorocarbonate (FEC). In the SPE, the plastic crystal electrolyte (PCE) is uniformly dispersed in the polymer matrix, forming a continuous 3D channel for efficient Li$^+$ transport. When the volume ratio of elastomer to PCE is 1 : 1, the ionic conductivity of the SPE at room temperature is as high as 1.9 mS cm^{-1}. In addition, the SPE has a shape memory effect, which can restore its shape after the temporary deformation induced by external stimuli such as temperature. The LMBs based on this SPE showed good rate performance and cycle reversibility over a wide temperature range.

Xu et al. [62] synthesized a highly ionic conducting hyper-elastic terpolymer SPE using a polar-spatial freedom design strategy for molecular orbitals. The SPE consisted of three functionalized side chains: poly 2,2,2-trifluoroethyl acrylate (PTFEA), PVC, and PEGMEA. Among them, fluorine-rich side chain PTFEA can improve the stability and interface compatibility; High polarity side chain PVC can enhance the effective dissociation and transport of ions. Flexible side chain PEGMEA with high spatial freedom can promote segment motion and interchain ion exchange. The in situ SPEs can adapt to the volume fluctuations of the Li anode and promote the formation of stable F-rich mixed SEI. Both the symmetric Li battery and the Li||NCM811 full battery based on the electrolyte showed excellent cycling stability. Yang et al. [63] designed a "polymer-in-salt" electrolyte system consisting of multiple Li$^+$ transport pathways and cross-linked polymer networks with high mechanical strength and ionic conductivity. The salt-rich structure achieves ultrafast ion transport, and the cross-linked structure compensates for the loss of mechanical strength at high salt concentrations.

Inorganic fillers are often used to modify PA-based SPEs, Wang et al. [64] designed and synthesized a mercaptan-branched solid polymer electrolyte (M-S-PEGDA), which was composed of a secondary surface modification of MOF (UIO-66-MET), tetra-(3-mercaptopropionate) pentaerythritol (PETMP), and a high molecular weight PEGDA (10,000 g mol^{-1}). UIO-66-MET containing the —NH$_2$ groups and PETMP were connected to PEGDA through a branched structure, which allowed the lithium-ion transport rate to reach 2.26×10^{-4} S cm^{-1} at room temperature. The photoinduced polymerization of mercaptan and olefin provided multiple C—S—C covalent bonds, which greatly improved the mechanical properties of the material. The stable 3D network formed by in situ polymerization on the electrode enabled good contact between the electrode and the electrolyte and greatly reduced the interface resistance. The SPE with mercaptan branching had good mechanical properties, resistance, and lithium-ion transport ability, and the assembled all-solid LMB also showed excellent cycling stability and electrochemical performance. Wang et al. [65] prepared a novel CPE by in situ polymerization strategy after mixing polyethylene glycol methacrylate (MPEGA), PEGDA, MMT, and IL. By adjusting the content of MMT, the ion conductivity and Li$^+$ transport number of SPE are significantly improved, and the mechanical strength of the SPE is also enhanced. Qi et al. [66] proposed a composite GPE consisting of a PEGDA polymer as a gel electrolyte component (PGPE) and a porous LLZO framework (PLF) as a support matrix. The PLF had an integrated 3D framework, which not only formed a continuous Li transport network, but also acted as a rigid support matrix to enhance the puncture strength and elastic modulus of the composite GPE. PLF exhibited an interconnected porous structure that provides stronger mechanical support for the soft gel phase. Combined with the reliable mechanical strength of PLF and the high ionic conductivity of the gel phase, PLF@GPE showed good interfacial compatibility with the Li anodes.

PA-based SPE is the most suitable electrolyte system for in situ polymerization preparation, it has unique advantages in the regulation and dispersion of the components of the system, interface contact, and compatibility regulation, and has great development potential in the field of solid battery in the future.

3A.4.4 Polyvinyl Alcohol

PVA-based electrolyte has excellent mechanical strength, low cost, nontoxicity, and chemical stability, but its low ionic conductivity due to its semicrystalline structure and high flammability limit its application in batteries. The ionic conductivity of PVA is relatively low, but it can be modified or combined with other materials to improve its performance. In addition, PVA SPE is often used in water batteries due to its water solubility.

Chen et al. [67] developed a boron-cross-linked PVA/glycerol GPE, in which glycerol interacted strongly with the PVA chain, effectively preventing the formation of ice crystals throughout the gel network. The freezing point of this GPE is below −60 °C, which enabled it to show high ionic conductivity of 10.1 mS cm^{-1} and good mechanical properties at −35 °C. The flexible quasi-solid water Zn-MnO$_2$ battery based on the antifreeze GPE shows excellent cycling stability and low-temperature performance. Ye et al. [68] blended PVA and carrageenan as a polymer matrix to reduce the crystallinity of PVA and obtained GPE with flame-retardant properties. The self-flame-retardant Car can form a hydrogen bond with the —OH group of PVA, break its regular molecular chain, improve its nuclear energy, and reduce its crystallinity. The GPE had high ionic conductivity, and the abundant sulfuric acid groups in the Car molecular chain provide excellent self-flame retardancy, greatly enhancing the performance and safety of the batteries. Lu et al. [69] prepared an organic GPE by combining Starch/PVA/glycerol/$CaCl_2$, which had good mechanical properties, frost resistance, and room temperature stability. The abundant hydrogen bonds formed between PVA, starch, glycerol, and water give GPE high stretchability (>790%) and good thermo-plasticity. Zhou et al. [70] report on a strategy to solve the CO_2 poisoning problem of electrolytes in zinc-air batteries by designing and preparing CO_2-ionized PVA-based GPE. They used the strong organic base tetramethylguanidine to fix CO_2 to PVA in the form of a polar branched chain -OCO_2-, and prepared a CO_2-resistant flexible GPE for zinc-air batteries. This strategy solved the problem of electrolyte CO_2 poisoning and reduced the deposition of zinc dendrites/ZnO, resulting in higher electrochemical performance and longer service life of zinc-air batteries.

Liu et al. [71] prepared a GPE using a combination of PVA, PEO, and the crosslinking agent N,N'-methylene bisacrylamide (MBA). They used PEO to reduce the crystallinity of PVA, thereby softening the PVA molecular chain to obtain a tight contact interface. The active site on the PVA chain was passivated by the cross-linking agent MBA in the GPE, which reduced the occurrence of hydrogen evolution reaction. After crosslinking, the long-distance transmission of Al in the GPE system was also realized, and OH^- ions achieved fast transmission, resulting in a stable gel size and strong flexibility. The electrochemical properties of the assembled Al-air battery were optimized. Ning et al. [72] used hydrogen bond interaction to compound PAA and PVA to form a double network (PAM-PVA) condensation to prepare GPE. After freeze-thaw crosslinking, PAM-PVA composite polymers can obtain higher mechanical strength without sacrificing ionic conductivity. In addition, the internal pore size of freeze-thaw cross-linked GPE was enlarged and higher ionic conductivity was obtained. Wang et al. [73] proposed a strategy to cross-link PVA and butyraldehyde complex with 3-glycidyl propyl trimethoxysilane (KH560) to prepare composite SPE with ionic conductivity. The prepared SPEs had good optical, mechanical, and thermal performances, including high visible light transmittance (>91%), strong bond strength (2.13 MPa), and excellent thermal stability.

Tian et al. [74] prepared GPE by combining tetramethylurea (TMU) additive with a PVA matrix and adjusting the solvation structure. The hydrophilic —C=O groups in the TMU structure had a strong affinity with PVA chains, which enhanced the mechanical strength of the PVA matrix. The $N(CH_3)_2$ groups at both ends of TMU exhibited strong hydrophobicity, resulting in local hydrophobicity and water activity. In addition, the rich oxygen-containing (electronegative) groups on PVA and TUM can adsorb Zn^{2+} and provide sites for its transport. Based on these advantages, the solvation structure and deposition behavior of Zn^{2+} were regulated. Xu et al. [75] developed a novel super-elastic GPE by introducing α-helix protein molecules into the PVA network, the transport efficiency of zinc-ions and mechanical strength of the GPE were significantly improved. The abundant nitrogen and oxygen elements in α-helix protein and its unique secondary structure can provide rapid transport channels for zinc-ions. The coordination effect between α-helix protein and zinc-ion can reduce the transmission energy barrier and shorten the transmission distance, achieving ultra-high ionic conductivity. In addition, this GPE was beneficial to reduce environmental pollution because it was also environmentally friendly and rapidly biodegradable. Zheng et al. [76] introduced a novel choline chloride/ethylene glycol polyvinyl alcohol (ChCl/EG-PVA) eutectic GPE for high-performance and wide temperature range flexible Zn-air batteries. Through the combination of ChCl and PVA, eutectic gel electrolytes with stronger hydrogen bonds and denser polymer networks were obtained. The GPE had excellent liquid retention and high ionic conductivity, and can maintain good performance at extreme temperatures.

Although the PVA-based electrolyte has the advantages of low cost and environmental protection, its wide application in lithium batteries is limited by its low mechanical strength and low thermal stability. It has been developed as an electrolyte for a variety of new battery systems, and its application will be more and more extensive with the development of battery technology.

3A.5 Application of Synthetic Polymers in Battery Separators

The application of a polymer separator in the battery is mainly reflected in its role as the "third electrode" of the battery, the main role is to block the cathode and anode electrode contact of the battery and effectively transport lithium-ions. Polymer separators need to have good dimensional stability, high porosity, excellent mechanical properties, good wettability, high chemical stability, and excellent heat and flame-retardant properties. Commonly used polymer separators include PO (such as PP and PE), PVDF, PAN, PA, and so on. These materials are prepared into nanofiber films by electrospinning and other processes to improve the ionic conductivity and electrochemical performance of batteries. In lithium-ion batteries, the application of polymer separators not only improves the safety and electrochemical stability of the battery, but also promotes the efficient transport of lithium-ions through its high porosity and excellent wettability, thereby improving the overall performance and cycle life of the battery. The application of polymer separators in batteries is the key to improving the performance and safety of batteries, through the selection of appropriate materials and fine processes, can effectively solve the shortcomings of traditional PO separators, to meet the development needs of high-performance lithium-ion batteries.

3A.5.1 Polyolefin

PO is the most commonly used separator material in batteries, it plays a role in isolating cathode and anode electrodes, preventing short circuits, and ensuring that lithium-ions can normally pass through the micropore channel during charge and discharge [77]. PO separators are mainly made of PE and PP materials, which have excellent mechanical properties and good chemical stability, and the cost is relatively low. However, the melting point of the PE separator is low, resulting in a lower melting breaking temperature, which may affect the safety of the battery. Although PP separator has a high fusing and breaking film temperature, it may also lead to high closed-cell temperature.

At present, the most widely used lithium-ion battery separators are PO-based on PE and PP. It includes single-layer PE, single-layer PP, and three-layer PP/PE/PP composite film. There are two main manufacturing processes for PO separator production: dry process (including unidirectional and bidirectional drawing) and wet process (bidirectional drawing). The dry separator forming process is mainly to mix the separator raw material and the film-forming additive, form the sheet crystal structure by melting extrusion, and then annealing treatment to get the dry separators. Using the principle of thermal phase separation, the plasticizers such as paraffin oil and PO resin are mixed and melted to form a uniform mixture. Then the plasticizer is extracted from the film with solvent for a certain period of time, so as to obtain interpenetrate microporous film materials of submicron size.

The thermal shrinkage characteristics of PO film directly affect the safety of the battery, and its performance can be improved by coating other materials on the surface of the PO film. Common coatings are PA, PVDF, PVDF-HFP, PVDF-CTFE, PEO, etc. The coating of these conventional polymers is beneficial to improve the electrolyte wetting of PO separators while preserving ion transport channels, and the batteries assembled with such separators have good rate performance. Coating the inorganic oxide particles on the surface of the PO separators can improve the wetting property of the separator surface and also improve the thermal performance of the separator based on the excellent temperature resistance of inorganic materials. The common coating materials include Al_2O_3 ceramic particles, SiO_2 and Al_2O_3 mixed ceramic slurry, ZrO_2 nanoparticles, etc. The former two coatings can greatly improve the thermal stability and electrolyte wettability of the separator, and the latter coating can effectively improve the electronic conductivity and electrolyte wettability of the separator. In the oxide coating process, polymer binders (such as PVDF, PVDF-HFP, etc.) are usually used to bind the particle coatings. Although the amount of binder is low, it also plays an important role in ensuring the stability of the coating, so the performance of the binder will also affect the thermal stability of the separator.

The traditional coating modification method of PO separators has been very mature, and some schemes have been applied commercially. In recent years, some new modification strategies have been proposed. Zhao et al. [78] prepared an advanced composite separator by easily scraping a thin functional layer onto a commercial 12 μm PE separator to stabilize the lithium anode. The functional layer coated on the polyethylene separator consists of reduced graphene oxide (rGO), tannic acid (TA), and VS_4. The composite separator improved both lithium-ion transport and deposition behavior, and had uniform lithium-ion distribution characteristics, which can realize dendrite-free Li deposition. Zou et al. [79] successfully prepared PP nanofibers with controllable size using nano layer coextrusion technology. The directional transformation of PP nanoribbons to nanofibers can be achieved by adjusting the number of superimposed layers and the traction speed. The PP nanofiber separator prepared by this technology has higher tensile strength, low interface impedance and high ionic conductivity, excellent thermal stability, and can achieve more efficient ion migration. In addition, due to the higher porosity and good 3D network structure, the absorption rate of the electrolyte was also significantly improved. Tu et al. [80] introduced a —NH_2 functionalized MOF materials with multiple functional units to commercial PO separators, which have high ionic conductivity, K^+ migration number, electrolyte wettability, and excellent puncture resistance. The inorganic rigid MOF skeleton in the modified layer improved the high elastic modulus and piercing strength of the separator, showing strong resistance to dendrite piercing. Through the use of this functional separator, K metal batteries based on different cathode materials and electrolytic fluids showed excellent rate and cycle performance.

Cheng et al. [81] designed and achieved a novel ceramic-PE composite separator for high-safety Li batteries. An optimized ceramic paste composed of Al_2O_3 and PVP was coated on one side of the PE separator to form a new Al_2O_3/PVP-PE separator. The heat-resistant Al_2O_3 ceramic powder was firmly adhered to the PE substrate by PVP, and the thermal stability of the optimized separator was significantly improved. The Al_2O_3/PVP-PE separator hardly shrinks after 30 minutes of high-temperature treatment at 180°C. In addition, the prepared Al_2O_3/PVP-PE separator had good wettability with the electrolyte, large porosity, and high liquid absorption capacity, which was conducive to improving the ionic conductivity and charge storage dynamics. Cheng et al. [82] used PVP as the binder, and coated inorganic modified material $ZnMoO_4$ onto the PE separator, preparing an ultra-thin PVP/$ZnMoO_4$@PE composite separator for high-security Li and Na batteries. PVP can provide strong adhesion between nanoparticles and between nanoparticles and the PE substrate, thus inhibiting the shrinkage of the PE substrate. The heat-resistant ceramic coating can withstand the internal stress from the PE substrate during heat treatment, which significantly improves the thermal stability of the modified separator.

The degradation of electrode and interface structure induced by hydrofluoric acid (HF) is one of the important reasons for the degradation of performance of high voltage LMBs. In order to solve this problem, Ding et al. [83] proposed a strategy for modifying PE separators using piperidine (PI)-impregnated molecular sieve. When the modified PE separator was used, the PI in the molecular sieve porous structure reacted with HF to significantly reduce the HF content in the electrolyte, thereby protecting the nickel-rich cathode from HF etching. Meanwhile, the N—H site within PI promoted the protonation reaction, effectively clearing HF and releasing F-anion to the anode. This led to the formation of a uniform rich LiF SEI on the

Li metal anode, which enhanced lithium-ion conductivity and Young's modulus, further reducing the formation of dendrites.

In addition to the application of traditional battery systems, the application of PO separator s in new battery systems, especially Li—S battery systems, has become more and more extensive in recent years. A lot of research work has been devoted to modifying PO separators to improve their ability to inhibit polysulfide shuttle [84–86].

With the increasing demand for Li batteries, especially for high-energy-density batteries, PO separators need to continuously improve their performance, including heat resistance, chemical stability, and mechanical strength. Functional separators, such as through the surface chemical composition and microstructure control, have become an effective way to solve the intermediate product shuttle, uncontrollable metal dendrite growth and ensure the safety of the battery, and are also the main development direction of high-performance battery separators. Coating the separator can further improve the safety performance and cycle life of the battery. Inorganic coating materials, such as ceramics, have become an important direction in improving separator performance because of their excellent high-temperature resistance. In addition, the diversity of coating materials and formulation innovation are also the focus of current research.

3A.5.2 Polyvinylidene Difluoride

PVDF as battery separators can withstand the electrolyte and high-temperature environment, extending the service life of the battery [87, 88]. Its high ion conductivity can promote the diffusion of lithium-ions, and improve the charge and discharge efficiency of the battery. Its good mechanical strength and durability can resist stress and deformation during the battery charging and discharging process, and maintain the integrity of the separator. Its high melting point and thermal stability can maintain performance in high-temperature environments, improving the safety and stability of the battery. It can resist the corrosion of the electrolyte in the battery, improving the durability and reliability of the battery. However, the high crystallinity of PVDF results in low ionic conductivity and poor affinity for electrolytes. The high interface resistance limits its application in the field of lithium-ion battery separators, especially in applications requiring high ion conductivity.

Wang et al. [89] used continuous electrospinning technology to prepare Sb_2O_3 nanoparticles modified PVDF-CTFE fiber separators for lithium-ion batteries. The mechanical strength of the PVDF-CTFE separator can be significantly improved with the doping amount of 2% Sb_2O_3 nanoparticles, and the modified separator had excellent flame retardancy and thermal stability. Compared to the common PE separator, this composite separator exhibited higher electrolyte wettability, ionic conductivity, and electrochemical stability as well as lower interfacial resistance. Cai et al. [90] successfully prepared a novel two-component lithium-ion battery separator made from side-by-side PVDF-HFP/PI fibers with a cross-linked structure. The PVDF-HFP/PI two-component composite fiber achieved a high porosity of up to 85.9% while maintaining excellent mechanical strength. The PVDF-HFP/PI fiber separator showed excellent performance, including thermal structural stability, self-flame retardant, high electrolyte absorption rate, high ionic conductivity, wide electrochemical stability window, and good interface contact. Liu et al. [91] used electrospinning to prepare PVDF/PAN composite separators with uniform fiber morphology, and most of the fiber diameters were between 100 and 300 nm. The PVDF/PAN separator has excellent chemical stability and mechanical strength, and can be utilized as a separator for sodium-ion batteries, exhibiting a stable cycle performance.

Luiso et al. [92] investigated the preparation process of a melt-blown PVDF separator and studied its performance as a separator for the lithium-ion batteries. The advantages of the meltblow process were higher manufacturing capacity and solvent-free operation compared to electrospinning. The melt-blown PVDF had the potential to create unique pad structures with mechanical properties suitable for battery separator applications. Yang et al. [93] prepared a series of PVDF/PMMA battery separators by thermally induced phase separation (TIPS) method and introduced SiO_2 to improve the comprehensive properties of the composite separators. Both the porosity and electrolyte uptake rate of the composite separators increased upon the SiO_2 content, but an excessive loading will induce agglomerate, and 3% SiO_2 doping was the best for the porosity and electrolyte uptake rate.

In comparison, PVDF is not a mainstream battery separator material, but its good electrochemical stability and electrical conductivity create opportunities for its development as a high-performance separator material.

3A.5.3 Polyacrylonitrile

PAN separator has excellent electrochemical stability, which can effectively inhibit battery heating and reduce the possibility of over-charging and over-discharging. It has good ionic conductivity, which is conducive to the rapid transport of lithium-ions, thereby improving the charging and discharging efficiency of the battery. However, PAN separator has low mechanical strength and poor fatigue resistance makes their performance degrade faster during repeated charge and discharge processes, which can lead to more damage in situations such as piercing or squeezing.

Fang et al. [94] prepared a 69 μm PAN separator for zinc-ion battery by electrospinning. The inherent —CN group of PAN guided the orderly transport of zinc-ions through N—Zn bond, thus achieving uniform distribution of electric field on the surface of zinc anode electrode, and regulating the nucleation and growth of zinc-ions. Due to the strong interaction between Zn^{2+} and —CN, which acted as an inhibitory chain to bind zinc-ions firmly and guide their transport, the PAN separator has a high ion migration number of 0.85. Based on this, the PAN separator significantly promoted the uniform deposition of zinc-ions, and the separator effectively inhibited the production of V-based byproducts. Its high ion transport number can effectively reduce the formation of concentration gradients in the electrolyte. The results showed that using a separator to control ion deposition was an effective strategy to obtain a dendrite-free zinc anode electrode. Gu et al. [95] proposed a PAN-derived double-sided asymmetric separator to improve the performance of Li—S batteries.

In this double-sided asymmetric separator, the PAN nanofibers on the cathode side were carbonized at 800 °C and doped with titanium nitride. This composite layer had electrocatalytic activity, which can promote the rapid transformation of polysulfide and inhibit the shuttle effect of polysulfide. The anode side was formed by the oxidation of PAN at 280 °C, and the abundant electronegative groups on it can make the lithium-ion plating and stripping more uniformly. This asymmetric strategy was aimed at the different functional requirements of the positive side to inhibit the polysulfide shuttle effect and the negative side to promote the uniform deposition of lithium-ions. Through different treatment regulation and modification processes, the PAN separator had dual functions, while effectively optimizing the reaction process on both sides of the cathode and anode electrodes, and significantly improving the electrochemical performance of the Li—S battery.

Cheng et al. [96] used a composite solution of PAN and zwitterionic surfactant (sulfobetaine methacrylate) as precursors to prepare modified PAN nanofiber separators for zinc-ion batteries by electrospinning. Zwitterionic surfactants with a sulfonic acid group [SO_3^-] can improve the uniformity of fiber morphology obtained by the electrospinning process, and also reduce the concentration polarization of zinc-ions on the zinc foil surface in the battery. Lin et al. [97] constructed a polyethylenimidazole (PVIM) modified PAN nanofiber separator through an electrospinning strategy and applied it to Li—S batteries to solve the polysulfide shuttle problem. The electron-lacking imidazole group introduced on the surface of the PVIIM-PAN separator can be used as the polysulfide isolation layer, which can significantly inhibit its penetration into the anode and cause side reactions, thus extending the cycle life of Li—S batteries. The strong binding energy between the imidazole group and the anion in the lithium salt limited the anion migration, and the Li^+ transport number of the separator was significantly increased to 0.60. The inhibition of the shuttle effect of polysulfide and the increase of lithium-ion transfer number greatly enhanced the efficiency of lithium plating and stripping, so the performance and stability of the battery were significantly improved.

The research on PAN separators mainly focuses on improving their thermal stability, mechanical strength, and ionic conductivity. By employing advanced manufacturing techniques such as electrospinning and phase separation, the researchers were able to produce PAN separators with specific pore sizes and porosity to meet the high-energy-density and long-life requirements of lithium batteries. As a common processing method for PAN fiber separation separators, electrospinning is limited by its slow production speed and expensive equipment, so it is necessary to develop a new high-efficiency separator processing and preparation process to meet the needs of large-scale applications.

3A.5.4 Polyvinyl Alcohol

PVA separator is a kind of high-performance battery separator, it has excellent oxygen resistance under dry conditions, and its oxygen permeability coefficient is the lowest among all kinds of resin films. This allows the PVA separator to effectively prevent oxidation reactions inside the battery. The PVA separator has good resistance to vegetable oil, animal oil, mineral oil, and many organic solvents, which increases the stability of the battery in different operating environments. PVA is a material that can be completely degraded by microorganisms in the soil, and battery separators made of PVA help reduce environmental pollution. However, its water resistance is relatively poor due to the hydrophilic hydroxyl group in the PVA molecule. In humid environments, the performance of the PVA separator may be affected. The solubility and processability of PVA make it possible to require special treatment during processing, which increases manufacturing costs.

Zhang et al. [98] prepared a MOF-PVA composite separator by introducing MOF into the polymer separator in an electrospinning process. In the battery based on this MOF-PVA composite separator, MOF particles containing metal active centers can spontaneously adsorb anions while allowing efficient transport of lithium-ions in the electrolyte, resulting in significant improvements in lithium-ion transport number t_{Li+} (up to 0.79) and lithium-ion conductivity. In addition, the porous composite separator reduced the decomposition of the electrolyte, accelerated the electrode reaction kinetically, and reduced the interface resistance between the electrolyte and the electrode. The application of this composite separator can significantly improve the performance and cycle life of the batteries. Yu et al. [99] made a similar MOF-PVA composite separator, except that they used a bimetallic MOF.

Gong et al. [100] also used electrospinning technology to package natural nanotube Edalite (HNT) in a PVA substrate and further cross-linked to prepare a composite separator with inherent electronegativity. The composite separator had a smooth surface and interconnected 3D network structure, which was conducive to the absorption and infiltration of electrolytes and enhanced its thermal stability. Wang et al. [101] prepared the PVA/HNTs composite separator by the method of first pouring and then freeze-drying. The composite separator had high Young's modulus and ion conductivity, which can effectively delay the growth of lithium whisker and improve the safety of the LMBs. Lee et al. [102] used a collaborative self-assembly process to prepare a composite separator of PVA and porous arylon nanofibers (ANF), which has an extremely high volume porosity (>95%) and surface porosity (>48.6%), as well as a high ion affinity. The prepared separator had excellent ion conductivity (2.73 mS cm^{-1}) and lithium-ion transport number (0.78), which effectively inhibited the growth of lithium dendrites and formed a stable fluorine-rich SEI layer on the surface of lithium metal. NCM/Li full batteries assembled with this separator showed excellent electrochemical performance and cycling stability.

The PVA separator has good ion permeability and thermal stability, which can effectively isolate the positive and negative electrodes and prevent short circuits, thereby improving the safety and service life of the battery. With the global emphasis on sustainable energy and environmental protection technologies, the demand for new energy vehicles and energy storage systems continues to grow, which provides a strong impetus for the development of the PVA separator market.

3A.6 Application of Synthetic Polymers in Anodes

Polymers also have many applications in anodes and cathodes, and in recent years many polymers have been developed as active materials for batteries. In this chapter, we will only discuss the use of polymers as auxiliary materials in anodes. On the anode side of the battery, the polymer is mainly used to prepare various protective or functional coatings to inhibit the growth of metal electrode dendrites or to improve stability.

3A.6.1 Polyacrylonitrile

PAN with strong electron-absorbing groups (C≡N) interact with C=O in carbonate solvents to form a more stable SEI film, and PAN is also used as a protective layer for lithium metal negative electrodes to reduce side reactions between the electrolyte and lithium metal and achieve uniform lithium deposition. By combining PAN with a specific electrolyte additive, a uniform PAN artificial SEI film can be formed on the surface of the lithium metal, which helps to reduce uneven lithium deposition and dendrite growth, thereby improving the cycling stability and safety of the battery.

In the lithium battery anode PAN coating application field, Hong et al. [103] prepared nano-scale spinel structure (FeCoNiCrMn)$_3$O$_4$ HEO (NHEO) material as anode material by ball grinding and constructed a cyclized polyacrylonitrile (cPAN) coating on the surface of NHEO by in situ thermochemical cyclization reaction. This cPAN coating significantly improved the electrical conductivity of NHEO, and also significantly improved its structural stability and inhibited interface side reactions. Liu et al. [104] proposed to protect lithium metal anode electrodes by using a sustainable double-layer interface (SDI), which consists of a PAN nanofiber layer and an in situ Li$_x$M alloy phase (where M can be tin, zinc, cobalt, indium, etc.). During a long cycle, the SDI layer can even the lithium-ion flux distribution on the negative surface, thus promoting the dynamics of the interfacial reaction and achieving dendrite-free lithium deposition and stripping. It can also "self-repair" through metal ions released by the SDI film when the alloy phase is destroyed during the cycle. It is believed that the structure of SEI film can be changed by constructing a nitrogen-containing amorphous carbon layer on the surface of graphite, thus improving the electrochemical performance of graphite anode. Bao et al. [105] proposed a nitrogen-rich polymer PAN as a coating agent to construct nitrogenous amorphous carbon-coated graphite by a simple wet coating carbonization process. The results show that a thinner, smoother, and more stable inorganic-rich SEI film is obtained, and the electrochemical performance of the battery is significantly improved.

In the zinc battery anode PAN coating application field, Du et al. [106] constructed an organic/inorganic composite artificial interface layer (Zn@PAN-TS) using an organic material PAN and an inorganic material titanium silicon molecular sieve (TS) to protect the zinc anode electrode. The coordination effect between the cyanide group and zinc-ions in PAN improves the transport efficiency of zinc-ions.

The porous TS can regulate the zinc-ion flux through the anion confinement effect. The dual synergistic regulation provided by the PAN-TS organic/inorganic artificial protective layer offered a fast and orderly ion transport channel and promoted the uniform deposition of zinc. Yang et al. [107] prepared an artificial coating of cyclized cPAN on zinc anodes by spin coating and heat treatment. The cPAN layer is rich in nitrogen-containing groups and a delocalized π-conjugated structure, which endows Zn with an extremely low nucleation barrier and uniform electric field. Based on this, the nucleation overpotential of Zn was significantly reduced, and no obvious by-product dendrite production was observed on the Zn anode. Liu et al. [108] and Chen et al. [109] prepared an artificial PAN coating by dropping a PAN solution containing zinc trifluoromethanesulfonate $(Zn(TfO)_2)$ onto a zinc foil. The PAN coating combined with zinc salt had excellent hydrophilicity and Significantly decreased the interface resistance of the Zn anode electrode by adding $Zn(TfO)_2$. The coating improved the hydrophilicity of the zinc anode and the transport efficiency of zinc-ions, thus promoting the uniform deposition of zinc-ions and avoiding the formation of zinc-ions dendrites.

PAN as an anode electrode coating can effectively improve the performance of negative electrode materials through appropriate treatment methods, and it is suitable for different battery systems, and it is a new way to develop high-performance batteries.

3A.6.2 Polyacrylics

PA contains more oxygen-containing functional groups that can react with metal anodes to form a protective layer. It also has good bonding properties and is also used as a battery anode coating material.

Niu et al. [110] designed an acrylate-type hydrophilic binder polymer coating on the zinc anode electrode surface to balance bonding and hydrophilicity. The coating can enhance the transport efficiency of zinc-ions, and regulate the deposition/dissolution behavior of zinc-ions, thereby achieving excellent inhibition effect on dendrites and side reactions. Xie et al. [111] introduced a cost-effective bifunctional film-forming additive in the ternary lithium battery system, PEGDA PEGDA475, which is a long-chain organic molecule rich in oxygen-containing functional groups. This PA bifunctional film-forming additive can optimize the composition of SEI on the lithium anode and effectively inhibit the growth of lithium dendrites. It also helps to form a uniform CEI interface layer on the nickel-rich NCM cathode, which improves the cycle life of the battery.

As a battery anode coating material, polyacrylate provides good electrochemical performance and environmental protection, but also needs to overcome the problem of its possible consumption of active lithium and high cost.

3A.7 Conclusions and Outlook

This chapter reviews the application of synthetic polymers in battery auxiliary materials, including binders, electrolytes, separators, and anode functional coatings.

Polymer binders are an important part of rechargeable batteries, providing a guarantee for good contact and morphological integrity of the electrode structure, and are the basis for maintaining electron/ion transfer during the battery cycle. Traditional binders, such as PVDF, are difficult to meet the application requirements of high-energy-density batteries with thick electrodes and/or high capacity electrodes due to their low adhesion, poor mechanical strength, and single function. With the importance of bonding materials to the overall performance of the battery has been recognized, through the continuous efforts of researchers, binder technology has made great progress in recent years, and various types of binder systems have been proposed and studied. One of the main challenges facing current binder systems is to cope with the large volume changes in some electrode systems during charging and discharging. Large system changes of electrodes are prone to causing electrode cracking and morphological changes, which usually lead to decreased performance. Optimizing the properties of binders without increasing the number of binders is an effective strategy to solve these problems. Improving the flexibility of the binder and maintaining the mechanical toughness are conducive to maintaining the cyclic stability of the electrode. The design of functional groups in polymer binders, the combination of different polymer binders, and the introduction of organic or inorganic functional components to modify the binder have been proved to be effective methods to improve the comprehensive properties of binders. With the improvement of people's environmental awareness, binders are required to reduce the use of organic solvents while ensuring binder performance, and the development of water-soluble or solvent-free binders is the future development trend.

Electrolytes are an important part of conducting ions between positive and negative electrodes in lithium-ion batteries and are crucial to the performance of the entire battery. Traditional liquid electrolytes have safety risks and low energy density, and new solid-state electrolytes need to be developed to solve these problems. Spes are widely regarded for their strong molecular structure designability, flexibility, light weight, low cost, easy manufacturing, and no liquid leakage. In practical applications, the main performance requirements of SPE are as follows: high ion conductivity and cation migration number, good electronic insulation, excellent mechanical strength, and high thermal and chemical stability. The traditional linear polymer electrolyte (such as PEO) is easy to crystallize at room temperature and hinders ion transport, and the high-temperature melting state will react with the interface and affect the stability. Starting from the design of molecular structure, a new effective solution of SPE is prepared. The functional groups with different structures can integrate different functions such as self-healing and flame retardancy. In terms of improving the conduction ion efficiency, the current research is mainly achieved by adding plasticizers. In the future, more attention should be paid to the design of molecular structure and the regulation of composition, and the use of small molecular compounds should be reduced to achieve a true all-solid polymer system. In addition, SPE has good processability, and it is necessary to develop new and efficient large-scale SPE preparation and processing methods.

Polymer separators need to have good dimensional stability, high porosity, excellent mechanical properties, good wettability, high chemical stability, and excellent heat and flame-retardant properties. In lithium-ion batteries, the application of polymer separators not only improves the safety and electrochemical stability of the battery, but also promotes the efficient transport of lithium-ions through its high porosity and excellent wettability, thereby improving the overall performance and cycle life of the battery. The application of polymer separators in batteries is the key to improving the performance and safety of batteries, through the selection of appropriate materials and fine processes, which can effectively solve the shortcomings of traditional PO separators, to meet the development needs of high-performance lithium-ion batteries. Since each polymer separators have its own disadvantages, blends or copolymers are usually used, or inorganic additives are used to improve the physical properties of the separator. Surface coating modification, blending modification, gel filling, and crosslinking modification are commonly used polymer separator modification methods. These modification methods can effectively improve the performance of polymer separators, including improving thermal stability, improving electrolyte wettability, enhancing mechanical strength, etc., so as to meet the higher requirements of lithium batteries for separators. As the core component of lithium batteries, the development trend of polymer separators is mainly focused on improving safety, thermal stability, chemical stability, and reducing production costs.

Polymer anode coatings can improve the performance and stability of metal anode-based batteries by providing a stable interface, optimizing ion transport dynamics, reducing overpotential, inhibiting side reactions, and preventing volume expansion. The application range of polymer anode coating is limited at present, but with the deepening of research and technological progress, it is expected that polymer anode coating will play an increasingly important role in the field of metal batteries.

With the application of secondary batteries becoming more and more widespread, in order to reduce pollution and save resources, the recycling demand for failed battery components is becoming higher and higher. In the future, the design and development direction of polymer functional materials for batteries should consider more characteristics such as green pollution-free, biodegradable, recyclable, and sustainable.

References

1 Saal, A., Hagemann, T., and Schubert, U.S. (2020). Polymers for battery applications—active materials, membranes, and binders. *Advanced Energy Materials* 11: 2001984.
2 Zou, F. and Manthiram, A. (2020). A review of the design of advanced binders for high-performance batteries. *Advanced Energy Materials* 10: 2002508.
3 Srivastava, M., Anil Kumar, M.R., and Zaghib, K. (2024). Binders for Li-ion battery technologies and beyond: a comprehensive review. *Batteries* 10: 268.

4 He, Q., Ning, J., Chen, H. et al. (2024). Achievements, challenges, and perspectives in the design of polymer binders for advanced lithium-ion batteries. *Chemical Society Reviews* 53: 7091–7157.

5 Wang, Y.-B., Yang, Q., Guo, X. et al. (2021). Strategies of binder design for high-performance lithium-ion batteries: a mini review. *Rare Metals* 41: 745–761.

6 Zhong, X., Han, J., Chen, L. et al. (2021). Binding mechanisms of PVDF in lithium ion batteries. *Applied Surface Science* 553: 149564.

7 Qin, T., Yang, H., Li, Q. et al. (2024). Design of functional binders for high-specific-energy lithium-ion batteries: from molecular structure to electrode properties. *Industrial Chemistry & Materials* 2: 191–225.

8 He, Y., Jing, L., Feng, L. et al. (2022). A smart polymeric sol-binder for building healthy active-material microenvironment in high-energy-density electrodes. *Advanced Energy Materials* 13: 2203272.

9 Jing, S., Shen, H., Huang, Y. et al. (2023). Toward the practical and scalable fabrication of sulfide-based all-solid-state batteries: exploration of slurry process and performance enhancement via the addition of $LiClO_4$. *Advanced Functional Materials* 33: 2214274.

10 Gu, Z.-Y., Cao, J.-M., Guo, J.-Z. et al. (2024). Hybrid binder chemistry with hydrogen-bond helix for high-voltage cathode of sodium-ion batteries. *Journal of the American Chemical Society* 146: 4652–4664.

11 Liu, Z., Dong, T., Mu, P. et al. (2022). Interfacial chemistry of vinylphenol-grafted PVDF binder ensuring compatible cathode interphase for lithium batteries. *Chemical Engineering Journal* 446: 136798.

12 Lee, S., Koo, H., Kang, H.S. et al. (2023). Advances in polymer binder materials for lithium-ion battery electrodes and separators. *Polymers* 15: 4477.

13 Wang, X., Chen, S., Zhang, K. et al. (2023). A polytetrafluoroethylene-based solvent-free procedure for the manufacturing of lithium-ion batteries. *Materials* 16: 7232.

14 Wei, Z., Kong, D., Quan, L. et al. (2024). Removing electrochemical constraints on polytetrafluoroethylene as dry-process binder for high-loading graphite anodes. *Joule* 8: 1350–1363.

15 Dong, H., Liu, R., Hu, X. et al. (2022). Cathode-electrolyte interface modification by binder engineering for high-performance aqueous zinc-ion batteries. *Advanced Science* 10: 2205084.

16 Chang, W.J., Lee, G.H., Cheon, Y.J. et al. (2019). Direct observation of carboxymethyl cellulose and styrene-butadiene rubber binder distribution in practical graphite anodes for Li-ion batteries. *ACS Applied Materials & Interfaces* 11: 41330–41337.

17 Isozumi, H., Horiba, T., Kubota, K. et al. (2020). Application of modified styrene-butadiene-rubber-based latex binder to high-voltage operating $LiCoO_2$ composite electrodes for lithium-ion batteries. *Journal of Power Sources* 468: 228332.

18 Mao, Z., Shi, X., Zhang, T. et al. (2023). Ultrastable graphite-potassium anode through binder chemistry. *Small* 19: 2302987.

19 Lee, S., Cho, S., Choi, H. et al. (2024). Bottom deposition enables stable all-solid-state batteries with ultrathin lithium metal anode. *Small* 20: 2311652.

20 Li, Z., Zuo, Y., Zhang, H. et al. (2024). Towards industrial applications: ultra-stable silicon-based pouch cell conducted via a lithiated polymer binder over a wide temperature range from −25 °C to 25 °C. *Nano Energy* 125: 109619.

21 Huang, Q., Wan, C., Loveridge, M. et al. (2018). Partially neutralized polyacrylic acid/poly(vinyl alcohol) blends as effective binders for high-performance silicon anodes in lithium-ion batteries. *ACS Applied Energy Materials* 1: 6890–6898.

22 Parikh, P., Sina, M., Banerjee, A. et al. (2019). Role of polyacrylic acid (PAA) binder on the solid electrolyte interphase in silicon anodes. *Chemistry of Materials* 31: 2535–2544.

23 Preman, A.N., Vo, T.N., Choi, S. et al. (2023). Self-healable poly(acrylic acid) binder toward optimized electrochemical performance for silicon anodes: importance of balanced properties. *ACS Applied Energy Materials* 7: 749–759.

24 Wang, D., Zhang, Q., Liu, J. et al. (2020). A universal cross-linking binding polymer composite for ultrahigh-loading Li-ion battery electrodes. *Journal of Materials Chemistry A* 8: 9693–9700.

25 Yang, X., Liu, J., Pei, N. et al. (2023). The critical role of fillers in composite polymer electrolytes for lithium battery. *Nano-Micro Letters* 15: 74.

26 Lu, X., Wang, Y., Xu, X. et al. (2023). Polymer-based solid-state electrolytes for high-energy-density lithium-ion batteries – review. *Advanced Energy Materials* 13: 2301746.

27 Zeng, X., Liu, X., Zhu, H. et al. (2024). Advanced crosslinked solid polymer electrolytes: molecular architecture, strategies, and future perspectives. *Advanced Energy Materials* 14: 2402671.

28 Meng, N., Ye, Y., Yang, Z. et al. (2023). Developing single-ion conductive polymer electrolytes for high-energy-density solid state batteries. *Advanced Functional Materials* 33: 2305072.

29 Li, Z., Fu, J., Zhou, X. et al. (2023). Ionic conduction in polymer-based solid electrolytes. *Advanced Science* 10: 2201718.

30 Liu, Y., Zeng, Q., Li, Z. et al. (2023). Recent development in topological polymer electrolytes for rechargeable lithium batteries. *Advanced Science* 10: 2206978.

31 Park, E.J., Jannasch, P., Miyatake, K. et al. (2024). Aryl ether-free polymer electrolytes for electrochemical and energy devices. *Chemical Society Reviews* 53: 5704–5780.

32 Wang, Z., Chen, J., Fu, J. et al. (2024). Polymer-based electrolytes for high-voltage solid-state lithium batteries. *Energy Materials* 4: 400050.

33 Reinoso, D.M. and Frechero, M.A. (2022). Strategies for rational design of polymer-based solid electrolytes for advanced lithium energy storage applications. *Energy Storage Materials* 52: 430–464.

34 Huo, S., Sheng, L., Xue, W. et al. (2023). Challenges of polymer electrolyte with wide electrochemical window for high energy solid-state lithium batteries. *InfoMat* 5: 12394.

35 Li, Z., Ren, Y., and Guo, X. (2023). Polymer-based electrolytes for solid-state lithium batteries with a wide operating temperature range. *Materials Chemistry Frontiers* 7: 6305–6317.

36 Zhou, S., Zhong, S., Dong, Y. et al. (2023). Composition and structure design of poly(vinylidene fluoride)-based solid polymer electrolytes for lithium batteries. *Advanced Functional Materials* 33: 2214432.

37 Karabelli, D., Leprêtre, J.C., Cointeaux, L. et al. (2013). Preparation and characterization of poly(vinylidene fluoride) based composite electrolytes for electrochemical devices. *Electrochimica Acta* 109: 741–749.

38 Yang, W., Liu, Y., Sun, X. et al. (2024). Solvation-tailored PVDF-based solid-state electrolyte for high-voltage lithium metal batteries. *Angewandte Chemie International Edition* 63: 2401428.

39 Barbosa, J.C., Correia, D.M., Fidalgo-Marijuan, A. et al. (2023). High performance ternary solid polymer electrolytes based on high dielectric poly(vinylidene fluoride) copolymers for solid state lithium-ion batteries. *ACS Applied Materials & Interfaces* 15: 32301–32312.

40 Xu, F., Deng, S., Guo, Q. et al. (2021). Quasi-ionic liquid enabling single-phase poly(vinylidene fluoride)-based polymer electrolytes for solid-state $LiNi_{0.6}Co_{0.2}Mn_{0.2}O_2$||Li batteries with rigid-flexible coupling interphase. *Small Methods* 5: 2100262.

41 Zhang, D., Liu, Y., Yang, S. et al. (2024). Inhibiting residual solvent induced side reactions in vinylidene fluoride-based polymer electrolytes enables ultra-stable solid-state lithium metal batteries. *Advanced Materials* 36: 2401549.

42 Liu, J.F., Wu, Z.Y., Stadler, F.J. et al. (2023). High dielectric poly(vinylidene fluoride)-based polymer enables uniform lithium-ion transport in solid-state ionogel electrolytes. *Angewandte Chemie International Edition* 62: 2300243.

43 Wang, H., Yang, Y., Gao, C. et al. (2024). An entanglement association polymer electrolyte for Li-metal batteries. *Nature Communications* 15: 2500.

44 Yang, H.X., Liu, Z.K., Wang, Y. et al. (2022). Multiscale structural gel polymer electrolytes with fast Li^+ transport for long-life Li metal batteries. *Advanced Functional Materials* 33: 2209837.

45 Li, Z., Gao, Y., Wang, W. et al. (2024). In situ bridging soft polymer and robust metal-organic frameworks as electrolyte for long-cycling solid-state potassium-organic batteries. *Energy Storage Materials* 72: 103732.

46 Ma, Y., Qiu, Y., Yang, K. et al. (2024). Competitive Li-ion coordination for constructing a three-dimensional transport network to achieve ultra-high ionic conductivity of a composite solid-state electrolyte. *Energy & Environmental Science* 17: 8274–8283. https://doi.org/10.1039/d4ee03134b

47 Ma, Y., Wan, J., Yang, Y. et al. (2022). Scalable, ultrathin, and high-temperature-resistant solid polymer electrolytes for energy-dense lithium metal batteries. *Advanced Energy Materials* 12: 2103720.

48 Shen, Z., Zhong, J., Jiang, S. et al. (2022). Polyacrylonitrile porous membrane-based gel polymer electrolyte by in situ free-radical polymerization for stable Li metal batteries. *ACS Applied Materials & Interfaces* 14: 41022–41036.

49 Yang, J., Cao, Z., Chen, Y. et al. (2023). Dry-processable polymer electrolytes for solid manufactured batteries. *ACS Nano* 17: 19903–19913.

50 Zhang, S., Huang, L., Zhang, C. et al. (2024). Non-flammable polymer electrolyte with fast ion conductivity for high-safety Li batteries. *Energy Storage Materials* 71: 103581.

51 Chen, W.-P., Duan, H., Shi, J.-L. et al. (2021). Bridging interparticle Li$^+$ conduction in a soft ceramic oxide electrolyte. *Journal of the American Chemical Society* 143: 5717–5726.

52 Yao, M., Ruan, Q., Yu, T. et al. (2022). Solid polymer electrolyte with in-situ generated fast Li$^+$ conducting network enable high voltage and dendrite-free lithium metal battery. *Energy Storage Materials* 44: 93–103.

53 Li, Y., Yang, X., He, Y. et al. (2023). A novel ultrathin multiple-kinetics-enhanced polymer electrolyte editing enabled wide-temperature fast-charging solid-state zinc metal batteries. *Advanced Functional Materials* 34: 2307736.

54 Wang, T., Zhang, M., Zhou, K. et al. (2023). A hetero-layered, mechanically reinforced, ultra-lightweight composite polymer electrolyte for wide-temperature-range, solid-state sodium batteries. *Advanced Functional Materials* 33: 2215117.

55 Chai, Y., Ning, D., Zhou, D. et al. (2024). Construction of flexible asymmetric composite polymer electrolytes for high-voltage lithium metal batteries with superior performance. *Nano Energy* 130: 110160.

56 Liu, H., Liao, Y., Leung, C. et al. (2025). Ring-opening polymerization reconfigures polyacrylonitrile network for ultra stable solid-state lithium metal batteries. *Advanced Energy Materials* 15: 2402795.

57 Shi, X., Zhong, Y., Yang, Y. et al. (2024). Anion-anchored polymer-in-salt solid electrolyte for high-performance zinc batteries. *Angewandte Chemie International Edition* 64: 2414777.

58 Lee, M.J., Han, J., Lee, K. et al. (2022). Elastomeric electrolytes for high-energy solid-state lithium batteries. *Nature* 601: 217–222.

59 Han, J., Lee, M.J., Lee, K. et al. (2022). Role of bicontinuous structure in elastomeric electrolytes for high-energy solid-state lithium-metal batteries. *Advanced Materials* 35: 2205194.

60 Wang, H., Song, J., Zhang, K. et al. (2022). A strongly complexed solid polymer electrolyte enables a stable solid state high-voltage lithium metal battery. *Energy & Environmental Science* 15: 5149–5158.

61 Gong, Y., Wang, C., Xin, M. et al. (2024). Ultra-thin and high-voltage-stable bi-phasic solid polymer electrolytes for high-energy-density Li metal batteries. *Nano Energy* 119: 109054.

62 Xu, H., Yang, J., Niu, Y. et al. (2024). Deciphering and integrating functionalized side chains for high ion-conductive elastic ternary copolymer solid-state electrolytes for safe lithium metal batteries. *Angewandte Chemie International Edition* 63: 2406637.

63 Yang, J., Li, R., Zhang, P. et al. (2024). Crosslinked polymer-in-salt solid electrolyte with multiple ion transport paths for solid-state lithium metal batteries. *Energy Storage Materials* 64: 103088.

64 Wang, H., Wang, Q., Cao, X. et al. (2020). Thiol-branched solid polymer electrolyte featuring high strength, toughness, and lithium ionic conductivity for lithium-metal batteries. *Advanced Materials* 32: 2001259.

65 Wang, Y., Li, X., Qin, Y. et al. (2021). Local electric field effect of montmorillonite in solid polymer electrolytes for lithium metal batteries. *Nano Energy* 90: 106490.

66 Qi, S., Li, M., Gao, Y. et al. (2023). Enabling scalable polymer electrolyte with dual-reinforced stable Interface for 4.5 V lithium-metal batteries. *Advanced Materials* 35: 2304951.

67 Chen, M., Zhou, W., Wang, A. et al. (2020). Anti-freezing flexible aqueous Zn-MnO$_2$ batteries working at −35 °C enabled by a borax-crosslinked polyvinyl alcohol/glycerol gel electrolyte. *Journal of Materials Chemistry A* 8: 6828–6841.

68 Ye, T., Zou, Y., Xu, W. et al. (2020). Poorly-crystallized poly(vinyl alcohol)/carrageenan matrix: highly ionic conductive and flame-retardant gel polymer electrolytes for safe and flexible solid-state supercapacitors. *Journal of Power Sources* 475: 228688.

69 Lu, J., Gu, J., Hu, O. et al. (2021). Highly tough, freezing-tolerant, healable and thermoplastic starch/poly(vinyl alcohol) organohydrogels for flexible electronic devices. *Journal of Materials Chemistry A* 9: 18406–18420.

70 Zhou, Y., Pan, J., Ou, X. et al. (2021). CO$_2$ ionized poly(vinyl alcohol) electrolyte for CO$_2$-tolerant Zn-air batteries. *Advanced Energy Materials* 11: 2102047.

71 Liu, S., Ban, J., Shi, H. et al. (2022). Near solution-level conductivity of polyvinyl alcohol based electrolyte and the application for fully compliant Al-air battery. *Chemical Engineering Journal* 431: 134283.

72 Ning, L., Zhou, J., Xue, T. et al. (2023). Freeze-thawed polyacrylamide-polyvinyl alcohol double network with enhanced mechanical properties as hydrogel electrolyte for zinc-ion battery. *Journal of Energy Storage* 74: 109508.

73 Wang, X., Yang, Y., Jin, Q. et al. (2023). A scalable, robust polyvinyl-butyral-based solid polymer electrolyte with outstanding ionic conductivity for laminated large-area WO$_3$-NiO electrochromic devices. *Advanced Functional Materials* 33: 2214417.

74 Tian, H., Yao, M., Guo, Y. et al. (2024). Hydrogel electrolyte with regulated water activity and hydrogen bond network for ultra-stable zinc electrode. *Advanced Energy Materials* 2403683.

75 Xu, X., Li, S., Yang, S. et al. (2024). Superelastic hydrogel electrolyte incorporating helical protein molecules as zinc ion transport pathways to enhance the cycling stability of zinc metal batteries. *Energy & Environmental Science* 17: 7919–7931.

76 Zheng, Y., Wu, D., Wang, T. et al. (2024). Advanced Eutectogel electrolyte for high-performance and wide-temperature flexible zinc-air batteries. *Angewandte Chemie International Edition* 64: 2418223.

77 Jia, H., Zeng, C., Lim, H.S. et al. (2024). Important role of ion flux regulated by separators in lithium metal batteries. *Advanced Materials* 36: 2311312.

78 Zhao, Q., Wang, R., Hu, X. et al. (2022). Functionalized 12 μm polyethylene separator to realize dendrite-free lithium deposition toward highly stable lithium-metal batteries. *Advanced Science* 9: 2102215.

79 Zou, Z., Wei, Y., Hu, Z. et al. (2023). Synthesis of polypropylene nanofiber separators for lithium-ion batteries via nanolayer coextrusion. *Chemical Engineering Journal* 474: 145724.

80 Tu, L., Zhang, Z., Zhao, Z. et al. (2023). Polyolefin-based separator with interfacial chemistry regulation for robust potassium metal batteries. *Angewandte Chemie International Edition* 135: 2306325.

81 Cheng, S., Deng, R., Zhang, Z. et al. (2024). A novel Al_2O_3/polyvinyl pyrrolidone-coated polyethylene separator for high-safety lithium-ion batteries. *Journal of Power Sources* 614: 234964.

82 Cheng, S., He, Q., Deng, R. et al. (2024). Ultrathin PVP/$ZnMoO_4$@PE separator for high-safety sodium-ion batteries. *Chemical Engineering Journal* 500: 156778.

83 Ding, L., Chen, Y., Sheng, Y. et al. (2024). Eliminating hydrogen fluoride through piperidine-doped separators for stable Li metal batteries with nickel-rich cathodes. *Angewandte Chemie International Edition* 63: 2411933.

84 Lei, T., Chen, W., Lv, W. et al. (2018). Inhibiting polysulfide shuttling with a graphene composite separator for highly robust lithium-sulfur batteries. *Joule* 2: 2091–2104.

85 Wang, R., Qin, J., Pei, F. et al. (2023). Ni single atoms on hollow nanosheet assembled carbon flowers optimizing polysulfides conversion for Li-S batteries. *Advanced Functional Materials* 33: 2305991.

86 Jiang, Z., Jin, L., Jian, X. et al. (2022). Manufacturing N,O-carboxymethyl chitosan-reduced graphene oxide under freeze-dying for performance improvement of Li-S battery. *International Journal of Extreme Manufacturing* 5: 015502.

87 Costa, C.M. and Lanceros-Mendez, S. (2021). Recent advances on battery separators based on poly(vinylidene fluoride) and its copolymers for lithium-ion battery applications. *Current Opinion in Electrochemistry* 29: 100752.

88 Bicy, K., Gueye, A.B., Rouxel, D. et al. (2022). Lithium-ion battery separators based on electrospun PVDF: a review. *Surfaces and Interfaces* 31: 101977.

89 Wang, L., Wang, Z., Sun, Y. et al. (2019). Sb_2O_3 modified PVDF-CTFE electrospun fibrous membrane as a safe lithium-ion battery separator. *Journal of Membrane Science* 572: 512–519.

90 Cai, M., Yuan, D., Zhang, X. et al. (2020). Lithium ion battery separator with improved performance via side-by-side bicomponent electrospinning of PVDF-HFP/PI followed by 3D thermal crosslinking. *Journal of Power Sources* 461: 228123.

91 Liu, Z., Li, G., Qin, Q. et al. (2021). Electrospun PVDF/PAN membrane for pressure sensor and sodium-ion battery separator. *Advanced Composites and Hybrid Materials* 4: 1215–1225.

92 Luiso, S., Henry, J.J., Pourdeyhimi, B. et al. (2021). Meltblown polyvinylidene difluoride as a Li-ion battery separator. *ACS Applied Polymer Materials* 3: 3038–3048.

93 Yang, B., Yang, Y., Xu, X. et al. (2022). Hierarchical microstructure and performance of PVDF/PMMA/SiO$_2$ lithium battery separator fabricated by thermally-induced phase separation (TIPS). *Journal of Materials Science* 57: 11274–11288.

94 Fang, Y., Xie, X., Zhang, B. et al. (2021). Regulating zinc deposition behaviors by the conditioner of PAN Separator for zinc-ion batteries. *Advanced Functional Materials* 32: 2109671.

95 Gu, M., Wang, J., Song, Z. et al. (2023). Multifunctional asymmetric separator constructed by polyacrylonitrile-derived nanofibers for lithium-sulfur batteries. *ACS Applied Materials & Interfaces* 15: 51241–51251.

96 Cheng, L., Li, W., Li, M. et al. (2024). Zwitterion modified polyacrylonitrile fiber separator for long-life zinc-ion batteries. *Advanced Functional Materials* 34: 2408863.

97 Lin, C., Feng, P., Wang, D. et al. (2024). Safe, facile, and straightforward fabrication of poly(N-vinyl imidazole)/polyacrylonitrile nanofiber modified separator as efficient polysulfide barrier toward durable lithium-sulfur batteries. *Advanced Functional Materials* 35: 2411872.

98 Zhang, C., Shen, L., Shen, J. et al. (2019). Anion-sorbent composite separators for high-rate lithium-ion batteries. *Advanced Materials* 31: 1808338.

99 Yu, W., Shen, L., Lu, X. et al. (2023). Novel composite separators based on heterometallic metal-organic frameworks improve the performance of lithium-ion batteries. *Advanced Energy Materials* 13: 2204055.

100 Gong, Z., Zheng, S., Zhang, J. et al. (2022). Cross-linked PVA/HNT composite separator enables stable lithium-organic batteries under elevated temperature. *ACS Applied Materials & Interfaces* 14: 11474–11482.

101 Wang, W., Yuen, A.C.Y., Yuan, Y. et al. (2023). Nano architectured halloysite nanotubes enable advanced composite separator for safe lithium metal batteries. *Chemical Engineering Journal* 451: 138496.

102 Lee, D., Jung, A., Liu, P. et al. (2024). Coordinated self-assembly of ultra-porous 3D aramid nanofibrous separators with a poly(vinyl alcohol) sheath for lithium-metal batteries. *Energy Storage Materials* 65: 103107.

103 Hong, C., Tao, R., Tan, S. et al. (2024). In situ cyclized polyacrylonitrile coating: key to stabilizing porous high-entropy oxide anodes for high-performance lithium-ion batteries. *Advanced Functional Materials* 35: 2412177.

104 Liu, Y., Guan, W., Li, S. et al. (2023). Sustainable dual-layered interface for long-lasting stabilization of lithium metal anodes. *Advanced Energy Materials* 13: 2302695.

105 Bao, C., Liu, Q., Chen, H. et al. (2024). Catalytic decomposition of ethylene carbonate by pyrrolic-N to stabilize solid electrolyte interphase on graphite anode under extreme conditions. *Nano Energy* 127: 109775.

106 Du, Y., Li, R., Wang, T. et al. (2024). Dual effect of organic/inorganic artificial protective layer to construct high-performance zinc metal anode. *Chemical Engineering Journal* 486: 150139.

107 Yang, J., Wang, S., Du, L. et al. (2024). Thermal-cyclized polyacrylonitrile artificial protective layers toward stable zinc anodes for aqueous zinc-based batteries. *Advanced Functional Materials* 34: 2314426.

108 Liu, Z., Li, G., Xi, M. et al. (2024). Interfacial engineering of Zn metal via a localized conjugated layer for highly reversible aqueous zinc ion battery. *Angewandte Chemie International Edition* 63: 2319091.

109 Chen, P., Yuan, X., Xia, Y. et al. (2021). An artificial polyacrylonitrile coating layer confining zinc dendrite growth for highly reversible aqueous zinc-based batteries. *Advanced Science* 8: 2100309.

110 Niu, B., Li, Z., Cai, S. et al. (2022). Robust Zn anode enabled by a hydrophilic adhesive coating for long-life zinc-ion hybrid supercapacitors. *Chemical Engineering Journal* 442: 136217.

111 Xie, Y., Huang, Y., Chen, H. et al. (2024). Dual-protective role of PM475: bolstering anode and cathode stability in lithium metal batteries. *Advanced Functional Materials* 34: 2310867.

3B

Application of Synthetic Polymers in Batteries: Hetero-chain Polymers

3B.1 Introduction

Synthetic polymers are long chains of large molecules made from many small molecules connected by chemical reactions. They have unique properties and a wide range of uses and are an indispensable part of modern industry and scientific research. A polymer material is a material consisting of a large number of repeating units connected by covalent bonds. These macromolecular compounds usually have high molecular weights and are capable of forming complex structures. These materials are highly malleable and functional, and their physical and chemical properties can be regulated by adjusting the type, number, and connection of monomers. In 1920, German scientist Staudinger proposed the long-chain structure of polymers, forming the concept of polymers, which began the era of synthetic polymers prepared by chemical methods. In 1929, Wallace Hume Carothers studied a series of condensation reactions, validated and developed the macromolecular theory, and led to the creation of polyamide 66. After entering the 1960s, polymer science entered a turning point, from basic regularity research to design and creation, synthesis of a large number of high-performance polymers, such as ultra-high modulus, ultra-high strength, flame resistance, high-temperature resistance, oil resistance, and other materials, as well as biomedical materials, semiconductor or superconductor materials, and low-temperature flexible materials.

3B.2 Overview of Synthetic Polymers Materials

Synthetic polymer materials are widely used in batteries due to their diversity of structure and function, for example, they can be used as binders, electrolyte materials, separators, functional coatings, flame retardants, and active material carriers. These applications demonstrate the diversity and importance of synthetic polymers in battery technology, which can further optimize battery performance and drive further development of battery technology through molecular design and materials engineering.

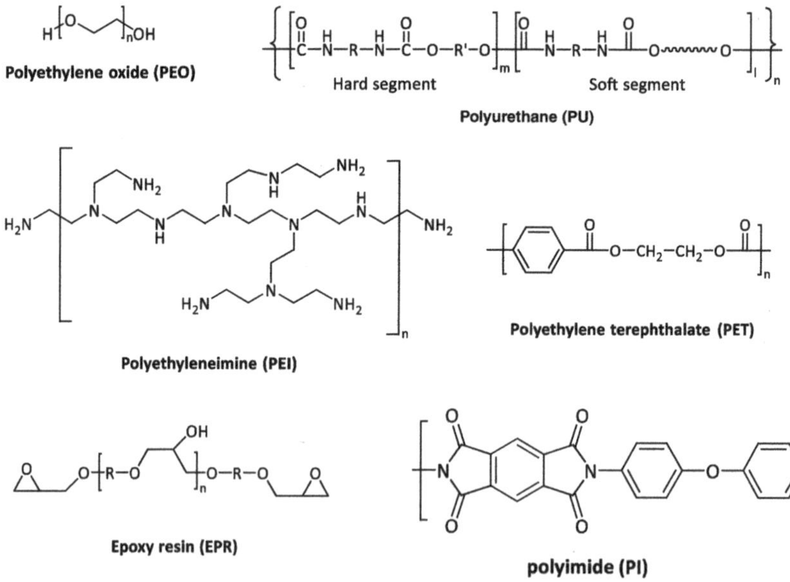

Figure 3B.1 Typical chemical structure of hetero-chain polymers commonly used in batteries.

The structure, physicochemical properties, processing methods, and application fields of polymers (their typical chemical structures are shown in Figure 3B.1) commonly used in batteries are summarized in the subsequent text.

3B.2.1 Epoxy Resin (EPR)

EPR mainly refers to a class of polymers containing more than two epoxy groups in the molecule, which is a kind of condensation reaction product of epichlorohydrin and bisphenol A or polyol. The molecular structure of EPR is characterized by active epoxy groups in the molecular chain, which can be located at the end and middle of the molecular chain, or in a ring structure. According to the molecular structure, they can be broadly divided into five categories: glycidyl ether EPR, glycidyl ester EPR, glycidyl amine EPR, linear fat EPR, and alicyclic EPR. The most used EPR variety in the composite material industry is the first type of glycidyl ether EPR, which is dominated by diphenol propane type EPR (referred to as bisphenol A type EPR). The second is glycidyl amine EPR.

Due to the high chemical activity of epoxy groups in the molecular structure, EPR as a thermosetting resin can be cross-linked with various types of curing agents containing active hydrogen to form insoluble and nonmelting polymers with three-dimensional (3D) network structures. The cured EPR has good physicochemical properties, it has excellent bonding strength on the surface of metal and nonmetallic materials, good dielectric properties, small fixed shrinkage, good dimensional stability of products, high hardness, good flexibility, and stability to alkali and most solvents.

At present, epoxy system products are widely used in the coating industry and the electronics industry. First, it can be made to various EPR coatings and inks, which are used for the protection of ships, automobile underframes, steel structures, pipeline containers, building floors and other aspects; Second, it can be made to a variety of binders, which are used for bonding between various materials, such as the repair of building cracks, bonding of interior decoration, bonding of automotive parts, and structural binders for aircraft and ships. Third, it can be made to a variety of castables, which are used for electrical components of the pooping, sealing, concrete structure manufacturing, or repair; Fourthly, it can be made to plastic sealing material, packaging all kinds of electrical equipment; Fifthly, it can be made to composite materials, for example, using EPR solution (impregnation) and glass cloth or glass fiber to prepare laminate or copper plate, which are widely used in the electrical industry as insulation board and printed circuit board; Sixthly, it can be made to organic conductive materials, which are used in resistance heating elements, resistors, antistatic materials, high-voltage shielding materials, electrode materials, sensitive element converters, conductive binders, etc.

In the battery field, EPRs are mainly used to combine with other curing agents as binder or sealant materials of batteries, improving the reliability and safety of the batteries.

3B.2.2 Polyethylenimine (PEI)

PEI is a stable water-soluble cationic polymer, which is prepared by cationic polymerization of ethylene imide catalyzed by acid or alkylating agent. According to the differences between synthetic raw materials and processes, it is divided into branched-chain PEI and straight-chain PEI. Branched-chain PEI was synthesized from ethylene imide, and linear was synthesized from oxazole derivatives. Branched-chain PEI contains both primary and secondary amines, and straight-chain PEI contains only secondary amines. Primary and secondary amines are active amines, which have strong nucleic acid binding ability. In addition, branched chain PEI is usually liquid; while straight chain PEI is solid at room temperature, it is insoluble in cold water, phenol, and ethyl ether, and more soluble in hot water and organic solvents such as methanol, ethanol, and chloroform. Branched-chain PEI has more advantages in mechanics, processability, film formation, and film light transmission, and is widely used in paper making, petroleum, textile, construction, and other fields. Straight-chain PEI is more effective in intracellular lysosome escape and DNA release, so it has more advantages in gene transfection of eukaryotic cells and has been widely used in gene transfection.

Amino groups in PEI can react with carboxyl groups to form hydrogen bonds. Amino groups can react with carboxyl groups to form ionic bonds and can react with carbonyl groups to form covalent bonds. It has a structure of hydrophilic groups (amino) and hydrophobic groups (vinyl) and can be combined with different substances. Using these comprehensive binding forces, it can be widely used in binders, inks, coatings, binders, and other fields. PEI exists in water as a polymer cation, which can neutralize and adsorb all anionic substances, and can also chelate heavy metal ions. With its high cation degree, it can be used in the fields of paper making, water treatment, electroplating solution, and dispersant. PEI is a highly reactive primary and secondary amine that can easily react with epoxies, acids, isocyanate compounds, and acidic gases. Using this property, it can be used as an EPR modifier, aldehyde adsorbent, and color-fixing agent.

In the battery field, PEI can be used to improve the electrochemical performance and cycling stability of batteries. For example, it can be used to construct hybrid cross-linked network binders to improve the mechanical properties and chemical stability of lithium-ion battery cathode materials, thereby improving the conductivity and specific capacity of lithium-ion batteries.

3B.2.3 Polyurethane (PU)

PU is a general term for large molecular compounds containing repeated urethane groups on the main chain, which is synthesized by the addition polymerization of organic diisocyanates or polyisocyanates with dihydroxyl or polyhydroxy compounds. Depending on the hydroxyl compounds used, PUs can be divided into polyester and polyether types. The polyester-type PU is prepared with diisocyanate and hydroxy-terminated polyester as raw materials. Polyether-type PU is prepared from diisocyanate and hydroxy-terminated polyether. In addition to urethane, PU macromolecules can also contain ether, ester, urea, biuret, allophanate, and other groups.

Two different products of thermosetting PUs and thermoplastic PUs can be prepared by adjusting the ratio of NCO/OH in the formulation, and they can be divided into linear type and body type according to their molecular structure. Due to the different cross-linking densities in the body structure, it can show the performance of hard, soft, or between the two, with high strength, high wear resistance, and solvent resistance. The lower density of PU makes it very popular in applications where weight reduction is required. At the same time, PUs have high tensile strength and tear strength, making them perform well in environments where they need to withstand pressure and tension. PU has excellent wear resistance and is suitable for use in environments that need to withstand friction and wear. Its molecular structure contains urethane groups, which gives the molecular chain strong toughness and wear resistance. PU can be made into products of various shapes and properties by adjusting the formulation and processing process. Which makes it highly flexible and adaptable in the manufacturing process.

In the battery field, PUs, as a very important binder material, play a vital role in the safety of batteries in potting binders, structural binders, and thermal conductivity binders. In addition, PU can also be used as a battery pack housing to provide cushioning, vibration isolation, and sealing of the battery.

3B.2.4 Polyethylene Oxide (PEO)

PEO is a crystalline, thermoplastic water-soluble polymer, which is formed by polyphase catalytic ring-opening polymerization of ethylene oxide. According to the molecular weight, it can be divided into two categories: molecular weight below 10^5 is polyethylene glycol, which is a viscous liquid or waxy solid. And molecular weight above 10^5 is solid PEO, which is a white free-flowing powder. The concentration of the active end group of these resins is low, and there is no obvious end-group activity. Due to the presence of C—O—C bonds, it is usually compliant and can form associated compounds with electron acceptors or certain inorganic electrolytes.

PEO has the following characteristics: It is soluble in water and some organic solvents, however, high molecular weight PEO is difficult to prepare into high concentration aqueous solution, due to the viscosity of high concentration solution; PEO with high molecular weight has a good aggregation effect on fine particles suspended in water, and its aggregation performance increases with the increase of molecular weight; It has a strong affinity for hydrogen bonding due to its ether-oxygen nonshared electron pairs, and can form complexes with many organic low molecular weight compounds, polymers, and some inorganic electrolytes; It is a soft, high strength thermoplastic resin, and can be processed by calendaring, extrusion, casting, etc. It is resistant to bacterial attack, and will not corrupt, its hygroscopicity is usually not large in the atmosphere. It has good compatibility with other resins, and its high molecular weight products have flocculation. The molecular chain of PEO has a large number of ether bonds, so it is easy to be attacked by oxygen and undergo oxidative degradation. As a kind of multipurpose polymer, PEO can be used in many fields such as medicine, new energy, electronics, and so on.

In the battery field, PEO is commonly used in the preparation of solid electrolytes. Among existing solid electrolytes, PEO-based solid electrolytes have attracted wide attention due to their advantages of low interface impedance, excellent electrode compatibility, flexibility, easy processing, and low price, especially when combined with lithium salts such as LiTFSI. They have several advantages over other types of solid electrolytes: They have relatively high ionic conductivity and good mechanical properties, which can improve the overall efficiency and durability of the battery. They also have good compatibility with lithium metal anodes, reducing the risk of dendrite formation and improving the overall safety of the battery. In addition, they are relatively inexpensive and easy to manufacture.

3B.2.5 Polyethylene Terephthalate (PET)

PET is an aliphatic polyester, which is prepared by transesterification of dimethyl terephthalate with ethylene glycol, or esterification of terephthalate with ethylene glycol to synthesize dihydroxy ethyl terephthalate and then polycondensation reaction. It is a milky white or light yellow, highly crystalline polymer with a smooth, shiny surface. PET plastic in terms of physical properties shows colorless transparent or opal opaque, with a certain luster. Its density is 1.30–1.33 and 1.33–1.38 g cm^{-3} in amorphous and crystalline states, respectively, which is easy to orient and crystallize.

PET has good creep resistance, fatigue resistance, friction resistance and dimensional stability, small wear, and high hardness, and has the greatest toughness among thermoplastics. It has good electrical insulation performance and is less affected by temperature, but its corona resistance is poor. It is nontoxic, weather resistant, has good stability against chemicals, low water absorption, and resistance to weak acids and organic solvents, but not heat resistant in water immersion, nor alkali resistance. PET resin has a high glass transition temperature, slow crystallization rate, long molding cycle, long molding cycle, large molding shrinkage, poor dimensional stability, brittle crystallization molding, and low heat resistance. PET has excellent properties (heat resistance, chemical resistance, strength and toughness, electrical insulation, safety, etc.), and the price is cheap, so it is widely used as fiber, film, engineering plastics, polyester bottles, etc.

In the battery field, PET is used as a separator material for lithium-ion batteries. Because PET has good mechanical and electrical properties, it is widely used to prevent electrodes from unfolding during assembly. PET battery separator plays an important role in improving the safety performance, heat resistance, and cycle performance of the battery, and is one of the important development directions of lithium-ion battery separator materials.

3B.2.6 Polyimide (PI)

PI is an aromatic heterocyclic polymer with repeated phthalimide groups in the molecular backbone, which is mainly prepared by ring-opening polymerization of dianhydride and diamine, and then cyclodehydration. It can be divided into thermoplastic PI and thermosetting PI two categories. The main chain of thermoplastic PI contains an imine ring and an aromatic ring, which has a stepped structure. This type of polymer has excellent heat resistance and thermal oxidation resistance, excellent mechanical properties, dielectric and insulating properties, and radiation resistance. Thermosetting PI overcomes the disadvantage that thermoplastic PI materials are not easy to process and form, and its processing performance is excellent. It not only has various excellent properties of thermoplastic materials, but also integrates excellent processing and forming performance and high performance in one.

PI has excellent high and low-temperature resistance, high-temperature resistance of >400 °C, and can still maintain good physical properties in a wide temperature range, long-term use temperature range of −200–300 °C. It has good electrical insulation properties, and mechanical properties, chemical stability, aging resistance, radiation resistance, low dielectric loss, and rich structural designability. The excellent properties and processing properties of PI materials enable them to be made into different products in a variety of ways, including films, resins, slurries, fibers, foams, orienting agents, binders, and many other types. It has a wide range of applications in electrical, electronic, microelectronics, aviation, aerospace, new energy, and other fields.

In the battery field, PI can be used as a positive electrode coating material, and the positive electrode active material can be modified by a specific preparation method. It can be used as separator material by virtue of its outstanding heat resistance, good electrolyte wettability, and large mechanical strength. It can be used as a binder with high adhesion, good mechanical properties, excellent thermal stability, and outstanding electrolyte wettability, which can maintain the structural integrity of the electrode. Its excellent mechanical strength, good electrolyte wettability, and excellent thermal stability allow it to be used as a solid electrolyte.

3B.3 Application of Synthetic Polymers in Binders

The main functions of the binder inside the battery include: improving the uniformity of the electrode components as a dispersant or thickener; Bonding active material, conductive agent, and fluid collector to maintain electrode structural integrity. Providing the required electron conduction within the electrode. Improving the wettability of electrolytes at the interface to enhance the transmission performance of Li^+ at the battery interface [1–5]. Although the proportion of binders in the batteries is small, the internal binder is an important part and the most easily ignored component of the battery. Protecting the electrode structure from damage is a key factor to ensure the long-term operation of the battery. With the further development of thick electrode technology, solid-state battery, and silicon anode material technology, the importance of the binder is further highlighted. At present, the binders used for battery products are mainly synthetic.

3B.3.1 Epoxy Resin

EPR is an ideal alkaline battery binder, due to the advantages of strong adhesion, low shrinkage, good electrical insulation, high mechanical strength, good alkaline resistance, and so on. EPR is a thermosetting resin whose curing reaction needs to add a curing agent, in order to improve some curing conditions and properties, an appropriate amount of accelerator, catalyst, coupling agent, diluent, and some additives also were needed to add.

Yan et al. [6] reported an EPR PEI-ER formed by in situ cross-linking between an epoxide compound and PEI, which was used as a cathode binder for lithium-sulfur (Li—S) batteries. Li—S batteries using this new binder showed high initial specific capacity along with remarkable cycling stability. Zhang et al. [7] prepared a new environmentally friendly multifunctional water-based binder for the Li—S batteries, by using PEI to cure an EPR compound triglycidyl isocyanurate (TGIC). This binder had a 3D cross-linked network, which was conducive to ion transport and accommodating volume changes during charge and discharge, and had excellent bonding force and mechanical properties. In addition, through experiments and theoretical simulations, the enhanced chemical adsorption, strong catalytic conversion ability, and high safety of these polymer binders on polysulfides were confirmed, and the flame-retardant mechanism of these polymer binders was clarified. This work is oriented toward practical application and provides a new idea for the development of a positive electrode binder for Li—S batteries with high performance and high safety [7].

Yu et al. [8] synthesized a novel epichlorohydrin-carboxymethyl cellulose (CMC) sodium binder with a 3D network crosslinking structure by a simple ring-opening reaction. This EPR-based binder can effectively bond silicon anodes with abundant covalent bonds and hydrogen, reducing the powder of silicon anodes. Benefiting from the advantages of this binder, the electrochemical properties were significantly enhanced compared to sodium cellulose binder alone. EPR also can be composed with inorganic materials to realize functionalization. Wang et al. [9] constructed an excellent integrated anode by using an epoxy-amine binder combined with ultrafine tin nanoparticles. The integrated electrode had high adhesion and tensile strength, which can effectively endure large volume changes of Sn electrode during the recharge and discharge cycle process, thus ensuring excellent lithium storage performance. Therefore, the development of these nanomaterials integrated binders provided a novel and promising strategy for improving the integration and mechanical strength of electrodes.

EPR battery binder had some limitations in acid resistance, formula complexity, curing temperature and process control, cost-effectiveness, etc. With new materials and technologies developing, these limitations may gradually be addressed.

3B.3.2 Polyethylenimine

PEI as a battery binder has the following significant characteristics: It has strong polar nitrogenous functional groups, which make it have good binding performance. This is conducive to the realization of a high electrode load, and its branch structure can maintain the stability of the electrode structure. The amine functional groups make it have a certain alkaline and cationic activity, which helps to interact with active substances such as polysulfides in the Li—S batteries. Through the interaction with polysulfide, PEI can effectively anchor polysulfide lithium, limiting its dissolution and diffusion, thereby inhibiting the shuttle effect and improving the cycling stability of the battery [1, 4].

Liao et al. [10] presented a water-soluble linear PEI binder for high sulfur loading cathodes in the Li—S batteries, which both have the functions of the strong binding performance and the trapping capability of soluble lithium polysulfide. Compared with traditional PVDF binders, it can effectively improve sulfur utilization of the cathode, reduce capacity decay, and extend the cycling life of the batteries. Huang et al. [11] reported a water-soluble N-cyanoethyl PEI binder for lithium-ion batteries based on a LiFePO$_4$ cathode, which was synthesized by conducting a Michael addition reaction of PEI and acrylonitrile. This binder was endowed with outstanding dispersion capability, excellent adhesion strength, and higher ionic conductivity by introducing polar cyano groups. LFP electrodes with this binder can maintain good mechanical integrity and low polarization during cycling, exhibiting high cycling stability and improved rate performance.

Liu et al. [12] prepared a novel multi-functional crosslinking binder by the ionic interaction of the amino group in polyethylenimide molecule and carboxyl group in PAA molecule and successfully applied it to the cathode material of Li—S battery. They demonstrated that the binder can significantly inhibit the dissolution shuttle problem of lithium polysulfide, and improve the sulfur loading and electrochemical cycling stability of Li—S batteries compared with commercial binder. In addition, the cross-linked polymer network in the binder can quickly dissolve in alkaline aqueous solutions, due to the ionic interaction between the amino and carboxyl groups being pH-sensitive. After the battery fails, the valuable components in the electrode material can be quickly dispersed in water and recovered by simple washing, then reprepared a new electrode with stable cycle capacity using the recovered materials. This study provides a new idea and method for the sustainable development of large-scale battery energy storage systems.

High-nickel layered oxide cathode is considered to be one of the most promising positive electrodes, which is expected to achieve the development requirements of high-energy-density and low-cost lithium storage batteries. However, with the increase of nickel content and state of charge, the electrochemical cycling performance of high nickel cathode will decrease. In order to stabilize the high nickel positive electrode, researchers have tried a variety of optimization and modification methods of positive electrode materials. Jin et al. [13] significantly improved the cyclic stability of high nickel positives by applying an in situ cross-linked binder, which is a functionalized terpolymer grafted by catechol groups. The preparation and characterization of the in situ crosslinking binder is shown in Figure 3B.2. The prepared binder can well maintain the integrity of the cathode structure, inhibit the cathode-anode chemical crossover, and improve the chemistry/structure of the electrode-electrolyte interface. This study demonstrates a rational structure and composition design strategy for polymer binders to mitigate the structural and interphase degradation of high nickel cathodes in lithium batteries.

PEI binder, because it contains a large number of active polar amino functional groups, can not only significantly improve the bonding ability and adsorption capacity of polysulfide in Li—S batteries, but also easily react with

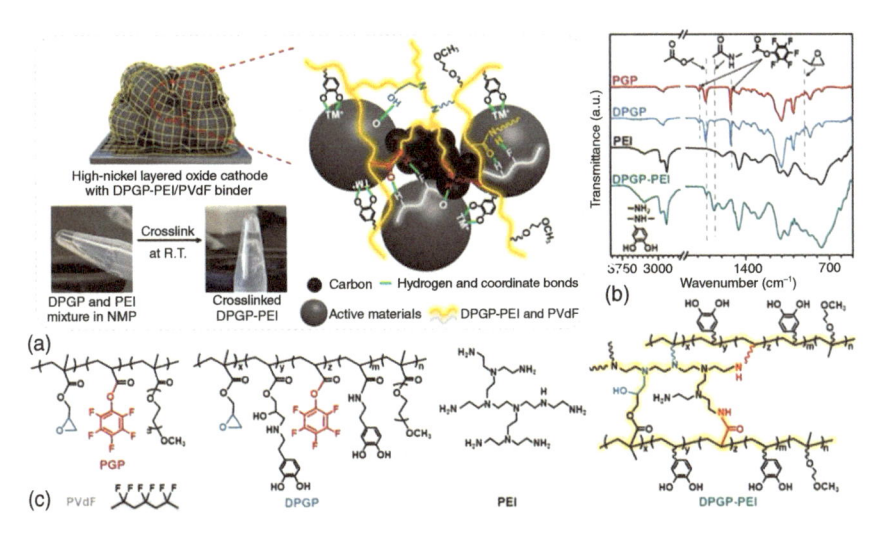

Figure 3B.2 Preparation of in situ crosslinking binder. (a) Chemical interactions between binder and electrode particles. In situ crosslinking process of the binder; (b) FTIR spectra; and (c) chemical structures of the binder composition. Source: Reproduced with the permission from Jin et al. [13]/John Wiley & Sons.

other functional materials for modification. Compared to traditional binders, it has a higher potential to meet the development needs of high-energy-density batteries.

3B.3.3 Polyurethane

The most commonly used PU binders are two-component, which are widely used in new energy batteries because of their high thermal conductivity, high adhesion, and good operating performance. The potting binder is suitable for the potting of battery packs, providing flame retardancy, shock absorption, low-temperature resistance, and good adhesion to the battery housing material. It can be applied to the potting of battery packs because it has the characteristics of flame retardant, shock absorption, low-temperature resistance, and good bonding with the battery housing material. It also can be used as an electrode binder due to its excellent thermal and electrochemical stability.

Sulfide-based all-solid-state batteries (ASSBs) offer high safety and potentially high energy density, especially with the recently proposed "anode-less" electrode strategy. This strategy creates a thin silver-carbon protective layer on the anode collector fluid, using silver particles to form a solid solution with lithium to reduce nucleation energy, thus homogenizing the lithium morphology to prevent the growth of lithium dendrites. Although this anode-less protective layer opens up a new research direction for ASSBs, the initial irreversibility and poor stability of the protective layer limit the sustainability of the battery. This shortcoming usually stems from silver particles (or other alloying metal particles), which will undergo large volume expansion during charge and discharge, thus reducing the integrity of the electrode. At the electrode level, repeated volume changes loosen the contact between particles, expanding side reactions and the formation of dead lithium in the protective layer, and subsequently compromising reversibility in the cycle. Therefore, the stability of the protective layer is crucial to improving the critical performance of anode-less ASSBs.

Oh et al. [14] proposed the use of highly elastic binders to solve the problems caused by changes in the volume of electrode materials. They developed a highly stable polymer binder for silver-carbon composite electrodes to promote electrode stability and reduce initial irreversibility, which is a highly elastic polymer named "Spandex." The Structure and bonding mechanism of the prepared Spandex polymer is shown in Figure 3B.3. The prepared PU-based binder consists of both hard and soft segments: the soft segment is composed of a poly(ethylene glycol) chain; the hard one is composed of a combination of urethane, methylene diphenyl diisocyanate, urea, and ethylene diamine. Hard segments support the entire network, while soft segments allow the polymer to stretch. The soft and hard segments of this binder work synergistically, the former performs strong hydrogen bonding with the active material and the latter facilitates elastic adjustment of the binding network. The design of this binder greatly improves the recharge-discharge reversibility and long-term cyclicity of anode-less ASSBs, and provides implications for elastic

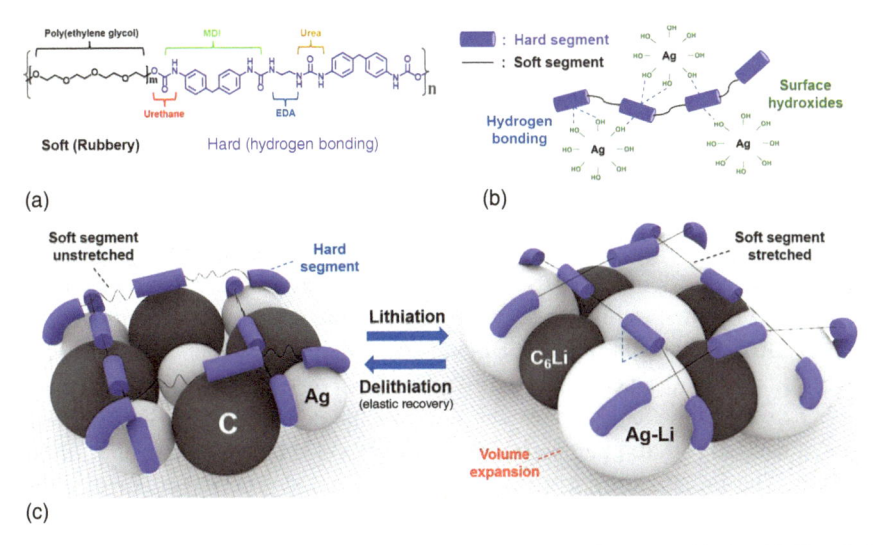

Figure 3B.3 Structure and bonding mechanism of prepared PU-based binder. (a) Chemical structure; (b) schematic illustration of hydrogen bonding; (c) 3D graphic of elastic recovery of the electrode. Source: Reproduced from Oh et al. [14]/with permission of American Chemical Society.

binder systems for high-capacity ASSB anodes, which can withstand large volume changes.

Silicon (Si) has been recognized as a promising anode material for lithium batteries due to its high theoretical capacity and cost-effectiveness. However, it will occur the significant volume changes that occur during repeated lithiation/delithiation processes, the traditional battery binders are not up to the task. Kim et al. [15] prepared an environmentally friendly PU binder using water as the dispersion solvent, which was obtained by introducing nonionic hydrophilic poly(ethylene glycol monomethyl ether)-based trimethylolpropane chains into the PU structure. In Si/binder systems, polar PU groups formed H-bonds with Si groups on Si particles, while stretchable and elastic PP oxide chains eliminated the stress induced by volume changes. These results suggest that sustainable and environmentally friendly aqua PU binders may be an attractive option for manufacturing environmentally friendly and stable silicon anodes. To deal with the expansion of silicon-based batteries, Sun et al. [16] also reported a novel water-soluble binder composed of waterborne PU and sodium CMC. The prepared binder can establish a cross-linked 3D network through hydrogen bonding, effectively maintaining the integrity of the electrode. It also can help to form a stable solid-electrolyte interphase (SEI) layer, thereby improving the cycling stability and durability of the batteries.

All these works have proved that PU-based battery binders have better mechanical properties and can meet the needs of ensuring the integrity of electrodes that will produce large volume changes during the cycle. With the construction and development of new electrode batteries, more advantages of PU binder will be discovered and developed.

3B.3.4 Polyimide

PI has high adhesion, good mechanical properties, excellent thermal stability, and outstanding electrolyte wettability. As a binder, it can effectively maintain the integrity of the electrode structure, especially in materials with large volume expansion and contraction [17, 18]. For example, in the silicon-based anode battery, PI can restrain the active particles, prevent the pulverization of silicon particles and the loss of electrical contact, maintain structural integrity and stability, and improve the cycling stability of the battery. PI binder has high mechanical strength and thermal stability, which can improve the safety of the battery at high temperatures. In addition, PI can be modified or modified by molecular structure design to improve the performance of the binder. For example, the introduction of ether-oxygen functional groups can improve the conductivity of lithium-ions. The synthesis of soft and hard segments of coembedded polymers can offer better elasticity. The introduction of strong bonding functional groups can enhance the bonding performance.

The high nickel layer oxide $LiNi_{0.8}Co_{0.1}Mn_{0.1}O_2$ (NCM811) cathode material is considered to be the most promising cathode material for the next generation of high-energy-density lithium batteries, due to its high reversible capacity of up to 280 mAh g^{-1}. It has great economical and practical feasibility to obtain a higher reversible capacity of nickel-based cathode materials by increasing the charging cut-off voltage. The high voltage will convert unstable Ni^{4+} to more stable Ni^{2+}, promote lattice oxygen oxidation, and cause surface phase transition and oxygen release. This will lead to a rapid decline in cathode material capacity and magnification performance, so more effective strategies are needed to stabilize cathode structures and surfaces at high voltages. Wang et al. [19] proposed a strategy of combining a PI with a rigid chain and a polysiloxane with a flexible chain to prepare a composite binder with a transportable ion channel through structural tailoring and copolymerization. The nanoscale composite binder layer formed in situ on the surface of NCM811 particles can provide resistance to extreme conditions and strong attractive interaction, which makes the cathode-electrolyte interface robust and promotes ion transfer during charge and discharge. The durable PI-polysiloxane composite binder can successfully inhibit the cathode-electrolyte interface side reaction, effectively inhibit the dissolution of the transition metal ions, and maintain the layered structure of NCM811, ultimately achieving an extremely long cycle life and rate performance at the high voltage. Subsequently, they prepared a PI-based gel binder using the same material to stabilize NCM811-based batteries and also achieved good results [20].

Silicon has a high theoretical weight energy density as an active negative electrode material, but it undergoes significant volume changes during the lithiation/de-lithiation process, resulting in rapid capacity decay. The performance of the silicon anode depends on the ability of the binder to bind the silicon particles, which is required to maintain the integrity of the anode during the repeated lithium process. Lusztig et al. [21] applied a novel PI binder P84 to a composite silicon anode composed of metallurgical micron silicon particles (80% silicon) and investigated the effect of heat treatment on the properties of the binder. It was found that after heat treatment at 400 °C, the performance of the electrode was improved the most. In addition, by comparing the performance of the composite silicon anode system, it is found that the performance of the PI binder is better than that of advanced binders based on polyacrylate and sodium alginate. Tan et al. [22] synthesized a copolyimide binder with adjustable rigidity and flexibility through a simple copolymerization reaction. The rigid chains can provide excellent force properties and interfacial interactions, and the flexible chains can relieve volume changes by enhancing the move ability and entanglements of the molecular chain. They found that appropriately increasing the flexibility of the PI binder can make it adapt to the volume change of the SI electrode, which is conducive to improving the cycling stability of the Si anode. Excessive proportion of flexible segments will lose mechanical and bonding properties, resulting in the inability to resist the volume expansion of Si electrodes and reducing electrochemical performance.

PI binder shows high thermal stability, excellent chemical stability, and mechanical strength in the battery system, but it still needs to solve the problems of dissolution processing, electrochemical performance, and cycling stability in practical applications. Future research can focus on solving these shortcomings, so as to promote the practical use of PI in the field of batteries.

3B.4 Application of Synthetic Polymers in Electrolytes

Polymer electrolyte is a material composed of a polymer matrix and electrolyte components that can conduct ions in batteries or other electrochemical devices. They are typically made from high molecular weight polymers that can combine with lithium salts to form a conductive network. Polymer electrolytes offer significant advantages for battery applications [23–33], including improved interface stability due to their good compatibility with electrode materials, which enhances cycling stability and efficiency while reducing interfacial resistance. Their ionic conductivity and mechanical strength can be increased by doping with various salts and incorporating liquid plasticizers or inorganic fillers. Compared to traditional liquid electrolytes, polymer electrolytes are generally more environmentally friendly and less toxic. They also have deformable properties that can be prepared into flexible batteries, making them suitable for flexible and wearable devices. Their mechanical toughness helps to withstand volume changes in electrodes during cycling, benefiting interface stability and charge transfer resistance. Additionally, the polymer matrix and functional filler can inhibit the growth of lithium dendrites, thereby improving the safety and cycling stability of the battery. Collectively, these attributes position polymer electrolytes as a promising solution for enhancing battery performance, safety, and environmental sustainability.

3B.4.1 Epoxy Resin

EPR-based electrolyte is a special material that combines the excellent properties of EPR with the functionality of electrolyte. EPR electrolyte has the characteristics of high ionic conductivity, high mechanical strength, excellent corrosion resistance, and environmental protection, which makes it have wide application potential in the field of electrical insulation materials. It provides good ion transport channels, ensuring that ions in electrochemical energy storage devices can move efficiently. It has high bending modulus and bending strength, can withstand mechanical stress, and is not easy to damage. It has good corrosion resistance, can resist chemical corrosion, and is suitable for harsh environmental conditions. In addition, the introduction of bio-based EPRs reduces the dependence on petroleum resources, while reducing environmental pollution. These characteristics make EPR show great application potential in the field of electrical insulation and energy storage devices.

EPR can be combined with inorganic materials to prepare composite electrolytes. Li et al. [23] used the traditional direct casting method to prepare the composite mesoporous EPR-based multi-layer electrolyte with lithium lanthanum titanate (LLTO) nanoparticles. LLTO doped EPR matrix electrolyte presents high ion conductivity (2.02×10^{-3} S cm^{-1}) and lithium-ion transfer number (0.82), a wide electrochemical stability window of up to 5.5 V (vs. Li$^+$/Li) at room temperature, and a small interfacial resistance. This multilayer electrolyte can fully infiltrate the electrode surface and maintain uniform Li deposition even at a high current density of up to 4 mA cm^{-2} with an area capacity of 12 mA h cm^{-2}. The unique mesoporous structure of EPR and the fast ion transport path endowed by LLTO facilitate the uniform deposition of lithium-ions. Zekoll et al. [24] proposed a method for the preparation of a 3D bi-continuous phase structure mixed electrolyte for ASSBs, which combines a ceramic solid electrolyte LAGP (Li$_{1.4}$Al$_{0.4}$Ge$_{1.6}$(PO$_4$)$_3$) and an insulating polymer. The test results show that the LAGP-epoxy electrolyte retains the same level of ionic conductivity of LAGP (1.6×10^{-4} S cm^{-1}), but its mechanical properties are better, and the bending failure strain of this electrolyte before rupture is five times that of LAGP. This shows that ordered ceramic and polymer hybrid electrolytes can achieve superior mechanical properties while retaining higher ionic conductivity, which solves one of the key challenges of ASSBs.

ILs are also commonly used to compound gel electrolytes with EPR, Demir et al. employed molecular dynamics simulations to study structural electrolytes composed of IL and epoxy polymer. It is found that the T_g, mechanical property, diffusivity, and ionic conductivity of high cross-linked EPR-based electrolyte change significantly with the variation of IL content, which is induced by the swelling effect of IL on the formed 3D polymer network. The computational results about the ionic conductivity, and thermal and mechanical properties of EPR-based SPE can help us to understand the phase behavior at the molecular level, which is critical for the design and optimization of SPE performance. Song et al. [25] demonstrated a simple and easy strategy to prepare a multifunctional SPE by adjusting the lithium salt concentration, which can be used to develop safe and flexible energy storage devices. The chemically cross-linked EPR-based SPE was obtained by ring-opening polymerization of oligoether solvated by Li$^+$, IL with high ionic conductivity, weakly bonded lithium salt. The polymerization causes discontinuous microphase separation, thus forming microscopic fast ion migration channels and macroscopic mechanical support cross-linked matrix. The combination of

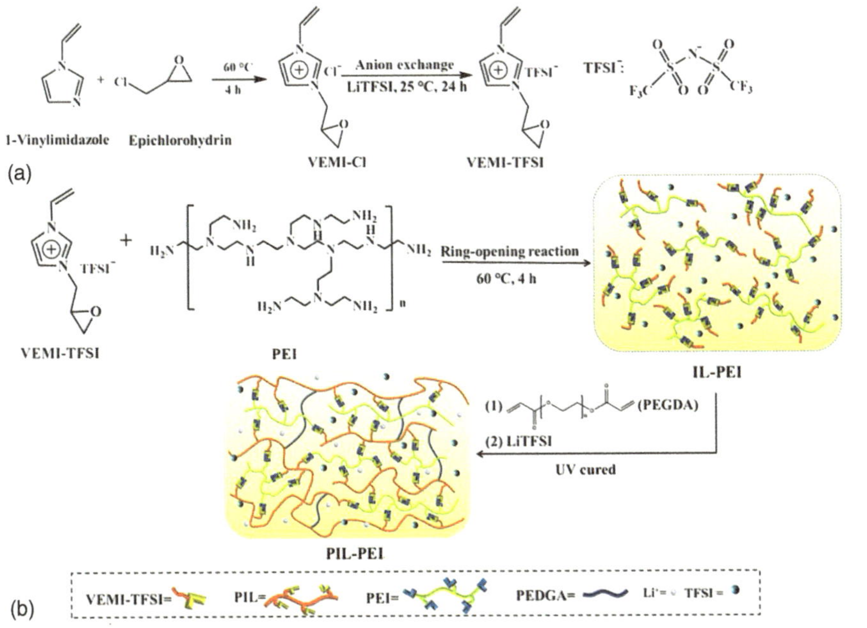

Figure 3B.4 Preparation mechanism of double cross-linked poly(ionic liquid) electrolyte (a) synthesis route of VEMI-TFSI and (b) synthesis route of PIL-PEI. Source: Reproduced from Liang et al. [26]/with permission of Elsevier.

electrochemical and mechanical stability enables excellent electrochemical performance to be maintained in bending and rolling tests. Liang et al. [26] synthesized a novel bi-functional IL and prepared a double cross-linked poly IL electrolyte (PIL-PEI) by ring-opening and radical cross-linking polymerization (Figure 3B.4.). The ring-opening cross-linking polymerization occurs between the epoxy groups of 1-vinyl-3-epoxypropylimidazolium bis(trifluoromethanesulfonyl)imide (VEIMTFSI) and amino groups of PEI. The radical cross-linking polymerization occurs between vinyl groups of VEIM-TFSI and acrylic groups of poly(ethylene glycol) diacrylate (PEGDA). This double-cross-linked network PIL-PEI electrolyte has good flexibility, thermal stability, and excellent electrochemical performance (room temperature ionic conductivity of 1.03×10^{-3} S cm^{-1}, lithium-ion transport number of 0.47). The pouch battery assembled with PIL-PEI electrolytes was able to successfully light the LED light in a variety of harsh situations such as bending and cutting, demonstrating the electrolyte's potential for applications in highly safe flexible electronic devices.

High ionic conductivity and excellent mechanical properties of SPEs are usually difficult to combine, Zeng et al. [27] proposed a reaction-controlled way to balance the ionic conductivity and force strength. The nanophase separation was induced by two-step polymerization to form an elastic epoxy polymer electrolyte with a special ion transport pathway and support matrix, which has excellent force strength (3.4 MJ m^{-3} toughness) and high ionic conductivity (3.5×10^{-4} S cm^{-1}, 25 °C). This elastic polymer electrolyte with both electrochemical and mechanical properties can be used in flexible, high energy density, and safe next-generation energy storage devices. Manarin et al. [28] prepared a bio-based GPE membrane using cashew phenol-based EPR as raw material by esterification with glutaric anhydride, succinic anhydride, and hexahydro-4-methylphthalic anhydride. This bio-based GPE has excellent electrochemical stability against potassium metal in the voltage range of −0.2-5 V and high ionic conductivity at room temperature (10^{-3} S cm^{-1}), which can be used in potassium ion batteries. Wang et al. [29] prepared a new EPR-based ionic gel electrolyte, in which epoxy monomer cured by using amine with soft segments is a matrix and IL is a plasticizer. The resulting GPE has a bi-continuous ion conduction path and thus has a high room temperature ionic conductivity of 0.05–1.69 mS cm^{-1}, which is a potential candidate gel electrolyte with high performance. Zhang et al. [30] prepared a bi-continuous phase structure SPE with high modulus (bending modulus ~1.0 GPa) and high ionic conductivity (6.7×10^{-1} mS cm^{-1}), through phase separation of EPR 5284 and IL-induced by in situ polymerization. The SPE can be used in battery systems with high energy density and high current density.

Because of its highly reactive epoxy functional group, EPR can react with other active curing agents to improve its performance. With the development of research, it is expected to be more widely used in the field of batteries.

3B.4.2 Polyurethane

PU-based electrolyte has excellent mechanical flexibility and electrochemical stability, which can effectively overcome the safety problems of liquid electrolytes, such as leakage and fire risks. It has the advantages of light weight, good film formation, good viscoelasticity, and high stability. Due to its special carbamate structure and a large number of hydrogen bond elements in the molecule, PU usually has the ability to self-heal solid/solid interface defects, thereby building a stable interfacial ion transport path. These characteristics make PU-based electrolyte one of the important research directions in the field of solid-state batteries [31].

High-performance lithium metal batteries that operate in high-temperature environments require stable and robust electrolytes, Andersson et al. [32] prepare a high-performance LMB with a PU-based SPE that can be cycled >2000 times at 80 °C and still remain stable. The strength provided by the hard PU segment makes it exhibit high stability under a low deformation rate, so that it can maintain mechanical stability at temperatures of 100 °C and above. Doping modification can further improve the performance of PU-based SPE, Gao et al. [33] prepared an organic inorganic composite SPE based on PU. The addition of oxide electrolyte $Li_{6.4}La_3Zr_{1.4}Ta_{0.6}O_{12}$ (LLZTO) increased the coordination ability of lithium-ions in the electrolyte, improved the electrochemical performance of PU-SPE, and made it free of lithium dendrite penetration after 500 h cycle. The composite electrolyte has the advantages of high room temperature ionic conductivity of 2.22×10^{-4} S cm^{-1}, high lithium-ion transfer number of 0.55, and excellent negative electrode contact interface. Huang et al. [34] combined an amino-modified metal-organic skeleton (UIO-66-NH$_2$) with PU-SPEs, which had abundant active defect sites, and the mechanical strength and room temperature ionic conductivity of the SPE were significantly improved. The abundant interfaces between UiO-66-NH$_2$ and polymer chains markedly reduce the crystallization rate of polymer chains and decrease lithium-ion transport impedance, which significantly improved the ionic conductivity (2.1×10^{-4} S cm^{-1}) and the Li$^+$ transference numbers (0.71). This organic and inorganic composite SPE design strategy is a simple and effective preparation method for high-performance and high-voltage LMBs.

It is also a common modification method to combine other organic components into PU to improve its overall performance, Hou et al. [35] prepared delignified wood-based PU composites by vacuum infusion of PU matrix with different ratios of polyethylene glycol/2,2-bis(hydroxymethyl)propionic acid into delignified woods. The obtained composite polymer matrix was mixed by vacuum perfusion method with electrolyte lithium perchlorate (LiClO$_4$) to prepare working GPE. The composite GPEs have efficient Li$^+$ transport pathways based on the abundant internal cavities, especially exhibiting good ion conductivity which presents a great potential in the field of electrochemistry. Shi et al. [36] prepared an ultrathin SPE with decreased phonon scattering by coating the composites of PU, IL, and lithium salt on the PE separator. The PE-based separator matrix with mechanical strength and flexibility can decrease the thickness of the electrolyte and enhance Li$^+$ mobility, offering a more regular thermal diffusion pathway for SPE to decrease the external phonon scattering. Therefore, the thermal conductivity of the as-prepared SPE is enlarged by about six times, which can effectively inhibit the thermal runaway of the battery. Wang et al. [37] prepared a hyperbranched PU (HPU) SPE by doping lithium salt and IL into a polymer matrix obtained from hyperbranched polyethylene glycol and isophoradione diisocyanate. The prepared composite HPU SPE has a wide electrochemical window and a high ionic conductivity at room temperature. This work exhibits a new way to develop the SPE with good practical application potential by using hyperbranched polymer electrolytes.

Other advanced functions for SPE, such as flame-retardant effect, interfacial self-healing ability, and mechanical energy dissipation capacity, also can be endowed with the PU by material structure design. Wu et al. [38] prepared a nonflammable and stretchable PU-based SPE by introducing a reactive flame-retardant group into the polymer chains through covalent bonding. The force strength of the polymer matrix is enhanced by grafting a rigid aromatic ring group, which can effectively inhibit the growth of lithium-dendrites. Rich oxygen-containing functional groups in the PU soft segment enhanced the ion conductivity and migration number of Li$^+$, and the PU hard segment can improve the force strength of the polymer matrix. It is a promising strategy to prepare SPE with advanced functions for high-energy-density and safe LMBs. Pei et al. [39] presented a poly(ether-urethane)-based SPE with a self-healing ability that can decrease the interfacial resistance and offer a high-performance solid-state LMB. By introducing dynamic covalent disulfide bonds and H-bonds, the prepared SPE has good interfacial self-healing capacity and exhibits sufficient interfacial contact. The Li||Li symmetric batteries based on this SPE presented stable long-term cycling of exceeding 6000 h, and the PU-based SPE Li—S battery exhibited a long cycling life

(700 cycles at 0.3 C). It is a promising way to construct high-performance LMBs by using self-healing SPEs with dynamic bonds.

Yan et al. [40] presented a supramolecular PU material enhanced by aromatic charge-transfer interactions for an electrolyte matrix with high strength, good toughness, and novel mechanical energy dissipation ability. The improved electrolyte matrix can observably fit the volume change of Li metal, and effectively eliminate the stress-concentrating effect at a deformed state. When the polymer chain is stretched by external forces, the internal stress is mainly concentrated in the deformation region. As the stress in some areas gradually increases to exceed the fixed strength limit, the polymer chain in the electrolyte film breaks and micro-defects appear. The internal stress caused by the repeated volume change of the lithium metal anode in the solid electrolyte will be further concentrated at the micro defect location, resulting in greater crack propagation. Figure 3B.5. is the schematic illustration of the PEO and PU samples at a tension state with serious stress concentration, which demonstrates the PU-based sample has better stress resistance. Therefore, the PEO-based electrolyte is more easily broking by lithium dendrites when the highly reactive Li metal cannot evenly deposit (Figure 3B.5.a), which will decrease the long-term cycling stability of LMBs.

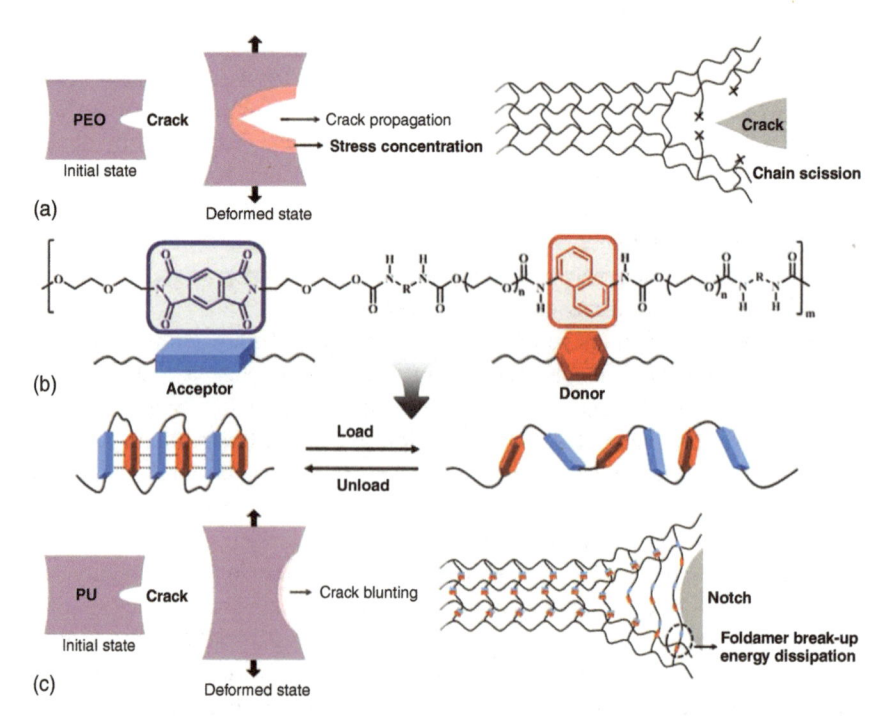

Figure 3B.5 Schematic illustration of PEO and PU sample at tension state with serious stress concentration. (a) Notched PEO sample at tension state with serious stress concentration. (b) Molecular structure of PU material with D-A charge-transfer interactions. (c) Notched PU sample eliminates external energy by dissociating the D-A self-assembly. Source: Reproduced from Yan et al. [40]/with permission of Wiley-VCH Verlag GmbH & Co. KGaA.

Aromatic charge transfer interactions usually occur between two chromophores of electricity-deficient acceptor (A) and electricity-rich donor (D), because of their inherent complementary properties at alternately arranged state. Sequenced alternating D-A segmental packs that exist in both inter- and intra-chains can enhance the microstructure of the soft matter, as the D-A self-assembly is achieved in a polymer material. The properties of D-A dynamic self-assembly that can reversibly form and dissociate endows the PU materials with good stretchability and excellent self-healing capacity (Figure 3B.5.b). In addition, absorbing and eliminating external impact mechanical energy by fracturing the noncovalent bonding is also beneficial to enhance the toughness. Therefore, the notched PU sample can eliminate external energy by dissociating the D-A self-assembly to blunt the crack (Figure 3B.5.c). This PU-based electrolyte has higher mechanical strength for better-resisting dendrites, and reversible D-A interactions can spontaneously recover to dynamically adapt to the shape-changing lithium anode, forming a stable SEI. This method of constructing a supramolecular system to improve the mechanical properties of SPE is of great significance for the preparation of solid LMBs with excellent comprehensive performances.

PU-based electrolyte shows great development potential in the battery field due to its characteristics of high conductivity, wide electrochemical window, stability and safety, and diversified applications. With the deepening of research and technological advances, PU electrolytes are expected to occupy an important position in the future battery market.

3B.4.3 Polyethylene Oxide

PEO-based electrolyte is considered one of the most promising polymer electrolytes for solid LMBs due to its good electrochemical stability and excellent solubility of lithium salt. Compared with traditional liquid electrolytes, PEO-based SPEs have greater superiority for improvement in thermal stability, battery energy density, operating temperature range, mechanical strength, and other aspects, which is one of the effective pathways to solve the safety problems of LMBs. In order to improve the high voltage resistance of PEO-based polymer solid electrolyte, the electrolyte or positive electrode interface can be modified by filler doping, molecular design, and interface construction, so as to improve the ionic conductivity, enhance the mechanical strength and inhibit the formation of lithium dendrites. Mixing PEO electrolytes with inorganic filler or other organic material is the most commonly used method to improve performance, which can enhance electrochemical performance, improve processing performance, or offer special functions [41, 42].

Lithium salts are essential for the performance of PEO-based SPEs, Li et al. [43] selected three lithium salts of LiFSI, LiTFSI, and LiPFSI to prepare SPEs with PEO and PI. The effects of lithium salts with different molecular sizes, F chemical environments, and structures on the performances of the LMBs were studied. The results demonstrated that the F-bond type is more important than the molecular size and F contents, and the performance of the battery based on LiPFSI is the best one. The SEI of SPE based on LiPFSI and LiTFSI has a large amount of LiF and

Li$_3$N, but the SPE based on LiFSI cannot product Li$_3$N. LiF-rich SEI can effectively restrain the formation of Li dendrites and improve the long-term cycling stability of LMBs, therefore, adjusting the composition of the formed SEI by using different lithium salts is an effective strategy to improve the performance of the SPEs. He et al. [44] prepared a double-layered SPE for LBMs, in which the PEO toward the cathode side had a high-concentration lithium salt (EO:Li = 4 : 1) and that toward the anode side had a low-concentration lithium salt (EO:Li = 16 : 1). Because the PEO with high-concentration lithium salt (EO:Li$^+$ ≤ 6 : 1) has high oxidation potentials (>5 V vs. Li/Li$^+$), this double-layered SPE strategy can markedly improve oxidation resistance of the PEO-based SPEs. He et al. [45] adopted a novel lithium salt lithium difluorobis(oxalato)phosphate (DFBOP) to prepare PEO-based SPE, which had excellent anode and cathode interfacial performances. Robust LiF, Li$_2$C$_2$O4, Li$_x$PO$_y$F$_z$, and Li$_3$N-enriched SEI can be formed in situ on the surface of Li anode under the coordination and synergistic decomposition of DFBP-LiTFSI. The HOMO level of PEO can be effectively reduced under the effect of the dipole–dipole interaction between PEO and DFBOP, which can help to inhibit the oxidative decomposition of PEO under high voltage. The collaborative decomposition of DFBP-LiTFSI can form a CEI with a novel organic–inorganic interpenetrating network. In this study, the interface design of EPO-based electrolytes is realized through electrolyte coordination modulation, which provides a new idea for the development of high-performance PEO-based LMBs.

The combination of PEO with other organic compounds is also a common method of electrolyte modification, Han et al. [46] used photo-controlled radical alternating copolymerization (photo-CRAP) of lithium fluoride monomer (electron acceptor, A) and PEO-substituted vinyl ether (electron donor, D) to prepare a PEO-based SPE. The precise positioning design of repeated units in the polymer sequence is realized. This special structure promotes the uniform distribution, the solvation, and the migration of lithium-ions, and significantly improves the lithium-ion conductivity. All-solid-state LMBs assembled with this PEO-based SPE have better cycling stability at both ambient and high temperatures, proving that fine programming of polymer sequences can confer more advanced functions on materials. Rong et al. [47] studied the performance differences of four kinds of PEO-based SPE with different terminal groups, including hydroxyl, propyl carbonate, methoxy terminal, and trifluoroethoxy. The results demonstrate that the terminal groups have great influences on the physical and electrochemical properties of PEO-based SPE, such as crystallization, viscosity, lithium salt dissociation ability, and so on. The hydroxyl terminal PEO exhibits less stability, propyl carbonate, and methoxy terminal PEO offer moderate ionic conductivity, and trifluoroethoxy terminal PEO has higher ionic conductivity and high voltage resistance.

Li et al. [48] introduced 1,1,2,2-tetrafluoroethyl-2,2,3,3-tetrafluoropropylether (TTE) into poly(ethylene oxide carbonates) to prepare composite PEO-based SPE with promoted ionic conduction performance. The synergistic effect between two polymers with different structures can significantly increase the transport rate of lithium-ions. Cheng et al. [49] presented an aqueous solution casting strategy to introduce zwitterionic cellulose nanofiber (ZCNF) into PEO-based SPE, addressing

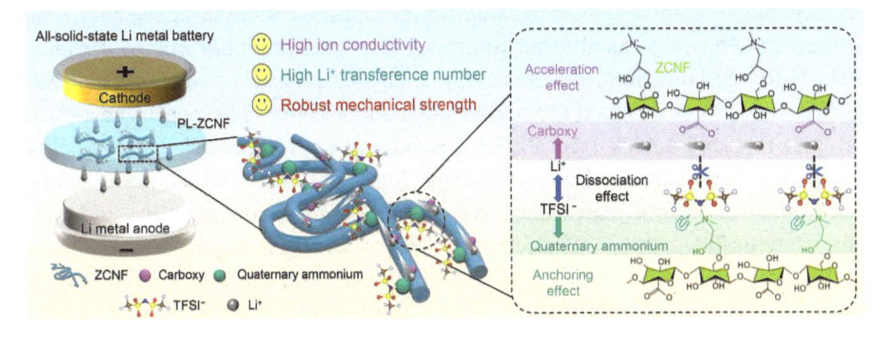

Figure 3B.6 Schematic illustration of the functions of ZCNF for PEO-based LMBs. Source: Reproduced from Cheng et al. [49]/with permission of Wiley-VCH Verlag GmbH & Co. KGaA.

the SPE defect including low ionic conductivity and Li$^+$ transference number as well as poor mechanical strength. The characterization data and theoretical calculations showed that the novel zwitterionic structure of ZCNF can Exert multiple positive effects in the LMBs based on this SPE. Schematic illustrations of the functions of ZCNF for PEO-based LMBs are shown in Figure 3B.6, by reducing PEO crystallinity, dissociating lithium salts, anchoring anions and promoting Li$^+$ migration, the ion conductivity, Li$^+$ transference number, and mechanical strength of the PEO-based SPE are significantly improved.

In addition to organic compounds, inorganic fillers are often used to modify PEO SPEs. Li et al. [50] synthesized a flexible SPE composed of TiO$_2$ nanotubes and PEO matrix, the composite SPE can offer outstanding interfacial conduction and rational electrode compatibility in the LMBs. A new lithium-ions transport pathway can be formed between the interfacial phase of TiO$_2$ and PEO, which can significantly reduce the interface resistance. In addition, TiO$_2$ can improve the dissociation of lithium salts and generate free lithium-ions, thereby increasing the ionic conductivity. Furthermore, the introduction of TiO$_2$ can achieve an ideal stable interphase layer with rich LiF, offering rapid lithium-ions transport and uniform Li deposition. Li et al. [51] introduced Cu-montmorillonite (MMT) in the PEO matrix to prepare PEO-based SPE, whose ability to inhibit dendrite growth was significantly improved. The unique spatial constraints of the vertical array structure inhibit the growth of PEO crystals and their distribution in the non-ionic conduction direction, reduce the ion migration crosstalk, and realize the rectification of the polymer chain. There are abundant copper ions and oxygen-containing groups in the matrix, which inhibit anion conduction and accelerate Li$^+$ migration at the nanometer scale, respectively, and increase the transport number of lithium-ions.

Zhou et al. [52] prepare a PEO-based SPE by using magnesium-aluminum fluoride (MAF) nanorods as the fillers, which can easily adjust the fluorine vacancy concentrations. Since MAF can immobilize the lithium-ions, and the F vacancy weakens the Li—F bond energy, the ionic conductivity and lithium-ions transport number of the SPE are significantly enhanced. The H-bond between MAF and PEO can weaken the reactivity of the —OH groups in the PEO molecules, widening the electrochemical window and increasing the mechanical strength of the SPE. The prepared PEO-based SPEs showed improved interfacial stability due to a robust LiF/Li$_3$N-rich SEI obtained, and the assembled batteries based on the SPE exhibited high discharge capacity, good cycling stability, and excellent rate performance. An et al. [53] reported a 4.8 V-level PEO-based battery with improved cyclic stability through a Lewis-acid coordinated strategy. Mg^{2+} and Al^{3+}, which have strong electron absorption ability, are introduced into the coordination structure, and their chelation with PEO reduces the electron density of the ether oxygen (EO) chain and weakens the local interaction between the EO chain and cathode surface at high charge. The Lewis-acid coordinated SPEs exhibit improved electrochemical oxidation potential (>5 V) and high ionic conductivity (0.51 mS cm^{-1}), and the assembled batteries based on the SPEs have excellent cycling stability. The electrolyte can be prepared as a rolled electrolyte film by a continuous process on an industrial scale, showing practical potential for the preparation of all-solid LMBs with high energy density.

PEO-based SPE is the earliest and most widely studied polymer electrolyte, but its application in LMBs faces challenges such as low ionic conductivity at low temperatures and poor electrochemical stability at high voltages. With the continuous efforts of researchers, these problems will be gradually solved, and the application range of PEO electrolytes is expected to be further expanded.

3B.4.4 Polyimide

PI-based electrolyte has the characteristics of high heat resistance, low thermal expansion coefficient, excellent mechanical properties, high voltage stability, good electrolyte infiltration, high chemical resistance, and stability, which make PI a popular material in high-performance lithium battery electrolytes and other electronic applications.

Cui et al. [54] prepared a composite SPE consisting of a porous PI matrix and a lithium-ion conductor filler. Among them, the matrix is a porous film made of the lightweight flame-retardant materials decabromo-diphenyl ethane (DBDPE) and PI, and the lithium-ion conductor filler is composed of polyethylene oxide/bistrifluoromethylsulfonyl lithium (PEO/LiTFSI). The use of high-strength porous PI film and the light weight of organic flame-retardant DBDPE can inhibit the generation of lithium crystal dendritic, and also solve the safety problem of battery short-circuit resulting in combustion. Li et al. [55] designed a novel composite SPE based on PI fiber frames and PEO electrolytes. In the PI-based SPE, the PI fiber matrix provided high elastic modulus and tensile strength. The strong interaction between electron-absorbing atoms in lithium salts and PI groups accelerated the dissociation of lithium salts and formed a more efficient ion transport network in the electrolyte system. The Li$^+$ transport number of PI composite SPE was higher than that of pure PEO electrolytes, which was conducive to inhibiting lithium dendrites and improving Li deposition. Its electrochemical window was expanded to 4.87 V, showing higher compatibility and stability with lithium metal anodes. Huang et al. [56] prepared an SPE with high electrical conductivity and thermomechanical stability by combining 1,3-dioxane (DOL), 1,2-dimethoxy-ethane (DME), LiTFSI, and LiPF6 in porous PI nanofiber films. In this system, LiPF6 can trigger the ring-opening reaction of DOL, LiTFSI can promote self-polymerization to form poly-DOL (PDOL), and DME acts as the plasticizer of PDOL. The PDOL had a strong affinity with PI and can form a stable SPE network. The prepared SPE films had high ionic conductivity, high lithium-ion transport number, wide electrochemical window, and excellent thermomechanical stability at room temperature.

Ma et al. [57] prepared composite SPE by pouring a PI/PVDF/LiTFSI mixture. In the system, the active conjugated carbonyl group in the PI polymer matrix facilitated Li$^+$ transport because the negatively charged N and O atoms accelerated the dissociation of lithium salts through electrostatic interactions, increasing the ionic conductivity and lithium-ion transfer number to 4.1×10^{-4} S cm^{-1} and 0.46, respectively. Compared to pure PVDF/LiTFSI electrolytes, the high thermal resistance PI-based SPE had a 68% reduction in thermal shrinkage and a 35% increase in mechanical strength. Both lithium-symmetric batteries and LiFePO$_4$ full batteries equipped with PI solid electrolytes can effectively homogenize Li + deposition at the electrolyte/anode interface, achieving high cycling stability. Wang et al. [58] adopted a polymer-coated polymer strategy, using strong and flexible PI as the main matrix material, and impregnated it with PEO-based electrolyte to prepare a high heat resistance PI-reinforced PEO-based SPE. The composite SPE films exhibited good mechanical properties and high ionic conductivity, which can effectively promote the interfacial stability of Li metal. Zhang et al. [59] developed a flame-retardant GPE for Li—S batteries using fluoro-substituted PI as a polymer matrix, which was gelated with carbonate electrolyte. The trifluoromethyl group in the PI chain provided a negative electric environment, which was conducive to the transport of Li$^+$, thus increasing the lithium-ion transport number (0.727). Aromatic benzene ring in PI acted as an H acceptor, and alkyl in solvent acted as an H donor, they formed a CH/π interaction. This can promote Li$^+$/solvent weak interaction, thereby reducing the parasitic reaction between metal Li and electrolyte and promoting the uniform deposition of lithium-ions. This PI-based SPE showed good cycling stability and high security in Li—S batteries. Wang et al. [60] prepared a flame-retardant high-strength SPE consisting of a high-strength PI fiber matrix, pentaerythritol tetracrylate (PPT) cross-linked network, and a LiNO$_3$/triethyl phosphate (TEP)/vinyl FEC as the plasticizer. During the formation of SEI, LiNO$_3$ can generate Li$_3$N and promote the generation of LiF from FEC. Therefore, a stable SEI rich in Li$_3$N and LiF was constructed, which effectively stabilized the Li metal anode interface and avoided side reactions between the SPE and Li metal anode. The SEI layer between the SPE and Li anode interface and the flame-retardant performance of the electrolyte were conducive to improving the cycling stability and safety of the battery.

PI-based electrolytes have great potential in the field of lithium batteries, especially in improving battery safety and high-temperature performance. However, its processing difficulty and cost issues are the key issues that need to be solved in order to achieve a wider commercial application.

3B.5 Application of Synthetic Polymers in Battery Separators

The application of a polymer separator in the battery is mainly reflected in its role as the "third electrode" of the battery, the main role is to block the cathode and anode electrode contact of the battery and effectively transport lithium-ions. Polymer separators need to have good dimensional stability, high porosity, excellent mechanical properties, good wettability, high chemical stability, and excellent heat and flame-retardant properties. Commonly used polymer separators include PO (such as PP and PE), PVDF, PAN, PA, and so on. These materials are prepared into nanofiber films by electrospinning and other processes to improve the ionic conductivity and electrochemical performance of batteries. In lithium-ion batteries, the application of polymer separators not only improves the safety and electrochemical stability of the battery, but also promotes the efficient transport of lithium-ions through its high porosity and excellent wettability, thereby improving the overall performance and cycle life of the battery. The application of polymer separators in batteries is the key to improving the performance and safety of batteries, through the selection of appropriate materials and fine processes, can effectively solve the shortcomings of traditional PO separators, to meet the development needs of high-performance lithium-ion batteries.

3B.5.1 Polyethylene Terephthalate

PET separator is a material with excellent physical properties and chemical stability [61]. PET separators can withstand higher temperatures without shrinkage or deformation due to their high thermal stability, which helps to improve the performance of the battery in high-temperature environments. It has good mechanical properties, including high tensile strength and good toughness, which helps to maintain the structural integrity of the battery. It has good electrical insulation properties, which can effectively isolate cathode and anode electrodes and prevent short circuits. Compared with other high-performance separator materials, the production cost of PET separators is relatively low, which is conducive to reducing the overall cost of the batteries. However, the porosity of the PET separator is relatively low, which may affect the efficiency of ion transport. The wetting of the electrolyte is relatively poor, which may affect the cycle life of the batteries. Despite the relatively low cost of PET separators, their manufacturing process is complex and requires specific processing techniques, which may limit their large-scale application.

Al_2O_3/PET composite separator has excellent thermal stability, but the Al_2O_3 particle shedding hinders the practical application of the separator. Cai et al. [62] developed a highly stable Al_2O_3/PET composite film by dipping the photoactive oligomer PEGDA into the Al_2O_3 coating and then in situ polymerization under ultraviolet UV irradiation. The composite separator exhibited outstanding thermal stability and good electrochemical performance. Chen et al. [63] used ion irradiation technology to prepare a novel PET separator with vertically aligned nanochannels with uniform channel size and distribution. This new PET separator has excellent electrolyte wettability, a high Li$^+$ transport number, and excellent thermal stability. Li et al. [64] studied the application of a two-sided asymmetric separator composed of PET-PTFE fiber in LMBs. The results showed that the cycle life of LMB can be significantly extended when the rigid PET side contacts the anode and the soft PTFE side contacts the cathode. The design strategy of asymmetric isolation separators has great potential in preparing high-performance separators for long-cycle-life LMBs.

PET battery separators show broad development prospects in the Li battery industry because of their cost effectiveness, good performance, and environmental friendliness. In the future, with further innovation in technology and further cost reduction, PET separators are expected to occupy a more important position in the Li battery market.

3B.5.2 Polyimide

PI separators have good thermal dimensional stability due to outstanding high-temperature resistance, and can be used at 300 °C for a long time, greatly improving the high-temperature use safety of the battery. The abundant polar groups in PI molecular structure can improve the infiltrability of the electrolyte, and help to improve the interface performance between the separator/electrolyte and the overall performance of the battery. In addition, PI materials have flame-retardant and self-extinguishing characteristics, providing a more powerful security guarantee for lithium-ion batteries. However, the film-forming process of PI material is difficult, and there are problems such as low productivity, poor reproducibility, and environmental pollution in the production process of the electrostatic spinning method, which are still facing many bottlenecks in industrial-scale manufacturing.

By combining chemical imide and freeze-drying, Jeong et al. [65] prepared a dry gel separator for lithium battery by hybridizing the surface-modified silica nanoparticle hollow microspheres (AHMS) with PI, which improved the heat resistance and ion permeability of the separator. Polyamide acid (PAA) was prepared by the reaction of diphenyl tetanic anhydride BPDA with diamine ODA, and an appropriate amount of chemically modified AHMS microspheres were added, and then the mixed solution was chemically imitated and freeze-dried to obtain PI-AHMS separator. The separator showed high porosity and uniform pore size, and had high ion conductivity and high affinity to the electrolyte, and the battery based on this separator showed high discharge capacity and high rate performance. Wang et al. [66] prepared the electrostatic spun PI nanofiber separator with uniform small pore size by blending PAA and polystyrene (PS). The fused PS phase can partially dissolve the PAA phase and cross-link the nanofibers with each other when the electrospun film is thermimidized at high temperature. The pore size and uniformity of nanofiber separators were controlled, and the PI separators with an average pore size of 0.78 µm and a porosity of 81% were obtained. Therefore, the PI nanofiber separator had good electrolyte absorption and retention ability, and the lithium battery assembled with it showed good rate performance and cycling stability under high current density.

Guo et al. [67] reported a mesoporous PI separator that can inhibit the formation of sharp lithium dendrites, based on its high energy storage modulus of 1.80 GPa. The mesoporous PI separator was prepared by slow pyrolysis of polylactic-polyimide-polylactic acid triblock copolymer at 280 °C. Polylactic acid was gradually removed by slow pyrolysis, and mesoporous pores of 21 nm were produced without disturbing the PI matrix. The combination of mesoporous structure and high modulus of mesoporous PI separator can effectively inhibit the formation of sharp lithium dendrites. Mu et al. [68] used electrospinning combined with laser irradiation to construct an integrated multifunctional separator consisting of a defective graphene (DG) layer and a PI layer. The abundant imide functional groups in PI nanofibers can promote the rapid diffusion of lithium-ions and induce the uniform deposition of lithium on the negative electrode surface. A large number of topological defects and vacancy defects in DG caused by transient thermal shock induced by laser are conducive to the capture and rapid reversible transformation of polysulfide lithium on the positive side. Moreover, the strong chemical interaction between PI layer and DG layer induces the electronic structure rearrangement of each other, which further strengthens the inhibition of the composite separator on the shuttle effect of lithium dendrites and polysulfide. Li—S batteries based on this DG-PI separator show excellent electrochemical performance and cycling stability.

Due to its excellent high-temperature resistance and electrolyte wettability, as well as the application of a variety of innovative technologies, PI separators are gradually becoming the preferred material for next-generation battery separators. Despite the challenges of cost and scale production, the industrialization of PI separators is expected to accelerate with formulation optimization, process innovation, and application practice.

3B.6 Conclusions and Outlook

This chapter reviews the application of synthetic polymers in battery auxiliary materials, including binders, electrolytes, separators, and anode functional coatings.

Polymer binders are an important part of rechargeable batteries, providing a guarantee for good contact and morphological integrity of the electrode structure, and are the basis for maintaining electron/ion transfer during the battery cycle. Traditional binders, such as PVDF, are difficult to meet the application requirements of high-energy-density batteries with thick electrodes and/or high capacity electrodes due to their low adhesion, poor mechanical strength, and single function. With the importance of bonding materials to the overall performance of the battery has been recognized, through the continuous efforts of researchers, binder technology has made great progress in recent years, and various types of binder systems have been proposed and studied. One of the main challenges facing current binder systems is to cope with the large volume changes in some electrode systems during charging and discharging.

Large system changes of electrodes are prone to causing electrode cracking and morphological changes, which usually lead to decreased performance. Optimizing the properties of binders without increasing the number of binders is an effective strategy to solve these problems. Improving the flexibility of the binder and maintaining the mechanical toughness are conducive to maintaining the cyclic stability of the electrode. The design of functional groups in polymer binders, the combination of different polymer binders, and the introduction of organic or inorganic functional components to modify the binder have been proved to be effective methods to improve the comprehensive properties of binders. With the improvement of people's environmental awareness, binders are required to reduce the use of organic solvents while ensuring binder performance, and the development of water-soluble or solvent-free binders is the future development trend.

Electrolytes are an important part of conducting ions between positive and negative electrodes in lithium-ion batteries and are crucial to the performance of the entire battery. Traditional liquid electrolytes have safety risks and low energy density, and new solid-state electrolytes need to be developed to solve these problems. Spes are widely regarded for their strong molecular structure designability, flexibility, light weight, low cost, easy manufacturing, and no liquid leakage. In practical applications, the main performance requirements of SPE are as follows: high ion conductivity and cation migration number, good electronic insulation, excellent mechanical strength, and high thermal and chemical stability. The traditional linear polymer electrolyte (such as PEO) is easy to crystallize at room temperature and hinders ion transport, and the high-temperature melting state will react with the interface and affect the stability. Starting from the design of molecular structure, a new effective solution of SPE is prepared. The functional groups with different structures can integrate different functions such as self-healing and flame retardancy. In terms of improving the conduction ion efficiency, the current research is mainly achieved by adding plasticizers. In the future, more attention should be paid to the design of molecular structure and the regulation of composition, and the use of small molecular compounds should be reduced to achieve a true all-solid polymer system. In addition, SPE has good processability, and it is necessary to develop new and efficient large-scale SPE preparation and processing methods.

Polymer separators need to have good dimensional stability, high porosity, excellent mechanical properties, good wettability, high chemical stability, and excellent heat and flame-retardant properties. In lithium-ion batteries, the application of polymer separators not only improves the safety and electrochemical stability of the battery, but also promotes the efficient transport of lithium-ions through its high porosity and excellent wettability, thereby improving the overall performance and cycle life of the battery. The application of polymer separators in batteries is the key to improving the performance and safety of batteries, through the selection of appropriate materials and fine processes, which can effectively solve the shortcomings of traditional PO separators, to meet the development needs of high-performance lithium-ion batteries. Since each polymer separators have its own disadvantages, blends or copolymers are usually used, or inorganic additives are used to improve the physical properties of the separator. Surface coating modification, blending modification, gel filling, and crosslinking modification are commonly used polymer separator modification methods. These modification methods can effectively improve the performance of polymer separators, including improving thermal stability, improving electrolyte wettability, enhancing mechanical strength, etc., so as to meet the higher requirements of lithium batteries for separators. As the core component of lithium batteries, the development trend of polymer separators is mainly focused on improving safety, thermal stability, chemical stability, and reducing production costs.

With the application of secondary batteries becoming more and more widespread, in order to reduce pollution and save resources, the recycling demand for failed battery components is becoming higher and higher. In the future, the design and development direction of polymer functional materials for batteries should consider more characteristics such as green pollution-free, biodegradable, recyclable, and sustainable.

References

1 Saal, A., Hagemann, T., and Schubert, U.S. (2020). Polymers for battery applications—active materials, membranes, and binders. *Advanced Energy Materials* 11: 2001984.
2 Zou, F. and Manthiram, A. (2020). A review of the design of advanced binders for high-performance batteries. *Advanced Energy Materials* 10: 2002508.
3 Srivastava, M., Anil Kumar, M.R., and Zaghib, K. (2024). Binders for Li-ion battery technologies and beyond: a comprehensive review. *Batteries* 10: 268.
4 He, Q., Ning, J., Chen, H. et al. (2024). Achievements, challenges, and perspectives in the design of polymer binders for advanced lithium-ion batteries. *Chemical Society Reviews* 53: 7091–7157.
5 Wang, Y.-B., Yang, Q., Guo, X. et al. (2021). Strategies of binder design for high-performance lithium-ion batteries: a mini review. *Rare Metals* 41: 745–761.
6 Yan, L., Gao, X., Wahid-Pedro, F. et al. (2018). A novel epoxy resin-based cathode binder for low cost, long cycling life, and high-energy lithium-sulfur batteries. *Journal of Materials Chemistry A* 6: 14315–14323.
7 Zhang, T., Li, B., Song, Z. et al. (2023). Ten-minute synthesis of a new redox-active aqueous binder for flame-retardant Li-S batteries. *Energy & Environmental Materials* 7: e12572.
8 Yu, L., Tao, B., Ma, L. et al. (2024). A robust network sodium carboxymethyl cellulose-epichlorohydrin binder for silicon anodes in lithium-ion batteries. *Langmuir* 40: 17930–17940.
9 Wang, Y.X., Xu, Y., Meng, Q. et al. (2016). Chemically bonded Sn nanoparticles using the crosslinked epoxy binder for high energy-density Li ion battery. *Advanced Materials Interfaces* 3: 1600662.

10 Liao, J., Liu, Z., Liu, X. et al. (2018). Water-soluble linear poly(ethylenimine) as a superior bifunctional binder for lithium-sulfur batteries of improved cell performance. *The Journal of Physical Chemistry C* 122: 25917–25929.

11 Huang, J., Wang, J., Zhong, H. et al. (2018). N-cyanoethyl polyethylenimine as a water-soluble binder for LiFePO$_4$ cathode in lithium-ion batteries. *Journal of Materials Science* 53: 9690–9700.

12 Liu, Z., He, X., Fang, C. et al. (2020). Reversible crosslinked polymer binder for recyclable lithium sulfur batteries with high performance. *Advanced Functional Materials* 30: 2003605.

13 Jin, B., Cui, Z., and Manthiram, A. (2023). In situ interweaved binder framework mitigating the structural and interphasial degradations of high-nickel cathodes in lithium-ion batteries. *Angewandte Chemie International Edition* 62: 2301241.

14 Oh, J., Choi, S.H., Chang, B. et al. (2022). Elastic binder for high-performance sulfide-based all-solid-state batteries. *ACS Energy Letters* 7: 1374–1382.

15 Kim, J.-O., Kim, E., Lim, E.Y. et al. (2024). Stress-dissipative elastic waterborne polyurethane binders for silicon anodes with high structural integrity in lithium-ion batteries. *ACS Applied Energy Materials* 7: 1629–1639.

16 Sun, X., Lin, X., Wen, Y. et al. (2024). A water-soluble binder in high-performance silicon-based anodes for lithium-ion batteries based on sodium carboxymethyl cellulose and waterborne polyurethane. *Green Chemistry* 26: 9874–9887.

17 Nimkar, A., Bergman, G., Ballas, E. et al. (2023). Polyimide compounds for post-lithium energy storage applications. *Angewandte Chemie International Edition* 62: 2306904.

18 Zhang, M., Wang, L., Xu, H. et al. (2023). Polyimides as promising materials for lithium-ion batteries: a review. *Nano-Micro Letters* 15: 135.

19 Wang, Y., Dong, N., Liu, B. et al. (2022). Enhanced electrochemical performance of the LiNi$_{0.8}$Co$_{0.1}$Mn$_{0.1}$O$_2$ cathode via in-situ nanoscale surface modification with poly(imide-siloxane) binder. *Chemical Engineering Journal* 450: 137959.

20 Wang, Y., Dong, N., Liu, B. et al. (2023). Self-adaptive gel poly(imide-siloxane) binder ensuring stable cathode-electrolyte interface for achieving high-performance NCM811 cathode in lithium-ion batteries. *Energy Storage Materials* 56: 621–630.

21 Lusztig, D., Luski, S., Shpigel, N. et al. (2024). Silicon anodes for lithium-ion batteries based on a new polyimide binder. *Batteries & Supercaps* 7: e202400255.

22 Tan, W., Liang, B., Chen, M. et al. (2024). Rigid-flexible mediated co-polyimide enabling stable silicon anode in lithium-ion batteries. *Chemical Engineering Journal* 496: 153822.

23 Li, M., Li, H., Lan, J.-L. et al. (2018). Integrative preparation of mesoporous epoxy resin-ceramic composite electrolytes with multilayer structure for dendrite-free lithium metal batteries. *Journal of Materials Chemistry A* 6: 19094–19106.

24 Zekoll, S., Marriner-Edwards, C., Hekselman, A.K.O. et al. (2018). Hybrid electrolytes with 3D bicontinuous ordered ceramic and polymer microchannels for all-solid-state batteries. *Energy & Environmental Science* 11: 185–201.

25 Song, Y.H., Kim, T., and Choi, U.H. (2020). Tuning morphology and properties of epoxy-based solid-state polymer electrolytes by molecular interaction for flexible all-solid-state Supercapacitors. *Chemistry of Materials* 32: 3879–3892.

26 Liang, L., Yuan, W., Chen, X. et al. (2021). Flexible, nonflammable, highly conductive and high-safety double cross-linked poly(ionic liquid) as quasi-solid electrolyte for high performance lithium-ion batteries. *Chemical Engineering Journal* 421: 130000.

27 Zeng, Z., Chen, X., Sun, M. et al. (2021). Nanophase-separated, elastic epoxy composite thin film as an electrolyte for stable lithium metal batteries. *Nano Letters* 21: 3611–3618.

28 Manarin, E., Corsini, F., Trano, S. et al. (2022). Cardanol-derived epoxy resins as biobased gel polymer electrolytes for potassium-ion conduction. *ACS Applied Polymer Materials* 4: 3855–3865.

29 Wang, Z., Shi, H., Zheng, W. et al. (2022). One-step preparation of epoxy resin-based ionic gel electrolyte for quasi-solid-state lithium metal batteries. *Journal of Power Sources* 524: 231070.

30 Zhang, J., Yan, J., Zhao, Y. et al. (2023). High-strength and machinable load-bearing integrated electrochemical capacitors based on polymeric solid electrolyte. *Nature Communications* 14: 64.

31 Lv, Z., Tang, Y., Dong, S. et al. (2022). Polyurethane-based polymer electrolytes for lithium batteries: advances and perspectives. *Chemical Engineering Journal* 430: 132659.

32 Andersson, R., Hernández, G., See, J. et al. (2022). Designing polyurethane solid polymer electrolytes for high-temperature lithium metal batteries. *ACS Applied Energy Materials* 5: 407–418.

33 Gao, Y., Wang, C., Wang, H. et al. (2023). Polyurethane/LLZTO solid electrolyte with excellent mechanical strength and electrochemical property for advanced lithium metal battery. *Chemical Engineering Journal* 474: 145446.

34 Huang, D., Wu, L., Kang, Q. et al. (2024). Amino-modified UiO-66-NH$_2$ reinforced polyurethane based polymer electrolytes for high-voltage solid-state lithium metal batteries. *Nano Research* 17: 9662–9670. https://doi.org/10.1007/s12274-024-6886-9

35 Hou, P., Gao, C., Wang, J. et al. (2023). A semi-transparent polyurethane/porous wood composite gel polymer electrolyte for solid-state supercapacitor with high energy density and cycling stability. *Chemical Engineering Journal* 454: 139954.

36 Shi, X., Jia, Z., Wang, D. et al. (2024). Phonon engineering in solid polymer electrolyte toward high safety for solid-state lithium batteries. *Advanced Materials* 36: 2405097.

37 Wang, H., Li, X., Zeng, Q. et al. (2024). A novel hyperbranched polyurethane solid electrolyte for room temperature ultra-long cycling lithium-ion batteries. *Energy Storage Materials* 66: 103188.

38 Wu, L., Pei, F., Cheng, D. et al. (2023). Flame-retardant polyurethane-based solid-state polymer electrolytes enabled by covalent bonding for lithium metal batteries. *Advanced Functional Materials* 34: 2310084.

39 Pei, F., Wu, L., Zhang, Y. et al. (2024). Interfacial self-healing polymer electrolytes for long-cycle solid-state lithium-sulfur batteries. *Nature Communications* 15: 351.

40 Yan, S., Wang, Z., Liu, F. et al. (2023). Aromatic donor-acceptor charge-transfer interactions reinforced supramolecular polymer electrolyte for solid-state lithium batteries. *Advanced Functional Materials* 33: 2303739.

41 Su, Y., Xu, F., Zhang, X. et al. (2023). Rational design of high-performance PEO/ceramic composite solid electrolytes for lithium metal batteries. *Nano-Micro Letters* 15: 82.

42 Jia, Z., Liu, Y., Li, H. et al. (2024). In-situ polymerized PEO-based solid electrolytes contribute better Li metal batteries: challenges, strategies, and perspectives. *Journal of Energy Chemistry* 92: 548–571.

43 Li, J., Hu, H., Fang, W. et al. (2023). Impact of fluorine-based lithium salts on SEI for all-solid-state PEO-based lithium metal batteries. *Advanced Functional Materials* 33: 2303718.

44 Xiong, Z., Wang, Z., Zhou, W. et al. (2023). 4.2 V polymer all-solid-state lithium batteries enabled by high-concentration PEO solid electrolytes. *Energy Storage Materials* 57: 171–179.

45 He, C., Ying, H., Cai, L. et al. (2024). Tailoring stable PEO-based electrolyte/ electrodes interfaces via molecular coordination regulating enables 4.5 V solid-state lithium metal batteries. *Advanced Functional Materials* 34: 2410350.

46 Han, S., Wen, P., Wang, H. et al. (2023). Sequencing polymers to enable solid-state lithium batteries. *Nature Materials* 22: 1515–1522.

47 Rong, Z., Sun, Y., Yang, M. et al. (2023). How the PEG terminals affect the electrochemical properties of polymer electrolytes in lithium metal batteries. *Energy Storage Materials* 63: 103066.

48 Li, R., Hua, H., Yang, X. et al. (2024). The deconstruction of a polymeric solvation cage: a critical promotion strategy for PEO-based all-solid polymer electrolytes. *Energy & Environmental Science* 17: 5601–5612.

49 Cheng, Y., Cai, Z., Xu, J. et al. (2024). Zwitterionic cellulose-based polymer electrolyte enabled by aqueous solution casting for high-performance solid-state batteries. *Angewandte Chemie International Edition* 63: 2400477.

50 Li, J., Cai, Y., Zhang, F. et al. (2023). Exceptional interfacial conduction and LiF interphase for ultralong life PEO-based all-solid-state batteries. *Nano Energy* 118: 108985.

51 Li, X., Feng, J., Li, Y. et al. (2024). Regulating Li$^+$ transport behavior by cross-scale synergistic rectification strategy for dendrite-free and high area capacity polymeric all-solid-state lithium batteries. *Energy Storage Materials* 72: 103107.

52 Zhou, M., Cui, K., Wang, T.-S. et al. (2024). Bimetal fluorides with adjustable vacancy concentration reinforcing ion transport in poly(ethylene oxide) electrolyte. *ACS Nano* 18: 26986–26996.

53 An, H., Li, M., Liu, Q. et al. (2024). Strong Lewis-acid coordinated PEO electrolyte achieves 4.8 V-class all-solid-state batteries over 580 Wh kg^{-1}. *Nature Communications* 15: 9150.

54 Cui, Y., Wan, J., Ye, Y. et al. (2020). A fireproof, lightweight, polymer-polymer solid-state electrolyte for safe lithium batteries. *Nano Letters* 20: 1686–1692.

55 Li, Y., Fu, Z., Lu, S. et al. (2022). Polymer nanofibers framework composite solid electrolyte with lithium dendrite suppression for long life all-solid-state lithium metal battery. *Chemical Engineering Journal* 440: 135816.

56 Huang, Y., Liu, S., Chen, Q. et al. (2022). Constructing highly conductive and thermomechanical stable quasi-solid electrolytes by self-polymerization of liquid electrolytes within porous polyimide nanofiber films. *Advanced Functional Materials* 32: 2201496.

57 Ma, Y., Jiao, Y., Yan, Y. et al. (2022). Polyimide-reinforced solid polymer electrolyte with outstanding lithium transferability for durable Li metal batteries. *Journal of Power Sources* 548: 232034.

58 Wang, Z., Sun, J., Liu, R. et al. (2023). Thin solid polymer electrolyte with high-strength and thermal-resistant via incorporating nanofibrous polyimide framework for stable lithium batteries. *Small* 19: 2303422.

59 Zhang, H., Chen, J., Liu, J. et al. (2023). Gel electrolyte with flame retardant polymer stabilizing lithium metal towards lithium-sulfur battery. *Energy Storage Materials* 61: 102885.

60 Wang, X., Xu, L., Li, M. et al. (2024). LiNO$_3$ regulated rigid-flexible-synergistic polymer electrolyte boosting high-performance Li metal batteries. *Energy Storage Materials* 73: 103778.

61 Adamson, A., Tuul, K., Bötticher, T. et al. (2023). Improving lithium-ion cells by replacing polyethylene terephthalate jellyroll tape. *Nature Materials* 22: 1380–1386.

62 Cai, B.-R., Cao, J.-H., Liang, W.-H. et al. (2021). Ultraviolet-cured Al$_2$O$_3$-polyethylene terephthalate/polyvinylidene fluoride composite separator with asymmetric design and its performance in lithium batteries. *ACS Applied Energy Materials* 4: 5293–5303.

63 Chen, L., Gui, X., Zhang, Q. et al. (2023). Direct fabrication of PET-based thermotolerant separators for lithium-ion batteries with ion irradiation technology. *ACS Applied Materials & Interfaces* 15: 59422–59431.

64 Li, J., Gao, Y., Duan, M. et al. (2024). Influence of the PET-PTFE separator pore structure on the performance of lithium metal batteries. *ACS Applied Materials & Interfaces* 16: 34902–34912.

65 Jeong, T.-Y., Lee, Y.D., Ban, Y. et al. (2021). Polyimide composite separator containing surface-modified hollow mesoporous silica nanospheres for lithium-ion battery application. *Polymer* 212: 123288.

66 Wang, Y., Guo, M., Fu, H. et al. (2022). Thermotolerant separator of cross-linked polyimide fibers with narrowed pore size for lithium-ion batteries. *Journal of Membrane Science* 662: 121004.

67 Guo, D., Mu, L., Lin, F. et al. (2023). Mesoporous polyimide thin films as dendrite-suppressing separators for lithium-metal batteries. *ACS Nano* 18: 155–163.

68 Mu, J., Zhang, M., Li, Y. et al. (2023). Laser irradiation constructing all-in-one defective graphene-polyimide separator for effective restraint of lithium dendrites and shuttle effect. *Nano Research* 16: 12304–12314.

4

Application of Nontraditional Organic Ionic Conductors in Batteries

4.1 Ionic Liquids

4.1.1 Introduction of Ionic Liquids

Traditional organic electrolytes are liquid organic electrolytes, which have some significant shortcomings in lithium-ion batteries (LIBs). They usually have low thermal stability and are easy to decompose at high temperatures to produce gas and heat, resulting in safety hazards. They are prone to leakage, increasing the safety risk of the battery during use. At high temperatures or overcharged states, they may decompose and produce gas, increasing the internal pressure of the battery, and causing the battery to expand or even rupture. Side reactions may occur on the electrode surface, forming solid electrolyte interface (SEI) layer instability, affecting the cycle life and safety of the battery. Some organic solvents are harmful to the environment, are not easy to degrade, and may cause pollution. These shortcomings limit the use of conventional organic electrolytes in certain applications with high safety requirements, so researchers are developing new organic electrolytes to provide higher safety and better ionic conductivity.

Ionic liquids (ILs) are a special type of liquid consisting mainly of positive and negative ions, which combine to form liquid compounds by coulomb forces. This liquid usually remains liquid at or near room temperature, so it is also called room-temperature ILs or room-temperature molten salt. Due to the large size difference between the anions and cations in their molecules, their structure is relatively loose, resulting in weak ionic interactions. This allows them to have a lower melting point than other salt compounds and exhibit liquid-like flow properties even at room temperature. They are considered to be the third class of solvents after water and organic solvents, with unique properties such as nonvolatility, high thermal stability, and high ionic conductivity. However, these properties are influenced by the Lewis pH of the cation/anion (i.e. the Coulomb interaction), the directivity of the interaction with the cation and anion, and the van der Waals interaction between the ions.

Due to their strong affinity for many compounds, ILs have been mainly used as a catalytic and synthetic solvent for a long time. Compared to other organic solvents, ILs have a very wide and stable liquid range, while most liquid liquids

Functional Auxiliary Materials in Batteries: Synthesis, Properties, and Applications, First Edition. Wei Hu.
© 2025 WILEY-VCH GmbH. Published 2025 by WILEY-VCH GmbH.

have a temperature range of about 90–400 °C. In addition, ILs have a very low vapor pressure, high thermal stability, and nonvolatility compared to common organic compounds, making it more suitable as a solvent. Another good feature of ILs is their nonflammability. The use of noncombustible organic solvents instead of traditional combustible organic solvents is an effective way to improve the safety of laboratory and chemical production processes, and their reusability can also indirectly save costs. ILs have a strong solubility, allowing a wide range of organic, inorganic, metallic, and polymer compounds to dissolve to a large extent, thereby extending their service range.

In addition, ILs have strong catalytic activity and can be recycled as a catalyst in organic synthesis to achieve homogeneous catalysis [1]. Gases such as hydrogen, oxygen, carbon monoxide, and carbon dioxide can also be dissolved in ILs, making them suitable for catalytic hydrogenation and hydroformylation reactions. They typically exhibit moderate viscosity, often higher than organic solvents, which allows them to be used as stationary phases in high-performance liquid chromatography. ILs also have good electrical conductivity and a relatively wide electrochemical window, allowing them to be electrolyzed at room temperature and used as an electrolyte in electrochemical research [2]. These characteristics have opened up a wide range of application fields for ILs [3, 4].

In summary, ILs have become an important field in scientific research and industrial applications because of their unique properties and wide application potential. With the development of research and technology, the application range of ILs will continue to expand.

4.1.2 Development of Ionic Liquids

ILs refer to organic salts that are completely composed of positively charged ions and negatively charged ions, typically existing in liquid form at room temperature. Theoretically, there could be trillions of different ILs, allowing scientists to select the specific IL needed for their work. The development of ILs can be summarized in the following stages [4].

Early explorations (1914–1948): In 1914, P. Walden reported that ethylamine nitrate ($[E_tNH_3][NO_3]$) with a melting point of 12 °C was the first organic salt to be liquid at room temperature, but it did not attract much attention due to its unstable and explosive properties. In 1948, Hurley and Wier added alkyl pyridine to $AlCl_3$, mixed and heated to become a colorless transparent liquid, that is, chloroaluminate IL, which laid the foundation for the study of ILs.

Functionalization and stability improvement (1976–1992): In 1976, Robert of Colorado State University in the United States found that the $AlCl_3/[n\text{-EtPy}]Cl$ electrolyte used room-temperature IL with good performance, can be miscible with organic matter, does not contain protons, and has a wide electrochemical window. In 1992, J.S. Wilkes et al. synthesized dialkyl imidazoles tetrafluoroboric acid and hexafluorophosphate plasma liquids, which have strong stability and water resistance, marking the birth of the second generation of water-resistant ILs.

Wide application and functionalization (1992–present): Since 1992, the research and application of ILs have expanded rapidly, including catalysis, electrochemistry, functional materials, and other fields. Researchers are committed to developing functional ILs, such as chiral ILs, acidic ILs, and alkaline ILs, to meet specific application needs.

Since the twenty-first century, the research of ILs has reached a new milestone, new ILs continue to emerge, the application field continues to expand, and the technology has become a new technology with great development prospects [5–8].

This evolution shows the transformation of ILs from initial exploration to functionalization and then to widespread application, providing powerful tools for scientific research and industrial applications [9]. The research and development of ILs align with the current advocacy for clean technologies and sustainable development, and they are increasingly recognized and accepted by the scientific community [10, 11].

4.1.3 Catalog of Ionic Liquids

ILs come in various types, and they can be categorized into three main classes based on the different combinations of anions and cations: organic cation–organic anion, organic cation–inorganic anion, and inorganic cation–inorganic anion. The most commonly used organic cations in ILs typically exhibit low symmetry and larger sizes, such as various alkyl-substituted imidazolium ions, pyridinium ions, and quaternary ammonium ions, structures of some typical cations and anions of ILs as shown in Figure 4.1 [12].

ILs can be divided into two categories based on the type of anion present. One category consists of aluminum chloride-based ILs, whose components can vary. The anions in this type of IL are typically $AlCl_4^-$ and $Al_2Cl_7^-$. When the molar fraction of $AlCl_3$ is ($x = 0.5$), the solution is neutral; when ($x < 0.5$), it is basic; and when ($x > 0.5$), it is acidic. Mixing solid halides with $AlCl_3$ usually yields a liquid IL, but the preparation process is highly exothermic. Therefore, a small amount of each solid is typically added alternately to the already prepared IL to facilitate heat dissipation. This type of IL has many advantages, but it is extremely sensitive to water and must be handled and applied under vacuum or inert atmospheres. The presence of proton and oxide impurities can have a decisive impact on the chemical reactions conducted in this type of IL. Moreover, when $AlCl_3$ encounters water, it releases HCl gas, which can be irritating to the skin.

The other category, also known as new ILs, has fixed compositions and can generally remain stable in air and water. The anions that primarily compose this type of IL include PF_6^-, BF_4^-, $CF_3SO_3^-$, $(CF_3SO_3)_2N^-$, and AsF_6^-. Additionally, functional groups with specific properties can be introduced to the cations or anions of ILs through organic chemical reactions such as substitution and addition, resulting in functionalized ILs. The introduced functional groups mainly include —NH_2, —OH, —SO_2H, and —CH_2OCH_3, and the positions of the cations and anions can also be adjusted to obtain new types of ILs.

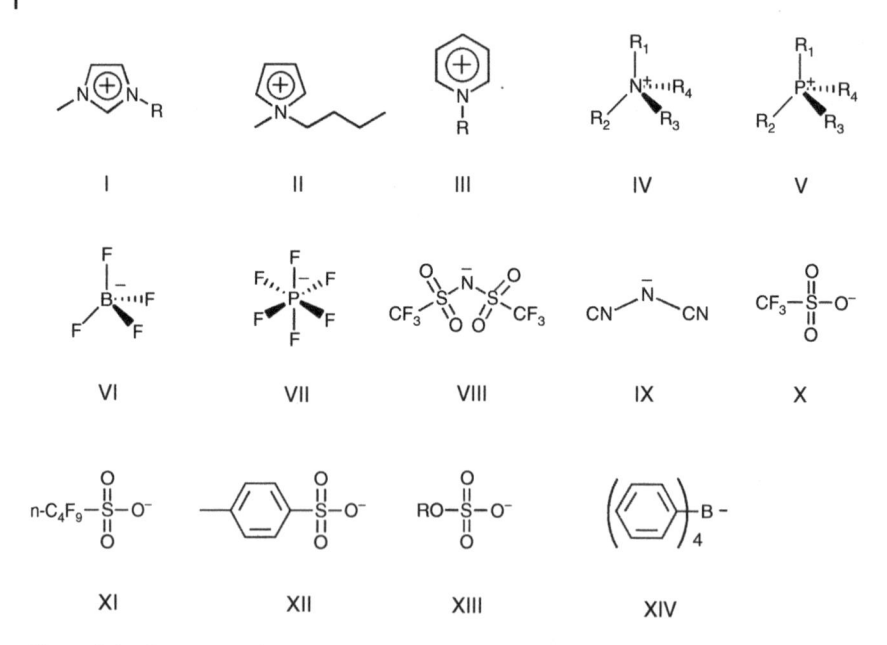

Figure 4.1 Structures of some typical cations and anions of ILs: I: 1-alkyl-3-methylimidazolium (Cnmim+); II: 1-butyl-1-methylpyrrolidinium (CnPyr+); III: 1-alkylpyridinium (CnPy+); IV: tetraalkylammonium; V: tetraalkylphosphonium; VI: tetrafluoroborate (BF−); VII: hexafluorophosphate (PF$_6^-$); VIII: bis(trifluoromethylsulfonyl)imide (Tf$_2$N−); IX: dicyanimide ((CN)$_2$N−); X: trifluoromethanesulfonate (TfO−); XI: nonafluorobutane sulfonate (NfO−); XII: tosylate (OTos−); XIII: alkylsulfate (CnOSO$_3^-$); XIV: tetraphenylborate (TPhB−). Source: Cui et al. [12]/with permission of John Wiley & Sons.

In addition, the anions, whether organic or inorganic, can be divided into mononuclear and polynuclear categories. Mononuclear anions are usually basic or neutral ions, including BF_4^-, PF_6^-, HSO_4^-, NO_3^-, CH_3COO^-, and X^-. Polynuclear anions, on the other hand, are generally unstable in water and air and include $Al_2Cl_7^-$, $Al_3Cl_{10}^-$, $Au_2Cl_7^-$, $Fe_2Cl_7^-$, and $Sb_2F_{11}^-$, among others. By varying the combinations of anions and cations, it is possible to design trillions of different ILs.

Based on the type of cation, ILs can be further divided into four categories: quaternary ammonium salts, quaternary phosphonium salts, alkyl-substituted imidazolium salts, and pyridinium salts. Among these, the most commonly used cations include: alkyl quaternary ammonium cations, represented as NR_4^+; alkyl quaternary phosphonium cations, represented as PR_4^+; alkyl-substituted imidazolium cations, such as 1-butyl-3-methylimidazolium cation (Bmim+) and 1-ethyl-3-methylimidazolium cation (Emim+); alkyl-substituted pyridinium cations, represented as Rpy+.

Furthermore, ILs also can be classified based on their structures into several categories: functionalized ILs, chiral ILs, switchable polarity solvents ILs, bio-ILs, poly-ILs, high-energy ILs, neutral ILs, acidic ILs, basic ILs, protonic ILs, metal ILs, and supported ILs.

4.1.4 Advantages of Ionic Liquids for Batteries

ILs have wide application potential in battery technology due to their unique physical and chemical properties [13, 14]. The characteristics of IL as a battery material mainly include:

1) High thermal and chemical stability: ILs are able to maintain their structure at higher temperatures and do not burn, which makes them safer for use in batteries.
2) High ionic conductivity: ILs have high ionic conductivity, which means that they transport ions efficiently, which is crucial for the performance of batteries.
3) Wide electrochemical window: The electrochemical window of ILs is wide, and the antioxidant/antireduction window of the electrolyte can be adjusted by adjusting the type of ions and even widened to >7 V.
4) Nonvolatile and noncombustible: The nonvolatile and noncombustible characteristics of IL reduce the safety risks during use and also avoid pollution to the environment.
5) Good solubility: ILs have good solubility to many inorganic salts and organics, which allows them to flexibly adjust their composition and structure in battery manufacturing.

These advantages make ILs an important research direction in the field of battery materials, although they also have some limitations, such as high viscosity and adverse effects on the diffusion process of ions in the electrolyte and the infiltration between materials inside the battery, these challenges are being overcome through technological innovation and material modification.

4.1.5 Synthesis and Characterization Method of Ionic Liquids

There are various synthesis methods for ILs, commonly including:

1) Direct method: The corresponding cation and anion are directly mixed according to a certain molar ratio and the IL is formed after room temperature or heating.
2) Reactive ion exchange method: The solid mixture of two ions is mixed and the solvent is added to carry out the reactive ion exchange reaction to generate an IL.
3) In situ ion exchange method: The cation and anion are synthesized separately and then mixed in accordance with a specific proportion, under the conditions of heating or pressurization, the ion exchange reaction produces an IL.
4) Density functional theory (DFT) design method: According to the molecular structure and chemical properties, the appropriate IL is designed and synthesized by computer simulation.

In addition, researchers have developed new green chemistry methods such as microwave heating technology, ultrasonic radiation technology, and solvothermal methods. These techniques help shorten reaction times and improve reaction conversion efficiency and have been applied in both inorganic and organic synthesis [15, 16].

The characterization methods of ILs mainly include the following [17–19]:

1) Nuclear magnetic resonance (NMR) spectroscopy: Used to determine the structure and relative conformation of organic molecular groups in ILs.
2) Fourier transform infrared (FTIR) spectroscopy: Through the adsorption spectrum and vibration spectrum of ILs, information such as chemical bonds and molecular structures can be determined in ILs.
3) Differential scanning calorimetry (DSC): It can determine the heat capacity, melting point, and glass transition temperature of ILs.
4) Solubility and heat of solution: Can reflect the solubility of ILs and different solutes.

These methods can be used in combination to comprehensively characterize ILs from multiple perspectives to explore their internal structure and adsorption properties.

4.1.6 Application of Ionic Liquids

ILs exhibit many attractive characteristics. For instance, their physical and chemical properties can be adjusted through the design of cations and anions, making them ideal candidate materials. Numerous materials based on ILs have been developed through chemical modifications (covalent, coordination, or ionic functionalization) or physical mixing of ILs with other functional materials. These materials can be applied in areas such as chemical synthesis, electrochemistry, separation technology, lubricants, catalysts, drug delivery, and environmental protection [20, 21].

1) Chemical synthesis: IL, as a green solvent, can replace traditional organic solvents for various chemical reactions, such as polymerization, alkylation reaction, and acylation reaction. They not only act as solvents, but also provide an ionic environment different from conventional solvents, thus optimizing the reaction mechanism. For example, the synthesis of nanomaterials. The successful synthesis of metal nanoparticles in IL media has been achieved due to their ability to stabilize metal nanoparticles through the cations and/or anions of ILs, resulting in monodispersity and nonaggregation behavior. The physicochemical properties of ILs greatly influence the size distribution of metal nanoparticles. For instance, ILs with longer side chains favor the production of smaller diameter and narrower size distribution metal nanoparticles. An increase in side chain length alters the physicochemical properties of ILs as well as the degree of interaction between the liquid and the nanoparticles. Generally, larger nanoparticle sizes can be attributed to unstable metal nanoparticles aggregating due to strong Coulombic attraction with the smaller anions of ILs. Anions with fewer coordination sites in ILs tend to bind with metal nanoparticles, leading to larger and more unevenly distributed nanoparticle sizes.
2) Electrochemistry: ILs have important applications in the field of electrochemistry, where they can be used as electrolytes in energy-storage and conversion devices such as batteries, supercapacitors, and fuel cells. In addition, ILs are also

used in electrochemical sensors and electroanalytical chemistry, showing good conductivity and electrochemical stability. For example, replacing aqueous solution systems with IL systems during electroplating can reduce the waste of precious metals; toxic precious metals and semiconductor materials can be deposited from contaminated water, and by selecting appropriate cations and anions to add to the electrolyte, the electrochemical performance of energy-storage batteries can be improved [22–31]. ILs demonstrate unique advantages in the field of electrochemistry. Currently, various types of ILs, including quaternary ammonium salts, phosphonium salts, alkyl-substituted imidazolium salts, and pyridinium salts, have been further developed and widely applied in electrochemical fields.

3) Separation technology: Due to their good solubility and selectivity, ILs are widely used in liquid-liquid extraction, extraction chromatography, membrane separation, and other separation technologies. These technologies are used to separate and purify organic compounds, metal ions, and gases from complex mixtures. For example, imidazole ILs can be directly used for the extraction and separation of natural products due to their good stability and designability. Techniques such as ultrasonic extraction, microwave-assisted extraction, two-phase aqueous extraction, enzyme-assisted extraction, and solid-phase extraction have been applied to the extraction and separation of natural components such as flavonoids, terpenes, alkaloids, and pigments. The application of ILs in extraction separation is not limited to laboratory research, they are also widely used in industrial processes, such as the extraction of organic matter, metal ions, and distillation separation of azeotrope. These ILs become efficient separation solvents because of their excellent physical and chemical properties, such as low melting point, low saturated vapor pressure, nonvolatilization, and high thermal stability. The combination of ILs with advanced separation technologies, such as catalysts for reactive distillation and entrainment agents for extractive distillation, enhances the distillation process. In addition, ILs can also be prepared into supporting IL membranes, IL composite polymer membranes, etc., for membrane separation processes, these technologies provide efficient separation solutions for petrochemical and other industries.

In addition, ILs can absorb carbon dioxide, which allows them to separate carbon dioxide from other gases [32]. Traditional ILs absorb CO_2 through physical interactions, while functionalized ILs do so via chemical interactions. By adjusting the structure of the functional groups, the interaction between the functional sites and CO_2 can be modulated, enabling high-capacity CO_2 absorption, low-energy desorption, and the recycling of functionalized ILs.

4) Lubricants: Some ILs have excellent lubrication properties and can be used as an alternative to high-performance lubricants. They are used as friction improvers and antiwear additives for greases because of their low viscosity and ability to form a moderate friction protective layer on the surface of components. Through self-assembly technology and regulation of IL alkyl chain length, the scientists successfully prepared a single-layer ordered fringe structure of IL, which achieved a very low friction coefficient (order of 0.001) in the very low friction velocity range. This super lubricity provides a new possibility

for the development of micro and nano devices. For example, new ILs based on pyridine and pyridine cations show excellent performance in cycloalkyl greases, especially under high-temperature conditions, which can effectively improve the performance of greases and reduce wear. The IL is considered an environmentally friendly lubricant additive because of its low toxicity and excellent lubricity. For example, specific ionic liquid formulations have been developed that, while reducing wear losses, also ensure smooth and uniform lubrication films, showing good environmental friendliness. These applications not only demonstrate the extensive potential of ILs in the field of lubrication, but also provide new ideas and methods for solving lubrication problems in modern industry.

5) Catalyst: ILs can be used as a catalyst or a component of the catalytic reaction system to participate in organic synthesis, catalytic conversion, and catalytic hydrogenation reactions. In the field of catalysis, modifying the structural parameters of the cations and anions of ILs can be used to adjust the acid strength and the solubility of reactants when using ILs or ILs combined with inorganic acids as homogeneous catalysts. The immobilization of ILs is an effective method for regulating the adsorption/desorption properties and the distribution of acid strength on solid acid catalytic surfaces, offering significant potential to reduce catalyst deactivation and enhance the activity of solid acid catalysts.

6) Drug delivery: ILs can effectively increase the solubility of insoluble drugs. ILs can also promote the transdermal penetration of drugs by destroying the integrity of skin cells, making skin cells laminar fluid, and establishing diffusion channels. Hydrophilic ILs promote intercellular transport primarily by enhancing fluidization within protein and lipid regions, opening tight connections within the skin. However, hydrophobic ILs can be inserted into the skin lipid bilayer to disrupt the orderly arrangement of the phospholipid bilayer and provide channels to improve the distribution of epithelial cell membranes, thus promoting the transcellular transport of lipid regions. Some ILs themselves have antibacterial, anti-cancer, and other biological activities, and can be used as active ingredients of drugs; ILs can also be used to prepare special percutaneous dosage forms.

7) ILs are considered an environmentally friendly solvent due to their low volatility and high thermal stability. They have potential applications in environmental protection technologies such as wastewater treatment, gas adsorption, and carbon capture.

With the continuous design, synthesis, and research of new ILs, their application fields will be more and more extensive.

4.2 Application of ILs in Batteries

The application of ILs in batteries is mainly reflected in the following aspects:

1) Lithium battery electrolyte: ILs are widely used in lithium battery electrolytes because of their low volatility, noncombustible, high thermal stability, and

chemical stability, as well as wide electrochemical window and high ionic conductivity. These features help improve the safety and stability of the battery. For example, ILs can be used as solvents for lithium batteries, effectively solving the flammability and safety problems of traditional organic liquid electrolytes. In addition, by combining heat-stable organic solvents and lithium salts, as well as searching for low-viscosity ILs, the viscosity of ILs can be effectively reduced and its applicability in lithium battery electrolyte systems can be improved.

2) Solid state battery electrolyte: Solid state battery electrolytes based on ILs can effectively solve safety problems such as battery electrolyte leakage and flammability, but also improve the energy density of the battery. ILs can improve the ionic conductivity of solid electrolytes as the interface wetting agent between different component grain boundaries. For example, ILs can be used as interfacial infiltrators in solid-state lithium batteries (SSLBs), and through specific structural designs, such as polyionic liquid gel electrolytes based on hydroxyl function, excellent crystallization resistance, and electrochemical properties can be achieved at ultra-low temperatures (−80 °C). Utilizing the property of ILs being liquid at room temperature, incorporating ILs into polymer solid-state electrolytes can effectively reduce the crystallinity of the polymer, enhance the mobility of polymer chain segments, and provide conductive pathways to improve the ionic conductivity of the electrolyte. The role is similar to that of carbonate plasticizers, but compared to conventional carbonate plasticizers, ILs offer several advantages: (i) they can conduct electricity themselves and have good solubility for inorganic salts and polymers, significantly enhancing the ionic conductivity of polymer electrolytes; (ii) they have high thermal decomposition temperatures and are nonflammable, which effectively improves flame resistance compared to traditional organic electrolytes, reducing safety risks; (iii) they possess good chemical stability and stable electrochemical windows, allowing for a broader range of battery operating environments; (iv) they are nonvolatile, environmentally friendly, and align with the principles of green chemistry [33–37].

3) IL electrolyte additive: IL electrolyte additive has the characteristics of small amount and strong specificity, and which is considered to be an economical and effective additive to improve battery performance. These additives can be divided into three categories: (i) SEI/CEI film-forming additives; (ii) additives to improve high and low-temperature performance; and (iii) other functional additives. For example, the unique advantages of ILs can make them more suitable for complex battery systems, such as forming electrostatic shielding layers and removing trace amounts of water and HF.

However, ILs also have some obvious disadvantages when applied to batteries. They usually have a high viscosity, which leads to a decrease in ionic conductivity at low temperatures, affecting the cycle performance and charge and discharge rate of the battery. Although the conductivity of ILs can be improved in some cases by adding specific diluents or adjusting the anionic and cationic structures, in general, the conductivity of ILs is generally lower than that of conventional organic solvent electrolytes. Compared with traditional LIB electrolyte materials, the production cost of ILs is higher, which limits its promotion in large-scale applications. Although there are many issues related to the application of IL electrolytes, if these problems can be effectively resolved, LIBs will make significant advancements in safety performance and high-temperature environments. Consequently, many studies are dedicated to optimizing the performance of IL electrolytes.

The application forms of ILs in LIB electrolytes typically fall into three categories. First, pure ILs are used directly as solvents for lithium salts in LIBs. Second, they are mixed with organic solvents in specific ratios for use in LIB electrolytes. Additionally, ILs can serve as fillers in gel polymer electrolytes (GPEs) or be polymerized directly, resulting in the use of polymeric ILs in GPEs. The following sections will introduce the applications of IL electrolytes in LIBs based on these three application forms [38, 39].

4.2.1 Ionic Liquid Electrolyte

IL electrolytes have shown great development prospects in the field of LIBs, mainly due to their excellent thermal stability, low toxicity, nonflammability, and good electrochemical stability. The application research of IL electrolytes in batteries mainly focuses on the following aspects:

1) Improving safety and stability: Although IL electrolyte has many advantages, further improving their safety and stability is still the focus of research. This includes the development of novel IL derivatives to enhance the performance of batteries under extreme conditions.
2) Improving electrochemical performance: By adjusting the composition and structure of IL, its electrochemical performance can be further optimized, including improving the conductivity and improving the cycle stability of the battery. One of the research priorities is to develop IL with high ionic conductivity to support higher energy density battery designs.
3) Cost reduction: The high production cost of IL electrolytes is a major factor limiting their widespread use. By studying more efficient methods of IL synthesis and optimizing the production process, the production cost of IL electrolytes can be reduced, thus improving its market competitiveness.
4) Innovative applications: In addition to traditional LIB applications, IL electrolytes have the potential to find applications in other fields, such as fuel cells, supercapacitors, and solar cells. Exploring the application prospects of IL in these fields will be another important direction to promote its development.

Because the physical and chemical properties of ILs are closely related to their molecular structure, a variety of new ILs composed of different cations and ions have been prepared, and their characteristic properties are used to improve the performance of batteries. Basile et al. [40] achieved the goal of inhibiting the formation of lithium metal dendrites on the anode during cycling by immersing the lithium metal electrode before assembly into an IL electrolyte for a certain time. During this immersion process, a durable lithium-ion permeable SEI is formed on the lithium metal surface, enabling commercial lithium iron phosphate (LiFePO$_4$) batteries to safely perform for 1000 cycles with a coulomb efficiency >99.5%. The SEI layer was prepared by immersing the lithium metal anode (LMA) in a room-temperature IL based on n-propyl-n-methylpyridinium bis(fluorosulfonyl)imide containing lithium salts. It has been found that the formation of lithium-ions changes dynamically with time and the type of lithium salt, and LMA treated under optimal conditions can significantly inhibit the formation of dendrites and the consumption of electrolytes during the cycle. The formed SEI after pretreatment was composed of LiF, Li_2CO_3, $LiSO_2F$, LiOH, and cation-breakdown products. This excellent performance of the SEI layer is the key to the long-term cycle stability of lithium metal batteries (LMBs).

The slow transport of Li^+ due to ion aggregation caused by coulomb interaction is one of the obstacles to the application of IL electrolytes in LMB materials with high safety and high energy density. By modulating the competitive coordination between solvent–cationic and anion–cationic interactions in the electrolyte, Zou et al. [41] developed anion-enhanced solvated ionic liquid electrolytes (ILE) to significantly improve the transport capacity of Li^+ and stabilize the electrode–electrolyte interface (EEI) (Figure 4.2). The designed anion-reinforced solvating ILEs (ASILEs), which contain chlorinated hydrocarbons and two types of anions, FSI^- and $TFSI^-$, aim to enhance Li^+ transport capabilities, and stabilize the interface of high-nickel cathode materials ($LiNi_{0.8}Co_{0.1}Mn_{0.1}O_2$, NCM811), and maintain flame-retardant properties. With these ASILEs, the Li/NCM811 batteries demonstrated a high initial specific capacity (203 mAh g^{-1} at 0.1 C), excellent capacity retention (81.6% after 500 cycles at 1.0 C), and outstanding average coulombic efficiency (CE) (99.9% after 500 cycles at 1.0 C). Moreover, Ah-level Li/NCM811 pouch batteries achieved a significant energy density of 386 Wh/kg, indicating the practical feasibility of this electrolyte. This study provides some reference value for the development of advanced ILE with enhanced Li^+ transport and CEI stability in high-voltage cathode systems.

Currently, the electrolytes used in lithium batteries typically consist of at least two compounds: an organic solvent, such as cyclic carbonates, and a lithium salt, such as $LiPF_6$. Guzmán-González et al. [42] demonstrated the concept of using a single-component electrolyte in lithium batteries based on a novel boron-containing lithium IL (LiIL) at room temperature. The design concept of this type of lithium IL is centered around a tetra-coordinated boron atom with asymmetric substitutions, including oligomeric ethylene glycol groups, fluorinated electron-withdrawing groups, and an alkyl group. The optimized borate-based LiILs exhibit a high ionic conductivity of >10^{-4} S cm^{-1} at 25 °C, a high lithium transfer number (t_{Li^+} = 0.4–0.5), and excellent electrochemical stability (>4 V). Some of these LiILs demonstrate good compatibility with lithium metal electrodes, showing stable polarization profiles in plating/stripping tests. The selected LiIL was investigated as a single-component electrolyte in LMBs, achieving discharge capacities of 124 and 75 mAh g^{-1} for Li/LiIL/LiFePO$_4$ and Li/LiIL/Li$_4$Ti$_5$O$_{12}$ batteries, respectively, under low capacity loss conditions at 0.2 and 65 °C.

Due to its low volatility and lack of expensive fluoride properties, cyanide ILs are an ideal choice for cheaper and safer ILs for batteries. Karimi et al. [43] designed n-methyl-n-butylpyridinium (Pyr14)-based ILs that incorporate two

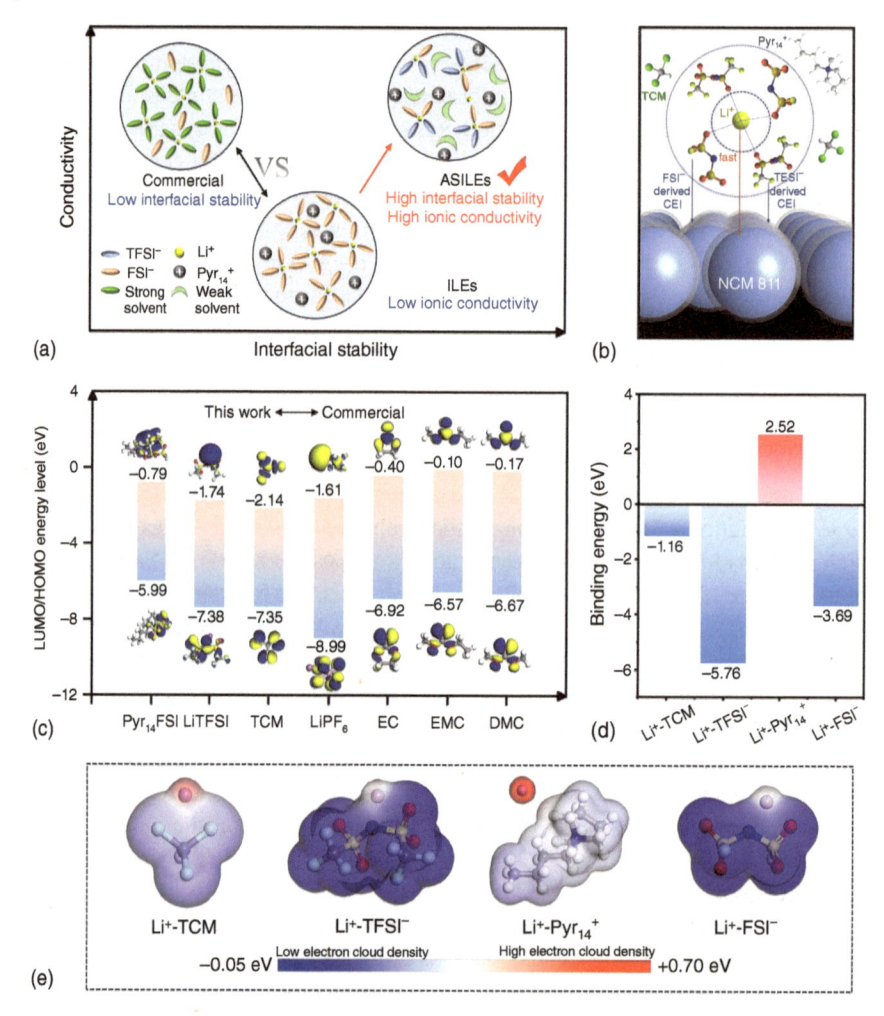

Figure 4.2 (a) Anion-reinforced effect on the ionic conductivity and electrochemical stability; (b) electrode–electrolyte interphase; (c) LUMO/HOMO energy level. (d) Binding energy; (e) electrostatic potential mappings. Source: Zou et al. [41]/with permission of John Wiley & Sons.

different cyanide anions, namely dicyanamide (DCA) and tricyanomethane (TCM), along with their respective mixtures with lithium salts (at a 1 : 9 salt: IL molar ratio) and their combination (DCA-TCM). These combinations exhibit significant ionic conductivity (5 mS cm^{-1}) at room temperature, an electrochemical stability window of up to 4 V, and high cycling stability. The formed SEI is dominated by a polymer-rich layer that includes carbon-nitrogen single bonds, double bonds, and triple bonds, providing high ionic conductivity and mechanical stability, which contributes to the aforementioned cycling stability. In the TCM-based electrolyte, SEI has a polymer-rich outermost layer and a more inorganic innermost layer. In the mixed electrolyte, the SEI layer shows a more uniform composition in the thickness direction. Nonfluorinated IL electrolytes exhibit great potential as safe, sustainable, and relatively low-cost electrolytes for LMBs. However, their properties are not as well known as fluorinated compounds, and this project provides important insights into the electrochemical and interfacial properties of novel F-free electrolytes.

Li et al. [44] developed an organic cage ionic conductor, Li-RCC1-ClO$_4$, for the preparation of high-performance solid-state cathodes. The ionic cage structure not only facilitates high ionic conductivity and ion transfer numbers but also provides options for solution processing in cathode fabrication, such as slurry coating. This molecular cage cathode is fully compatible with current cathode manufacturing processes, demonstrating significant application potential in SSLBs. Future efforts can focus on introducing additional advantages, stability, enhanced mechanical properties, and improved ionic conductivity by utilizing structured organic molecular additives, such as organic cages. Porous solids, such as metal-organic frameworks (MOFs) and covalent organic frameworks (COFs), have been explored extensively for their ion conduction properties. Unlike insoluble frameworks MOFs and COFs, organic cages have discrete molecular covalent structures and can be solution processable. These molecules can aggregate to form crystals with a highly interconnected three-dimensional (3D) network of pores. The discrete nature of the cage molecules makes them processable by being soluble in different solvents and can be modified by mixing with other soluble compounds. In the solid Li$^+$ conductor of this obtained organic cage, the ionizing functional groups in it provide an environment with efficient dielectric shielding that enables the dissociation of dissolved Li salts (such as LiClO$_4$) into mobile ions. In addition to its ideal room-temperature ionic conductivity, a key feature is its solubility in polar solvents. Therefore, in the mixing step, the organic cage-type Li$^+$ conductor is easily incorporated into the solid cathode as a cathode. The caged cathode electrolyte is uniformly dissolved in the cathode slurry, crystallizes when the solvent evaporates and grows on the surface of the cathode particles during the coating process, thus forming an effective ion conduction network inside the cathode. This method of using organic cages in SSLMs minimizes the number of ionic additives required for the cathode and results in excellent room-temperature cycling performance.

Solid polymer electrolyte (SPE) is one of the widely studied electrolytes for SSLBs, which has the potential to meet the evolving needs of lib. However, in order to meet the increasing power and energy density requirements, improvements in transmission performance and electrochemical stability are essential. It should meet the following requirements: high lithium-ion conductivity at room temperature ($>10^{-3}$ S cm^{-1}), high ion transfer number, strong electrolyte absorption capacity, good thermal stability, good electrochemical performance, good mechanical properties, able to withstand the impact of the battery assembly process, to prevent the growth of lithium dendrites during use. However, due to the crystalline behavior of the polymer, the ionic conductivity of SPE at room temperature is very low. Taking advantage of the fact that ILs are liquid at room temperature, the addition of ILs to the SPE can effectively reduce the crystallinity of the polymer, enhance the mobility of the polymer segments, and provide a conductive pathway to improve the ionic conductivity of the electrolyte [45]. This effect is similar to that of carbonate plasticizers, but compared with conventional carbonate plasticizers, ILs have the following advantages: (i) It can conduct electricity itself, has good solubility to inorganic salts and polymers, and significantly improves the ionic conductivity of polymer electrolytes. (ii) High thermal decomposition temperature, nonflammable, compared with traditional organic electrolytes, can effectively improve flame retardant and reduce safety risks. (iii) They have good chemical stability and a stable electrochemical window, allowing the battery to operate in a wider range of environments. (iv) Nonvolatile, environmental protection, in line with the principles of green chemistry.

SPEs have many advantages over liquid electrolytes, and much research has focused on developing SPEs with enhanced mechanical properties while maintaining high ionic conductivity. The recently developed phase separation polymer induced phase separation (PIPS) technique provides a simple method for fabricating bicontinuous nanostructured materials with independently adjustable mechanical properties and electrical conductivity. Melodia et al. [46] utilized digital light processing 3D printing to create SPEs with tunable mechanical properties and conductivity, employing the PIPS process to achieve nanostructured ionic conductive materials for energy-storage applications. A rigid cross-linked poly(isobornyl acrylate-stat-trimethylpropane triacrylate) scaffold provided materials with room-temperature shear modulus above 400 MPa, while soft poly(oligo(ethylene glycol) methyl ether acrylate) (POGMEA) domains containing the IL 1-butyl-3-methylimidazolium bis-(trifluoromethyl sulfonyl)imide endowed the material with ionic conductivity up to $1.2\,\text{mS}\,\text{cm}^{-1}$ at 30 °C. These features make 3D-printed SPEs highly competitive in all solid-state energy-storage devices, including supercapacitors.

Demarthe et al. [47] found that a poly(vinylidene fluoride) (PVDF) matrix rich in lone pairs reduced the aggregation expected to occur in IL-based electrolytes and increased the proportion of free ions. At high salt concentrations, the diffusion rate of Li^+ in PVDF-based ion gels is higher than that of EMIMTFSI, indicating the interaction between metal cations and polymers. Factors such as glass transition temperature, macroscopic ionic conductivity, self-diffusion coefficient, and coordination number show that the activation energy of the confined ion interface is lower than that of the unconfined ion interface. Therefore, the transport performance of PVDF-based ion gels in the liquid phase is improved by limiting IL. The mixed electrolyte has high ionic conductivity and can be used to prepare SSLBs with high safety and performance.

In addition to the introduction of IL into the polymer electrolyte to form IL Solid polymer electrolyte (ILSPE), IL reactivity is made through structural design, and IL can be directly polymerized to form PIL-SPEs. PILs consist of repeated motifs of monomer units, forming dimers, trimers, oligomers, and eventually polymers or copolymers. PILs combine the unique properties of IL with the flexibility of its macromolecular structure to produce compounds with novel functions, including solid ionic conductors, strong dispersants, stabilizers, absorbers, carbon material precursors, and porous polymers. They have great potential applications in polymer chemistry and materials science. So far, the preparation of IL based on

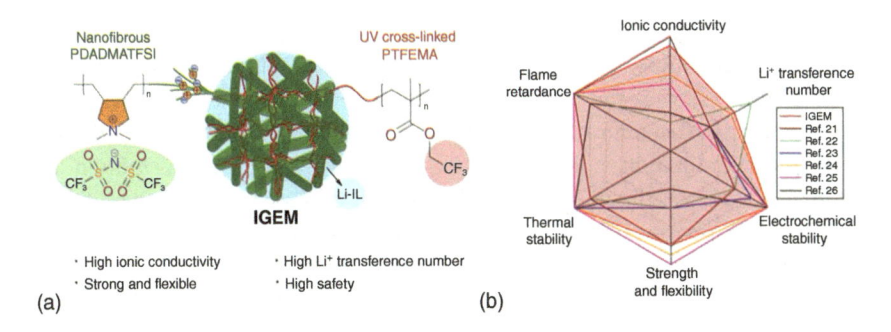

Figure 4.3 (a) Structure and composition of the IGEM. (b) Comparison between IGEM and previously reported ionogel electrolytes. Source: Yu et al. [48]/with permission of John Wiley & Sons.

various cationic and anionic forms has mainly focused on the conventional radical polymerization of IL monomers.

Monolithic ionic gel electrolyte membranes (IGEMs) have garnered significant attention due to their wide processing compatibility, nonflammability, and favorable thermal and electrochemical properties. However, IGEMs lack the high mechanical strength and Li$^+$ transportable properties offered by functional stands, limiting the battery's power and safety. Yu et al. [48] designed a task-specific IGEM monolith with high Li$^+$ conductivity and excellent thermal stability by using electrospun positively charged poly(ionic liquid) nanofibers as a thermally stable scaffold (Figure 4.3).

By adjusting the lithium-ion environment in IGEM, the transition from slow lithium-ion transport mode to fast structural lithium-ion transport mode can be achieved. Based on this unique IGEM, solid-state Li||LiFePO4 batteries exhibit excellent magnification capabilities and good cycle stability over a wide temperature range of 0–90 °C.

Adding healing agents to the material may be an effective and responsive material self-healing method, which can repair defective interfaces immediately after stimulation. Bis(fluorosulfonyl)imide (FSI$^-$) anions were considered excellent healing agents for LMA due to their outstanding passivation capacity to lithium metal, leading to the formation of rich inorganic compounds such as LiF and Li3N. The solubility of LiFSI salts in carbonates and ethers is high, but the performance of high-concentration electrolytes is poor. While the vinyl composition can be polymerized to form a stable SEI on the LMA, and the imidazole IL is compatible with the electrode. Qin et al. [49] obtained a self-healing SEI by UV light polymerizing 1-vinyl-3-methyli-midazolium bis(fluorosulfonyl)imide (VMI-FSI) monomer under and cross-linking it with polyethylene oxide (PEO). FSI$^-$ anions exchanged from the film were electrochemically decomposed into inorganic salts to improve the SEI film. Due to the self-healing properties of the film, a Li/LiCoO$_2$ battery with a loading of 16.3 mg cm^{-2} achieved an initial discharge capacity of 183.0 mAh g^{-1}, exhibiting stable operation over 500 cycles within the voltage range of 3.0–4.5 V, with a capacity retention of 81.4% and an average CE of 99.97%.

Fu et al. [50] proposed a polymer electrolyte based on PILs and IL plasticizers. The polymer matrix was composed of poly(diallyldimethylammonium) bis(fluorosulfonyl)imide and n-butyl-n-methylpyridinium TFSI$^-$ as a plasticizer, combined with LiFSI as the lithium salt. They all had excellent chemical stability and wide electrochemical stability windows. Compared with the traditional lithium-ion coordination polymer matrix, the positively charged lithium-ion chain reduces the coordination of lithium-ions and promotes the high mobility of lithium-ions within the polymer. The binding energy between LiFSI and Li$^+$ is low, and a stable interface can be formed when in contact with lithium metal. As a result, the PILSPE exhibited high ionic conductivity and a wide electrochemical stability window, enabling high cycling stability of LMAs with high-voltage cathodes at room temperature.

Wang et al. [51] synthesized redox pyridine-based PILs (PILs-py-400) by conducting a trimerization reaction with cyano groups using pyridine ILs as the raw material at an appropriate temperature (400 °C). The positively charged framework of PILs-py-400, along with its extended conjugated system, rich microporosity, and amorphous structure, can enhance the utilization efficiency of redox sites. At a current density of 0.1 A g^{-1}, a high capacity of 1643 mAh g^{-1} (96.7% of the theoretical capacity) was obtained, indicating that 13 Li$^+$ redox reactions occurred per repeat unit consisting of one pyridine ring, one triazine ring, and one methylene group. Furthermore, PILs-py-400 demonstrated excellent cycling stability, retaining a capacity of approximately 1100 mAh g^{-1} after 500 cycles at 1.0 A g^{-1}, with a capacity retention rate of 92.2%.

LMA is considered an ideal anode material for batteries due to its low electrode potential and the highest theoretical energy density. However, lithium dendrite growth and low interfacial stability hinder the practical application of LMAs. To overcome these drawbacks, Zhou et al. [52] proposed a GPE that contains imidazolium IL end groups with perfluoroalkyl chains (F-IL) (Figure 4.4). This structural design significantly changes the solubility of the salt electrolyte, and high ionic conductivity and lithium-ion transfer numbers are achieved through the Lewis acid segment in the polymer main chain. The presence of the F-IL backbone reduces the binding affinity of the lithium-ion with the glycol chain, allowing the lithium-ion to transfer rapidly within the gel network. These structures effectively anchor anions to the IL segment, reduce the space charge effect of ions, and promote the enhancement of anion coordination and the weakening of cationic coordination in Lewis acidic polymers. The obtained GPE had lithium-ion conductivity of 9.16×10^{-3} S cm^{-1}, lithium-ion transport number of 0.69, good electrochemical stability (4.55 V), and dendrite suppression performance. Based on the GPE, Li||Li symmetric batteries demonstrated excellent cycling stability, achieving stable cycling performance for over 1800 h with an areal capacity of 9 mAh cm^{-2}. Furthermore, the lithium-sulfur (Li—S) full batteries maintained 86.7% of their original capacity after 250 cycles.

In addition to the common applications in the LIBs field, ILs also have widespread applications in other battery systems. For zinc metal batteries, artificial interface layer engineering is an effective modification strategy to prevent the growth of

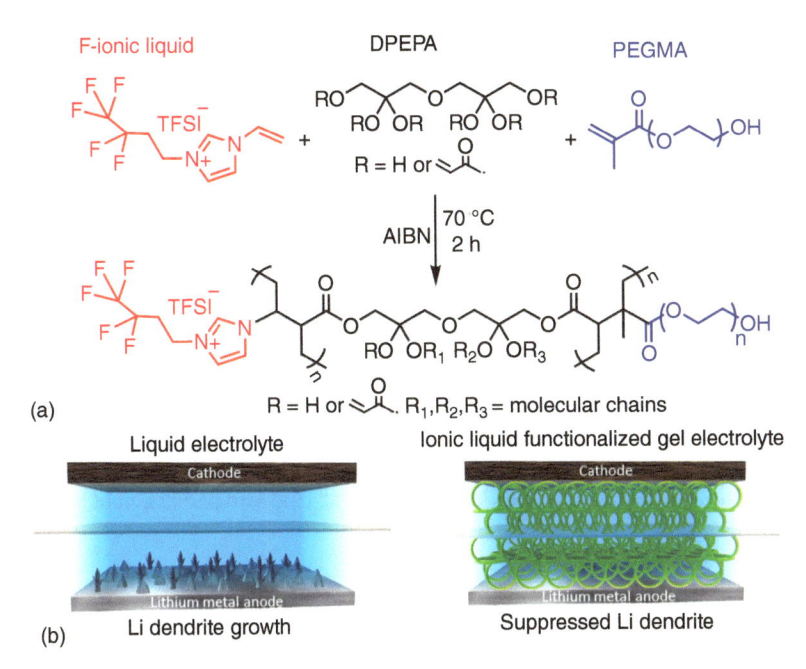

Figure 4.4 (a) Synthesis route of GPE; (b) Li plating mechanism in LE and F-IL-GEL electrolyte. Source: Zhou et al. [52]/CC BY-NC-ND 4.0/with permission of John Wiley & Sons.

zinc dendrites and the formation of side products. However, the high bulk ionic conductivity of most artificial interface layers is primarily contributed by the movement of anions (SO_4^{2-}), which is the source of parasitic reactions on the zinc anode. Ke et al. [53] designed a high zinc ion donor transition imidazolium polymer IL interface layer (1-carboxymethyl-3-vinylimidazoli-lium bromide monomer, CVBr) for zinc metal protection. The N^+ atom of the imidazolium ring forms cavities through chain connectivity, and the anions are confined within these cavities. Consequently, the hindrance of surrounding units to the anions leads to a sub-diffusion state, suppressing the diffusion of SO_4^{2-} at the interface and increasing the transfer number of Zn^{2+}. Moreover, the cation–anion coordination mechanism of PolyCVBr ensures the accelerated hopping of Zn^{2+}, facilitating rapid internal migration pathways for Zn^{2+}. Therefore, the Zn@CVBr||AM symmetric battery simultaneously exhibits high bulk ionic conductivity (4.42×10^{-2} S cm^{-1}) and a high transfer number for Zn^{2+} ($t_{Zn^{2+}} = 0.88$). The Zn@CVBr||AM-NaV$_3$O$_8$ pouch battery demonstrated an 88.9% capacity retention rate after 190 cycles under 90° bending, validating its potential for practical applications.

ILs with imidazolium cation structures exhibit higher ionic conductivity compared to other cation types. The anion part, TFSI$^-$, possesses high hydrophobicity and lower coordination ability, allowing it to integrate better with lithium-ions compared to other commonly used anions such as tetrafluoroborate (BF_4^-) and PF_6^-. COFs have structural designability, good porosity, and suitable pore size distribution, which can promote the movement of ions and provide a way for ion transport.

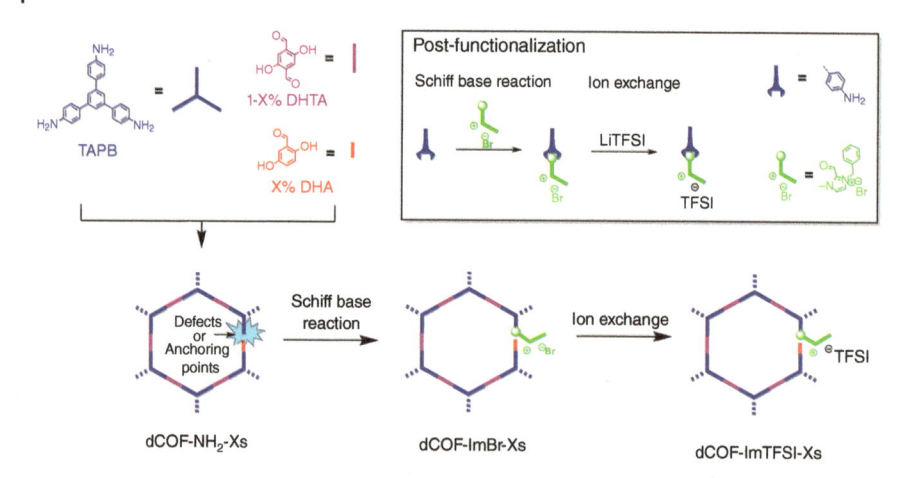

Figure 4.5 Synthesis route of dCOF-NH$_2$-Xs. Source: Li et al. [54]/with permission of John Wiley & Sons.

Zhen et al. [54] introduced defects into COFs by using a three-component condensation strategy (Figure 4.5). The resulting defective COFs (dCOF-NH$_2$-Xs, where X = 20, 40, and 60) exhibited good crystallinity and porosity, along with reactive amine functional groups serving as anchor points for further postfunctionalization. By introducing imidazole functional groups, the COF pore walls reacted to yield dCOF-ImBr-Xs and dCOF-ImTFSI-Xs materials, which function as solid electrolytes with lithium-ion conductivity over a wide temperature range (from 303 to 423 K). Notably, the dCOF-ImTFSI-60 ionic electrolyte achieved a conductivity of 7.05×10^{-3} cm^{-1} at 423 K. To date, this is the highest value reported for all polymer crystal porous materials based on solid-state electrolytes. Furthermore, the Li/dCOF-ImTFSI-60@Li/LiFePO$_4$ all-solid LIB demonstrated satisfactory performance at 353 K. This work provides a new approach to constructing accurately controlled postfunctionalized defects for SSLB materials.

It can be seen from these application examples, that IL electrolytes have broad development prospects in the field of LIBs, and its research and application will continue to deepen to meet the growing demand for energy storage.

4.2.2 Ionic Liquid/Organic Solvent Electrolyte

Due to the low concentration of lithium salt in pure IL system and poor wettability with electrode material, the cycle performance and coulomb efficiency of the battery will be adversely affected. In recent years, researchers have improved IL performance by adding specific organic electrolyte components. Mixing ILs with organic carbonate can combine the best of both worlds. But this is not in line with the ultimate goal of replacing existing organic solvents, as their coexistence could pose safety issues for batteries. However, if the IL content in the mixture is high enough, the electrolyte can become nonflammable. Adding 50% IL to the conventional electrolyte can significantly reduce the flammability of the organic electrolyte while maintaining stable battery performance. In addition, mixing il with organic carbonate produces better viscosity and conductivity, and the ionic conductivity of the resulting mixture may exceed that of pure organic carbonate [55].

Mixed electrolytes do not necessarily require a comparable volumetric ratio of the two components, as one component can serve as an additive. For instance, a small amount of organic carbonate (e.g. 5%) can significantly enhance battery performance. Adding 10–20% organic carbonate can improve the viscosity and ionic conductivity of the IL while still keeping the electrolyte nonflammable. Molecular dynamics simulations indicate that organic additives increase ion mobility by reducing lithium coordination. In addition to adding organic carbonates to improve the performance of ILs, incorporating ILs into traditional organic electrolytes is also a practical approach. In these cases, ILs act not as liquid electrolytes but as organic ionic salts that introduce large ions into the electrolyte matrix, serving as flame retardants for organic electrolytes.

Severe side reactions between LMAs and traditional carbonate electrolytes lead to the formation of SEI films with poor stability and significant lithium dendrite growth. While ether-based electrolytes have demonstrated excellent reductive stability on LMAs, they are generally unsuitable for high-voltage LMBs due to their poor oxidative stability (close to 4.0 V vs. Li^+/Li). ILEs have nonflammability, low volatility, wide electrochemical windows, and excellent interfacial film-forming ability. However, the high viscosity of ILEs limits the transport kinetics of lithium-ions, affecting the battery's high-rate performance and low-temperature performance. To address these issues, researchers proposed combining inert diluents with ILEs to form locally concentrated IL electrolytes (LCILEs). This strategy effectively reduces the viscosity of ILEs while preserving the solvation structures of contact ion pairs (CIPs) and anion aggregates (AGGs), significantly improving the compatibility of LCILEs with high-voltage LMBs. Despite extensive research on LCILEs, the mechanism by which diluents affect the lithium-ion solvation structure remains unclear, and whether "inert diluents" are truly "inert" warrants further investigation.

Tu et al. [56] used a chlorinated alkane 1,3-dichloropropane (DCP13) as diluent to develop a wide-temperature range DCP13-LCILE, investigating the nonequilibrium solvation structure of the electrolyte under an external electric field. Unlike the traditional understanding of solvation structures, considering external electric field conditions in constructing the lithium-ion solvation structure revealed that the diluents participate in the solvation structure of lithium-ions under the influence of the electric field, thereby affecting battery performance. In the nonequilibrium solvation structure, diluent molecules with strong lithium-ion affinity exhibited a higher capability to stabilize the solvation structure of lithium-ions, which can enhance the oxidative stability of the electrolyte. Simultaneously, the lithium-affinitive sites of the diluents can improve the transport kinetics of lithium-ions. The study on the nonequilibrium solvation structure of electrolytes under an electric field and the action of diluent was analyzed from the kinetic point of view, which provided a new idea for the development of high-performance electrolytes for LIBs.

The SEI plays a crucial role in high-performance LMBs, due to its excellent mechanical properties and high ionic conductivity, effectively preventing the growth of the notorious lithium dendrites. Tu et al. [57] prepared a modified IL electrolyte by diluting IL with 1,2-difluorobenzene (DFB) diluent. The diluent DFB not only created a crowded electrolyte environment, but promoted the interaction between Li$^+$ and FSI ions, resulting in the formation of AGGs; It also participated in reduction reactions to build robust SEI with high ionic conductivity. Based on this M-ILE, Li/LiFePO$_4$ batteries achieved a remarkable 96% capacity retention over 250 cycles with a mass loading of 9.5 mg cm^{-2}. Similarly, Li/LiNi$_{0.5}$Co$_{0.2}$Mn$_{0.3}$O$_2$ batteries delivered a discharge capacity of 132 mAh g^{-1} after 100 cycles, maintaining an impressive retention rate of 88%. Therefore, the use of diluents is proved to be an effective strategy for constructing advanced SEIs on lithium anodes.

Huang et al. [58] developed a novel ILE using lithium bis(trifluoromethanesulfonyl)imide (LiTFSI) as the lithium salt, 1-ethyl-3-methylimidazolium nitrate ([EMIm][NO$_3$]) IL and fluoroethylene carbonate (FEC) as functional solvents, and 1,2-dimethoxyethane (DME) as a diluent. The use of [EMIm][NO$_3$] IL as a solvent component promoted the formation of a unique Li$^+$-coordinated NO$_3^-$ solvation structure, allowing the solvated NO$_3^-$ to undergo continuous electrochemical reduction and resulting in a very stable and conductive SEI. The inclusion of FEC as an additional functional solvent and DME as a diluent enhances the electrolyte's oxidative stability and ionic conductivity while improving the kinetics of electrochemical reactions. The results demonstrate that this electrolyte exhibits excellent reversible and stable lithium stripping/deposition behavior, with a high average CE of 98.8% and ultra-long cycling stability of 3500 h. This IL-based electrolyte has been shown to have enhanced lithium deposition behavior and cycling properties, enabling reversible and stable lithium stripping/plating properties.

Wang et al. [59] investigated the nanostructures and physicochemical properties of LCILs formed by mixing HMIM TFSI, HMIM FAP, and BMIM FAP with TFTFE in weight ratios of 2 : 1, 1 : 1, and 1 : 2, as well as a 2 : 1 mixture of HMIM TFSI-TFTFE with 0.25, 0.5, and 0.75 m LiTFSI. The results were compared with those of pure ILs, parent ILEs, and PC-diluted ILEs. Rheological and conductivity measurements indicated that the addition of the diluent reduced viscosity by over 50% and increased conductivity by >40%.

Highly reactive electrodes often exhibit poor interface compatibility with conventional electrolytes, leading to limited cycling stability. Liu et al. [60] designed an LCILE composed of LiFSI, 1-ethyl-3-methylimidazolium bis(fluorosulfonyl)imide (EmimFSI), and DFB to overcome this challenge. As a co-solvent, DFB not only facilitated the transport of Li$^+$ in the solely ILE, but also had beneficial effects on the EEIs of LMAs and NMC811 cathodes. The developed LCILE enabled dendrite-free cycling of LMAs, achieving a CE of up to 99.57% at 0.5 mA cm^{-2}. Additionally, the Li/NMC811 batteries demonstrated high cycling stability, with 93% capacity retention after 500 cycles at C/3 charge and 1 C discharge (1 C = 2 mA cm^{-2}) at 4.4 V. In contrast, the electrolyte without DFB only achieved lithium stripping/deposition CE, with capacity retention rates of 98.22% and 16% for the Li/NMC811 batteries under the same conditions, respectively.

Wang et al. [61] designed an LCILE composed of LiFSI salt, n-methyl-n-propylpiperidinium bis(fluorosulfonyl)imide ([PP13][FSI]) as the IL solvent, and

1,1,2,2-tetrafluoroethyl-2,2,3,3-tetra-fluoropropyl ether (HFE) as the diluent. The introduction of HFE significantly reduced the viscosity and cost of the pure IL electrolyte, improved its ionic conductivity, and enhanced its ability to wet the surface of separators. The obtained LCILE exhibited nonflammability, low viscosity, improved room-temperature ionic conductivity, and good electrochemical performance in LMBs. The diluent HFE does not participate in the solvation sheath of Li^+ cations, which enhances the interaction probability between Li^+ cations and FSI^- anions. More FSI^- anions are preferentially reduced to form a dense inorganic SEI layer. The transmission of Li^+ is accelerated, and a uniform Li^+ flux is formed on the EEI. These characteristics effectively inhibit the growth of lithium dendrites and induce the uniform deposition of lithium.

Wang et al. [62] reported a novel ionic regulation strategy that combined dual ILs with organic solvents to achieve synergistic controlling cations and anions, improving lithium-ion transport characteristics and in situ generating stable SEIs to suppress lithium dendrite growth, thereby enhancing the stability of LMAs. Specifically, the cation 1-benzyl-3-methylimidazolium ($Bzmim^+$) in the IL had a large steric hindrance, which can promote the migration of lithium-ions within the SEI. In contrast, the cation 1-ethyl-3-methylimidazolium ($Emim^+$) had a higher conductivity, enhancing lithium-ion transport in the electrolyte. Simultaneously, the synergistic effect of FSI^- from the lithium salt and $TFSI^-$ from the IL promoted the formation of a LiF-rich SEI layer. Additionally, optimizing the components of ILs and ether-based electrolytes significantly reduced the electrolyte viscosity, accelerating the diffusion of lithium-ions in the liquid phase and within the SEI. This coupling mechanism has been shown to effectively suppress lithium dendrite growth and enhance the electrochemical performance of LMAs in symmetric and full batteries. The double salt addition strategy of BzmimTFSI and EmimTFSI enables the operation of dendrite-free LMA, where the integration of anionic FSI and TFSI creates a stable SEI. The hybrid cations guarantee fast Li^+ transport within the in situ formed SEI and the liquid solvent.

Lee et al. [63] presented an LCILE with a nonsolvating, nonflammable diluent HFE. The LCILE compost of LiTFSI, 1-methyl-1-propyl pyrrolidinium bis(fluorosulfonyl)imide (P13FSI), and 1,1,2,2-tetrafluoroethyl 2,2,3,3-tetrafluoropropyl ether (TTE). The addition of TTE made the LCILE have low viscosity and good separator wettability, which is conducive to the transfer of lithium-ions to the LMA. The nonflammability of TTE enabled the electrolyte to have excellent thermal stability. In addition, the synergy between the dianions (FSI/TFSI) contributed to the formation of ideal SEI, improved Li coulomb efficiency, and reduced lithium dendrite formation.

In summary, LCILEs can significantly improve the performance of LMBs. By combining an inert diluent and an IL, LCILEs reduced viscosity while retaining compatibility with high-voltage LMBs. This strategy helps to achieve faster lithium-ion transport dynamics and enhanced oxidation stability, thereby improving the coulomb efficiency and cycle life of the battery. Traditional ILEs are flammable and highly volatile, while locally concentrated ILs reduce these risks through the use of inert diluents. This improvement makes LCILEs a safer electrolyte option, particularly suitable for lithium battery applications with high energy density. The design of LCILEs allows for excellent electrochemical performance over a wide temperature range, including low temperatures. This is essential for the development of LMBs suitable for low-temperature environments, which can expand the range of applications of lithium batteries, especially in the transportation and industrial sectors. By optimizing the combination of ILs and diluents, LCILEs are able to provide a more stable electrolyte/electrode interface. This is important to prevent dendrite growth and reduce the risk of battery short circuits, thereby improving battery reliability and safety. In conclusion, the locally concentrated ILE provides an efficient, safe and environmentally friendly solution for the development of lithium batteries through its unique chemical and physical properties.

4.2.3 Organic-Inorganic Composite Ionic Liquid Electrolyte

The main advantages of inorganic fillers to enhance ILE include increased ionic conductivity, increased mechanical strength, and improved voltage resistance. These advantages are mainly achieved in the following ways: Improving ionic conductivity The ionic conductivity of ILE can be significantly improved by introducing inorganic fillers. The crystal structure of these fillers can provide additional Li^+ channels, making ion migration smoother. Adding inorganic fillers, such as MOFs, into the polymer electrolyte can significantly improve the mechanical strength of the electrolyte and effectively inhibit the growth of lithium dendrites. This enhanced mechanical strength helps prevent the electrolyte from cracking due to changes in internal pressure during charging and discharging. The use of inorganic fillers can improve the voltage resistance of ILE, so that they can work stably at higher voltages. This is important for developing applications such as high-voltage lithium batteries. In short, the introduction of inorganic fillers not only improves the electrochemical performance of the ILE, but also enhances its mechanical stability and voltage resistance, which is of great significance for improving the safety and energy density of lithium batteries.

Inorganic ceramics can significantly promote ILE, mainly in improving ionic conductivity, stability, and safety. The addition of inorganic ceramics can create more ion transport paths, thereby improving the ionic conductivity of the electrolyte. The introduction of inorganic ceramics can enhance the chemical and electrochemical stability of the electrolyte, inhibit the growth of lithium dendrites, and thus improve the cycle stability of the battery. The high melting point and high thermal stability of inorganic ceramics help to improve battery safety and reduce the risk of fire and explosion. An efficient ion transport network is formed by wrapping ion gels around ceramic particles. The solvated IL is introduced into the nano-porous inorganic ceramics to achieve high ionic conductivity and good cyclic stability at room temperature.

Electrolytes that can operate over a wide temperature range are crucial for sustainable advanced energy systems. In general, the low-temperature cycle performance of batteries is mainly due to insufficient ion transport in the electrolyte and charge transfer dynamics at the electrolyte/electrode interface, resulting in changes in the structure of the SEI. For polymer electrolytes at low temperatures, these problems are even more serious. The introduction of low melting point IL into the polymer can significantly improve the conductivity of ions at low temperatures. In addition, the incorporation of il can effectively improve the electrolyte/electrode interface, thus accelerating the ion transport at the interface [64]. Zhang et al. [65] prepared a layered IL composite electrolyte (L-ILCE) by confining IL to a two-dimensional (2D) ordered interlayer nanostructure of vermiculite. The microstructure inside the nanostructure can induce IL rearrangement and crystallinity, giving L-ILCE the combined advantages of both liquid and solid electrolytes. The obtained L-ILCE exhibited continuous and dense site distribution characteristics similar to liquids, and excellent thermal and physicochemical stability. This L-ILCE had a high transport number of 0.89 and a wide electrochemical window of 0–5.3 V. LiFePO$_4$||Li and NMC811||Li batteries assembled with L-ILCE exhibited highly stable electrochemical performance at −20 to 60 °C. This work provides a new idea for developing electrolyte materials with a wide temperature range for LMBs.

Limiting IL in the SPEs can improve the performance of the LMBs, Hu et al. [66] prepared restricted SPE with IL-grafted SiO$_2$, LiTFSI, and PEO, explored the effect of restricted IL on the properties and properties of the electrolyte, and clarified the transport mechanism of Li$^+$. Compared to unrestricted IL SPEs, IL-restricted SPEs decrease the crystallinity of the polymer due to increased dissociation of lithium salts. It shows higher ionic conductivity, higher lithium-ion transport number, wider electrochemical window, and more stable cycle performance. IL@ SiO$_2$ provides an additional Li$^+$ transport route that accelerates ion transfer and alleviates lithium dendrites. As a result, IL confinement is an effective strategy to improve the performance of LMBs.

Jin et al. [67] prepared a composite SPE by integrating Laponite (LAP)-IL-TFSI multilayer particles with a PEO matrix on LMBs. LAP-IL-TFSI had a unique structure of repulsive face-to-face interaction and attractive face-to-face interaction, resulting in partial separation of its multilayer structure, thus enhancing its compatibility with the PEO matrix. The cation at the edge of the LAP particle and the nitrile group in grafted IL can promote the dissociation of lithium salt. Thus, the obtained composite SPE exhibited decreased low crystallinity, improved ionic conductivity (1.5×10^{-3} S cm^{-1}), and a high lithium-ion transport number (0.53). The integration in the Li/SPE interface significantly reduced parasitic reactions, and the stable operation at 1000 h demonstrated the achievement of ultra-long dendrite-free Li deposition behavior.

Lithium garnet is a promising inorganic ceramic solid electrolyte for LMBs, demonstrating good electrochemical stability under lithium anodes [68]. The instability of nickel-rich cathodes at high cycle rates and the narrow operating temperature range hinder the development of LMBs with high energy density. To address this issue, Deng et al. [69] incorporated IL-grafted ceramic electrolyte Li$_{6.4}$La$_3$Zr$_{1.4}$Ta$_{0.6}$O$_{12}$ (LLZTO) into a poly(vinylidene fluoride-co-hexafluoropropylene) (PVDF-HFP) polymer matrix to prepare a functional ILSPE. Grafting IL on the surface of LLZTO can enhance the organic/inorganic compatibility of LLZTO and PVDF-HFP, and reduce the reaction between LLZTO and

PVDF-HFP. The obtained ILSPE has the ability to anchor the solvent and promote the dissolution of Li$^+$, which prevents the cracking of the nickel-rich cathode and inhibits the dissolution of Ni^{2+}, resulting in a stable and high-rate cycle.

Garnet electrolytes have excellent ionic conductivity and resistance to high pressures, but their brittleness and rigidity limit their close contact with the two electrodes, resulting in high interfacial resistance to ion migration. To address this issue, Pervez et al. [70] employed a strategy of using an ILE thin interlayer at the EEI, which overcame the barriers to ionic transport. The chemically stable ILE enhances the contact between the electrode and the solid electrolyte and significantly reduces the interfacial resistance at the cathode and anode interfaces. This makes the deposition of lithium metal in the anode more uniform and significantly inhibits lithium dendrite growth even at a high current density of 0.3 mA cm^{-2}. In addition, the improved lithium/electrolyte interface reduces the overpotential of the symmetric lithium/lithium battery from 1.35 to 0.35 V.

At present, the development of aluminum batteries is mainly limited by electrochemical stability, corrosion, and moisture sensitivity. Aluminum chloride is a commonly used electrolyte in rechargeable aluminum batteries because it can reversibly electrodeposit aluminum at room temperature. Leung et al. [71] developed SPE based on 1-ethyl-3-methylimidazolium chloride, PEO, and gas-phase silica, which exhibited higher electrochemical stability than ILs while maintaining a high ionic conductivity (~13 mS cm^{-1}). When gaseous SiO$_2$ (<1 wt%) is added to the polymer electrolyte, the content of Al$_2$Cl$_7^-$ increases due to the reaction between SiO$_2$ and AlCl$_4^-$ in the presence of PEO. In aluminum–graphite batteries, the SPE can be charged to 2.8 V, achieving a maximum specific capacity of 194 mAh g^{-1} at 66 mA g^{-1}. With long-term cycling at 2.7 V, a reversible capacity of 123 mAh g^{-1} was observed at 360 mA g^{-1}, and a CE of 98.4% was achieved after 1000 cycles. The results of this study reveal the unique synergistic effect of PEO and gas phase SiO$_2$ in the EMIMCl-AlCl$_3$-based polymer electrolyte mixture, which provides a new way for the development of solid electrolytes for high-performance aluminum batteries.

Dong et al. [72] imbibed IL by repacking it on nanosheets and investigated the ionic conductivity of IL in nanoconfinement. The layered nanostructured channels of 2D materials such as graphene oxide (GO) and molybdenum disulfide (MoS$_2$) provide an efficient way for the rapid transport of ions. Ions migrate significantly faster in the confined liquid than in the free-state sample. The ionization of GO and MoS$_2$ systems increased from 0.65 to 0.86 and 0.83, respectively, demonstrating the effective promotion of il dissociation by nano confinement. In this kind of system, the interaction between ions is obviously inhibited, the free volume is obviously increased, and the nano bound ions are distributed in layers, which is conducive to the increase of ion transport speed. Therefore, the corresponding ionic conductivity is drastically enhanced under confinement, proving the promising efficiency of 2D nanochannels in ion transport. The corresponding ionic conductivity is greatly improved under constrained conditions, which proves the great advantage of 2D nanochannels in ion transport.

MOFs have a significant promotion effect on ILEs, which is mainly reflected in the following aspects including improved ionic conductivity and electrolyte stability, adjustment of electrochemical properties, and enhanced interface contact. MOFs have high porosity and large specific surface area, which can effectively adsorb and fix ions in ILs. This structural property facilitates the rapid transport of ions in MOF channels, thereby improving the ionic conductivity of ILs as an electrolyte. They can be used as a carrier for ILs to enhance the stability of the entire electrolyte system through the stability of its structure. Their porosity and rigidity can inhibit the volatilization and leakage of ILs and improve the safety of electrolytes.

By choosing different combinations of MOF materials and ILs, it is possible to fine-tune the electrochemical properties of the electrolyte, such as conductivity, ion migration number, etc. For example, MOFS-loaded ILs can achieve high ionic conductivity and good electrochemical stability in a specific IL and MOF selection. The surface characteristics of MOFs can improve the interface contact between the electrolyte and the electrode, reduce resistance, and thus improve the charge and discharge efficiency of the battery. In short, MOFs can effectively promote the performance of ILE through their special structure and physical and chemical properties, and provide a strong support for the development of high-performance electrochemical energy-storage systems.

To address the interfacial issues and extend the cycling life of SSLBs, Liu et al. [73] constructed the GPE by mixing a MOF UIO-66 particle with SIL ([Li(G4)$_1$][TFSI]). UIO-66, as a carrier of IL, had a uniform 3D channel and porosity, which realized the limitation of SIL. SIL was limited to UIO-66, and the coordination environment of Li$^+$ was weakened by the interaction of metal ions with TFSI, which resulted in the improvement of ionic conductivity and transfer number of Li$^+$. In addition, according to Lewis acid–base theory, TFSI is used as a Lewis acid, while Zr^{4+} in UIO-66 makes a Lewis base. Therefore, UIO-66 can inhibit the release of more Li$^+$ ions from SIL, promote the participation of Li$^+$ ions in the ion transport process, and produce uniform lithium-ion flux, thus inhibiting the formation of lithium dendrites and realizing stable lithium deposition.

Zhang et al. [74] reported a novel family of proton conductors based on MIL-101 and protic IL polymers (PILPs) containing IL different anions. Protic IL monomers were first introduced into the layered pores of the highly stable MIL-101, followed by in situ polymerization to synthesize a series of PILP@MIL-101 composites. The resulting PILP@MIL-101 composites not only retained the nanoporous cavities and water stability of MIL-101, but also provided numerous opportunities for enhanced proton transport due to the interwoven structure of the PILPs compared to MIL-101 alone. The obtained PILP@MIL-101 composite not only had the nanoporous cavity of MIL-101 and excellent water stability, but also improved proton transport through hydrogen bonding and interaction with water molecules.

Ho et al. [75] incorporated IL@BUIL fillers into a PEO matrix, significantly enhancing the performance of composite SPE, including increased ionic conductivity, lithium-ion transport number, electrochemical stability, and long-cycle stability of LMAs (Figure 4.6). Lithium salt was added to the filler to dissociate the IL and made it play the role of active filler. The incorporation of MOF fillers with high Lewis acid activity was designed to enhance the migration of Li$^+$ ions by interacting with lithium salts and PEO substrates. This interaction increased the dissociation of

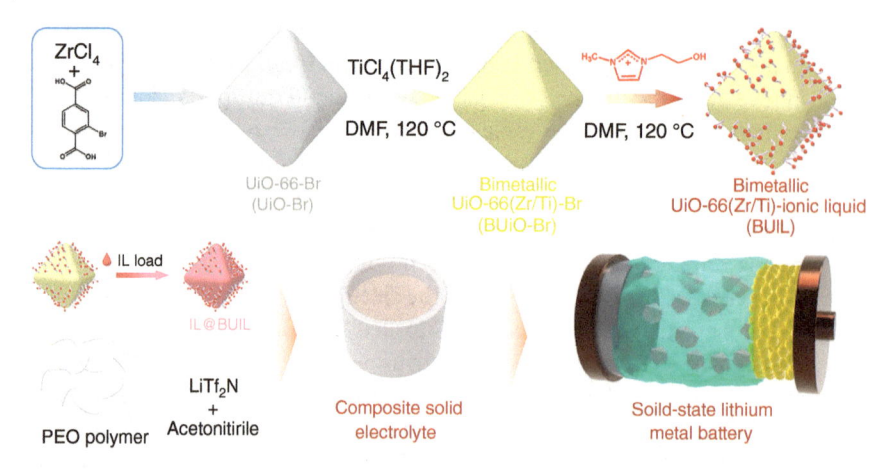

Figure 4.6 Diagram of the procedure for UiO-66-IL based SSLMB [76] Source: Ho et al. [75]/with Permission of John Wiley & Sons.

lithium salts, improved anion capture, and reduced the crystallinity of the polymer. In addition, the addition of MOF fillers expanded the electrochemical stability window. The optimized SPEs exhibited an enhanced ionic conductivity (0.458 mS cm^{-1}, 30 °C), increased lithium-ion transport number (0.668), and an expanded electrochemical stability window (4.5 V). In addition, in continuous plating/stripping tests, a lithium metal symmetric battery using the composite SPEs was stably cycled for 500 h at 0.2 mA·cm^{-2}. Finally, in the LiFePO$_4$/IL@BUIL/Li battery configuration, a capacity of 148 mAh g^{-1} was achieved at a rate of 1 C. This method of simultaneously modifying the metal and ligand sites of MOFs to enhance the performance of SPEs can serve as a universal strategy for designing MOF fillers in LIBs.

Nguyen et al. [77] prepared a composite SPE composed of zeolitic imidazolate frameworks (ZIF-67) as fillers and ILEs as plasticizers and PEO/LiTFSI as the matrix. By optimizing the preparation process, an integrated ultra-thin PEO/LiTFSI-IL-ZIF-67 electrolyte film with a thickness of 32 μm was prepared. It had high ionic conductivity (1.19 × 10^{-4} S cm^{-1} at 25 °C), wide electrochemical stability (5.66 V), and high lithium-ion transport number (0.8). The assembled LMBs exhibited excellent cycle stability at both low and high temperatures, presenting an initial specific discharge capacity of 166.4 mAh g^{-1} and a capacity retention of 83.7% after 1000 cycles at 3 C under 60 °C, owning a low fading rate of 0.0163% per cycle. In addition, the prepared composite SPEs demonstrated high safety performance.

Zhang et al. [78] constructed an IL-confined MOF/Polymer 3D-porous membrane, which can promote in situ electrochemical conversion of LiF/Li$_3$N 3D-Janus structure SEI films on nanofibers. The SEI-incorporated into the separator provided fast Li$^+$ transport pathways, exhibiting high room-temperature ionic conductivity (8.17 × 10^{-4} S cm^{-1}) and Li$^+$ transport number (0.82).

The composite electrolytes of ILs and other materials have great application potential in the fields of energy, electronics, and materials science. With the deepening of research and technological advances, these materials are expected to

4.3 Single-Ion Conductive

4.3.1 Introduction of Single-Ion Conductive

LIBs have achieved significant success in powering portable devices and are increasingly used in large-scale applications such as (hybrid) electric vehicles and stationary energy storage. Nonetheless, further improvements are needed to enhance energy and power density, as well as safety [79]. Traditional electrolytes, whether liquid or solid, can be considered as composed of "lithium salt + solvent." Both lithium-ions and anions contribute to ionic conductivity during charge and discharge processes, hence they are referred to as "bivalent ionic conductors." In practice, however, lithium-ions are often restricted within a "coordination cage" due to interactions with Lewis basic sites in the matrix material, which limits their migration. Consequently, the lithium-ion transfer number in bivalent ionic conductors is only about 0.2–0.3, meaning that the primary contribution to conductivity comes from anions. Anions are highly mobile during charge and discharge, leading to their accumulation at the electrodes and causing concentration polarization, which degrades battery performance [76].

Ionic concentration polarization can slow down the diffusion rate of lithium-ions at the electrode surface, thereby increasing the internal resistance of the battery and reducing its charge and discharge efficiency. Changes in the migration rate of lithium-ions in the electrolyte affect their transport efficiency between electrodes, which may lead to decreased charging and discharging performance. Additionally, under high-concentration polarization conditions, the deposition rate of lithium-ions on the electrode surface may become uneven, increasing the risk of lithium dendrite formation. These dendrites can penetrate the separator, causing battery short-circuits and compromising safety. The uneven lithium deposition and dissolution caused by ionic concentration polarization can accelerate the aging of electrode materials, reducing the cycling life of the battery. Ionic concentration polarization may also prevent the battery from fully utilizing all of its active materials during charge and discharge, thereby lowering its actual capacity and energy density. Due to obstructed ion transport, excess heat may be generated within the battery, potentially affecting its temperature stability and thermal management. Under high-concentration polarization conditions, the electrolyte may decompose on the electrode surface, forming unstable byproducts that could further impact battery performance and lifespan [80–83].

Two approaches have been considered to be effective in reducing the migration rate of anions, thereby alleviating concentration polarization. The first method involves adding anion receptors to the electrolyte can slow down the migration of anions, allowing more cations to migrate. This method optimizes the transport of cations by adjusting the ionic coordination structure in the polymer electrolyte. The second method is to fix the anion on the polymer main chain, that is, to prepare a cationic single-ion conductor (SIC). Cationic SICs have a lithium-ion migration number close to 1, which means that lithium-ions conduct very fast in the conductor with little interference from other ions. This high selectivity allows SICs to effectively reduce side reactions and concentration polarization in battery applications, thereby improving battery performance and stability. Low concentration gradient: Due to the high migration number of lithium-ions, SICs are able to form a low concentration gradient in the electrolyte. This helps to slow down dendrite growth on the metal lithium anode, thereby preventing short circuits on the electrode surface and extending the battery life. Cationic SICs have particular application potential in the field of solid-state batteries. Since the solid electrolyte does not contain liquid solvents, the problem of anion migration in conventional two-ion conductors is mitigated. SICs can provide better ion conductivity and chemical stability, helping to improve the safety and energy density of solid-state batteries.

Although SICs have many advantages, they still face some challenges in practical applications, mainly due to their low ionic conductivity and processing properties at low temperatures. Researchers are addressing these issues through chemical modification and structural design to advance the application of SICs in fields such as solid-state batteries [84, 85].

4.3.2 Catalog of Single-Ion Conductive

SICs can be roughly categorized based on the type of anion into several groups: carboxylate, sulfonate, borate, and bis(sulfonyl)imide types.

Carboxylate-Based SICs: These electrolytes have high dissociation energy, making it difficult for lithium-ions to dissociate, resulting in generally low electrical conductivity (around 10^{-7} S cm^{-1}). Consequently, there have been fewer reports on these in recent years.

Sulfonate-based SICs: The delocalized negative charge of the sulfonate anion has long attracted the attention of researchers, and the synthesis of sulfonate-based polymers is relatively simple and economical. The researchers improved their performance by introducing electron-withdrawing groups to facilitate the dissociation of lithium-ions from sulfonates.

Sulfonimide-based SICs: Due to the strong interaction between carboxylate ($-CO_2^-$) and sulfonate ($-SO_3^-$) anions and Li$^+$, the conductivity remains at a low level. Attention therefore turned to the sulfonimide anion ($-SO_2N-SO_2^-$) and its derivatives, such as the trifluoromethylsulfonimide anion ($-N(CF_3SO_2)_2$, TFSI$^-$). Compared with the sulfonate anions, the sulfonimide anions have a higher degree of negative charge delocalization and a larger conjugated structure, resulting in lower dissociation energy of these polymer lithium salts. These electrolytes are common and can be synthesized by copolymerizing a lithium salt monomer containing the sulfonimide anion with a polymer monomer [such as ethylene oxide (EO) and methacrylate], or by modifying the side chain so that the anion group is added to the flexible main chain.

Organic borate-based SIC: Lithium borate is centered on boron atoms and combines with oxygen-containing ligands to form a conjugated system, which effectively disperses negative charges and reduces the interaction between cations and anions. This results in higher solubility and ionic conductivity, as well as good thermal stability and a wide electrochemical window, making it a promising organic life salt. Therefore, many researchers have fixed borate functional groups to polymer frameworks to prepare various borate-based SICs [42].

4.4 Application of Single-Ion Conductive in Batteries

SICs are ideal electrolyte materials for batteries because they can provide ionic transport performance comparable to traditional liquid electrolytes while avoiding the safety hazards associated with liquid electrolytes. In LMBs, SICs can suppress the growth of lithium dendrites, thereby enhancing the cycling stability and safety of the battery. Additionally, SICs can also be used in sensors and electronic devices as ion-selective transport layers [86, 87].

However, the application of SICs in batteries also faces some challenges: including interfacial compatibility, low ionic conductivity at room temperature, processing difficulty, and long-term stability. The compatibility between SICs and electrode materials is critical for achieving high-efficiency batteries and requires further optimization. Their ionic conductivity at room temperature is usually low, which limits their performance in practical applications. Processability is another challenge, as the synthesis costs of some SICs can be high, and processing can be difficult, limiting their commercial application in practical applications. SICs must maintain stable ionic transport performance over extended periods of operation [88].

In general, the application of SICs in batteries shows great potential, especially in terms of improving battery safety and cycle life. However, to overcome the current challenges, further research and technological innovation are needed.

4.4.1 Organic Single-Ion Conductor Electrolyte

Traditional polymer electrolytes contain dissolved lithium salts, allowing both cations and anions to move freely, which can lead to uneven lithium deposition and associated safety concerns. SICPEs effectively address this issue by covalently fixing anions in the polymer backbone, allowing only lithium cations to be mobile. SIC electrolytes have high room-temperature ionic conductivity, which ensures that lithium-ions can be transported quickly, thus supporting high-rate charge and discharge performance. The lithium-ion transport number in this type of electrolyte is close to 1, which helps to eliminate concentration polarization and maintain the stable performance of the battery at different current densities. Especially in the process of fast charging, this is particularly important. In addition, the SIC electrolyte can also maintain good cyclic stability under high current density. Therefore, the SIC electrolyte is very suitable for SSLBs supporting fast charge due to its high ion conductivity, high lithium-ion migration number, excellent cycling performance, and stability.

Wang et al. [89] developed a SICPE with high ionic conductivity (1.1×10^{-3} S cm^{-1}) and lithium-ion transport number (0.92) at room temperature. The polymer network structure in SICPEs not only promoted the rapid jump of lithium-ions, enhancing the ion dynamics, but also improved the dissociation ability of lithium-ions, making the transport number of lithium-ions close to 1. LIBs assembled with the SICPEs, lithium metal, and various cathodes (such as LiFePO$_4$, S, and LiCoO$_2$) exhibited excellent high-rate cycling performance and fast charging capability.

Li et al. [90] reported a novel electrospun SICPE composed of nanoscale mixed PVDF-HFP and lithium poly(4,4'-diaminodip-henyl sulfone, bis(4-carbonylphenyl)-sulfone imide (LiPSI) (Figure 4.7). This design overcame the shortcomings of polyolefin-based separators (which have low porosity, poor electrolyte wettability, and thermal dimensional stability) and LiPF$_6$ salt (which has poor thermal stability and moisture sensitivity). The electrospun nanofiber membrane exhibited high porosity and adequate mechanical strength. The fully aromatic polyamide backbone endowed the es-PVPSI membrane with high thermal dimensional stability even at 300 °C, while its high polarity and porosity ensure rapid wettability of the electrolyte. By soaking the membrane in a solvent mixture of ethylene carbonate (EC) and dimethyl carbonate (DMC) (v : v = 1 : 1), a SICPE with broad electrochemical stability, good ionic conductivity, and high lithium-ion transfer number was obtained. Based on these advantages, Li/LiFePO$_4$ batteries using this SICPE demonstrated excellent rate capacity and remarkable electrochemical stability for at least 1000 cycles, indicating that this electrolyte can replace the traditional liquid electrolyte-polyolefin combination in LIBs. Furthermore, long-term stripping and plating cycling tests, combined with scanning electron microscopy

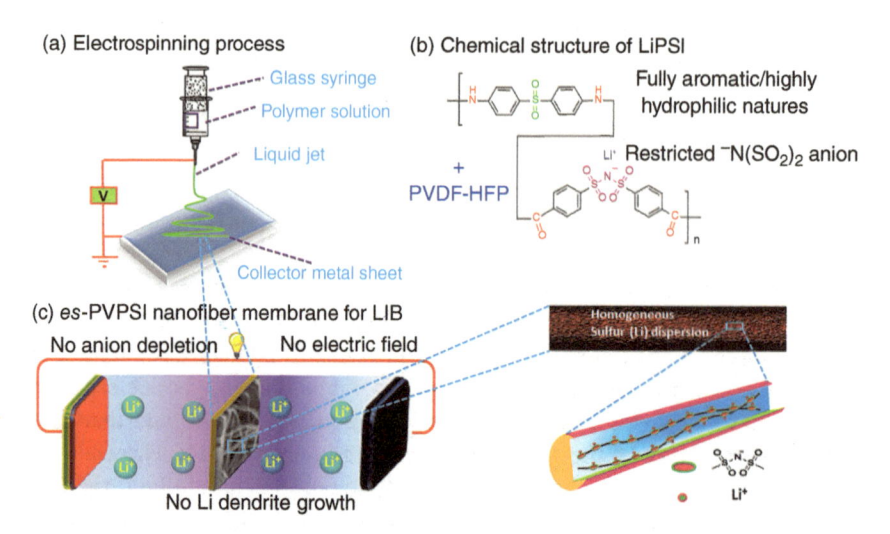

Figure 4.7 Diagram of the fabrication (a), composition (b), and operation (c) of SICPE. Source: Li et al. [90]/with permission of John Wiley & Sons.

(SEM) images of lithium foils, clearly confirmed that the es-PVPSI membrane can suppress the growth of lithium dendrites, establishing its application foundation in high-energy LMBs.

Zhang et al. [91] developed a self-supporting SICPE, designed through the synergistic interaction between anionic receptors and dissolved ILs. This innovative chemical synergy significantly enhances the complete dissociation of lithium salts while fixing the anions, thereby facilitating the rapid transport of Li$^+$. As a result, the ionic conductivity of the SICPE was improved to 8.0×10^{-4} S cm^{-1}, and the Li$^+$ transference number increased to 0.75. These properties effectively alleviate concentration polarization and dendrite growth, ensuring the long-term stability of the battery. Furthermore, the Li||PBSIL||NCM811 batteries demonstrated a cycle life of up to 1300 cycles, with a discharge capacity of 183 mAh g^{-1} and a capacity retention rate of up to 75%. Additionally, SICPE was successfully integrated for the first time into the production of rolled semi-solid cylindrical and Z-type stacked pouch LMBs. Through the synergistic regulation of Li$^+$ transport and anion fixation, SICPE provides an effective design strategy for self-supporting SICPE, showcasing outstanding electrochemical performance and contributing to the development and commercialization of long-term cycle LMBs.

The addition of high dielectric constant organic carbonates, such as EC or propylene carbonate (PC), can effectively improve the room-temperature conductivity of SICPEs, as these molecules can cooperate with lithium-ions and promote them to jump from one anionic site to another. Nevertheless, this can lead to a reduction in the mechanical properties of the electrolyte, necessitating blending with a second polymer, such as PVDF or PVDF-HFP. Dominic Bresser et al. [92] proposed an advanced SIC block copolymer electrolyte with a main chain of lower fluorine content, which can dramatically reduce costs compared to previous systems. Furthermore, this electrolyte can provide highly stable cycling performance for LMBs based on LiNi$_{0.6}$Co$_{0.2}$Mn$_{0.2}$O$_2$ (NCM622) and NCM811 cathodes. Additionally, this research indicates that small molecules with high mobility and high dielectric constants can effectively facilitate Li$^+$ transport, which is important for achieving high-performance LMBs. Specifically, the transition from pure EC to a mixture of EC and PC improved the electrochemical stability against oxidation and increased the limiting current density, thereby improving the rate performance and cycling stability of Li||NCM batteries at ambient temperatures.

Deng et al. [93] synthesized a new type of environment-friendly multifunctional CO$_2$-based polycarbonate SICPE, which has good electrochemical performance. Polycarbonate propylene glycol ether (PPCAGE) with different allyl content was prepared by the trimerization of CO$_2$, propylene oxide (PO) and allyl glycidyl ester (AGE) under the catalysis of zinc glutarate (ZnGA). The glass transition temperature (T_g) of the obtained trimers is <110 °C. By efficient click reaction, the terpolymer was functionalized with 3-mercaptopropionic acid and treated with lithium hydroxide to obtain SICPE with different lithium contents. The obtained SPE achieved an ionic conductivity of 1.61×10^{-4} S cm^{-1} at 80 °C and a lithium-ion transport number of 0.86. It also possessed electrochemical stability of up to 4.3 V vs. Li$^+$/Li.

Porcarelli et al. [94] prepared a novel SICPE, which was composed of polylithium (1-[3-(methylpropenoxy)-propyl sulfonyl]-1-(trifluoromethyl sulfonyl) imide) and polyethylene glycol methyl methacrylate (PEGMMA) blocks. The obtained SICPE exhibited a low T_g value (−61 to 0.6 °C), high ionic conductivity (2.3×10^{-6} S cm^{-1} at 25 °C and 1.2×10^{-5} S cm^{-1} at 55 °C), a wide electrochemical stability window (4.5 V relative to Li$^+$/Li), and a lithium-ion transport number close to 1 (0.83). Based on the synergistic effect of these properties, the prepared SICPE material can be applied to LMBs, which can achieve high charge and discharge efficiency and high specific capacity.

SICPE are ideal for inhibiting dendrite lithium deposition, but they are unstable at high potentials, making them incompatible with high-energy cathode materials such as nickel-rich cathodes. Chen et al. [79] prepared SCIPE, which can be used for high-energy LMBs based on NCM811 cathodes, using a combination of a polyblock copolymer (aryl ether sulfone) and a suitable "molecular transporter" such as PC. These batteries can be cycled at high reversible capacity at a variety of temperatures, including 20 °C or even 0 °C, and when optimized charging protocols are adopted, >500 cycles can be achieved without significantly reducing capacity.

Dong et al. [95] prepared three novel SICPEs composed of different perfluorinated ionic side chains: poly(1,4-phenylene ether sulfone)-Li, polysulfone-Li, and hexafluorinated polysulfone-Li. The effects of the chemical structure of the main polymer chain and the concentration of ionic groups in the side chain on the electrochemical properties of the polymer electrolyte were investigated. It was found that trifluoromethyl (-CF$_3$) in the main chain and higher concentrations of ion side chains are closely related to charge transport and electrochemical stability. Based on these factors resulted in excellent ionic conductivity (approximately 2.5×10^{-4} S cm^{-1}) and outstanding anode stability (>4.8 V), the Li∥NCM622 batteries with a high capacity retention rate were achieved.

Shin et al. [96] reported a cross-linked polymer with weakly coordinating anion nodes that serve as a high-performance SICPE with minimal plasticizer presence. This electrolyte exhibits a wide electrochemical stability window, a high room-temperature conductivity of 1.5×10^{-4} S cm^{-1}, and a special selectivity for lithium-ion conduction ($t_{Li^+} = 0.95$). Importantly, this material is also flame-retardant and highly stable when in contact with lithium metal. LMB prototypes containing this quasi-solid electrolyte demonstrated superior performance compared to traditional batteries with polymer electrolytes. As shown in Figure 4.8, this polymer forms a diamondoid network consisting of weakly coordinating borate anions connected through butanediol linkers. The borate nodes display a weak affinity for lithium cations, promoting Li$^+$ mobility, while the alkene units enable postsynthetic cross-linking to generate robust membranes. With minimal plasticizer, this electrolyte exhibits remarkable selectivity for Li$^+$ ion conduction and high room-temperature conductivity, together with flame retardancy and stability toward Li metal and high-potential cathode materials. Battery cycling tests reveal outstanding power performance and cycling stability, suggesting that the material can serve as a functional electrolyte for next-generation lithium batteries. The synthesized materials presented in this chapter have tight spacing and weak

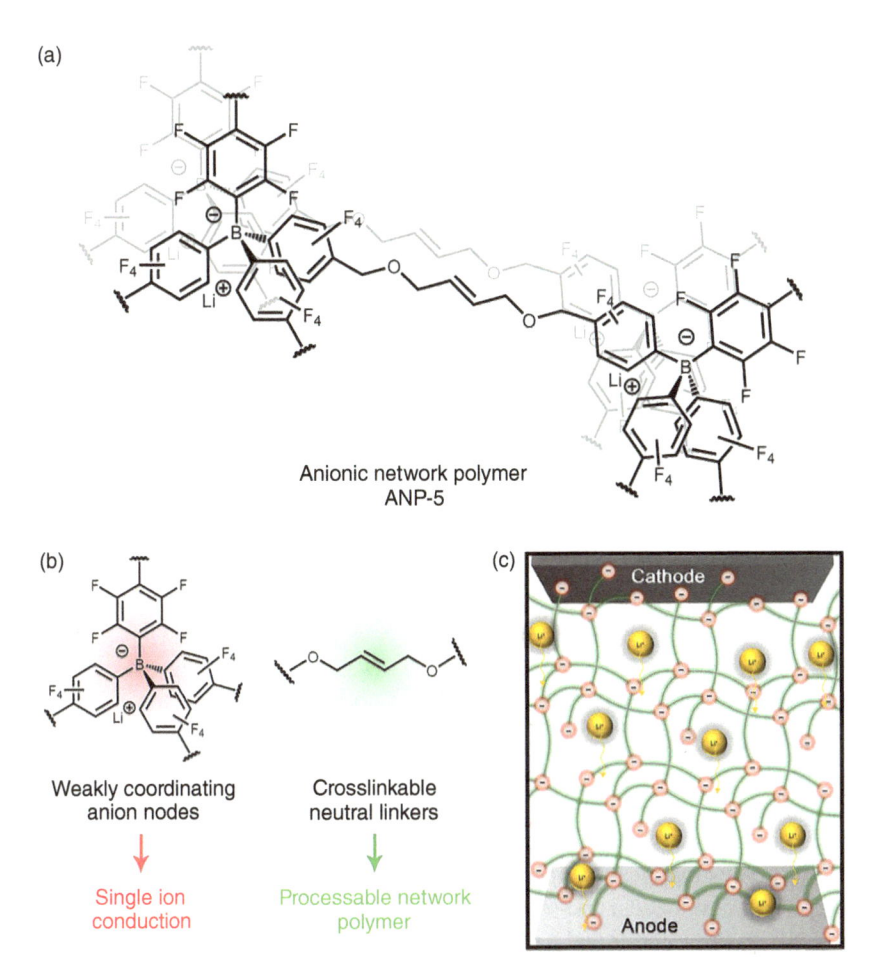

Figure 4.8 Structure and composition of anionic borate network polymer. (a) The SIC anionic borate network polymer, ANP-5. (b) Tetrafluorophenyl borate anion nodes (red), cis-2-butene-1,4-diol linker (green). (c) Schematic of ANP-5 as the electrolyte. Source: Shin et al. [96]/with permission of John Wiley & Sons.

coordination anions in interpenetrating network polymers and can be used in high energy density Li—S or Li—Br batteries after design optimization.

Wu et al. [97] synthesized a novel SICPE using minimally plasticized 3-sulfonyl (trifluoromethanesulfonyl) lithium imidazolium acrylate (MASTFSILi) as the starting material (Figure 4.9). This SICP allows for precise modulation of the ion–dipole interactions between Li$^+$ and carbonyl/cyano groups. The SIPE exhibits extremely high selectivity for lithium-ion conduction (with a lithium-ion transference number of up to 0.93), a high ionic conductivity of approximately 10^{-4} S cm^{-1} at room temperature, and a wide electrochemical stability window (>4.5 V). The resulting SICPE demonstrates excellent electrochemical stability with lithium metal during long-term cycling at room temperature and 60 °C. Within a wide temperature range of −20 to 90 °C, the LiFePO$_4$-based SSBs containing the SICPE show good rate and

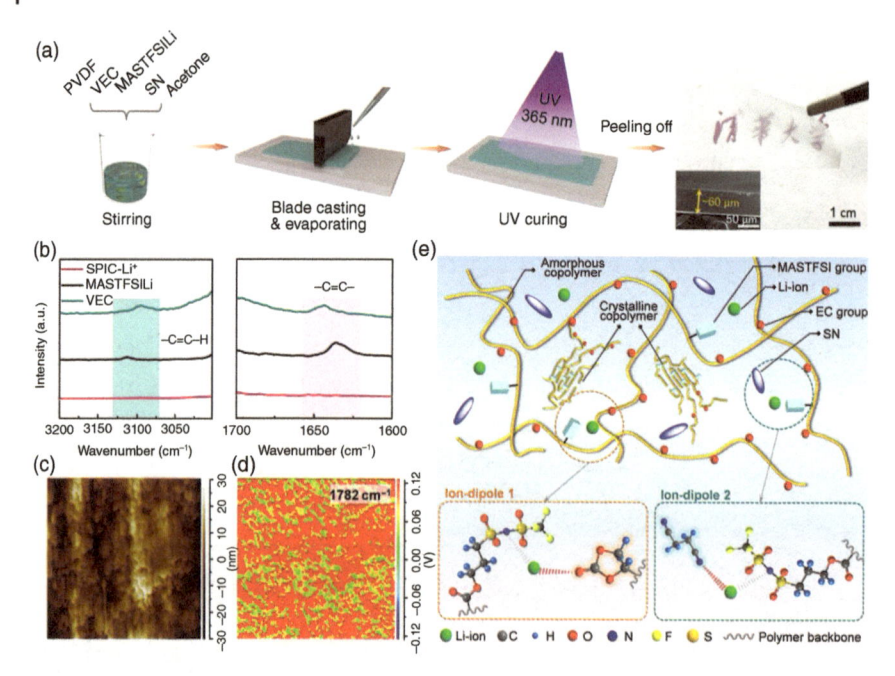

Figure 4.9 (a) Diagram of SICPE membrane synthetic processes; (b) FTIR spectra of the SICPE; (c) AFM topography image and (d) corresponding IR; (e) diagram of the ion–dipole interactions. Source: Reproduced with permission from Wen et al. [97]/John Wiley & Sons.

cycling performance. These experimental results show that the electrochemical performance of single-ion conducting polymer electrolytes can be effectively improved by adjusting the ion–dipole interaction.

Cui et al. [98] developed a two-salt system polymer electrolyte that contains a SIC polymer (SICP) lithium salt and a conventional bivalent lithium salt (LiTFSI). The prepared SICP can provide the polyanion to reduce the mobility of the free anion (TFSI) by repulsive force and increase the lithium-ion transport number to 0.75. The traditional bivalent lithium salt can effectively dissociate sufficient Li$^+$ to ensure a high lithium-ion conductivity (σ_{Li}^+, 0.87 mS cm^{-1}). The double salt system improves the low conductivity of SICP lithium salt and the low lithium-ion migration number of traditional bivalent lithium salt, and effectively alleviates the concentration polarization. The ions in the electrolyte can be evenly distributed, the transport environment of lithium-ions is improved, and the high-rate and long-term stable cycle of LMB is finally realized.

Li et al. [99] prepared an interpenetrating SCIPE PTF-4EO by cross-linking tetrabutylborate lithium with tetraethylene glycol. PTF-4EO had unique anion weak interaction and coordination ether-oxygen segment structure, room-temperature conductivity ($3.53 \times 10 \times 10$ S cm^{-1}), lithium-ion transport number (0.92), electrochemical window width (>4.8 V), and good mechanical properties. In addition, the generated SSPE can help form a stable SEI, further enhancing the interface stability of the LMA. The assembled LiFePO$_4$ LIBs exhibited high cycling stability, CE, and capacity retention for over 200 cycles under 2.50–4.25 V.

Zhang et al. [100] prepared a novel SICPE membrane with a high lithium-ion transport number, excellent mechanical strength, and high ionic conductivity by using a one-step photoinitiated click reaction on an electrospun PVDF carrier membrane. This process involved the reaction of lithium diallylborate (LiBAMB), pentaerythritol tetrakis(2-mercaptoacetate) (PETMP), and 3,6-dioxy-1,8-octanedithiol (DODT). The optimized SICPE exhibited a high ionic conductivity (1.32×10^{-3} S cm^{-1}, 25 °C), a high lithium-ion transport number (0.92), and a wide electrochemical window (6.0 V). The SICPE exhibited a tensile strength of 7.2 MPa and a breaking elongation of 269%. Based on these superior properties, The SICPE can inhibit the growth of lithium dendrites by constant current plating/stripping cycle test and surface morphology analysis of lithium metal electrode after cycle. The Li|LPD@PVDF|Li symmetric battery maintained very stable and low overpotential without short-circuiting over a period of 1050 h. Compared to batteries based on traditional LE and Celgard separators, the Li|LPD@PVDF|LiFePO$_4$ battery exhibited excellent rate and cycling performance.

Deng et al. [101] prepared a novel gel state SICPE with a high lithium-ion transport number, good mechanical strength, and excellent ionic conductivity. In the presence of gamma-butylactone (GBL), a 3D network structure was formed in the electrospun PVDF membrane by a one-step photo-triggered in situ mercaptoyl clicking reaction of LiBAMB, pentaerythritol tetras(2-mercaptoacetate) (PETMP) and DODT. The electrospun PVDF offered high mechanical strength as the supporting framework. The anions BAMB$^-$ were covalently tethered into the polymer network to limit the anionic movement, which resulted in a high lithium-ion transport number of 0.92. Charge delocalization of boron atoms in BAMB$^-$ was achieved through covalent bonding with electron-withdrawing groups, which resulted in weak electrostatic interactions between lithium-ions and anions. The gel SICPE exhibited improved ionic conductivity (>10^{-3} S cm^{-1} at 25 °C). Its high lithium-ion transfer number and good mechanical strength can inhibit the growth of lithium dendrites. Li|LPD@PVDF|Li symmetric battery achieved a highly stable and low overpotential without a short-circuiting cycle over 1050 h. Li|LPD@PVDF|LiFePO$_4$ batteries had excellent rate and cycle performance.

Porcarelli et al. [94] synthesized sulfonated polysulfone (SPSU(X)Li)) polymeric lithium salts through ion exchange after sulfonation. By curing poly(ethylene glycol) diglycidyl ether (PEGDGE) with 4,4′-diaminodiphenyl sulfone (DDS) within the SPSU(X)Li matrix, a novel SIC PE was prepared. The relationships between ionic conductivity, thermal stability, and tensile properties with the degree of sulfonation and PEGDGE concentration were investigated. The introduction of sulfonate lithium groups into the polysulfone enhanced the compatibility between SPSU(X)Li and PEGDGE in the SPE. AFM analysis indicated that as the degree of sulfonation increased, the phase morphology became more heterogeneous, and the size of the dispersed PEGDGE phase decreased. The interaction between sulfonate lithium and the polyether epoxy resin improved the thermal stability of the epoxy resin network. Compared to neat SPSU(X)Li, the enhanced compatibility also led to an increase in elongation at break. Higher lithium-ion concentration and the segmental mobility of polymer chains above the glass transition temperature (T_g) contributed to the high ionic conductivity of the SIC SPE at elevated temperatures.

Kwon et al. [102] designed a composite layer (S-CE/S-GE) based on SICs on lithium metal electrodes, which significantly reduced the loss of liquid electrolytes by adjusting the solvation environment for Li$^+$ movement within the layer. A Li||Ni$_{0.5}$Mn$_{0.3}$Co$_{0.2}$O$_2$ pouch cell, featuring a thin lithium metal (N/P ratio of 2.15), a high-loading cathode (21.5 mg cm^{-2}), and a carbonate electrolyte, achieved 400 cycles at an electrolyte-to-capacity ratio of 2.15 g Ah^{-1} (2.44 g Ah^{-1} including the mass of the composite layer). Alternatively, under a battery stack pressure of 280 kPa, it maintained 100 cycles at 1.28 g Ah^{-1} (1.57 g Ah^{-1} including the mass of the composite layer). The rational design of the composite layer based on SICs demonstrated in this work provides a method for constructing high-energy-density rechargeable LMBs with minimal electrolyte content.

To address the issue of lithium dendrite penetration, Hu et al. [103] developed a novel 3D cross-linked siloxane-based SIC membrane using an in situ sol–gel method and a nonsolvent induced phase separation (NIPS) process. The constructed PA-SICPE porous membrane, with a high porosity of 73.1%, achieved excellent liquid electrolyte absorption of 468.6%. At 25 °C, it demonstrated a remarkable ionic conductivity of 1.72×10^{-3} S cm^{-1}. The dense 3D interpenetrating network also endows the PA-SICPE membrane with enhanced thermal stability, ensuring battery safety. Crucially, the inherent SIC behavior of the PA-SICPE membrane allows the obtained electrolyte to have a high transference number t_{Li^+}, effectively suppressing the growth of lithium dendrites. Consequently, lithium/lithium symmetric cells using this membrane exhibited a very stable and reversible plating/stripping process with low overpotentials, without short circuits exceeding 400 h. Compared to cells using a PP separator, the LiPO$_4$/Li cells assembled with the cell membrane also exhibited excellent rate capability and cycling stability. Thus, the simple approach of using the PA-SICPE porous membrane offers great potential for practical applications in high-performance layered membranes.

The uncontrollable growth of lithium dendrites leads to low CE and severe safety issues, significantly hindering the practical application of LIBs using lithium metal as the anode. Fraile-Insagurbe et al. [104] developed a novel SCIPE, which is based on an ionomer with a poly (ethylene-maleimide) skeleton and a phenylsulfonyl lithium (trifluoromethane sulfonyl) imide overhang group, which is then blended with PEO and polyethylene glycol dimethyl ether (PEGDME). These SICPEs exhibited ionic conductivity of about $\sim 7 \times 10^{-6}$ S cm^{-1} at 70 °C, lithium transport number close to 1, and excellent mechanical strength (fracture toughness >30 J cm^{-3}). Additionally, the SCIPE had very high resistance against lithium dendrites growth, achieving cycling for >1200 h in Li symmetric batteries at 0.1 mA cm^{-2}.

Zeng et al. [105] have designed and fabricated a novel SIC cross-linked polymer gel electrolyte using PVDF-HFP as a reinforcing material, which contains negatively charged delocalized borate structures and abundant EO units. A new type of lithium borate was synthesized by covalently bonding the anion to the polymer backbone, while a vinyl monomer rich in EO units was chosen to construct the polymer framework, ensuring rapid ion transport. As expected, the custom boron-centered SIC polymer gel electrolyte exhibited a high ionic conductivity of up to 1.03×10^{-3} S cm^{-1} at 32 °C, an excellent oxidation potential of up to 5.05 V vs. Li$^+$/Li at 1 mV s^{-1}, and near single-ion conductive behavior (with a lithium-ion transfer number of 0.65). The lithium metal symmetric batteries assembled with the novel boron-centered SIC polymer gel electrolyte demonstrated long-term stable cycling at room temperature for over 700 hours without short-circuiting, indicating good stability of the mixed membrane during lithium metal deposition and stripping processes. Moreover, lithium/LiFePO$_4$ batteries assembled with the boron-centered SIC polymer gel electrolyte exhibited excellent rate capability and cycling performance. At a rate of 0.1 °C, the initial discharge capacity of the battery was 161.3 mAh g^{-1}, and after 100 cycles, the capacity remained at 127.7 mAh g^{-1}, with CE approaching 100%. Additionally, the lithium/LiFePO$_4$ battery provided nearly 100% stable CE after 300 cycles at 0.5 °C.

COFs based SICs have garnered significant attention due to their unique structures and chemical diversity, making them potential alternatives to inorganic ion conductors. However, the slow lithium-ion conductivity has limited their practical applications. Li et al. [106] proposed a solvent-free COF SIC (Li-COF@P) based on weak ion–dipole interactions, in contrast to traditional strong ion–ion interactions. The ion–dipole interaction between the lithium-ions from the COF and the oxygen atoms of polyethylene glycol diacrylate (PEGDA) embedded in the COF pores facilitates the ion dissociation and the migration of Li$^+$ through oriented ion channels. Driven by this single-ion transport behavior, Li-COF@P demonstrated reversible lithium deposition/stripping and stable cycling performance (maintaining 88.3% after 2000 cycles) in a LMAs|| 5,5'-dimethyl-2,2'-biphenol quinone (Me$_2$BBQ) cathode organic battery under ambient operating conditions, highlighting its electrochemical feasibility in all-solid-state organic batteries.

4.4.2 Organic–Inorganic Composite Single-Ion Conductor Electrolyte

SICPEs have great potential in the field of solid-state batteries, but they still face some challenges and problems in practical applications [88]. SICPEs currently have relatively low room-temperature ionic conductivity, which limits their performance in high-performance battery applications and requires them to be further enhanced. Its ability to inhibit the growth of lithium dendrites also needs to be improved, which is essential to prevent battery short circuits and improve safety.

By combining inorganic filler with a polymer matrix, the ionic conductivity of organic and inorganic composite electrolytes can be significantly improved. The design of this composite electrolyte allows the lithium-ions to move quickly through the polymer network, thereby increasing the efficiency of the battery's charging and discharging. Reasonable selection and design of inorganic fillers can effectively reduce the surface resistance of composite electrolytes, which is very important for the cycle stability and rate performance of SSLBs. The lower surface resistance helps to reduce the energy loss of the battery during use. The flexibility and viscoelasticity of the organic–inorganic composite electrolyte enable it to form close contact with the lithium metal negative electrode, reduce the interface resistance, and thus reduce the phenomenon of local electric field concentration. In addition, the introduction of inorganic ceramic similar fillers can improve the thermal stability of the electrolyte, which can effectively block the growth of lithium dendrites and prevent short circuits of the battery. This composite electrolyte can match the high-pressure positive electrode and the lithium metal negative electrode, which is conducive to achieving a significant increase in the energy density of SSLBs. The wide electrochemical window means that the battery can operate stably over a higher voltage range, which is very advantageous for the development of battery systems with high energy density. In addition, the organic inorganic composite electrolyte has the characteristics of small thickness and light weight, which helps to reduce the mass and volume of SSLBs, thus significantly improving the energy density of the battery. This is important for applications that require high energy density, such as electric vehicles. In short, the organic–inorganic composite SIC electrolyte is regarded as an important development direction in the field of SSLBs in the future because of its advantages of high conductivity, low surface resistance, high safety, wide electrochemical window, and lightweight.

Organic–inorganic composite SICs mainly utilize the method of grafting organic SIC components onto the surfaces of inorganic materials. This approach leverages the nanostructured inorganic materials to influence the crystallization of the polymer electrolyte, improving the mechanical properties and stability of the composite electrolyte. Additionally, the acid–base groups present on the surface interact with the anions of the SICs, thereby immobilizing these anions and improving lithium-ion migration. The inorganic precursors commonly used include alumina, silica, and other inorganic ceramic nanoparticles, as well as novel inorganic sources employed as SICs, such as modified inorganic materials and complexes [107, 108].

The LMA faces severe interfacial issues that significantly hinder its practical application. An unstable interface directly leads to low cycling efficiency, lithium dendrite deposition, and even serious safety concerns. Xu et al. [109] designed an advanced artificial protective layer with single-ion pathways, which is expected to achieve a uniform distribution of ions and electric fields across the lithium metal surface, thereby effectively protecting the LMA under long-term operating conditions. By rationally integrating doped garnet, a robust biphasic artificial interface was constructed that exhibits single-ion conductivity, high mechanical rigidity, and considerable deformability. This artificial SEI significantly stabilizes the repeated charge and discharge processes of the battery by regulating the facile transfer of lithium-ions and the dense lithium plating behavior, which helps to improve the CE of LMBs and significantly enhances their cycling stability. This work emphasizes the importance of rationally manipulating the interfacial characteristics of working LMAs and provides new insights into achieving dendrite-free lithium deposition behavior in operational batteries.

SICPEs may reduce polarization and the growth of lithium dendrites, but these materials can be mechanically too rigid, necessitating the use of ion mobilizers, such as organic solvents, to facilitate the transport of lithium-ions. Uneven distribution of mobilizers and the emergence of preferential lithium transport pathways can ultimately create sites that favor lithium-ion conduction, applying additional mechanical stress and potentially leading to premature short circuits in the battery. Overhoff et al. [110] explored a ceramic-polymer hybrid electrolyte composed of a blend of SIC polymers and PVDF-HFP, incorporating EC and PC as plasticizers, along with silane-functionalized LATP particles. The mixed electrolyte features an oxide-rich layer that significantly stabilizes the interface of lithium metal, allowing lithium to deposit at a current density of 0.1 mA cm^{-2} for over 700 hours. The addition of oxide particles markedly reduced the natural solvent absorption from 140 to 38 wt%, while still maintaining a relatively high ionic conductivity. The electrochemical performance was evaluated in NMC622 LMBs, demonstrating impressive capacity retention over 300 cycles. Notably, a very thin LiNbO$_3$ coating on the cathode material further enhanced cycling stability, with a total capacity retention of 78% over >600 cycles, highlighting the potential of the hybrid electrolyte concept.

Oh et al. [111] have demonstrated that solid-state soft electrolyte (SICSE) is a novel quasi-solid electrolyte strategy for practical semi-solid LMBs (SSLMBs), enabling the batteries to be manufactured and operated under ambient conditions. SICSE consists of a nonflammable coordinating electrolyte and an ion-flexible framework (cation copolymer/Ti-SiO$_2$@Al$_2$O$_3$). By combining SICSE slurry with a UV-cured assisted multilevel printing process, seamless integrated SSLMBs can be fabricated without the need for high-temperature/high-pressure sintering steps. The soft characteristics of SICSE address concerns regarding grain boundary resistance and the instability of the electrolyte–electrode interface, which have long been challenged for inorganic solid electrolytes. The single-ion conductive properties of SICSE facilitate the formation of a stable interface with the LMA and NCM811 cathode. Driven by SICSE and the monolithic battery structure, the SSLMB demonstrates stable cycling performance, rate capability, adjustable voltage, and high specific/volumetric energy density under ambient operating conditions. Furthermore, the low-temperature performance, mechanical flexibility, and safety (nonflammability) of the SSLMB significantly surpass those of previously reported solid-state LMBs (LMBs).

Zhang et al. [112] developed a polymer for solid-state sodium metal batteries, polymer-in-MOF SICPE, in which the polymer segment was partially confined to nanopore ZIF-8 particles by Lewis acid–base interaction. This unique nano-confinement effectively weakened the coordination between sodium ions and anions, and promoted the dissociation between sodium ions and salts. Simultaneously, the nanopore inside the ZIF-8 particle also provides a directional and ordered channel for the migration of sodium ions. This SCIPE achieved a Na ion transport number of 0.87, Na ion conductivity of 4.01×10^{-4} S cm^{-1}, and an improved electrochemical voltage window up to 4.89 V vs. Na/Na$^+$. The assembled batteries (with Na$_3$V$_2$(PO$_4$)$_3$ as the cathode) exhibited dendrite-free Na-metal deposition, achieving excellent rate and cycling performance with 96% capacity retention over 300 cycles. This polymer-in-MOF design offered an effective strategy for preparing metal batteries with high performance and safety.

Hui et al. [113] anchored anions in dissociated zinc salts by introducing deletion-linked MOFs and stimulated the preparation of SCIPEs, a copolymer matrix with a low-energy Zn^{2+} diffusion pathway. The conduction of Zn^{2+} in high-performance zinc metal batteries was accelerated, thereby regulating both anion and cationic movement in SCIPEs. The carboxyl ferrocene (Fc) modulated deletion ligand MOF was designed with an open Co^{2+} site. It bonded to OTF^- in strong Lewis acid–base interactions, and promoted the dissociation of $Zn(OTF)_2$ salts and the single Zn^{2+} conduction of the anchoring OTF^- anion. The crystallinity of the polymer matrix was decreased by the introduction of MOF-Fc, and the mobility of local polymer chains and coordination Zn^{2+} ions was enhanced. The competitive coordination of —O— in PEGDA and NH_2—C=O in PAM significantly reduced the dissolution energy of Zn^{2+}. Based on the synergistic interaction of these groups, the obtained all-solid electrolyte achieved both excellent ionic conductivity (1.52 mScm^{-1}) and high Zn ion transport number (0.83) at room temperature. The SCIPE enables an integrated electrolyte–electrode interface with excellent compatibility and stability, enabling 1000 cycles of dendrite-free galvanizing/stripping and 2000 cycles of ultra-stable Zn//VO_2 full batteries.

Xu et al. [114] proposed a strategy for regulating ion transport in composite electrolytes using 2D MOF. The 2D MOF was used to create a fast ion permeation channel and promote the Li^+ dynamics in the electrolyte. Substituents in the 2D-MOF enhance the electron-donating effect and limit the movement of $ClO4^-$, resulting in improved mechanical properties and lithium-ion transport number (0.64). The ionic conductivity of the PEO/MOFs-NH_2 electrolyte at room temperature is 6.5×10^{-5} S cm^{-1}, which is nearly 10 times that of the SPE without MOFs. In addition, different modifications on the surface of MOF can make the composite electrolyte have different properties, in which the amino functional group of the electron donor in MOFs-NH_2 acts as an ion screen, which can enhance the selective ion transport, while MOFs or MOFs-NO_2 cannot. The interface behavior was significantly improved, showing a stable stripping/plating electrochemical curve at 0.2 mA cm^{-2} in LMBs.

From the aforementioned conclusions, we can see that organic and inorganic composite SICs show great application potential in battery technology, especially in improving the safety and performance of batteries.

4.5 Conclusions and Outlook

In the past decade, ILs have garnered significant attention as safer candidates for solid electrolytes in batteries due to their excellent ionic conductivity. Several key scientific issues arise in the study of the physicochemical properties of ILs:

1) The determination of any substance's properties assumes a purity of 100%. However, the purity of ILs has consistently posed a challenge for scientists. The nonvolatility of ILs makes it difficult to purify them through distillation, and being liquid makes it challenging to achieve purification via crystallization. Therefore, the purification methods for ILs require standardization and convenience to ensure reproducibility in physicochemical property measurements. Many of the physicochemical properties of ILs are closely related to their purity, and the impact is significant. Thus, the purity of ILs is fundamentally important for the reliability of physicochemical property data.

2) Many ILs are hydrophilic, and even hydrophobic ILs have a much higher hygroscopicity in the air compared to many other organic solvents. Therefore, the influence of the environment is crucial when measuring ILs, and measurements must be conducted in a dry environment or within a glove box.
3) While a substantial number of publications report on the physicochemical properties of ILs, systematic studies of these properties are still quite limited, particularly concerning the determination of specific properties needed for certain reactions. From a large-scale production or industrial perspective, there remains a vast amount of knowledge to be explored regarding the nature of ILs.
4) The ultimate goal of research on the physicochemical properties of ILs would be to select ideal ILs based on simple experimental data or convenient computational methods.

Ionic conductors play a crucial role in the REDOX reaction kinetics of electrochemical energy-storage systems, which facilitates the exploration of advanced ionic conductors with high ionic conductivity and electrochemical stability by using electrode materials. The application of IL-based electrolytes in LIBs shows improved safety and electrochemical performance, especially in terms of high energy density and long cycle life. Although IL-based electrolytes have many advantages, they still face the challenges of high cost and mass production technology. The development of new ILs and mixed electrolytes to improve electrochemical performance and reduce costs is a key direction for future development. IL battery technology is in a stage of rapid development, with huge market potential and technological innovation space. With the deepening of research and technological progress, IL batteries are expected to play a more important role in the field of energy-storage technology in the next few years.

On the other hand, although commercial liquid electrolytes are widely used in LIBs, the freely moving anions and organic solvents in these electrolytes often lead to uneven ion fluxes and may react adversely with electrode materials, leading to reduced battery performance and safety failures. In order to solve these problems, people have been working on the research of single lithium-ion conductors for all-SSLBs. SIC batteries use a single ion as a conducting ion, which makes them superior to conventional LIBs in terms of safety because there is no problem with the coexistence of two ions that can cause short circuits. The design of the SIC battery allows the use of high-energy materials as electrodes, thereby increasing the energy density of the battery. This is useful for applications that require prolonged use or high energy output. Compared with traditional LIBs, SIC batteries use materials and production processes that are more environmentally friendly, helping to reduce their environmental impact.

At present, the research of SIC batteries is still in the development stage, but the existing research results show that through the progress of materials science and electrochemical technology, the performance of batteries is constantly improving. For example, by improving the design of electrode materials, increasing ionic conductivity, and optimizing the battery structure and manufacturing process, the performance of SIC batteries can be further improved. Single ion conductor battery has special application potential in solid-state batteries, water batteries, and other fields. Especially in areas that require high safety and fast charging capabilities, such as electric vehicles and energy-storage systems, SIC batteries show huge market application prospects. SIC batteries are becoming an important direction of battery technology development because of their high safety, fast charging ability, high energy density, and environmental protection characteristics. With the deepening of research and technological progress, it is expected that SIC batteries will achieve commercial application in the near future.

References

1 Francis, C.F.J., Kyratzis, I.L., and Best, A.S. (2020). Lithium-ion battery separators for ionic-liquid electrolytes: a review. *Advanced Materials* 32: 1904205.
2 Avila, J., Corsini, C., Correa, C.M. et al. (2023). Porous ionic liquids go green. *ACS Nano* 17: 19508–19513.
3 Ding, F.W., Li, Y.X., Zhang, G.X. et al. (2024). High-safety electrolytes with an anion-rich solvation structure tuned by difluorinated cations for high-voltage lithium metal batteries. *Advanced Materials* 36: 2400177.
4 Huang, Z., Song, W.L., Liu, Y.J. et al. (2022). Stable quasi-solid-state aluminum batteries. *Advanced Materials* 34: 2104557.
5 Hyun, W.J., Thomas, C.M., Luu, N.S. et al. (2021). Layered heterostructure ionogel electrolytes for high-performance solid-state lithium-ion batteries. *Advanced Materials* 33: 2007864.
6 Li, L.L., Wang, X.W., Gao, S.N. et al. (2024). High-toughness and high-strength solvent-free linear poly(ionic liquid) elastomers. *Advanced Materials* 36: 2308547.
7 Liu, X., Mariani, A., Diemant, T. et al. (2023). Reinforcing the electrode/electrolyte interphases of lithium metal batteries employing locally concentrated ionic liquid electrolytes. *Advanced Materials* 36: 2309062.
8 Muralidharan, N., Essehli, R., Hermann, R.P. et al. (2020). Lithium iron aluminum nickelate, LiNi$_x$Fe$_y$Al$_z$O$_2$-new sustainable cathodes for next-generation cobalt-free Li-ion batteries. *Advanced Materials* 32: 2002960.
9 Wang, R.Y., Jeong, S., Ham, H. et al. (2022). Superionic bifunctional polymer electrolytes for solid-state energy storage and conversion. *Advanced Materials* 35: 2203413.
10 Watanabe, M., Thomas, M.L., Zhang, S.G. et al. (2017). Application of ionic liquids to energy storage and conversion materials and devices. *Chemical Reviews* 117: 7190–7239.
11 Fan, X.T., Liu, S.Q., Jia, Z.H. et al. (2023). Ionogels: recent advances in design, material properties and emerging biomedical applications. *Chemical Society Reviews* 52: 2497–2527.
12 Cui, J.C., Li, Y., Chen, D. et al. (2020). Ionic liquid-based stimuli-responsive functional materials. *Advanced Functional Materials* 30: 2005522.

13 Hunt, P.A., Ashworth, C.R., and Matthews, R.P. (2015). Hydrogen bonding in ionic liquids. *Chemical Society Reviews* 44: 1257–1288.
14 Wang, B., Qin, L., Mu, T. et al. (2017). Are ionic liquids chemically stable? *Chemical Reviews* 117: 7113–7131.
15 Qian, W.J., Texter, J., and Yan, F. (2017). Frontiers in poly(ionic liquid)s: syntheses and applications. *Chemical Society Reviews* 46: 1124–1159.
16 Amarasekara, A.S. (2016). Acidic ionic liquids. *Chemical Reviews* 116: 6133–6183.
17 Zhang, S.-Y., Zhuang, Q., Zhang, M. et al. (2020). Poly(ionic liquid) composites. *Chemical Society Reviews* 49: 1726–1755.
18 Weiss, M., Simon, F.J., Busche, M.R. et al. (2020). From liquid- to solid-state batteries: ion transfer kinetics of heteroionic interfaces. *Electrochemical Energy Reviews* 3: 221–238.
19 Shaplov, A.S., Marcilla, R., and Mecerreyes, D. (2015). Recent advances in innovative polymer electrolytes based on poly(ionic liquid)s. *Electrochimica Acta* 175: 18–34.
20 Yiming, B.R.B., Han, Y., Han, Z.L. et al. (2021). A mechanically robust and versatile liquid-free ionic conductive elastomer. *Advanced Materials* 33: 2006111.
21 Lee, J.H., Kim, S., Park, K. et al. (2023). Contrasting miscibility of ionic liquid membranes for nearly perfect proton selectivity in aqueous redox flow batteries. *Advanced Functional Materials* 33: 2306633.
22 Wang, Y., Yang, Y., Li, N. et al. (2022). Ionic liquid stabilized perovskite solar modules with power conversion efficiency exceeding 20%. *Advanced Functional Materials* 32: 2204396.
23 Wang, D., Hwang, J., Chen, C.Y. et al. (2021). A β''-alumina/inorganic ionic liquid dual electrolyte for intermediate-temperature sodium-sulfur batteries. *Advanced Functional Materials* 31: 2105524.
24 Reber, D., Borodin, O., Becker, M. et al. (2022). Water/ionic liquid/succinonitrile hybrid electrolytes for aqueous batteries. *Advanced Functional Materials* 32: 2112138.
25 Liu, Z., Cheng, H., He, H. et al. (2021). Significant enhancement in the thermoelectric properties of ionogels through solid network engineering. *Advanced Functional Materials* 32: 2109772.
26 Ye, L., Liao, M., Zhang, K. et al. (2024). A rechargeable calcium-oxygen battery that operates at room temperature. *Nature* 626: 313–318.
27 Wang, S., Jiang, Y.J., and Hu, X.L. (2022). Ionogel-based membranes for safe lithium/sodium batteries. *Advanced Materials* 34: 2200945.
28 Zhang, J., Chen, Z.Y., Zhang, Y. et al. (2021). Poly(ionic liquid)s containing alkoxy chains and bis(trifluoromethanesulfonyl)imide anions as highly adhesive materials. *Advanced Materials* 33: 2100962.
29 Zhang, J., Li, H., Zhou, X. et al. (2024). Adhesive Zwitterionic poly(ionic liquid) with unprecedented organic solvent resistance. *Advanced Materials* 36: 2403039.
30 Zhang, P.P., Guo, W.B., Guo, Z.H. et al. (2021). Dynamically crosslinked dry ion-conducting elastomers for soft iontronics. *Advanced Materials* 33: 2101396.

31 Hyun, W.J., Thomas, C.M., and Hersam, M.C. (2020). Nanocomposite ionogel electrolytes for solid-state rechargeable batteries. *Advanced Energy Materials* 10: 2002135.

32 Liu, Q.H., Wang, S.Y., Zhao, Z.Y. et al. (2022). Electrically accelerated self-healable polyionic liquid copolymers. *Small* 18: 2201952.

33 Chen, M., Wu, J.D., Ye, T. et al. (2020). Adding salt to expand voltage window of humid ionic liquids. *Nature Communications* 11: 5809.

34 Kim, O., Kim, K., Choi, U.H. et al. (2018). Tuning anhydrous proton conduction in single-ion polymers by crystalline ion channels. *Nature Communications* 9: 5029.

35 Kim, S.Y., Kim, S., and Park, M.J. (2010). Enhanced proton transport in nanostructured polymer electrolyte/ionic liquid membranes under water-free conditions. *Nature Communications* 1: 88.

36 Meng, J.S., Hong, X.F., Xiao, Z.T. et al. (2024). Rapid-charging aluminium-sulfur batteries operated at 85°C with a quaternary molten salt electrolyte. *Nature Communications* 15: 596.

37 Sun, H., Zhu, G.Z., Xu, X.T. et al. (2019). A safe and non-flammable sodium metal battery based on an ionic liquid electrolyte. *Nature Communications* 10: 3302.

38 Zhan, X., Li, M., Zhao, X.L. et al. (2024). Self-assembled hydrated copper coordination compounds as ionic conductors for room temperature solid-state batteries. *Nature Communications* 15: 1056.

39 Zhang, J.Q., Sun, B., Zhao, Y.F. et al. (2019). A versatile functionalized ionic liquid to boost the solution-mediated performances of lithium-oxygen batteries. *Nature Communications* 10: 602.

40 Basile, A., Bhatt, A.I., and O'Mullane, A.P. (2016). Stabilizing lithium metal using ionic liquids for long-lived batteries. *Nature Communications* 7: 11794.

41 Zou, W.H., Zhang, J., Liu, M.Y. et al. (2024). Anion-reinforced solvating ionic liquid electrolytes enabling stable high-nickel cathode in lithium-metal batteries. *Advanced Materials* 36: 2400537.

42 Guzmán-González, G., Alvarez-Tirado, M., Olmedo-Martínez, J.L. et al. (2022). Lithium borate ionic liquids as single-component electrolytes for batteries. *Advanced Energy Materials* 13: 2202974.

43 Karimi, N., Zarrabeitia, M., Mariani, A. et al. (2020). Nonfluorinated ionic liquid electrolytes for lithium metal batteries: ionic conduction, electrochemistry, and interphase formation. *Advanced Energy Materials* 11: 2003521.

44 Li, J., Qi, J.Z., Jin, F. et al. (2022). Room temperature all-solid-state lithium batteries based on a soluble organic cage ionic conductor. *Nature Communications* 13: 2031.

45 Correia, D.M., Fernandes, L.C., Martins, P.M. et al. (2020). Ionic liquid-polymer composites: a new platform for multifunctional applications. *Advanced Functional Materials* 30: 1909736.

46 Melodia, D., Bhadra, A., Lee, K.Y. et al. (2023). 3D printed solid polymer electrolytes with bicontinuous nanoscopic domains for ionic liquid conduction and energy storage. *Small* 19: 2206639.

47 Demarthe, N., O'Dell, L.A., Humbert, B. et al. (2024). Enhanced Li and Mg diffusion at the polymer-ionic liquid interface within PVDF-based ionogel electrolytes for batteries and metal-ion capacitors. *Advanced Energy Materials* 14: 2304342.

48 Yu, L., Yu, L., Liu, Q. et al. (2022). Monolithic task-specific ionogel electrolyte membrane enables high-performance solid-state lithium-metal batteries in wide temperature range. *Advanced Functional Materials* 32: 2110653.

49 Qin, Y.P., Wang, H.F., Zhou, J.J. et al. (2024). Binding FSI to construct a self-healing SEI film for Li-metal batteries by in situ crosslinking vinyl ionic liquid. *Angewandte Chemie-International Edition* 63: e202402456.

50 Fu, C.Y., Homann, G., Grissa, R. et al. (2022). A polymerized-ionic-liquid-based polymer electrolyte with high oxidative stability for 4 and 5 V class solid-state lithium metal batteries. *Advanced Energy Materials* 12: 2200412.

51 Wang, Y., Yang, G., Wang, G. et al. (2023). Superlithiation performance of pyridinium polymerized ionic liquids with fast Li^+ diffusion kinetics as anode materials for lithium-ion battery. *Small* 19: 2302811.

52 Zhou, T.H., Zhao, Y., Choi, J.W. et al. (2021). Ionic liquid functionalized gel polymer electrolytes for stable lithium metal batteries. *Angewandte Chemie-International Edition* 60: 22791–22796.

53 Ke, J.Q., Wen, Z.P., Yang, Y. et al. (2023). Tailoring anion association strength through polycation-anion coordination mechanism in imidazole polymeric ionic liquid-based artificial interphase toward durable Zn metal anodes. *Advanced Functional Materials* 33: 2301129.

54 Li, Z., Liu, Z.W., Li, Z.Y. et al. (2020). Defective 2D covalent organic frameworks for postfunctionalization. *Advanced Functional Materials* 30: 1909267.

55 Li, J., Cai, Y., Wu, H. et al. (2021). Polymers in lithium-ion and lithium metal batteries. *Advanced Energy Materials* 11: 2003239.

56 Tu, H., Wang, Z., Xue, J. et al. (2024). Regulating non-equilibrium solvation structure in locally concentrated ionic liquid electrolytes for wide-temperature and high-voltage lithium metal batteries. *Angewandte Chemie International Edition* e202412896.

57 Tu, H.F., Li, L.G., Wang, Z.C. et al. (2022). Tailoring electrolyte solvation for LiF-rich solid electrolyte interphase toward a stable Li anode. *ACS Nano* 16: 16898–16908.

58 Huang, G.X., Liao, Y.Q., Zhao, X.M. et al. (2023). Tuning a solvation structure of lithium ions coordinated with nitrate anions through ionic liquid-based solvent for highly stable lithium metal batteries. *Advanced Functional Materials* 33: 2211364.

59 Wang, J.A., Buzolic, J.J., Mullen, J.W. et al. (2023). Nanostructure of locally concentrated ionic liquids in the bulk and at graphite and gold electrodes. *ACS Nano* 17: 21567–21584.

60 Liu, X., Mariani, A., Diemant, T. et al. (2022). Difluorobenzene-based locally concentrated ionic liquid electrolyte enabling stable cycling of lithium metal batteries with nickel-rich cathode. *Advanced Energy Materials* 12: 2200862.

61 Wang, Z.C., Zhang, F.R., Sun, Y.Y. et al. (2021). Intrinsically nonflammable ionic liquid-based localized highly concentrated electrolytes enable high-performance Li-metal batteries. *Advanced Energy Materials* 11: 2003752.

62 Wang, T.H., Chen, C., Li, N.W. et al. (2022). Cations and anions regulation through hybrid ionic liquid electrolytes towards stable lithium metal anode. *Chemical Engineering Journal* 439: 135780.

63 Lee, S., Park, K., Koo, B. et al. (2020). Safe, stable cycling of lithium metal batteries with low-viscosity, fire-retardant locally concentrated ionic liquid electrolytes. *Advanced Functional Materials* 30: 2003132.

64 Li, Z., Yu, R., Weng, S.T. et al. (2023). Tailoring polymer electrolyte ionic conductivity for production of low-temperature operating quasi-all-solid-state lithium metal batteries. *Nature Communications* 14: 482.

65 Zhang, Y.F., Huang, J.J., Liu, H. et al. (2023). Lamellar ionic liquid composite electrolyte for wide-temperature solid-state lithium-metal battery. *Advanced Energy Materials* 13: 2300156.

66 Hu, H., Li, J., Wu, Y. et al. (2024). Revealing the role and working mechanism of confined ionic liquids in solid polymer composite electrolytes. *Journal of Energy Chemistry* 99: 110–119.

67 Jin, B., Wang, D., He, Y. et al. (2023). Composite polymer electrolytes with ionic liquid grafted-Laponite for dendrite-free all-solid-state lithium metal batteries. *Chemical Science* 14: 7956–7965.

68 Xu, F., Deng, S., Guo, Q. et al. (2021). Quasi-ionic liquid enabling single-phase poly(vinylidene fluoride)-based polymer electrolytes for solid-state LiNi$_{0.6}$Co$_{0.2}$Mn$_{0.2}$O$_2$||Li batteries with rigid-flexible coupling interphase. *Small Methods* 5: 2100262.

69 Deng, Y., Zhao, S., Chen, Y. et al. (2024). Wide-temperature and high-rate operation of lithium metal batteries enabled by an ionic liquid functionalized quasi-solid-state electrolyte. *Small* 20: 2310534.

70 Pervez, S.A., Kim, G., Vinayan, B.P. et al. (2020). Overcoming the interfacial limitations imposed by the solid-solid interface in solid-state batteries using ionic liquid-based interlayers. *Small* 16: 2000279.

71 Leung, O.M., Gordon, L.W., Messinger, R.J. et al. (2024). Solid polymer electrolytes with enhanced electrochemical stability for high-capacity aluminum batteries. *Advanced Energy Materials* 14: 2303285.

72 Dong, M.Y., Zhang, K.Y., Wan, X.Y. et al. (2022). Stable two-dimensional nanoconfined ionic liquids with highly efficient ionic conductivity. *Small* 18: 2108026.

73 Liu, Z., Hu, Z., Jiang, X. et al. (2022). Metal-organic framework confined solvent ionic liquid enables long cycling life quasi-solid-state lithium battery in wide temperature range. *Small* 18: 2203011.

74 Zhang, S.L., Xie, Y.X., Somerville, R.J. et al. (2023). MOF-based solid-state proton conductors obtained by intertwining protic ionic liquid polymers with MIL-101. *Small* 19: 2206999.

75 Ho, J.W., Choi, J., Kim, D.G. et al. (2024). Bimetallic UiO-66(Zr/Ti)-ionic liquid grafted fillers with intensified Lewis acidity for high-performance composite solid electrolytes. *Advanced Functional Materials* 34: 2308250.

76 Zhu, J.D., Zhang, Z., Zhao, S. et al. (2021). Single-ion conducting polymer electrolytes for solid-state lithium-metal batteries: design, performance, and challenges. *Advanced Energy Materials* 11: 2003836.

77 Nguyen, M.C., Nguyen, H.L., Duong, T.P.M. et al. (2024). Highly safe, ultra-thin MOF-based solid polymer electrolytes for superior all-solid-state lithium-metal battery performance. *Advanced Functional Materials* 2406987.

78 Zhang, X., Su, Q., Du, G. et al. (2023). Stabilizing solid-state lithium metal batteries through in situ generated Janus-heterarchical LiF-rich SEI in ionic liquid confined 3D MOF/polymer membranes. *Angewandte Chemie* 135: 2304947.

79 Chen, Z., Steinle, D., Nguyen, H.-D. et al. (2020). High-energy lithium batteries based on single-ion conducting polymer electrolytes and Li[Ni$_{0.8}$Co$_{0.1}$Mn$_{0.1}$]O$_2$ cathodes. *Nano Energy* 77: 105129.

80 Zhang, H., Li, C., Piszcz, M. et al. (2017). Single lithium-ion conducting solid polymer electrolytes: advances and perspectives. *Chemical Society Reviews* 46: 797–815.

81 Shan, X.Y., Zhao, S., Ma, M.X. et al. (2022). Single-ion conducting polymeric protective interlayer for stable solid lithium-metal batteries. *ACS Applied Materials & Interfaces* 14: 56110–56119.

82 Du, D., Hu, X., Zeng, D. et al. (2019). Water-insoluble side-chain-grafted single ion conducting polymer electrolyte for long-term stable lithium metal secondary batteries. *ACS Applied Energy Materials* 3: 1128–1138.

83 Cao, C., Li, Y., Chen, S.S. et al. (2019). Electrolyte-solvent-modified alternating copolymer as a single-ion solid polymer electrolyte for high-performance lithium metal batteries. *ACS Applied Materials & Interfaces* 11: 35683–35692.

84 Shan, X.Y., Morey, M., Li, Z.X. et al. (2022). A polymer electrolyte with high cationic transport number for safe and stable solid Li-metal batteries. *ACS Energy Letters* 7: 4342–4351.

85 Liang, H.P., Zarrabeitia, M., Chen, Z. et al. (2022). Polysiloxane-based single-ion conducting polymer blend electrolyte comprising small-molecule organic carbonates for high-energy and high-power lithium-metal batteries. *Advanced Energy Materials* 12: 2200013.

86 Zhao, S., Song, S.H., Wang, Y.Q. et al. (2021). Unraveling the role of neutral units for single-ion conducting polymer electrolytes. *ACS Applied Materials & Interfaces* 13: 51525–51534.

87 Deng, H., Qiao, Y., Wu, S. et al. (2018). Nonaqueous, metal-free, and hybrid electrolyte Li-ion O$_2$ battery with a single-ion-conducting separator. *ACS Applied Materials & Interfaces* 11: 4908–4914.

88 Borzutzki, K., Nair, J.R., Winter, M. et al. (2022). Does cell polarization matter in single-ion conducting electrolytes? *ACS Applied Materials & Interfaces* 14: 5211–5222.

89 Wang, Y.Y., Sun, Q.Y., Zou, J.L. et al. (2023). Simultaneous high ionic conductivity and lithium-ion transference number in single-ion conductor network polymer enabling fast-charging solid-state lithium battery. *Small* 19: e2303344.

90 Li, C.C., Qin, B.S., Zhang, Y.F. et al. (2019). Single-ion conducting electrolyte based on electrospun nanofibers for high-performance lithium batteries. *Advanced Energy Materials* 9: 1803422.

91 Zhang, J.P., Zhu, J., Zhao, R.Q. et al. (2024). An all-in-one free-standing single-ion conducting semi-solid polymer electrolyte for high-performance practical Li metal batteries. *Energy & Environmental Science* 17: 7119–7128.

92 Dong, X., Mayer, A., Chen, Z. et al. (2024). Advanced single-ion conducting block copolymer electrolyte for safer and less costly lithium-metal batteries. *ACS Energy Letters* 5279–5287.

93 Deng, K., Wang, S., Ren, S. et al. (2016). A novel single-ion-conducting polymer electrolyte derived from CO_2-based multifunctional polycarbonate. *ACS Applied Materials & Interfaces* 8: 33642–33648.

94 Porcarelli, L., Shaplov, A.S., Bella, F. et al. (2016). Single-ion conducting polymer electrolytes for lithium metal polymer batteries that operate at ambient temperature. *ACS Energy Letters* 1: 678–682.

95 Dong, X., Chen, Z., Gao, X.P. et al. (2023). Stepwise optimization of single-ion conducting polymer electrolytes for high-performance lithium-metal batteries. *Journal of Energy Chemistry* 80: 174–181.

96 Shin, D.M., Bachman, J.E., Taylor, M.K. et al. (2020). A single-ion conducting borate network polymer as a viable quasi-solid electrolyte for lithium metal batteries. *Advanced Materials* 32: e1905771.

97 Wen, K.H., Xin, C.Z., Guan, S.D. et al. (2022). Ion-dipole interaction regulation enables high-performance single-ion polymer conductors for solid-state batteries. *Advanced Materials* 34: e2202143.

98 Cui, M.Y., Qin, Y.Y., Li, Z.C. et al. (2024). Retarding anion migration for alleviating concentration polarization towards stable polymer lithium-metal batteries. *Science Bulletin* 69: 1706–1715.

99 Li, H., Du, Y.F., Zhang, Q. et al. (2022). A single-ion conducting network as rationally coordinating polymer electrolyte for solid-state Li metal batteries. *Advanced Energy Materials* 12: 2103530.

100 Zhang, J.W., Wang, S.J., Han, D.M. et al. (2020). Lithium (4-styrenesulfonyl) (trifluoromethanesulfonyl) imide based single-ion polymer electrolyte with superior battery performance. *Energy Storage Materials* 24: 579–587.

101 Deng, K.R., Qin, J.X., Wang, S.J. et al. (2018). Effective suppression of lithium dendrite growth using a flexible single-ion conducting polymer electrolyte. *Small* 14: 1801420.

102 Kwon, H., Choi, H.J., Jang, J.K. et al. (2023). Weakly coordinated Li ion in single-ion-conductor-based composite enabling low electrolyte content Li-metal batteries. *Nature Communications* 14: 4047.

103 Hu, Z., Zhang, Y., Fan, W. et al. (2023). Flexible, high-temperature-resistant, highly conductive, and porous siloxane-based single-ion conducting electrolyte membranes for safe and dendrite-free lithium-metal batteries. *Journal of Membrane Science* 668: 121275.

104 Fraile-Insagurbe, D., Boaretto, N., Aldalur, I. et al. (2023). Novel single-ion conducting polymer electrolytes with high toughness and high resistance against lithium dendrites. *Nano Research* 16: 8457–8468.

105 Zeng, X., Dong, L., Fu, J. et al. (2022). Enhanced interfacial stability with a novel boron-centered crosslinked hybrid polymer gel electrolytes for lithium metal batteries. *Chemical Engineering Journal* 428: 2003521.

106 Li, Z.P., Oh, K.S., Seo, J.M. et al. (2024). A solvent-free covalent organic framework single-ion conductor based on ion-dipole interaction for all-solid-state lithium organic batteries. *Nano-Micro Letters* 16: 265.

107 Nguyen, H.-D., Kim, G.-T., Shi, J. et al. (2018). Nanostructured multi-block copolymer single-ion conductors for safer high-performance lithium batteries. *Energy & Environmental Science* 11: 3298–3309.

108 Wan, J.J., Liu, X., Diemant, T. et al. (2023). Single-ion conducting interlayers for improved lithium metal plating. *Energy Storage Materials* 63: 103029.

109 Xu, R., Xiao, Y., Zhang, R. et al. (2019). Dual-phase single-ion pathway interfaces for robust lithium metal in working batteries. *Advanced Materials* 31: e1808392.

110 Overhoff, G.M., Ali, M.Y., Brinkmann, J.P. et al. (2022). Ceramic-in-polymer hybrid electrolytes with enhanced electrochemical performance. *ACS Applied Materials & Interfaces* 14: 53636–53647.

111 Oh, K.S., Kim, J.H., Kim, S.H. et al. (2021). Single-ion conducting soft electrolytes for semi-solid lithium metal batteries enabling cell fabrication and operation under ambient conditions. *Advanced Energy Materials* 11: 2101813.

112 Zhang, J., Wang, Y., Xia, Q. et al. (2024). Confining polymer electrolyte in MOF for safe and high-performance all-solid-state sodium metal batteries. *Angewandte Chemie International Edition* 63: 2318822.

113 Hui, X., Zhan, Z., Zhang, Z. et al. (2024). Missing-linker defect functionalized metal-organic frameworks accelerating zinc ion conduction for ultrastable all-solid-state zinc metal batteries. *ACS Nano* 18: 25237–25248.

114 Xu, L., Xiao, X., Tu, H. et al. (2023). Engineering functionalized 2D metal-organic frameworks nanosheets with fast Li$^+$ conduction for advanced solid Li batteries. *Advanced Materials* 35: 2303193.

5

Application of Self-Healing Materials in Batteries

5.1 Introduction

5.1.1 The Need for Battery Innovation

With the increasing global demand for energy, innovations in battery technology have become a focal point across various fields. Batteries, as key devices for energy storage and supply, are widely used in consumer electronics, electric vehicles, renewable energy systems, and other sectors. However, the limitations of existing battery technologies urgently need to be overcome, driving the demand for innovative solutions. Innovation in batteries is a crucial factor in tackling future energy challenges. By enhancing energy density, reducing costs, improving safety and sustainability, and meeting the needs of emerging applications, innovations in battery technology can significantly advance technological progress and the implementation of global energy strategies. Looking to the future, interdisciplinary collaboration and new materials science research will provide important support for battery innovation. By concentrating innovative forces and resources, the prospects for battery technology will become brighter, serving a wider range of people and applications.

Among these, battery safety is one of the top concerns for consumers and manufacturers. Lithium-ion batteries can easily undergo thermal runaway when overcharged, short-circuited, or exposed to high temperatures, leading to fires or explosions. Therefore, enhancing the intrinsic safety of batteries is essential. This includes developing safer electrolytes (such as solid-state electrolytes, SSEs), optimizing battery structures to prevent thermal runaway, and embedding smart monitoring systems to detect and adjust battery temperatures and voltages in real time. The development of internet of things devices, wearable tech, and drones imposes new requirements on batteries. Small-sized, lightweight, flexible, and high-efficiency battery solutions will greatly enhance the performance and commercial value of these devices.

5.1.2 Overview of Self-Healing Materials

The concept of self-healing materials can be traced back to early twentieth-century materials science research when scientists began focusing on materials' natural healing abilities. When the material is stressed, causing microcracks, the microcapsules

Functional Auxiliary Materials in Batteries: Synthesis, Properties, and Applications, First Edition. Wei Hu.
© 2025 WILEY-VCH GmbH. Published 2025 by WILEY-VCH GmbH.

break, releasing healing agents that chemically react with curing agents in the matrix to heal the cracks. Microencapsulation technology is relatively simple and moderately priced, marking an important milestone in the field of self-healing materials.

Entering the twenty-first century, dynamic chemical bonds and supramolecular compounds inspired by biological healing mechanisms began to be applied to self-healing materials. Dynamic chemical bonds, such as reversible covalent bonds and hydrogen bonds, allow materials to self-repair through chemical reactions when damaged. These materials can typically reheal after changes in environmental conditions (such as temperature or light) and can be reused multiple times. With ongoing research, self-healing materials are gradually evolving toward multifunctionality and intelligence. These materials not only have self-healing capabilities but also respond to external stimuli such as temperature, humidity, and pH levels, enabling functionality under various conditions.

Despite the promising prospects of self-healing materials, they still face many challenges, such as balancing material strength and healing ability, durability, and stability under extreme conditions. In conclusion, as a highly promising smart material, self-healing materials are expected to achieve more widespread and in-depth applications in more fields as science and technology advance. Their development is not only a progression in materials science but also an essential driver for promoting environmental sustainability and resource conservation.

5.1.3 Benefits of Self-Healing Technologies in Batteries

The application of self-healing technology in batteries is a significant innovation in the battery field. This technology greatly enhances battery performance, safety, and lifespan by enabling them to automatically repair after damage, thus providing remarkable advantages for various battery applications.

Firstly, self-healing technology can significantly enhance the cycle life of batteries because batteries inevitably suffer internal damage during repeated charge-discharge cycles, such as structural damage and cracks caused by the expansion and contraction of electrode materials during the charge-discharge process. Traditional batteries usually experience rapid capacity decline and eventually fail when these problems occur. Self-healing materials can autonomously repair and heal these cracks, restoring the integrity and functionality of the electrode structure. This not only extends the battery's lifespan but also enhances its reliability and durability in high-frequency usage environments.

Secondly, self-healing technology also shows significant value in battery safety. When batteries are impacted by external forces, overcharged, or heated due to internal short circuits, traditional battery materials are prone to cracking, leaking, and even catching fire or exploding. Self-healing materials, particularly self-healing polymers applied to electrolytes and electrodes, can effectively repair damages caused by heat or mechanical stress due to their unique physical or chemical self-healing properties, thus enhancing battery safety under harsh conditions. This improvement is particularly suitable for applications requiring high safety standards, such as electric vehicles, avionics, and military energy storage systems.

Thirdly, self-healing technology also helps improve battery performance. Self-healing materials are often combined with high-performance conductive materials, which not only provide self-repair functions but also improve battery conductivity and electrochemical stability. Applying self-healing materials can reduce battery internal resistance, and increase energy density and power density, meeting modern electronic devices' demands for high efficiency and fast charging. Particularly in the development of solid-state batteries, using self-healing electrolytes has been proven to significantly enhance ionic conductivity and battery performance, thereby advancing the commercialization process of this future battery technology.

Fourthly, self-healing technology also helps improve battery performance levels. Self-healing materials are often combined with high-performance conductive materials, which not only provide self-repair functions but also enhance battery conductivity and electrochemical stability. Applying self-healing materials can reduce battery internal resistance, and increase energy density and power density, meeting the demands of modern electronic devices for high efficiency and fast charging. Particularly in the development of solid-state batteries, using self-healing electrolytes has been proven to significantly enhance ionic conductivity and battery performance, thus advancing the commercialization of solid-state battery technology, which is considered the future of batteries.

Furthermore, the application of self-healing technology in batteries also promotes environmental protection and sustainable development. Traditional batteries inevitably require frequent replacement due to damage, leading to increased electronic waste. Self-healing batteries, due to their significantly extended lifespan, can reduce the number of discarded batteries, thus lowering resource consumption and environmental burden. Moreover, many self-healing materials are polymers derived from renewable resources, having a reduced environmental impact during both manufacturing and disposal, thus meeting green sustainability criteria.

It is noteworthy that in the field of flexible and wearable electronics, self-healing technology offers distinct advantages. These devices are continually subjected to deformation and bending, placing greater demands on the fatigue resistance of battery materials. Self-healing materials can maintain the integrity of electrical performance under dynamic strain, making them suitable for a variety of innovative products, including smartwatches and health monitoring devices, providing long-lasting and reliable energy support for people's smart lifestyles.

With the advancement of technology and deepening research, it is foreseeable that self-healing technology will gradually secure a place in the global battery technology and materials science field. By extending battery life, enhancing safety, improving performance, and promoting environmental protection, it not only brings innovation to existing electronic devices but also lays a solid foundation for the development of future battery technologies. The proliferation of self-healing batteries will also usher in a safer, longer-lasting, and more environmentally friendly energy storage era.

5.1.4 Challenges in Scaling and Commercializing Self-Healing Materials

Self-healing materials, particularly their application in the battery field, as an innovative technology, show extremely attractive prospects. However, this technology still faces numerous challenges on the path toward commercialization and large-scale application. Thorough understanding and overcoming these challenges are key to truly realizing the potential of self-healing materials in the battery field.

Firstly, the complexity of the technology and cost issues are the main obstacles to the commercialization of self-healing materials. The sophistication of self-healing materials lies in their ability to autonomously repair themselves when damaged, which typically involves complex chemical structures and reaction mechanisms. These materials are often composed of high polymer polymers or nanocomposite materials, requiring expensive chemical reagents and complex manufacturing processes, which greatly increase production costs. For mass production and widespread application, simplifying the synthesis process of self-healing materials without sacrificing their excellent self-healing properties is an important research direction. This step requires a deep understanding of material science and the advancement of new manufacturing processes to achieve more efficient and cost-effective manufacturing workflows.

Secondly, the stability and durability of self-healing materials in long-term use is another issue that needs to be addressed. Although laboratory research has shown the initial effectiveness of self-healing materials, it remains to be verified whether their self-healing performance can be sustained during long-term use and multiple cycles. Batteries undergo complex and variable conditions during use, such as temperature changes, mechanical stress, and chemical environments. Self-healing materials must maintain their functionality under these conditions without rapidly degrading or failing. This requires researchers to develop self-healing materials that can adapt to various application environments and are durable, rather than products that only perform under ideal experimental conditions.

Another significant challenge is the issue of material performance and compatibility. Application scenarios like electric vehicles and wearable devices are increasingly demanding higher battery performance, such as higher energy density, faster charging speeds, and long cycle life. In these aspects, self-healing batteries must maintain or surpass the performance of traditional batteries. Furthermore, since self-healing materials often have physical and chemical properties different from traditional materials, incorporating them into existing battery systems to ensure good compatibility and performance optimization is also challenging. Developing self-healing materials that can seamlessly integrate with existing battery technology is another key to realizing their practical application.

During the commercialization process, the standardization and quality control of self-healing materials are issues that cannot be overlooked. As an emerging technology, the performance evaluation standards for self-healing materials have not been fully established. How to measure the performance, durability, and reliability of self-healing materials, and develop relevant international standards, is crucial for their large-scale application. At the same time, it involves strict quality control to ensure the consistency and safety of performance for each batch of materials, avoiding failures or accidents caused by individual differences.

Overall, the commercialization and large-scale application of self-healing materials in the battery field face challenges from multiple aspects, including technology, economy, compatibility, standardization, and market. However, with ongoing research and technological development, the widespread application of self-healing technology in batteries will not only revolutionize the reliability and safety of energy storage devices but, more importantly, provide crucial technological support for building a sustainable energy future.

5.2 Types of Self-Healing Materials for Battery Applications

5.2.1 Physically Bonded Self-Healing Materials

Physically bonded self-healing materials are typically based on noncovalent interactions, such as hydrogen bonds, π–π interactions, and van der Waals (vdW) forces. These materials do not require chemical reactions to achieve repair but are self-healed through dynamic physical cross-linking networks. They have relatively low cost, easy handling, and can perform well over multiple cycles, although they may have limited repair effectiveness under extreme conditions due to poor stability in high-temperature environments.

For example, a physical network formed by intertwined polymer chains can withstand high energy during bending or stretching because they can quickly respond to external mechanical stimuli and self-repair. This type of material performs excellently in the cycle life and safety of batteries, making them suitable for wearable devices and flexible electronics that need to withstand repeated cycles. In battery applications, physically bonded self-healing materials exhibit good flexibility and deformability because their particles or chain segments can rearrange through physical interactions when damaged. When the material undergoes mechanical damage, new physical bonds are rapidly formed, restoring the integrity and function of the battery.

5.2.2 Chemically Bonded Self-Healing Materials

Chemically bonded self-healing materials achieve repair through reversible chemical reactions. In these materials, dynamic covalent bonds (such as ester bonds and disulfide bonds) are usually introduced; these chemical bonds can break and reform after the material is damaged.

In battery applications, chemically bonded self-healing materials can achieve more lasting repair effects. The self-healing ability of these materials is often influenced by the chemical reaction rate, so selecting appropriate chemical groups and reaction conditions can adjust their repair speed and efficiency. They have high repair efficiency and can maintain stable performance in high temperatures and complex environments, enhancing battery safety. The synthesis and preparation process of the materials is relatively complex, and the corresponding production costs are higher. For example, many self-healing electrolytes use cross-linked polymers, where self-healing mechanisms are achieved through dynamic chemical reactions [such as the Diels–Alder (D-A) reaction]. These chemical bonds can be repaired at room temperature, providing better cycle stability and safety for batteries. Compared with physically bonded materials, these chemically bonded materials are more resilient in the face of high temperatures or extreme conditions, making them suitable for energy storage devices that require high reliability.

5.2.3 Composite Self-Healing Materials with Multiple Repair Mechanisms

Composite self-healing materials combine multiple repair mechanisms to provide stronger self-healing capabilities. These materials are typically composed of a combination of physically bonded and chemically bonded materials, or a matrix combined with fillers, enhancing repair performance through synergistic effects.

In battery applications, composite self-healing materials use a combination of two or more self-healing mechanisms, allowing the materials to rapidly and effectively self-repair when faced with different types of damage. For instance, a certain composite electrolyte possesses both hydrogen bond cross-linking and dynamic covalent cross-linking, enabling it to swiftly repair damage through both mechanisms when the battery suffers from different forms of cracking or delamination. With comprehensive performance and strong adaptability, it can perform excellently in various environments, effectively handling complex working conditions. However, the design and manufacturing process of such materials is relatively complex, potentially increasing production costs and technical requirements. Research indicates that composite self-healing materials generally display higher ionic conductivity and superior mechanical properties in terms of electrochemical performance. For example, combining conductive nanoparticles with a polymer matrix can enhance the conductivity of the electrolyte while maintaining its continuity through self-healing mechanisms, improving the battery's cycling performance. The design approach of composite materials can effectively deal with complex working environments and enable multifunctional integration to meet the demands of future high-performance batteries.

5.3 Applications of Self-Healing Materials in Batteries

5.3.1 Gel Polymer Electrolytes

In the field of zinc-ion batteries, traditional electrolytes face issues such as zinc dendrite growth and side reactions at the electrode/electrolyte interface during application, which constrain their performance and stability. To address this, Pan et al. [1] developed a self-healing hydrogel polymer electrolyte (SHGPE). By introducing carboxyl-modified polyvinyl alcohol (PVA-COOH) cross-linked with Fe^{3+}, and prepared in the environment of $Zn(NO_3)_2$ and $MnSO_4$, this electrolyte achieved self-healing through dynamic bonding between carboxyl groups and Fe^{3+}. When the electrolyte was mechanically damaged, the carboxyl groups reformed bonds with Fe^{3+}, thereby repairing the damaged area and restoring its mechanical strength and conductivity. This mechanism effectively suppressed zinc dendrite growth and reduced side reactions at the electrode/electrolyte interface. Experimental results showed that when this hydrogel electrolyte was applied to zinc-manganese batteries, it maintained a capacity of 177 mAh g^{-1} after 1000 cycles at a rate of 1 C, demonstrating excellent capacity retention and efficient self-healing abilities. Additionally, the battery maintained efficient electrochemical performance even after multiple cycles of mechanical damage and repair. After five cut/repair cycles, the battery's capacity and cycling stability showed almost no change, demonstrating excellent self-healing capabilities.

Jin et al. [2] further explored the self-healing mechanism by immobilizing imidazolium-based ionic liquids (ILs) in fluoropolymer gels. Imidazolium cations formed strong ion–dipole interactions with fluorine atoms in the matrix, enabling the electrolyte to rapidly self-heal after mechanical damage, enhancing mechanical strength and thermal stability, and providing a wide electrochemical operating window (up to 4.5 V vs. Li$^+$/Li), which was highly beneficial for high-voltage cathode applications. By immersing a scratched P(VDF-HFP) film with fixed IL in a dimethyl ether (DME)/1,3-dioxolane (DOL) solution of LiTFSI, the scratches almost completely disappeared after 12 hours, demonstrating excellent self-healing capability. Simultaneously, regarding electrochemical performance, this gel electrolyte exhibited an ionic conductivity as high as 8.8×10^{-4} S cm^{-1} at room temperature, far exceeding traditional liquid electrolytes. In tests of symmetric Li/Li batteries, batteries using LiTFSI-IL-P(VDF-HFP) stably cycled for 1000 hours at a current density of 0.5 mA cm^{-2} and maintained for 230 hours at 2.0 mA cm^{-2}, clearly demonstrating its effective suppression of lithium dendrite growth. In LiFePO$_4$/Li batteries using this gel electrolyte, the battery demonstrated remarkable cycling stability and rate performance, with an initial discharge capacity of 132 mAh g^{-1} at 1 C, maintaining 96.5% capacity retention after 200 cycles, and a Coulombic efficiency of 99.8%. Additionally, in Li—S battery applications, this electrolyte effectively suppressed the polysulfide shuttle effect, significantly enhancing capacity retention and cycle life, particularly maintaining 867 mAh g^{-1} capacity after 200 cycles at 0.2 C, with a Coulombic efficiency close to 100%.

On the other hand, to enhance the stability of zinc anodes in zinc-based aqueous batteries, Huang et al. [3] introduced a SHGPE with a rigid-flexible framework, synthesized through cross-linking polymerization and composite reinforcement. The self-healing hydrogel electrolyte relied on dynamic hydrogen bonds formed by the hydroxyl groups in its polymer framework, achieving self-repair and stretchable properties. When the hydrogel electrolyte undergoes mechanical fracture, these hydrogen bonds enable the material to automatically repair, restoring its structural integrity and electrochemical performance. The self-healing electrolyte possessed high ionic conductivity (23.1 mS cm^{-1}, 25 °C) and excellent mechanical strength, promoting self-healing and stretchability through dynamic hydrogen bonds, thereby improving electrode–electrolyte interface compatibility and facilitating orderly zinc metal plating/stripping. Additionally, the introduction of carboxymethyl cellulose enhanced the film-forming ability of the composite electrolyte, creating a cross-linked network with higher porosity. This structure aided in improving water absorption capacity, providing rapid ion transport pathways.

In terms of electrochemical performance, this SHGPE enhanced the performance of zinc-based aqueous batteries with its excellent ionic conductivity and interface compatibility. The hydrogel electrolyte exhibited a high ionic conductivity of 23.1 mS cm^{-1} at 25 °C, significantly better than traditional pure polyacrylamide (PAM) electrolytes, which helped to conduct zinc ions more efficiently, reducing electrochemical polarization and overpotential during battery operation. Using this electrolyte, zinc symmetric batteries underwent 300 hours of plating and stripping cycles at a current density of 5 mA cm^{-2}, with voltage hysteresis remaining below 100 mV throughout, demonstrating excellent stability and antidendrite growth capability. In flexible zinc/manganese dioxide batteries, this electrolyte also supported a significant capacity of up to 304 mAh g^{-1} and exhibited excellent durability and cycle life at high current densities, with its capacity maintaining 83.1% after 1500 charge-discharge cycles. Meanwhile, self-healing performance played a crucial role in maintaining battery life and functional stability. Experiments proved that batteries using this hydrogel electrolyte can quickly restore their normal function and electrochemical performance after mechanical cuts through the self-repair process, maintaining good healing efficiency and battery capacity after five complete cut/heal cycles. This characteristic not only improved battery lifespan and durability but also showed its attractiveness and practicality when facing inevitable physical damage in everyday use, providing a reliable basis for developing durable and high-performance wearable energy storage devices.

To address the issues of short lifespan and difficulty in repairing existing zinc-air batteries (ZABs), Sun et al. [4] designed an SHGPE synthesized via photopolymerization of acrylamide and polyethylene glycol (PEG) monomethyl ether acrylate in 1-ethyl-3-methylimidazolium dicyanamide and used it as an electrolyte in the fabrication of self-healing ZABs for the first time (Figure 5.1). Dynamic hydrogen bonds are weak interactions that allow materials to restore their structure and function after mechanical damage or fracture through self-reorganization.

Additionally, these batteries could repeatedly self-heal and restore their performance after damage, providing durable and reliable power sources for wearable devices. This hydrogel electrolyte could restore its electrochemical performance to over 95% of its prebreak level after five complete cut and heal cycles. Testing the assembled ZAB at a current density of 0.1 mA cm^{-2} indicated a cycling duration of 340 hours with a charge-discharge voltage gap of 0.96 V at 25 °C, while showing outstanding environmental and electrochemical stability with cycling times of 400 hours at −20 °C and 275 hours at 40 °C.

Figure 5.1 Photograph of the PAM-PEGMA-IL ionogel and schematic illustration of its internal structure. Source: Reproduced with permission from Li et al. [4]/John Wiley & Sons.

Cheng et al. [5] designed a SHGPE featuring physically cross-linked dopamine (DA)-grafted sodium alginate and hydrophobic associative networks (DA-Alg/PAAm-O), wherein the hydrophobic groups spontaneously reentered the hydrophobic microdomain and rebound, restoring the electrolyte to its original state upon breaking. The noncovalent bonds and ionic interactions of DA-Alg/PAAm-O GPE enabled it to completely heal within 24 hours after breaking, showing a self-healing efficiency of 72.4% (after the fifth break-heal cycle). DA-Alg/PAAm-O GPE exhibited excellent electrochemical performance. The Zn//Zn symmetric battery cycled for over 2000 hours at a current density of 2.0 mA m^{-2}, maintaining stable voltage polarization. Additionally, the Zn//MnO$_2$ battery based on DA-Alg/PAAm-O GPE achieved a specific capacity of 290 mAh g^{-1} at 0.2 A g^{-1}, with a capacity retention rate of 91% after 4000 cycles. These excellent performances were mainly attributed to the GPE's high ionic conductivity (32.3 mS cm^{-1}), good interfacial adhesion (80 kPa), and the ability to inhibit zinc dendrite growth.

Among the safety challenges faced by lithium metal batteries (LMBs), traditional liquid electrolytes introduce significant safety hazards due to their low flash point and risk of leakage. To mitigate these issues, Wang et al. [6] developed a self-healing gel polymer electrolyte (IL-SHGPE) based on ILs, with main components including polymers with quadruple hydrogen bond units, 1-butyl-3-methylimidazolium bis(trifluoromethylsulfonyl)imide (BMIM-TFSI) IL, and polyethylene glycol diacrylate (PEGDA) as a cross-linker. This electrolyte incorporated polymers capable of forming quadruple hydrogen bonds, coupled with IL components, granting the electrolyte self-healing capabilities. Experimentally, this gel electrolyte could achieve 95% self-healing efficiency within 30 minutes at room temperature, coupled with outstanding thermal stability (up to 410 °C) and mechanical strength (41.3% elongation, 1.9 MPa tensile strength). The electrolyte exhibited an ionic conductivity of 1.29×10^{-3} S cm^{-1} at room temperature, and an electrochemical stability window of 5.18 V (relative to Li/Li$^+$). For battery performance testing, the LMB utilizing IL-SHGPE showed an initial discharge capacity of 139 mAh g^{-1} a 1 C rate, with a capacity retention rate of 88.3% after 200 cycles. These results indicated that IL-SHGPE exhibits excellent performance in self-healing ability, mechanical properties, and electrochemical properties, providing important reference and insight for the development of high-performance LMBs.

Quartarone et al. [7] developed a gel electrolyte formulated from a highly cross-linked PEGDA network blended with ureidopyrimidinone (UPy) units, which allows the gel to autonomously repair mechanical damages. The dynamic multiple hydrogen bonding among UPy units was pivotal in facilitating this self-healing capability, which restored the gel's ionic conductivity even after significant physical fractures. The PEGDA-UPy gels demonstrated solid-like mechanical stability while maintaining good ionic transport properties. The gels' ability to heal was notably enhanced at elevated temperatures, achieving substantial recovery of conductivity after damage in a short time. Electrochemical tests revealed that the PEGDA-UPy 67, containing a higher liquid electrolyte concentration, offered superior ionic conductivity and a stable lithium transference number, alongside preventing mossy lithium formation, which was crucial in avoiding dendrite growth commonly seen in lithium metal anodes. Furthermore, the electrochemical stability window of over 5.5 V indicated promising compatibility for lithium-ion applications, showing advantages over traditional liquid electrolytes. This self-healing attribute, combined with the solid-like structural benefits of a cross-linked gel, positions it as a safer, more reliable component for future lithium batteries. The gels' multifunctional nature extended beyond self-repair; they effectively immobilized the salt anions to enhance lithium-ion transport, a feature that could be indispensable in high-energy battery designs.

In addition to lithium-ion batteries, the self-healing ability of electrolytes is equally important in the field of zinc-based batteries. To address the challenges of zinc dendrite growth and mechanical durability, Cheng et al. [8] prepared a PAAm-O-B electrolyte in an aqueous solution through a simple micelle copolymerization reaction. They mixed hydrophobic octadecyl methacrylate (OMA) and hydrophilic acrylamide (AAm) to form an emulsion, then added N,N'-methylenebisacrylamide (BIS) as a chemical cross-linking agent to enhance the mechanical strength and maintain the hydrogel framework. Zinc and manganese ions were then introduced, and finally, by adding potassium persulfate (KPS) to initiate free radical polymerization, the cross-linked PAAm-O-B hydrogel electrolyte was obtained (Figure 5.2). Relying on the three-dimensional network structure spontaneously formed by hydrophobic chains in an aqueous environment, when the electrolyte was broken, the hydrophobic groups spontaneously re-entered the hydrophobic microdomains and recombined, restoring the electrolyte to its initial state. Experimental results showed that the PAAm-O-B electrolyte can completely heal within 24 hours after breaking and maintain up to 91.9% of its ionic conductivity. This self-healing ability mainly stemmed from the synergistic effect of dynamic physical cross-linking and chemical cross-linking, ensuring that the hydrogel maintains its structure and function after multiple damages and repairs.

The $Zn//MnO_2$ battery using this electrolyte achieved a specific capacity of 298 mAh g^{-1} at a 1 C discharge rate and maintained 75% capacity after 2000 cycles at 8 C. Furthermore, the battery exhibited excellent electrochemical performance

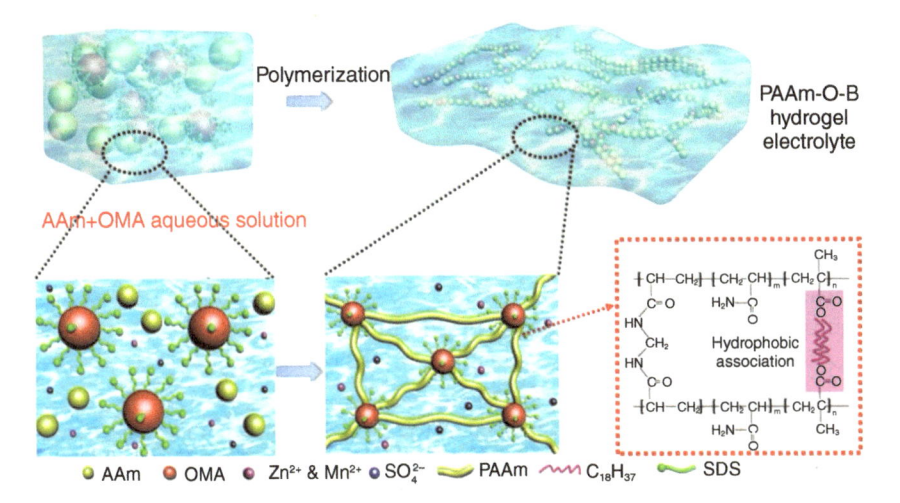

Figure 5.2 Schematic illustration of the synthesis of PAAm-O-B hydrogel electrolyte. Source: Wang et al. [8]/with permission of John Wiley & Sons.

and mechanical stability after multiple bending and break-healing cycles, showing good adaptability to deformation and self-healing capability. In particular, after several self-healing processes, the battery's ionic conductivity remained at 91.9% of its initial value, and the self-healed battery exhibited an 83% capacity retention rate after 1000 cycles at 8 C.

Wang et al. [9] synthesized a dual-network structured self-healing gel polymer electrolyte (DN-SHGPE) with enhanced mechanical and ionic conductive properties suitable for high-performance LMBs. Central to this system was the use of a dual-network structure, which combined the dynamic UPy dimers functionalities to enable excellent self-healing properties and a mechanically robust framework via PEGDA cross-linking. The urethane-pyrimidine units formed a quadruple hydrogen bonding network, providing both reversible self-repair capabilities and high dimensional stability. This combination ensured that the DN-SHGPE retained its mechanical integrity and continued to facilitate efficient ion transport under conditions that would typically degrade battery performance.

From a mechanistic standpoint, the UPy dimer system, with a remarkably high binding constant, was pivotal in dynamically recreating intermolecular connections through hydrogen bonds once damaged. This quadruple hydrogen bonding system was designed to act autonomously, reestablishing physical connections when polymer disruptions occurred, thus restoring the material's integrity. Meanwhile, the chemical cross-links formed by PEGDA supported the structure and facilitated the electrolyte's high mechanical strength and thermal stability, essential for sustaining operational effectiveness and protecting against dendrite penetration.

Quantitative tests demonstrated that DN-SHGPE exhibited tensile stress of up to 17.24 MPa, significantly higher than traditional gel polymer electrolytes, indicating superior mechanical resilience. The dynamic self-healing capability was assessed by deliberately inducing fractures, which the DN-SHGPE was able to repair autonomously at ambient conditions within five hours, without external stimuli. Furthermore, DN-SHGPE showed an ionic conductivity of up to 8.47×10^{-4} S cm^{-1} at room temperature, along with a wide electrochemical stability window reaching 4.93 V.

Sodium dendrite growth presents significant challenges in sodium metal batteries (SMBs) by compromising safety and performance. Huang et al. [10] designed a self-healing polymer electrolyte (SPE) that leverages an elastic interface to suppress dendrite formation. A self-healing polymer matrix (pBAA) was synthesized via free radical polymerization of butyl acrylate and acrylic acid (AA), which was then immersed in a liquid electrolyte to fabricate the SPE. This SPE exhibited non-porous and compact properties that play a crucial role in preventing sodium-potassium alloy (NaK) permeation and dendrite piercing.

The self-healing mechanism is enabled by the dynamic coordinate bonds between aluminum ions (Al^{3+}) from aluminum triacetylacetone (AlACA) and the carboxyl groups of AA. When mechanical damage occurs, these bonds can autonomously reform, promoting the reconnection of fractured polymer segments. This capacity for instant self-repair, occurring without external stimuli, was demonstrated by the recovery of 63.5% elongation in the polymer after damage, a robust self-repair in SPE observed both visually and through tensile testing. This mechanism ensured that the elasticity and structural integrity of the SPE were preserved, facilitating a smooth sodium deposition surface.

The Na|MC SPE|Na symmetric battery demonstrated a prolonged life of up to 590 hours with a low polarization voltage of 192 mV, significantly outperforming traditional liquid electrolyte systems. The underlying mechanism involves the dense non-porous structure of the SPE, which ensures uniform sodium-ion transport and distribution across the battery's electrode interface. In contrast, conventional systems allow for uneven sodium distribution and dendrite growth due to porous interfaces. Additionally, the reconstructed elastic interface induced by self-healing capacity helps wrangle and suppress the development of sodium dendrite protrusions, maintaining a uniform ion flux and contributing to the battery's extended lifespan.

In wearable devices and zinc ion application fields, NiCo//Zn batteries have garnered attention due to their mechanical deformation sensitivity. Zhi et al. [11] developed an inherently self-healing NiCo//Zn battery employing a novel SHGPE. The electrolyte was composed of sodium polyacrylate (PANa) hydrogel cross-linked with trivalent iron ions, which acted as a non-covalent cross-linker forming ionic bonds that reconnect broken surfaces when the hydrogel was cut, providing the electrolyte's self-healing ability (Figure 5.3). The PANa-Fe^{3+} hydrogel can self-heal without any external stimulus, maintaining its ionic conductivity. Electrochemical tests showed that the NiCo//Zn battery using PANa-Fe^{3+} electrolyte could maintain 87% of its initial capacity after self-healing and keep its typical charge-discharge profile even after multiple cutting and self-healing cycles.

This inherently self-healing battery demonstrated excellent electrochemical and mechanical performance. The NiCo||Zn battery had a discharge capacity of about 250 mAh g^{-1} at a current density of 0.2 A g^{-1}, which surpasses many reported

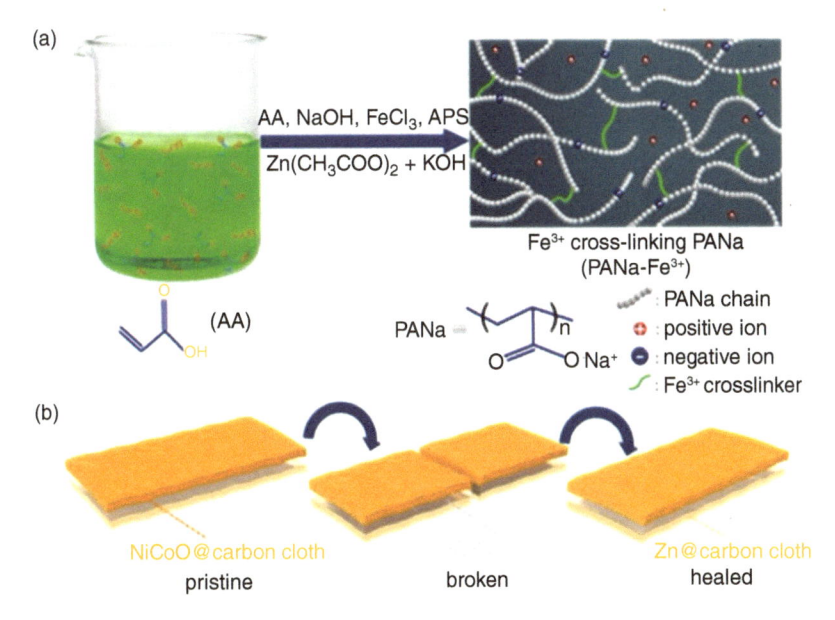

Figure 5.3 (a) Fabrication of the PANa-Fe^{3+} electrolyte from acrylic acid monomer (AA), NaOH (neutralizer), FeCl$_3$ (cross-linker), ammonium persulfate (APS, initiator), and zinc acetate and potassium hydroxide (ion sources). (b) Fabrication of the intrinsically self-healable NiCo||Zn battery. Source: Reproduced with permission from Huang et al. [11]/John Wiley & Sons.

solid-state zinc-based batteries. Furthermore, the battery's charge-discharge curve remained stable after multiple cutting and self-healing cycles, indicating that the battery's self-healing properties effectively restored its electrochemical performance. In practical applications, batteries using this self-healing electrolyte can reconnect and restore energy storage capability after breaking, showing promising application prospects in damaged electronic devices.

Wang et al. [12] employed silk fibroin's intrinsic properties and its interaction with PAM to create a semi-interpenetrating network structure within the hydrogel, resulting in a high-performance SHGPE. The supramolecular hydrogen bonding interactions between amide bonds in PAM and polypeptide chains in silk fibroin facilitated rapid self-repair by allowing the material to re-form these hydrogen bonds when broken. This capability was visually demonstrated by the SF-gel electrolyte's ability to heal completely within minutes and withstand substantial mechanical deformation post-repair. Quantitative assessments revealed that the SF-gel retains 68.2% of its maximum tensile strength and 79.1% of its break elongation even after five cutting-healing cycles.

The SF-gel electrolyte achieved a high ionic conductivity of 14.4 mS cm^{-1}, a significant improvement over the PAM-only hydrogel, due to its enhanced structural plasticity and low interface impedance with electrodes, which boosts ion diffusion. This self-healable electrolyte also demonstrated exceptional electrochemical performance in Ag—Zn batteries, with a notable capacity retention of 85.7% after 800 cycles at 0.5 A g^{-1}. Furthermore, its inherent mechanical strength and

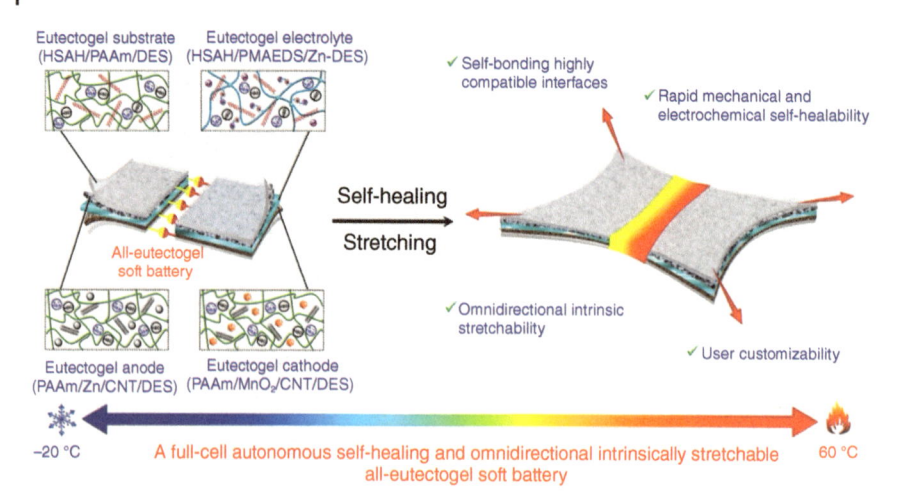

Figure 5.4 Design and performance of full-device autonomous self-healing and omnidirectional intrinsically stretchable all-eutectic gel soft battery (AESB). HSAH: (R)-12-hydroxystearic acid hydrazide; PAAm: polyacrylamide; DES: deep eutectic solvent; PMAEDS: poly[2-(methacryloyloxy)ethyl]diethy-(3-sulopropyl); CNT: carbon nanotube. Source: Gu et al. [14]/with permission of John Wiley & Sons.

hydrogen bonding capabilities provided substantial resilience against zinc dendrite formation, a common issue that leads to short-circuiting in metal anode batteries.

Additionally, Sun et al. [13] developed a SHGPE by integrating hydrogen-bonded supramolecular polymers with ILs. This electrolyte's self-healing mechanism relies on the quadruple hydrogen bonds between UPy groups and the charge interactions between ILs and PIL-UPy polymers. This material was endowed with self-repair ability due to the reversibility and dynamic characteristics of these noncovalent bonds, allowing it to repolymerize and restore its function after mechanical damage. In lithium-ion batteries, this self-healing electrolyte can effectively repair both external and internal damages, substantially increasing the battery's reliability and lifespan. Experimental findings demonstrated that the Li/LiFePO$_4$ battery employing Ionogel-3.5 ionogel film maintained a discharge capacity of 147.5 mAh g^{-1} and a coulombic efficiency of 99.7% after 120 cycles at a 0.2 C charge-discharge rate. Moreover, this electrolyte exhibited an ionic conductivity of up to 1.41×10^{-3} S cm^{-1} at room temperature, approaching the conductivity of traditional liquid electrolytes. The self-healed electrolyte could effectively restore the electrochemical performance of the battery, ensuring stability and safety after multiple cycles.

Liu et al. [14] developed a fully self-healing and inherently stretchable all-deep eutectic soft battery (AESB) aimed at realizing genuine full-battery healing and multidirectional stretch capability. By introducing self-adhesive eutectic gels into the battery, these gels utilized abundant noncovalent bonds (such as hydrogen bonding and ion–dipole interactions) to form strong self-adhesive interfaces between battery layers, enabling spontaneous healing after mechanical damage and restoring the battery's mechanical and electrochemical performance (Figure 5.4).

Batteries with SP-DN eutectic gel electrolyte exhibited a cycle life of over 300 hours at a current density of 1 mA cm^{-2} and an overpotential of merely 0.2 V. Moreover, the electrolyte achieved an ionic conductivity of 4.27 mS cm^{-1} at room temperature and retained a conductivity of 1.11 mS cm^{-1} even at a low temperature of −60 °C. After multiple damage and self-healing cycles, the electrolyte's conductivity remained largely unchanged, demonstrating excellent self-healing capability. The AESB battery had a discharge capacity of 257 mAh g^{-1} at a 0.1 A g^{-1} discharge rate, with a coulombic efficiency of over 99.9% after 500 cycles. The battery's capacity and cycle performance remained stable across various temperature conditions (−20 to 60 °C). Even when stretched to 500% and 1000% area strains, the battery still delivered a high capacity of 257.15 and 254.78 mAh g^{-1}, respectively, while maintaining stable cycling performance.

Xie et al. [15] synthesized a three-dimensional cross-linked network gel polymer electrolyte (CNGPE) through a thiol-ene click reaction. The DES introduced, consisting of LiTFSI and N-methylacetamide (NMA), significantly enhanced the ionic conductivity of CNGPE. The introduction of hydrogen bonds and dynamic disulfide bonds within the three-dimensional cross-linked network of CNGPE provided it with rapid self-healing capability at room temperature. Scratches on the surface of CNGPE disappeared within two hours without external stimulation, indicating its good self-healing properties. This was primarily due to the dynamic disulfide bonds and sacrificial hydrogen bonds in the three-dimensional cross-linked network; the hydrogen bonds broke first during damage, while the disulfide bonds undertook the main self-healing function.

The Li/CNGPE/Li symmetric battery demonstrated a low overpotential (0.07 V) and excellent interfacial stability while cycling for 800 hours at a current density of 0.1 mA cm^{-2}. At a higher current density of 0.2 mA cm^{-2}, the overpotential stabilized at 0.11 V, with the battery maintaining good performance even after 500 hours of cycling. The LiFePO4/CNGPE/Li battery exhibited an initial discharge specific capacity of 150 mAh g^{-1} at room temperature and 0.5 C, and retained a specific capacity of 126 mAh g^{-1} after 960 cycles, with a coulombic efficiency reaching 99.9%. Additionally, the NCM811/CNGPE/Li battery achieved a discharge-specific capacity of 154.9 mAh g^{-1} after 32 cycles at 0.1 C. Through X-ray photoelectron spectroscopy (XPS) analysis, it was found that after 300 cycles, the solid electrolyte interphase (SEI) formed on the lithium metal anode surface in the LiFePO$_4$/CNGPE/Li battery contained Li$_3$N and LiF, which effectively suppressed lithium dendrite growth and improved the battery's cycling performance.

Micro-cracks and breakage in wearable electronic devices impede ion transport and diminish battery performance. Yang et al. [16] utilized dicationic polymerized ionic liquids (PILs) as the structural backbone, with IL infill to form a sea-island structure. This structure facilitates efficient lithium-ion transfer through interlaced channels, significantly boosting the ionic conductivity of the electrolytes. The self-healing feature was driven by reversible ionic bonds formed between abundant cations and anions within the polymer chains and ILs, enabling the electrolytes to repair damages autonomously. The fabrication process involved synthesizing PVT polymers through nucleophilic substitution and ion exchange reactions, introducing TFSI anions to ensure higher electrochemical stability. Upon mechanical stress or cutting, the ionic interactions enabled the material to self-heal without external assistance, as the electrostatic forces reconstructed the ionic bonds, restoring the material's integrity and functionality. This self-healing process was evident from experiments demonstrating that the electrolytes regain their initial properties after being cut and subsequently healed, with optimal healing times of around 60 minutes.

D'Angelo and Panzer [17] designed a stretchable and self-healing gel electrolyte using fully zwitterionic polymer networks in solvate ionic liquids (SILs) for lithium-based batteries. The unique feature of these electrolytes was their ability to self-heal and maintain mechanical integrity and ionic conductivity, thanks to the dynamic interactions facilitated by zwitterionic components within the polymer network. They highlighted the synthesis of solvate ionogels via UV-initiated free-radical copolymerization of two zwitterionic monomers: 2-methacryloyloxyethyl phosphorylcholine (MPC) and sulfobetaine vinylimidazole (SBVI). These monomers formed a fully zwitterionic (f-ZI) polymer network that interacted with the SIL, [Li(G4)][TFSI], ensuring high ionic conductivity and mechanical robustness.

The mechanism of self-healing in these materials was primarily due to the dynamic physical cross-links created between the zwitterionic units. The MPC units, through phosphorylcholine groups, establish strong interactions with the lithium-containing SIL ions, while the SBVI groups form robust dipole–dipole interactions within the network. This configuration enabled the ionogels to self-heal effectively after being mechanically disrupted, as these zwitterionic cross-links can break and reform, bridging any fractures. The high mobility of the polymer chains, owing to the lack of covalent cross-links, allowed the network to reorganize and reestablish these interactions, facilitating the self-healing process. Tests conducted on the ionogels demonstrated their ability to recover their mechanical properties postdamage. For instance, puncture and slicing tests revealed complete healing when the materials were heated at 50 °C, showing no visible signs of the initial damage. Rheological measurements further confirmed this behavior, indicating that the ionogels exhibit solid-like properties even after substantial deformation.

Guo et al. [18] introduced a novel self-healing aqueous ammonium-ion micro battery (NH_{4^+} AMBs), featuring a PVA-NH_4Cl hydrogel electrolyte and an MXene-integrated perylene (PTCDI) anode. The development of this battery aimed to address the challenges of lowly rate performance and power density found in traditional metal-ion batteries, which were caused by the strong interactions between metal ions and electrode materials. Ammonium ions, with their smaller hydrated ionic radius and lower molar mass, offered enhanced reaction rates and reduced corrosion, making them an attractive alternative.

The self-healing capability was primarily achieved through the PVA-NH_4Cl hydrogel electrolyte, which was rich in hydroxyl side groups and hydrogen bonds, allowing it to self-repair when damaged. This hydrogel, modified via a freeze/thaw strategy, ensures good contact and ion conductivity between the electrolyte and electrodes. When the battery was mechanically damaged, the hydrogel's hydrogen bonds quickly reformed, restoring mechanical strength and ion transport capabilities, while maintaining consistent electron pathways, resulting in excellent self-healing properties. This battery demonstrated outstanding high-rate performance, energy density, and power density. It maintained 81.67% of its capacity after over 3000 charge/discharge cycles. The PTCDI-Ti$_3$C$_2$Tx MXene anode enhances material stability and conductivity by integrating with highly conductive MXene. MnO$_2$-CNT composites were used as the cathode, combining excellent conductivity and electrochemical performance. The overall design ensured flexibility, safety, and low cost, making it an ideal solution for next-generation wearable and portable smart devices.

Chen et al. [19] introduced self-healing functionality using polyvinyl alcohol/zinc trifluoromethanesulfonate (PVA/Zn(CF$_3$SO$_3$)$_2$) hydrogel electrolyte. This hydrogel was prepared through a freeze/thaw strategy, endowing the ZIBs with excellent ionic conductivity and stable electrochemical properties. It can autonomously self-heal through hydrogen bonding without external intervention. The hydrogel enabled the integration of the cathode, separator, and anode directly into its matrix, forming an all-in-one battery structure that prevents physical dislocation under bend conditions. The neutral hydrogel electrolyte overcame challenges associated with alkaline solutions such as poor Coulombic efficiency and capacity fading, providing a stable electrode–electrolyte interface even during fractures.

PVA's inherent self-healing properties due to its hydroxyl side groups and hydrogen bonding, make it ideal for Zn^{2+} ion integration, thus achieving a self-healing ZIB. The PVA/Zn(CF$_3$SO$_3$)$_2$ hydrogel showed optimal performance at a certain concentration and retained its conductivity after repeated cutting/healing cycles. This made it a robust choice for ZIB applications. Moreover, the electrochemical performance of these batteries was demonstrated to be excellent. The integrated ZIBs maintain high specific capacities and Coulombic efficiency after extensive cycling, showcasing remarkable durability and self-healing ability. Practical applications were demonstrated by powering LED arrays, emphasizing the battery's robustness and functional integrity even after mechanical damage and repair. The research offered a promising approach to developing advanced self-healing energy storage devices for flexible and wearable electronics.

Lithium-ion batteries face significant issues due to microcracks and mechanical damages, particularly at the electrode/electrolyte interfaces during charging and discharging cycles, which can lead to capacity loss and diminished battery reliability. Liu et al. [20] incorporated dissociative covalent adaptable networks (CANs) into GPEs, utilizing the thermally reversible D-A reaction between furan and maleimide groups. The dissociative CANs facilitated autonomous recovery of structural integrity by enabling the de-cross-linking and re-cross-linking reactions, providing the GPEs with dynamic self-healing capabilities.

By conducting in situ polymerization of trifunctional furan (TF) and trifunctional maleimide (TMI) in the presence of liquid electrolytes, the resulting GPE showed advanced self-healing properties. This mechanism allowed for effective mending of microcracks formed early in the cycle, thereby preventing further damage propagation. Experimental results demonstrated that a lithium-ion battery with these self-healing GPEs can withstand up to 500 cycles at 2 C with stable capacity retention. The GPE with dissociative CANs, specifically the GPE_EC/DEC-10 formulation, achieved a high ionic conductivity of 1.07 mS cm^{-1} and a favorable lithium transference number of 0.60. Additionally, the self-healing behavior did not compromise the electrochemical stability since the GPE maintains stability up to 4.9 V. The self-healing mechanism driven by DA reactions significantly contributes to the recovery of battery capacity after capacity fading due to mechanical or thermal stress. This property allowed the GPEs to restructure themselves at elevated temperatures (e.g. 120 °C), fully restoring functionality with the same ionic conductivity as the pristine state after repair.

Addressing the inherent challenge in batteries, which struggled to balance self-healing properties and thermopower efficiency, Huang et al. [21] integrated poly[2-(methacryloyloxy)-ethyl]dimethyl-(3-sulfo-propyl) ammonium hydroxide (PSBMA) within hydrogels to facilitate dual functionality. The self-healing mechanism hinged on strong electrostatic interactions between the zwitterions' anionic and cationic components. This hydrogel demonstrated a remarkable 90% self-healing efficiency and 325% stretchability post-recovery, owing to these dynamic ionic interactions that facilitate crack repair and structural maintenance.

On the other hand, the thermopower enhancement – or Seebeck coefficient – was achieved through selective cationic interactions involving the zwitterions and the redox couple, rearranging hydration structures and altering entropy. This intricate process resulted in heightened thermopower, reaching up to 3.5 mV K^{-1}, significantly outperforming conventional thermocells. The inclusion of LiCl further enhanced these properties by providing ionic-thermoelectric effects that contribute additional thermoelectric potential, evident even at low temperatures (−10 °C).

To corroborate these findings, the researchers subjected the hydrogel thermocells to rigorous mechanical and thermal tests, confirming their resilience and operational integrity under demanding conditions, such as low temperature and mechanical deformation. The adaptability of this self-healing hydrogel, encapsulated in a thermocell structure, was illustrated by its ability to convert low-grade heat (e.g. body heat) into electricity efficiently, showcased through powering an LED lamp consistently despite physical damages and environmental challenges. This study sets a precedent for future explorations into multifunctional materials that combine mechanical robustness and high-performance energy conversion, enlarging the scope of practical applications for thermocells in wearable and flexible electronics.

5.3.2 Solid Polymer Electrolytes

The recent advancements in self-healing materials have paved the way for their use in energy storage devices, aimed at enhancing the longevity and reliability of such technologies. Xu et al. [22] designed a polyurethane-based SPE utilizing a D-A dynamic network that embodies both self-healing and recyclability attributes. These polymer electrolytes were synthesized through a one-pot method using methoxy polyethylene glycol maleimide (mPEG-MAL) and N,N'-(4,4'-diphenylmethane) bismaleimide (BMI), exploiting the D-A reaction for cross-linking. The reversible nature of the D-A reaction allows the SPE to heal itself and maintain functionality, where the furan-maleimide adducts disassemble and reassemble at varying temperatures, thus enabling intrinsic self-repair. The hydrogen bonding within the polyurethane matrix further facilitates rapid surface repair at ambient temperatures, while the D-A bonds provide deeper repair capabilities upon mild heating.

The ionic conductivity of the recycled and self-healed PUB1SPE is comparable to the pristine material, validating the effectiveness of the self-healing and recycling processes. Specific thermomechanical characterizations showed that PUB1SPE maintains a wide electrochemical stability window (>5.0 V) and can withstand severe thermal and mechanical conditions, exhibited by the successful operation of Li|PUB1SPE|LiFePO$_4$ batteries in lighting applications under stress. Despite a moderate ionic conductivity at high temperatures, reaching up to 1.8×10^{-6} S cm^{-1} at 60 °C, the mechanical resilience and self-healing property of the polymer may compensate for operational challenges in real-world applications.

To solve the issue of lithium dendrite growth in solid electrolytes, Guo et al. [23] developed a self-healing solid polymer electrolyte (SHSPE), utilizing amino-terminated polyethylene glycol (NH$_2$-PEG-NH$_2$) as the supramolecular framework, enabling rapid self-healing through dynamic hydrogen bonds (Figure 5.5a). Elastic thermoplastic polyurethane (TPU) was used as a physical cross-linker, enhancing the mechanical strength of the solid electrolyte through intramolecular hydrogen bonds between urethane and ester groups. These dynamic and reversible intermolecular or intramolecular hydrogen bonds impart rapid self-healing ability to the solid electrolyte at the molecular level. Additionally, the ionic Coulombic force within the lithium salt further promoted rapid self-healing. Under optimized conditions, the solid electrolyte was capable of fully self-healing within 60 seconds after being deeply cut by a blade (Figure 5.5b).

Ye et al. [24] designed SPEs featuring self-healing capabilities based on dynamic imine networks, suitable for lithium-ion battery applications. The synthesis of SPEs involves a condensation polymerization process between terephthalaldehyde (TPA) and polyetheramines, specifically Jeffamine D2000, to form a flexible polymer matrix that facilitates lithium-ion transport. The incorporation of dynamic covalent imine bonds confers the material with self-healing properties, enabling the repair of cuts and tears at room temperature without external intervention.

Mechanistically, the imine bonds, characterized by reversible covalent interactions, allow the SPE to reconstruct its network structure after being physically damaged. This self-healing process was evidenced by the material's ability to recover 74% of its original tensile strength after damage. FT-IR confirmed the successful formation of imine bonds, while thermal analyses, including DSC and TGA, indicated that the SPEs maintain thermal stability up to 300 °C. The study found that decreasing the cross-linking density of the polymer leads to lower glass transition temperatures (T_g), thereby enhancing chain mobility which is beneficial for ion diffusion. Furthermore, the self-healing properties were visually validated as cuts in the SPE film autonomously sealed within 12 h at room temperature,

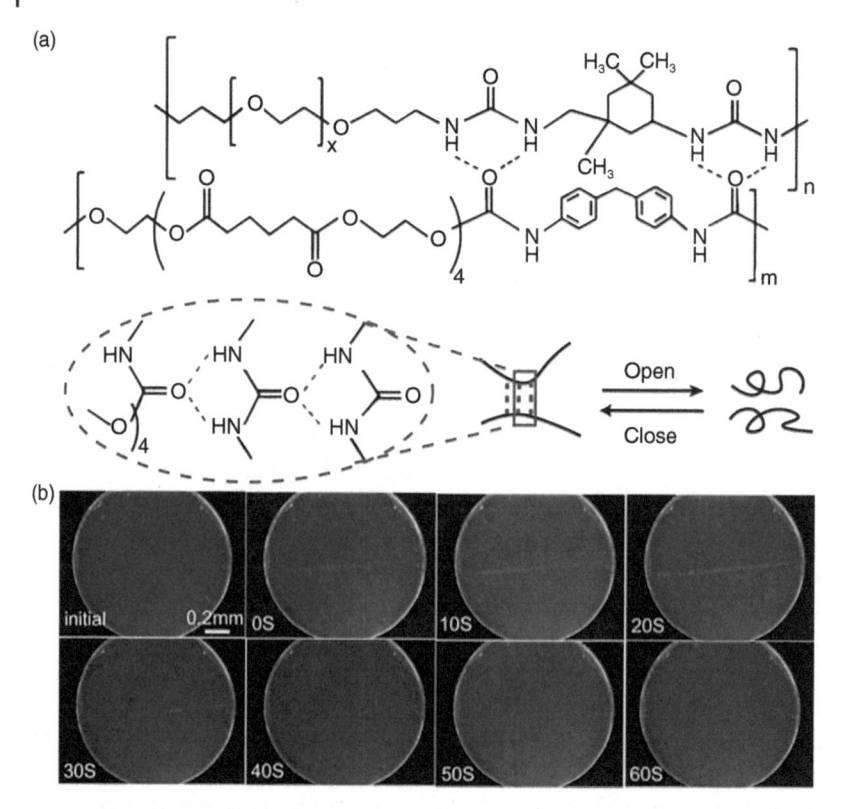

258 | *5 Application of Self-Healing Materials in Batteries*

Figure 5.5 (a) Supramolecular structure and dynamic hydrogen bonding for self-healing solid polymer electrolyte. (b) The self-healing process of SHSPE after being cut. The SHSPE could heal the cut in <60 s. Source: Reproduced with permission from Wu et al. [23]/John Wiley Sons.

restoring mechanical integrity sufficient to support weights up to 50 g. The intrinsic self-healing ability, coupled with the SPE's high ionic conductivity, positions these imine-based networks as promising candidates for robust, long-lasting solid-state battery electrolytes, highlighting their potential to reduce battery degradation and risk through effective autonomous repair.

Cui et al. [25] developed an SPE constructed from a comb-like polymer, poly-[propylene oxide-co-2-(2-methoxyethoxy)ethyl glycidyl ether] [P(PO/EM)], combined with a newly developed lithium salt, lithium trifluoro(perfluoro-tert-butyl-oxyl)borate (LiTFPFB). The supramolecular interactions in this study primarily involve noncovalent interactions, such as hydrogen bonding and vdW forces, between the highly fluorinated anions of LiTFPFB and the hydrogen atoms in the P(PO/EM) polymer. These interactions were confirmed and analyzed using FT-IR, solid-state NMR, and mechanical tests like stripping and self-healing experiments. The FT-IR spectra and NMR results indicate that the supramolecular interactions lead to stronger bonding between the salt and polymer matrix in the LiTFPFB/P(PO/EM) electrolyte compared to traditional systems like LiTFSI/P(PO/EM). The supramolecular bonds allow the SPE to exhibit self-healing properties, demonstrated by its ability to recover adhesion strength and maintain structural integrity after mechanical damage. Peel strength tests showed that the LiTFPFB/P(PO/EM) system could restore up to 81% of its original adhesive strength after mechanical separation and recombination, outperforming the comparative LiTFSI/P(PO/EM) system. Additionally, the self-healing rate was faster for the LiTFPFB/P(PO/EM) electrolyte, underscoring the effective self-repair facilitated by robust supramolecular interactions.

At room temperature, the self-healing electrolyte reaches an ionic conductivity of $1.9 \times 10^{-4}\,S\,cm^{-1}$, a lithium-ion transference number of 0.44, and only $1.87 \times 10^{-8}\,S\,cm^{-1}$ in electronic conductivity. These features provided it with outstanding electrochemical performance in battery applications. In Li/Li symmetric batteries, with the use of a self-healing electrolyte, the battery underwent cycling tests at a current density of $1\,mA\,cm^{-2}$, showing that it could maintain stable cycling performance even after 80 cycles. SEM and XPS analyses showed that the lithium metal surface became smoother and more uniform after cycling, indicating that self-healing electrolytes effectively suppressed the growth of lithium dendrites. The Li/SHSPE/LiFePO$_4$ (LFP) full battery had an initial discharge capacity of $147.9\,mAh\,g^{-1}$ at a 0.1 C rate, maintaining a capacity retention rate of nearly 90% after 100 cycles, with a coulombic efficiency of approximately 100%. Additionally, at high temperatures (200 °C), it exhibited good thermal stability and high resistance to combustion, further enhancing the safety of the battery. The self-healing electrolyte not only possesses excellent self-healing capability and electrochemical performance but also significantly enhances the cycle life and safety of LMBs.

Solid polymer electrolytes are considered the most promising materials for the fabrication of solid-state lithium batteries due to their light weight, good flexibility, and ease of film formation. However, solid polymer electrolytes inevitably suffer from external forces (e.g. stretching, twisting, and bending) during the assembly and operation of lithium batteries, which may cause microcracks or fractures, thus hindering Li$^+$ transport within the solid electrolyte and severely affecting the performance and longevity of lithium batteries. Self-healing polymers can significantly enhance cyclic stability and safety, and prolong the lifespan of lithium batteries. Xie et al. [26] reported a novel cross-linked network solid polymer electrolyte (PDDP) prepared via a thiol-ene click reaction. The combination of hydrogen bonds and dynamic disulfide bonds endowed the electrolyte with excellent mechanical and self-healing properties. It could repair cutting damage and restore its mechanical strength and electrochemical performance within one hour at 28 °C. Tensile tests showed that the PDDP film exhibits high mechanical stretchability, capable of withstanding volume changes during Li deposition/dissolution processes without incurring irreparable damage.

Compared to PDDP-0, the glass transition temperature (T_g) of PDDP-5 was lowered, indicating that the addition of EMIMTFSI improves segmental mobility, which was beneficial for Li$^+$ transport. The self-healing electrolyte could self-heal and support its own weight after three seconds of contact at 28 °C, and after 24 hours, it could withstand a weight of 100 g without external stimuli. The outstanding self-healing performance was attributed to the combination of dynamic disulfide bonds and hydrogen bonds. The assembled Li/PDDP/LFP battery had an initial discharge capacity of 139.6 mAh g^{-1} at 0.1 C, maintaining a capacity of 141.4 mAh g^{-1} after 150 cycles, showing good long-cycle performance. The Li/PDDP/LFP battery exhibited initial specific capacities of 160, 150, 130, 100, and 80 mAh g^{-1} at current densities of 0.1, 0.2, 0.5, 1, and 2 C, respectively, indicating excellent rate performance. The discharge capacity fully recovered to its initial state when the current rate was reverted to 0.1 C, indicating good rate performance. With the mixed electrolyte composed of PDDP-impregnated liquid electrolyte, the lithium metal electrode sustained a uniform, smooth, and compact deposition structure after repeated cycling, without experiencing dendrite growth or pulverization.

Lithium-ion batteries are extensively applied in the flexible electronics field due to their high volumetric energy density, superb capacity retention rate, and longevity. However, the rigid electrolytes and electrode materials used in current lithium-ion batteries readily suffer from sudden capacity drops or even battery failure when subjected to repeated deformation and external stresses. Therefore, developing flexible electrodes, electrolytes, and packaging materials to accommodate volume changes or fractures in each part is crucial. Xue et al. [27] designed and developed a polyurethane-based SPE possessing shape memory properties, which allowed the control of a temporary shape and the ability to revert to its original shape via polycaprolactone (PCL) soft segments. The abundance of disulfide bond exchange and hydrogen bonds equips it with the ability to self-heal readily when subjected to thermal stimuli.

The tensile stress of the polyurethane solid electrolyte film undoped with lithium salt reaches 19.83 MPa, and upon doping with 40 wt% LiClO$_4$, the electrolyte film changed from a plastic deformation behavior to an elastic one, showing that the crystallization behavior of PCL was completely inhibited. Given that the shape memory features of the material were principally afforded by the crystallization transition of PCL, 20 wt% LiClO$_4$ content was the optimal selection. Regarding the influence of different PCL molecular weights on mechanical properties, the larger the molecular weight of the PCL diol soft segment, the higher the crystallinity of Net-PU4000SS33, resulting in tensile stress dropping from 6.73 MPa for Net-PU6000SS33 to 0.63 MPa for Net-PU2000SS33. At a PCL molecular weight of 4000, the electrolyte film's flexibility was optimal, making it suitable for ion conduction. The tensile strength of Net-PU4000CC33 (5.38 MPa) is notably higher than that of Net-PU4000SS33 (3.37 MPa), whereas the elongation at break of Net-PU4000SS33 (518%) is longer, suggesting that the disulfide bonds with exchange capability enhanced the flexibility of the molecular chains.

The polyurethane solid electrolyte reached a self-healing efficiency of 91% within 12 hours at 60 °C, primarily attributed to the dynamic reversibility of the disulfide bonds and hydrogen bonds in polyurethane. The rapid exchange reactions of dynamic disulfide bonds at high temperatures significantly enhanced the self-healing properties of the material. Furthermore, the ionic conductivity of the polyurethane solid electrolyte attained 2.49×10^{-4} S cm^{-1} at 60 °C and 8.80×10^{-4} S cm^{-1} at 80 °C, exhibiting outstanding electrochemical stability. The

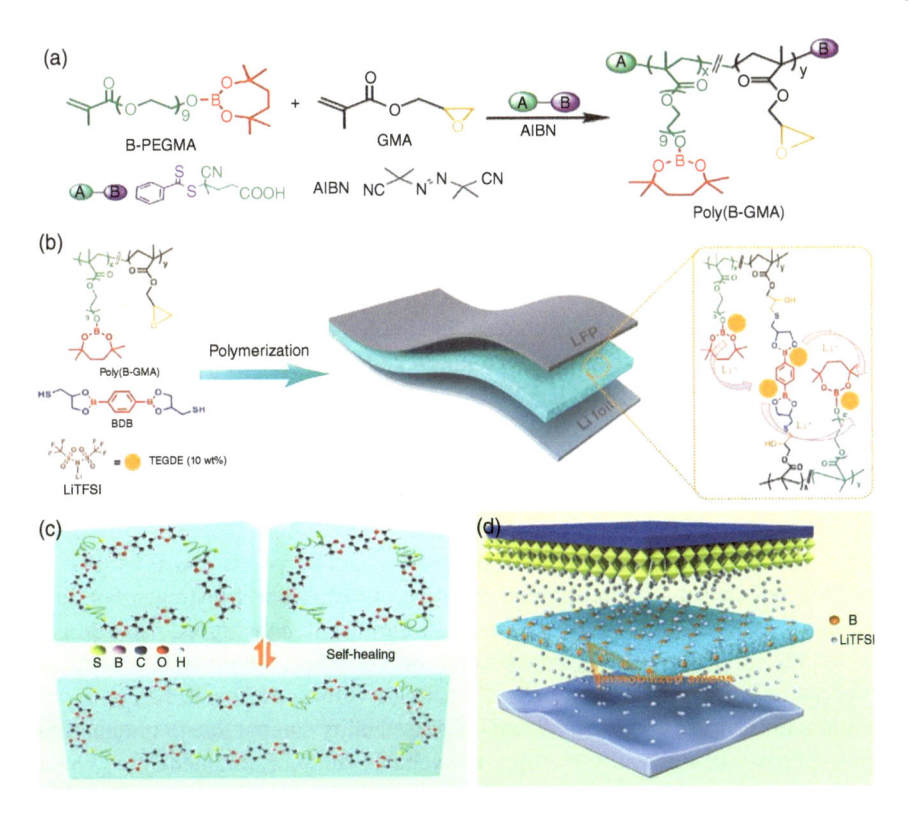

Figure 5.6 (a) Schematic of synthesizing poly(B-GMA) by RAFT polymerization. (b) Schematic illustration of the fabrication of the multifunctional DB-SHPE. (c) Schematic diagrams showing the self-healing mechanism in the DB-SHPE. (d) The contribution of the boron moieties in DB-SHPE to the homogeneous deposition of Li ions in LMBs. Source: Zhou et al. [28]/with permission of John Wiley & Sons.

lithium metal symmetric battery using polyurethane solid electrolyte maintained stability after 500 hours of cycling at a 0.05 mA cm^{-2} current density, reflecting its good interfacial stability and compatibility.

Their team [28] also utilized dynamic covalent chemistry to develop a dynamic self-healing polymer electrolyte based on boronate ester bonds (DB-SHPEs), prepared via a thermally initiated ring-opening reaction involving thiols and epoxy groups. These electrolytes demonstrated outstanding mechanical properties and interfacial stability. The introduction of borate ester bonds not only imparted self-healing ability to the electrolyte but also facilitated uniform lithium-ion deposition through Lewis acid–base interaction (Figure 5.6). Moreover, the boronate ester bonds provide the electrolyte with reprocessability and recyclability.

The self-healing mechanism of DB-SHPEs primarily relies on the dynamic covalent bond characteristics of borate ester bonds. When DB-SHPE is cut into two parts, the borate ester bonds exposed at the damaged interface can recombine through exchange reactions, restoring the original state. This process is achieved by rearranging the topology of borate ester bonds, thereby giving DB-SHPEs significant self-healing ability. Experiments indicated that after undergoing a three-hour self-healing process at 60 °C, DB-SHPE can endure a weight of 500 g and maintain 87% of its original mechanical strength.

DB-SHPEs exhibited outstanding performance improvements in real-world battery applications. The Li/dB-SHPE/Li symmetric battery maintained a stable voltage platform after 1200 hours of cycling, indicating exceptional long-cycle stability. Furthermore, the LiFePO$_4$/dB-SHPE/Li battery achieved a specific capacity of 140.0 mAh g^{-1} at 0.2 C during the first cycle, with a Coulombic efficiency of 99.5%. After 150 cycles, it retained 93% of its capacity, with a Coulombic efficiency still at 99.4%. In contrast, batteries with conventional polymer electrolytes had a capacity of only 126.5 mAh g^{-1}, with a capacity retention rate of 85.2%. XPS analysis showed that the SEI formed by DB-SHPE contains LiF and Li$_3$N, aiding in the formation of a uniform and dense SEI layer that prevents side reactions between lithium metal and the electrolyte, thereby enhancing interfacial stability. Additionally, the Lewis acidity of boron atoms in DB-SHPE assists in capturing anions of the Li salt, further aiding the uniform distribution and deposition of lithium-ions.

They [29] also explored SPEs that integrate disulfide bonds and urea groups into their polymer matrix, leveraging disulfide metathesis and hydrogen bonding to facilitate self-repair both at ambient and elevated temperatures. This design enabled the solid polymer electrolyte to not only self-heal autonomously at room temperature but also achieve rapid self-healing at higher temperatures, essential for maintaining ionic conductivity and cycling stability comparable to undamaged states. The study innovatively utilized reversible addition-fragmentation chain transfer (RAFT) polymerization, combining poly(ethylene glycol) methyl ether acrylate (PEGA) with cross-linkers that provide dynamic covalent and noncovalent bonding interactions. Among the tested configurations, 3PEG-SSH and 3PEG-CCH demonstrate significant self-healing efficiencies, with 3PEG-SSH achieving up to 95.7% self-healing efficiency at 30 °C. Further, the cyclic healing experiments underscored the SPEs' ability to restore mechanical and conductive properties effectively after damage. These SHSPEs showed a high initial ionic conductivity which is crucial to battery operation, decreasing negligibly after multiple damage-healing cycles, affirming their resilience. Electrochemical tests revealed that the self-healed SHSPEs maintain high voltage stability and can be integrated into lithium-ion batteries without compromising the electrochemical performance. The batteries employing these electrolytes exhibited commendable cycling performance, retaining high capacity even after extended use, which indicated their potential for practical applications.

Chen et al. [30] designed a SHPE through the in situ polymerization of a mixture of butyl acrylate (BA) and UPy with SN and LiTFSI, where SN served as a plastic crystal and DES to improve ionic conductivity while quadruple hydrogen bonding by UPy enhances self-healing and mechanical properties.

The formation of strong quadruple hydrogen bonds provided by low doses of UPy units promotes the spontaneous repair of structural defects such as cracks and voids that develop at the Li/electrolyte interface during battery operation. This process facilitated the inhibition of lithium dendrite growth by mending dendrite-induced damage, thus prolonging the battery's cycling life and enhancing safety. An optimal formulation with 3 mol% UPy demonstrated high mechanical strength and stretchability (up to 1500%), alongside a considerable ionic conductivity of 0.46 mS cm^{-1} at 30 °C and a broad electrochemical stability window of 4.7 V.

The investigation into the molecular interactions via FT-IR and Raman spectroscopy affirmed the coordination between lithium-ions and nitrile groups from SN within the DES phase, alongside the interaction of carbonyl groups with cations, both of which collectively enhanced ionic conductivity. The assertion that the microphase separation in conjunction with tailored hydrogen bonding interactions significantly contributed to the self-healing capability and electrochemical performance of SHPE, positioning it as a promising solution for advanced high-performance LMBs.

Besides, they [31] introduced an SHSPE designed for flexible LMBs through a dual-network structure combining quadruple hydrogen bonding and chemical bonding. This structure enhanced the mechanical robustness and self-healing abilities of the polymer matrix, addressing traditional challenges in polymer electrolytes related to mechanical strength and ionic conductivity. The self-healing mechanism primarily relies on the dynamic universal interactions provided by UPy dimers, which enable the continuous breaking and reformation of bonds, facilitating the spontaneous repair of damage. These UPy dimers form the first network, providing excellent self-repairing properties by allowing the polymer to recuperate its integrity after mechanical disruptions such as cuts or scratches. Upon thermal activation, specifically at 60 °C, the dynamic nature of these interactions supported quick and efficient self-healing, as demonstrated by the restoration of mechanical strength to 84% of the original post-repair. The introduction of a second, chemically cross-linked network composed of polyethylene glycol-bis-carbamate dimethacrylate (PEGBCDMA) enhances the mechanical strength and stability of the polymer electrolyte. This second network ensures that the polymer maintains its structural integrity and dimensional stability, even under deformation. The combination of these dual networks allows the electrolyte to sustain both flexibility and durability, crucial for prolonged cycle life and reliability in battery applications.

Rheological studies indicated that the dual-network structure contributes significantly to the mechanical properties, with improved storage modulus (G') compared to conventional single-network systems. The elastomeric properties from the cross-linked networks mitigate the inherent trade-off between ionic conductivity and mechanical strength typically observed in solid polymer electrolytes. The urethane segments in the PEGBCDMA are instrumental in enhancing ionic conductivity by promoting ionization, thus maintaining a delicate balance between charge mobility and polymer rigidity. The electrochemical assessment revealed that the SHPE possessed a wide electrochemical window of up to 5.2 V, alongside high thermal stability up to 350 °C, confirming its suitability for high-performance and safe battery operations. Additionally, the abundant hydrogen bonding in the SPHE matrix ensures strong adhesion and excellent interfacial stability between the electrolyte and lithium metal, thereby enhancing the prevention of lithium dendrite growth and ensuring consistent cycling performance.

The introduction of PIL block copolymers presents a viable solution by leveraging their intrinsic ionic conductivity and mechanical robustness. Liu et al. [32] integrated poly(vinylidene fluoride-co-hexafluoropropylene) (PVDF-HFP) into PIBCPs, leveraging dual ion–dipole interactions. The imidazole cations of PIL blocks interact with both the oxygen atoms in the poly(ethylene oxide) (PEO) blocks and the -CF$_3$ dipoles in PVDF-HFP. This dual interaction not only facilitates rapid lithium-ion transport but also imparts exceptional self-healing properties. The reversible nature of these interactions allows PIBCPEs to heal without external stimulation at ambient temperatures, a capability confirmed by experiments showing complete self-healing and restoration of mechanical and conductive properties within one hour.

Electrochemical and thermal tests revealed that PIBCPE exhibited significant ionic conductivity improvements due to the decreased crystallinity of PEO, reaching values of 3.7×10^{-4} S cm^{-1} at room temperature. Furthermore, density functional theory (DFT) calculations demonstrated an attractive binding energy of -53.09 kJ mol^{-1} for ion–dipole interactions, reinforcing the mechanism of self-healing at the molecular level. The presence of a strong electrostatic attraction, with a calculated binding energy of -343.00 kJ mol^{-1} between imidazole cations and bis(trifluoromethylsulfonyl)imide (TFSI) anions, increased the lithium-ion transference number to 0.57. Thermal analysis using TGA showed no degradation below 300 °C, signifying excellent thermal stability, vital for operational safety. The study further underscores the PIBCPE's potential through successful long-term cycling tests in lithium battery assemblies, demonstrating stable cycling performance with notable capacity retention and coulombic efficiency after extensive cycles.

Fan et al. [33] explored a novel SHSPE for all-solid-state ZIBs. The SPE was designed using a coordination polymer that incorporates the ligand 2,6-bis((propylimino)methyl)-4-chlorophenol (Hbimcp) grafted onto a poly(propylene oxide) (PPO) polymer chain. The self-healing mechanism is facilitated by the inherent reversibility and mobility of the zinc ion coordination bonds. These coordination bonds act as dynamic cross-linkers between polymer chains, which can spontaneously reform after being disrupted, for example, by a mechanical cut. When the polymer is split, zinc ions can migrate quickly to the broken areas and reestablish coordination bonds due to the high association constant of the zinc-ligand complex. This rapid coordination rebinding allows the polymer to repair itself efficiently without external stimuli, such as heat or pressure, making it ideal for maintaining functionality under operational stresses. Tensile tests demonstrated the efficacy of the self-healing process, where a cut sample can recover nearly 97.3% of its original fracture elongation within six hours. This ability to self-repair contributed significantly to the durability and resilience of the SPE, especially in flexible and wearable electronic applications. The large elongation at break, over 1000% in some samples, underscored the electrolyte's mechanical robustness. Another important aspect of the SPE is its chemical recyclability, enabled by the imine linkages in the polymer, which can degrade into their precursors under acidic conditions. This feature allows the spent electrolyte to be fully reclaimed and recycled, presenting an environmentally friendly approach to material lifecycle management.

Zhao et al. [34] synthesized the SPEs through free radical photopolymerization involving PEG methyl ether methacrylate and borate-containing monomers, leveraging the reversible transesterification reaction of borate ester which facilitates superior self-healing and mechanical properties. The reaction of boric acid with diols can rearrange through reversible transesterification. This process empowered the SPEs to repair themselves autonomously when damaged, effectively restoring their mechanical strength and integrity. Specifically, when the polymer matrix undergoes structural disruption, the borate bonds dissociate and reform through dynamic bond exchange, thereby reestablishing the SPE's original configuration without external intervention.

The SPE's internal comb-shaped polymer architecture is instrumental in enhancing lithium-ion conductivity and durability. This structure, combined with borate-induced complexation, supports an amorphous state favorable for ion mobility. Borate groups function as electron acceptors with strong interactions with anions, generating effective single-ion conduction pathways and reducing polarization. Correspondingly, the borate-based SPEs achieved an ionic conductivity of $0.272\,mS\,cm^{-1}$ at room temperature, alongside a high lithium-ion transference number of 0.76, indicating robust ion transport characteristics. Electrochemical evaluations revealed that these SPEs maintain a stable working voltage window of 5.1 V and exhibit resilient interface stability with lithium metal, evidenced by minimal overpotential increases during extensive cycling in symmetrical Li|PPBSPE|Li batteries. Additionally, the borate groups prove crucial in forming a stable SEI on lithium anodes, preventing dendrite formation and enhancing cycle life.

Feng et al. [35] enhanced safety and stability by leveraging the self-healing capability of imine bonds, which can undergo reversible chemical reactions, such as imine condensation/hydrolysis and imine exchange/metathesis, without the need for catalysts. This reversible bonding is key to the material's self-healing ability, allowing it to repair itself upon mechanical stress or fracture by re-establishing the imine bonds. The introduction of diglycidyl ether of bisphenol A (DGEBA) and TPA creates a cross-linked network that endows the electrolyte with excellent mechanical strength and adhesion properties, crucial for maintaining intimate contact with electrodes.

They described a polymer network synthesized through a simple mixing and heating process, where the active NH_2 groups of PEG reacted with the aldehyde groups of TPA to form imine bonds. These bonds provide the polymer with intrinsic self-healing properties. Upon experiencing internal or external forces, the network swiftly restores its original shape and functionality due to the dynamic reversible nature of the imine bonds. The self-healing mechanism is further enhanced by the epoxy groups in DGEBA, which react with amino groups to form a robust covalent structure, ensuring high mechanical durability.

Empirical results demonstrated that this SHSPE could autonomously self-heal at room temperature within 30 minutes, without any external interventions, whereas higher temperatures accelerate the healing process – complete recovery is achieved in just 10 minutes at 60 °C. XRD and TGA confirmed the amorphous nature and high thermal stability of the ShSPE, indicating its suitability for high-temperature applications in lithium batteries. The polymer's low glass transition temperature further signifies enhanced polymer chain mobility, contributing to better ionic conductivity.

To suppress the growth of lithium dendrites in lithium-sulfur batteries (LSBs), Kim et al. [36] designed and synthesized a bifunctional SHSPE by co-grafting ion-conductive and self-healing polymer segments onto a rigid aromatic polymer backbone. This PSE achieves a combination of high ionic conductivity and high mechanical strength by co-grafting PEG and PU or disulfide bond polymers (PUS) onto the polyarylether sulfone (PAES) backbone. The PAES-g-(PU/2PEG) and PAES-g-(PUS/2PEG) membranes form and recombine hydrogen bonds and/or disulfide bonds, allowing these reversible bonds to quickly self-repair when the membrane is damaged, restoring its original mechanical and electrochemical performance. After self-healing, the membranes had ionic conductivities of 0.659 and 0.694 mS cm^{-1} at room temperature respectively, maintaining 94.8% of their initial conductivity. The lithium transference number was 0.412, showing a high ion migration capability. Mechanical testing revealed that the self-healed membranes possessed a tensile strength of 2.14 MPa and an elongation at break exceeding 50%, nearly equivalent to the original membrane.

LSBs using PAES-g-(PUS/2PEG) membranes achieved an initial discharge specific capacity of 929.8 mAh g^{-1} at a 0.2 C rate, with a capacity retention rate of up to 98.7% after 200 cycles, and a Coulombic efficiency of 99.5%. This high performance was primarily due to the PSE membrane's superior interfacial stability and its capability to suppress lithium dendrite growth, ensuring battery safety and stability over prolonged cycling. Furthermore, these membranes demonstrated excellent thermal stability and high flame resistance at high temperatures (200 °C), further enhancing the battery's safety.

In the continuous innovation of lithium battery technology, self-healing mechanisms play a key role in optimizing and addressing challenges related to battery safety and performance durability. Pu et al. [37] designed and synthesized a PEO-based SHSPE. At room temperature, without external stimuli, it could completely recover its mechanical properties within 24 hours after breaking, showing excellent self-healing capability (initial tensile stress is 137 kPa, and break strain is 524%; self-healed tensile stress is 114 kPa, and break strain is 542%).

The ionic conductivity of SHSPE at 25 °C was 7.48×10^{-4} S cm^{-1}, which was significantly higher than that of traditional PEO-based SPEs. Assembled Li/SHSPE/Li symmetric batteries displayed a very low overpotential (319 mV) after 1200 hours of cycling at a current density of 0.1 mA cm^{-2}, indicating excellent interfacial stability. Furthermore, LiFePO4/SHSPE/Li batteries achieved an initial discharge specific capacity of 150 mAh g^{-1} at 0.1 C and 27 °C, and after 300 cycles, the specific capacity remains at 126.4 mAh g^{-1}, with a capacity retention rate of 84.3%. High electrochemical stability of SHSPE is demonstrated through EIS and LSV Throughout prolonged cycling, Li/SHSPE/Li batteries maintained stable interfacial impedance, indicating superior interfacial compatibility and electrochemical stability.

Meng et al. [38] developed a polymer electrolyte (PBPE) with high ionic conductivity, self-healing properties, and nonflammability based on dynamic metal coordination bonds, for use in dendrite-free LMBs. PBPE was synthesized through a

Schiff base reaction between PEG diamine and 1,3,5-triformyl benzene, forming a cross-linked polymer network structure. PBPE primarily relies on the combination of dynamic imine bonds and hydrogen bonds for self-healing, which can spontaneously break and reform at room temperature, thus achieving self-repair. This PBPE exhibited a self-healing efficiency of up to 87.9% after repair and can reach 100% self-healing efficiency after 30 minutes of repair at 60 °C. Additionally, PBPE showed no cracks after repeated bending to 180°, showcasing excellent flexibility and good mechanical performance. TGA showed that the 5% weight loss temperature of PBPE was 334 °C, indicating its excellent thermal stability.

At 30 °C, PBPE had an ionic conductivity of 4.79×10^{-3} S cm^{-1}, making it the highest among reported SPEs. PBPE can promote the formation of a LiF-rich SEI, effectively suppressing dendrite growth on the lithium metal anode. PBPE can promote the formation of a LiF-rich SEI, effectively suppressing dendrite growth on the lithium metal anode. Li/PBPE/Li symmetric batteries maintained stable voltage and exhibited no dendrite formation after cycling for 1000 hours at a current density of 0.5 mA cm^{-2}, illustrating excellent interfacial stability and dendrite suppression ability. The assembled LiFePO$_4$ (LFP) battery delivered a discharge capacity of 118.2 mAh g^{-1} at a 5 C rate, with a capacity retention rate of 97.8% after 125 cycles; importantly, the repaired PBPE could fully restore its performance in the LFP battery, significantly improving the battery's reliability.

To enhance the self-healing speed and conductivity of electrolytes, Yang et al. [39] designed a SHSPE based on a six-armed divalent cationic polymer ionic liquid (DPIL-6). They first synthesized a six-armed monovalent cationic polymer ionic liquid (MPIL-6) and the DPIL-6, then prepared the SPE using a solution casting method.

The rich non-covalent bonds and ionic interactions in DPIL-6-SPE allowed it to fully self-repair within two hours at room temperature, with exceptional tensile properties (elongation >1500%) and stress resistance (>490 kPa), as well as excellent adhesion to the lithium anode (supporting >200 g). Compared to the monocations MPIL-6-SPE, DPIL-6-SPE exhibits higher ionic conductivity (exceeding 10^{-5} S cm^{-1} at room temperature), a broader electrochemical window (4.9 V), and a lithium-ion transference number of 0.46. DPIL-6-SPE exhibited complete recovery in mechanical performance and conductivity after self-healing, and maintained stable output capabilities during repeated angular bending and folding. LiFePO4/Li batteries assembled with DPIL-6-SPE show an initial discharge-specific capacity of 152.6 mAh g^{-1} at a 0.1 C rate, achieving a capacity retention rate of 94% after 50 cycles and a coulombic efficiency of 96%.

Yang et al. [40] integrated a poly(HFBM-co-SBMA) network containing imidazole-based ionic liquid (EMI-TFSI) and LiTFSI into the electrolyte, forming an SPE with remarkable self-healing capability. This self-healing capability primarily relies on the stable ion–dipole interactions between imidazole cations and fluorine atoms, granting SHSPE the ability to recover within 60 minutes at room temperature. The electrolyte could achieve an elongation of over 4000% and stress over 130 kPa and could fully restore its mechanical properties and conductivity at 60 °C. Furthermore, the self-healing mechanism significantly improves the ion migration efficiency and interfacial compatibility of the battery, allowing for long-term stable operation with high discharge capacity and high coulombic efficiency. At a current density of 0.2 C, the Li/SHSPE3/LiFePO4 battery achieved a discharge capacity of 144.8 mAh g^{-1}, with a capacity retention rate of 82% after 100 cycles and a coulombic efficiency of 97%.

Furthermore, to improve the performance of conventional PEO-based solid electrolytes, Wong et al. [41] developed a new self-healing PEO-based solid electrolyte by incorporating 2 wt% amorphous three-dimensional carbon (3DC). The strong vdW interactions between the 3DC and PEO matrix and the 3D structure of 3DC offer abundant porosity and strong adhesion, enabling PEO chains to reconnect when broken, thereby repairing cracks. The PEO-2 wt% 3DC membrane could completely self-heal within two hours at 60 °C, retaining its ionic conductivity and mechanical strength. The ionic conductivity of the self-healed PEO-2 wt% 3DC electrolyte is 9.1×10^{-4} S cm^{-1} at 60 °C, much higher than the 4.5×10^{-4} S cm^{-1} of pure PEO. Additionally, the electronic conductivity of the PEO-2 wt% 3DC electrolyte at 60 °C is 6×10^{-9} S cm^{-1}, similar to pure PEO, demonstrating good electrical insulation.

The Li/Li symmetric battery with PEO-2 wt% 3DC electrolyte retained stable cycling performance after 5000 hours at a current density of 0.1 mA cm^{-2}, demonstrating its exceptional electrochemical stability. In full cell testing, the LiFePO$_4$/PEO-2 wt% 3DC/Li battery maintained 84% discharge capacity after 850 cycles at a 1 C rate, with nearly 100% coulombic efficiency. By contrast, batteries with pure PEO electrolyte showed a rapid capacity drop and a significant decrease in coulombic efficiency. Additionally, the PEO-2 wt% 3DC electrolyte demonstrated excellent thermal stability and good mechanical strength at high temperatures (60 °C), effectively suppressing lithium dendrite growth and enhancing the safety and lifespan of the battery.

Conventional electrolytes are susceptible to mechanical damage during battery operation, resulting in reduced ionic conductivity and decreased battery life. To address this challenge, Wang et al. [42] designed a single-ion conducting polymer electrolyte with dynamic anion exchange capability (Figure 5.7). When the electrolyte material experiences mechanical damage, the dynamic anions can rearrange within the material via interactions and fill the damaged areas, thus restoring the structure and function of the material. Electrolytes that had undergone self-healing treatment showed significant improvements in cycling performance and ionic conductivity, with a capacity retention rate of 85% after 500 cycles at 1 C, compared to only 60% for control group batteries without self-healing electrolytes under the same conditions. Electrolytes that had undergone self-healing treatment showed significant improvements in cycling performance and ionic conductivity, with a capacity retention rate of 85% after 500 cycles at 1 C, compared to only 60% for control group batteries without self-healing electrolytes under the same conditions. Furthermore, the self-healing electrolyte achieved an ionic conductivity of 1.2×10^{-3} S cm^{-1} at room temperature, significantly higher than that of traditional electrolytes.

Lu et al. [43] synthesized an ion-conducting elastomer (ICE) through in situ chain scission of polylactic acid precursor (PAP) triggered by lithium metal and

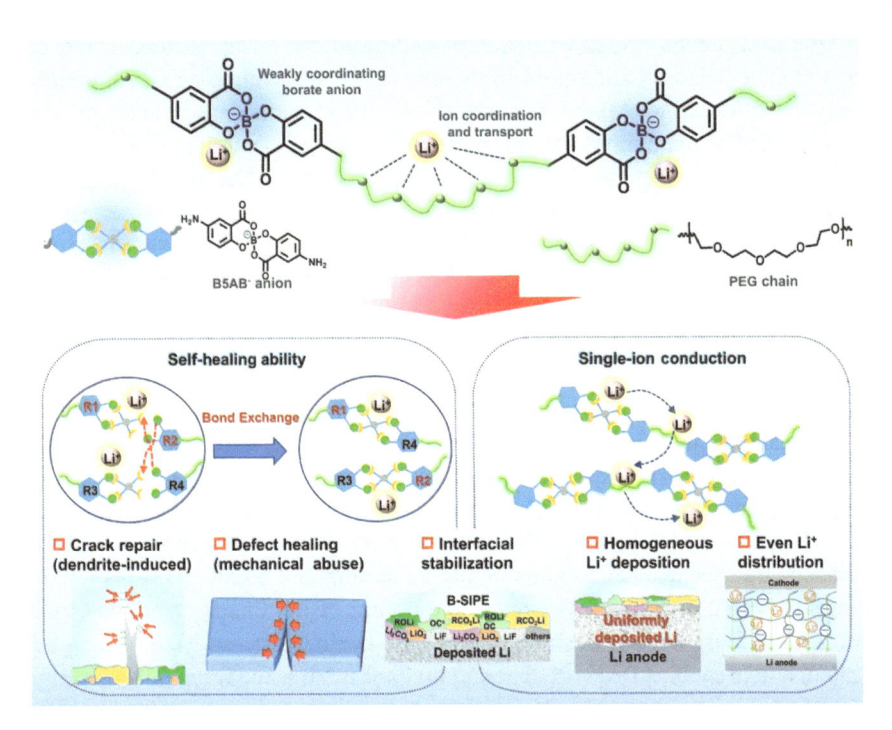

Figure 5.7 Schematic illustration of structures and functions of the self-healing single-ion conductor polymer electrolyte (B-SIPE). Source: Duan et al. [42]/with permission of John Wiley & Sons.

integrated it into the SEI to form ICE-SEI. The self-repair mechanism of ICE is grounded on lithium-ions quickly generating ionic groups to form an adaptive structure, endowing it with dynamic reversibility and superior elasticity. The self-repair mechanism of ICE is grounded on lithium-ions quickly generating ionic groups to form an adaptive structure, endowing it with dynamic reversibility and superior elasticity. After over 20,000 cycles, the cycle life of ICE-SEI surpassed 22,700 cycles (approximately 9110 hours) at a current density of 5 mA cm^{-2}, greatly exceeding existing records. Additionally, under high current density, the ICE-SEI@Li electrode exhibited extremely low voltage hysteresis and remarkable stability. In terms of electrochemical performance, the interfacial resistance of the ICE-SEI@Li electrode was significantly lower than that of the bare lithium electrode, and the single-layer pouch cell maintained a reversible specific capacity of about 140 mAh g^{-1} after 100 cycles, with a capacity retention rate of 87%. In comparison, the specific capacity of the bare lithium electrode rapidly decreased to 43 mAh g^{-1}, with a capacity retention rate of just 29%.

Solid electrolytes typically face challenges due to brittleness, which can lead to cracking, especially when the separator layer is thin. Lee et al. [44] fill voids in the solid electrolyte with an organic polymer, forming a cross-linked polymer network in situ. Specifically, a malleable thermoset polymer, when combined with a Li$_2$S-P$_2$S$_5$ inorganic electrolyte, formed a stand-alone membrane with a thickness of 64 μm and a high inorganic material loading of 80%. This method achieved near-theoretical density and significantly enhanced the gravimetric and volumetric energy densities of the membrane without inhibiting conduction capabilities. The developed SEPM maintains essential properties for solid electrolyte contact while significantly improving membrane mechanical properties. This novel membrane processing replaced thicker, traditional separators with a much thinner 64 μm variant, enhancing conductance and demonstrating stable cycling over 200 cycles in a battery configuration using a FeS_2-based cathode. The study utilized scanning electron microscopy and energy-dispersive X-ray spectroscopy to verify the structure of the self-healing polymer matrix dispersed through a densified solid electrolyte, highlighting the polymer's effective penetration and distribution.

Zhang et al. [45] investigated a SHSPE designed for lithium-ion batteries, using the DA reaction as its self-healing mechanism. In this research, the self-healing electrolyte (DASHPE) was synthesized using a DA cycloaddition reaction with furan groups as dienes and maleimide groups as dienophiles, enhancing the mechanical properties and self-repair capabilities of the polymer matrix. The resultant DASHPE demonstrated excellent self-healing ability by closing and mending cracks within 30 minutes at 80 °C through the reversible nature of DA bonds, combined with hydrogen bonding interactions that support the healing process. These reversible covalent bonds enabled DASHPE to maintain its structural and functional integrity, thereby reducing risks associated with electrolyte failure.

Additionally, DASHPE exhibited robust thermal stability, enduring temperatures over 300 °C, which underscored its resilience in thermal conditions typically challenging for battery materials. The cross-linked network stabilized by DA bonds also broadens the electrochemical window to 5.0 V, crucial for compatibility with diverse battery chemistries, while offering sufficient ion conductivity at elevated temperatures which supports efficient electrochemical performance.

Kim et al. [46] utilized sulfonyl imide-based anionic monomers and DA thermoset polymers to create a series of recyclable thermosets via thermo-reversible cross-linked networks. The incorporation of furan-containing monomers and the use of bismaleimide cross-linkers creates a DA reaction, which facilitates thermo-reversibility, allowing the polymer network to heal and reprocess efficiently.

The self-healing capability is the DA chemistry, which enables reversible cross-linking. Under thermal conditions, these cross-linked networks dissociate and recombine, providing a platform for healing without losing mechanical integrity. The reversible DA reaction and the entailing changes in solubility with temperature are evidenced by FT-IR and DSC analyses, confirming the formation of DA adducts and their reversible nature. Moreover, these polymers exhibited high ionic conductivity, achieving $7.07 \times 10^{-5}\,S\,cm^{-1}$ at 80 °C, attributable to enhanced ion mobility aided by PEO units embedded in the polymer matrix, which supports lithium-ion transport.

The self-healing process involved thermal treatment at 140 °C for 60 minutes, effectively repairing physical damage such as scratches, while maintaining electrochemical performance integrity throughout multiple reprocessing cycles. Test data showed that repeated thermal cycling (up to 30 cycles) did not significantly degrade ionic conductivity or mechanical properties, with the polymers retaining a high lithium-ion transference number close to 1, indicating singular ion conducting efficiency.

Zeng et al. [47] integrated boroxine-based dynamic covalent bonds into the polymer matrix through an amidation reaction between NH_2-PEG-NH_2 and ortho-triphenylboroxine, enhanced with lithium salt. The dynamic and reversible nature of boroxine bonds is facilitated by N—B coordination. Boroxines are notable for their electron-withdrawing properties, and when coordinated with nitrogen, they create a robust, reversible structure that can spontaneously mend damage without external stimuli, thus maintaining film integrity and prolonging lifespan.

DFT calculations revealed the structural advantages of ortho-triphenylboroxine for effective N—B coordination, as the planarity deviations of benzene rings in this configuration improve the cleavability of boroxine linkages. Experimentally, the ONBSPE-2000 with ortho boroxine exhibited an ionic conductivity of $0.168\,mS\,cm^{-1}$ at room temperature, a lithium-ion transference number of 0.60, and demonstrated rapid self-healing abilities characterized by the complete mending of damage at 60 °C within 5 min. This is significantly attributed to the enhanced chain mobility facilitated by the N—B dynamic bonds.

Traditional LMBs, although known for their high energy density, often face safety concerns due to irregular lithium deposition and dendrite formation, which can puncture separators. Kuang et al. [48] tackle these issues by integrating self-healing capabilities into the electrolyte design, leveraging 2-ureido-4-pyrimidinone (UPy) units that introduce self-healing properties through reversible hydrogen bonding. These bonds allow the electrolyte to repair itself autonomously at room temperature, mitigating mechanical damage and ensuring consistent electrolyte performance. A fluorinated polymer deep eutectic electrolyte (FPDE) combines the benefits of both deep eutectic solvents and fluorinated polymers. The deep eutectic solvents contribute to high ionic conductivity and a broad electrochemical window, while the fluorinated polymers enhance thermal stability and form a lithium fluoride (LiF)-rich SEI, crucial for inhibiting lithium dendrite growth.

The self-healing mechanism in FPDE primarily involves quadruple hydrogen bonding from UPy units and additional hydrogen bonds between hydrogen and fluorine atoms within the polymer chain, providing a flexible yet robust structural network that can recover from physical disruptions within an hour without external stimuli. Experimentally, the FPDE demonstrated excellent ionic conductivity of $0.97\,mS\,cm^{-1}$ at room temperature and maintained a stable lithium plating and stripping process over 6000 hours in symmetric Li batteries. Furthermore, the designed FPDE resulted in significantly improved fire-retardant properties with a self-extinguish time of zero seconds, as no combustion occurred within 10 seconds of exposure to flame.

These attributes not only contribute to safer electrolyte operation but also ensure the long-term electrochemical performance stability of LMBs. Extensive testing revealed that a Li/FPDE/LFP battery maintains 80.4% capacity retention after 600 cycles at 0.2 C, underscoring the endurance imparted by the self-healing and flame-retardant design.

5.3.3 Composite Electrolytes

Liu et al. [49] proposed a solid-liquid mixed electrolyte composed of a composite polymer electrolyte (CPE) impregnated with a liquid electrolyte. The CPE membrane is composed of a self-healing polymer and Li$^+$ conductive nanoparticles. Compared to conventional separators, Li metal electrodes with mixed electrolytes exhibited uniform, smooth, and dense electro-deposition structures rather than dendritic and powdery structures.

At room temperature, the CPE membrane could repair cut damage within one hour and restore its mechanical strength and electrochemical performance. Tensile experiments revealed that the CPE film possesses high mechanical stretchability, enabling it to endure volume changes during Li deposition/dissolution without sustaining irreparable damage. CPE facilitated the generation of a LiF-rich SEI, effectively inhibiting dendrite growth on the lithium metal anode. Li foil symmetric batteries using mixed electrolytes maintained very low voltage hysteresis while cycling 500–1500 times with a capacity of 1 mAh cm^{-2} at a current density range of 3–20 mA cm^{-2}. Even at an ultra-high current density of 20 mA cm^{-2}, the voltage hysteresis was just 240 mV, the lowest recorded so far. The Li|LTO half-cell based on mixed electrolytes exhibits excellent cycling performance, with a capacity retention rate of 99.4% after 120 cycles at a 0.2 C rate.

The issue with conventional PVA-based acidic electrolytes is their low self-healing efficiency and insufficient tensile capability. These conventional electrolytes struggle to recover after mechanical damage and have poor performance under large strain, thereby limiting their use in portable and wearable energy storage devices. Zhi et al. [50] introduced a polyacrylic acid electrolyte (VSNPs-PAA) with dual cross-linking through hydrogen bonds and vinyl-hybridized silicon nanoparticles (SiNPs), which not only has adjustable ionic conductivity but also self-healing and high tensile properties. The self-repair mechanism predominantly depends on reversible hydrogen bond cross-linking and the synergistic action of SiNPs. When the electrolyte is damaged, the hydrogen bonds can dynamically reassemble, thereby restoring mechanical and electrochemical properties. Experiments demonstrated that the capacitance retention rate is close to 100% after 20 cycles of damage and repair. As the water content increased, the ionic conductivity of the VSNPs-PAA electrolyte improved significantly, attaining approximately 1×10^{-2} S cm^{-1} conductivity at 60 wt% water content. This electrolyte maintained integrity at strains reaching 3700%, showcasing extremely high tensile capabilities. Further electrochemical tests showed that at 600% strain, the capacitance value increased 3.5 times, demonstrating excellent stress transfer and energy dissipation capabilities.

Traditional liquid electrolytes possess high ionic conductivity and good compatibility with electrode interfaces, but they encounter problems like leakage, dendrite growth, and electrolyte evaporation in real-world applications. Such problems not only impact battery performance but might also lead to safety risks. To solve these challenges, Yan et al. [51] introduced a novel leakage-responsive and self-repairing electrolyte system (LRE), integrating the benefits of both solid and liquid electrolytes to improve battery safety and stability.

When ethyl α-cyanoacrylate (ECA) is added to carbonate electrolytes, it rapidly polymerizes in the presence of air, forming a polymer layer that prevents further evaporation and oxidation of the electrolyte. This polymerization reaction is initiated by water molecules, forming a carbon anion through the nucleophilic attack on the β-carbon, which then rapidly initiates anionic polymerization. ECA not only rapidly repaired physical battery damage, but also prevented electrolyte leakage and lithium oxidation. Moreover, ECA helps form a robust polymer SEI layer, inhibiting lithium dendrite growth and enhancing battery stability. Experimental results showed that in Li||Li symmetric batteries, batteries using LRE had a cycle life of over 500 hours at a current density of 1.0 mA cm^{-2}, with an overpotential of only 45 mV. In LCO||Li batteries, batteries using LRE could still maintain 86% of their capacity after 400 cycles. With 30 vol% ECA, the ionic conductivity of LRE was comparable to that of traditional EC/DEC electrolytes (around 7.2 mS cm^{-1}), showing good Li plating/stripping behavior in Li||Li symmetric batteries. Additionally, in LCO||Li batteries, batteries using LRE had an initial capacity of 164 mAh g^{-1} at a 1 C rate, with a capacity retention rate of 84.6% after 400 cycles. These results indicated that LRE has significant advantages in improving battery performance and safety, making it a promising alternative to traditional liquid and solid electrolytes, potentially advancing practical applications of LMBs.

Current SSEs (such as polymer electrolytes, inorganic ceramic electrolytes, and CPEs) possess advantages in safety and energy density, but are susceptible to mechanical damage in practical applications. Therefore, developing composite electrolytes with self-healing properties is particularly crucial. Cheng et al. [52] embedded UPy-functionalized SiO$_2$ nanoparticles containing quadruple hydrogen bonds into a polymer matrix. These UPy-functionalized SiO$_2$ nanoparticles form dynamic cross-links with polymer chains through quadruple hydrogen bonds, thereby imparting self-healing capabilities to the electrolyte.

The different morphologies of SiO$_2$ fillers had a significant impact on the ionic conductivity and self-healing properties of composite electrolytes, among which the composite electrolyte filled with hollow mesoporous SiO$_2$ (hSiO$_2$) exhibits the best electrochemical performance. At room temperature, the ionic conductivity of HCPE reached 2.37×10^{-5} S cm^{-1}, and at 60 °C, it reached 9.77×10^{-4} S cm^{-1}. Furthermore, the sodium ion transference number of the HCPE electrolyte is 0.23, which is better than that of other morphologies of SiO$_2$ fillers. Sodium ion batteries using HCPE electrolyte exhibit an initial discharge capacity of 110.5 mAh g^{-1} at a rate of 0.1 C, with a capacity retention rate of 76.4% after 200 cycles. These data suggested that self-healing electrolytes not only enhanced the mechanical properties of batteries but also significantly improved the electrochemical performance and stability of the electrolytes.

The current polymer electrolytes have deficiencies in mechanical strength and ionic conductivity, which restrict their application in lithium-ion batteries. In order to solve these problems, Xue and colleagues [53] proposed the combination of UPy-functionalized SiO$_2$ nanoparticles (SiO$_2$-UPy) with a polymer matrix containing UPy units, forming a supramolecular network. The self-healing function improved the ionic conductivity and mechanical strength of the electrolyte, while also giving the electrolyte self-repairing abilities. For instance, the ionic conductivity of SHCPE at 30 °C reached 8.01×10^{-5} S cm^{-1}, roughly four times higher than that of the composite electrolyte with nonfunctionalized SiO_2. After 60 cycles, the discharge capacity was maintained at 139 mAh g^{-1}, with a coulombic efficiency of 97.9%.

Liu et al. [54] designed a unique cross-linked structure that combines poly(ethylene glycol) methoxy acrylate (PEGMA) and 2-hydroxyethyl methacrylate-g-4-formylbenzoic acid (CHO@HEMA) with $Li_{6.4}La_3Zr_{1.4}Ta_{0.6}O_{12}$ (LLZTO) nanoparticles, functionalized by 3-aminopropyltriethoxysilane (APTES). This configuration utilizes the formation of strong imine bonds between the amino groups on APTES@LLZTO and the aldehyde groups on CHO@HEMA. These covalent bonds act as dynamic connecting points within the polymer matrix, significantly enhancing the organic/inorganic interface. When the material is damaged, the imine bonds can break and reform, allowing the polymer network to self-repair without the need for external stimuli or conditions. This self-healing process restores the integrity and functionality of the electrolyte, crucial for maintaining performance and extending the lifetime of LMBs.

Furthermore, the introduction of APTES@LLZTO nanoparticles not only improves the dispersion and reduces agglomeration within the polymer matrix but also aids in maintaining a low crystallinity of the polymeric phase. The reduced crystallinity allows for enhanced flexibility and ionic conductivity, which are vital for efficient lithium-ion transport. The covalent imine network also offers superior mechanical strength, allowing for effective energy dissipation in response to mechanical stresses. By ensuring uniform dispersion of ceramic nanofillers and facilitating self-repair, this composite electrolyte represents a promising direction for the future of solid-state lithium batteries.

Mi et al. [55] accomplished this by adding fluoroethylene carbonate (FEC) into $Li_{0.35}La_{0.55}TiO_3$-polyethylene glycol (LLTO-PEO) composite SSE. Under the drive of lithium-ions, FEC migrates to the damaged interface to form a new LiF-rich interfacial layer, thereby enabling self-healing of the lithium metal and CSSE interface, suppressing the growth of lithium dendrites, and enhancing the electrolyte's electrochemical stability. The LLTO-(PEO-FEC) composite solid-state electrolyte (CSSE-1115) with added FEC exhibited a lithium-ion conductivity of 1.13×10^{-4} S cm^{-1} at 25 °C and 3.77×10^{-4} S cm^{-1} at 50 °C, while the electronic conductivity was 1.68×10^{-9} S cm^{-1} at 25 °C and 4.31×10^{-9} S cm^{-1} at 50 °C. Additionally, CSSE-1115 demonstrated a wide electrochemical window (5.2 V vs. Li$^+$/Li) and long-term stable lithium plating performance, operating stably for 800 hours at a current density of 0.2 mA cm^{-2}.

Repeated mechanical deformation (such as stretching and bending) can cause mechanical failure of electrolytes, leading to the loss of functionality in wearable devices and thereby limiting their lifespan. Kim et al. [56] introduced a self-healing mechanism based on the development of reversible nanostructures, utilizing the behavior of polymer scaffold structures in ILs. The nanophase separation of long-chain alkyl side chains enables the aggregates formed by these side chains to spontaneously reorganize after the removal of external forces, thereby restoring the original cross-linked structure. This self-healing mechanism can be realized at room temperature without external triggers. The sample with the greatest self-healing capability (TA10C) almost completely recovered its mechanical and electrochemical properties after 16 hours of contact at room temperature, achieving a self-healing efficiency of about 97%. Meanwhile, the TA10C sample exhibited a toughness of approximately 13.2 MJ m^{-3} in its initial state, which was significantly higher than previously reported SPE composites. Additionally, the ionic conductivity of the TA10C sample was about 0.36 mS cm^{-1}, which was comparable to or slightly higher than the self-healing PECs with low IL content reported in the literature.

5.3.4 Electrode Binders

The silicon anode undergoes a volume change of 300–400% during charge and discharge, leading to particle fracture, electrode detachment, and interface instability, resulting in rapid capacity decay. Traditional binders such as PVDF underperform in dealing with the substantial volume changes of silicon, failing to maintain the structural integrity and electrochemical performance of the electrode.

Bao et al. [57] introduced an ion-conductive self-healing polymer (SHP-PEG) binder, which incorporates PEG groups into the self-healing polymer. These PEG groups form hydrogen bonds to provide self-healing capability while also promoting lithium-ion conduction within the binder (Figure 5.8). The design of the SHP-PEG binder includes dynamic hydrogen bonding for self-healing and the ionic conductivity of PEG groups, aimed at enhancing the interface stability and electrochemical performance of silicon particulate electrodes during cycling. The silicon microparticle anode utilizing the SHP-PEG750(40) binder could still maintain a capacity of 1300 mAh g^{-1} after 150 cycles at a 0.5 C current density, with an initial coulombic efficiency (ICE) reaching 83%. Additionally, the binder significantly reduced the charge transfer resistance (Rct), showing 35 Ω in EIS testing, which is much lower than the 71 Ω of traditional SHP binders. This research showcases the tremendous potential of self-healing polymer binders in addressing the mechanical and electrochemical challenges presented by the volumetric changes of silicon anodes.

Silicon is regarded as a potential material for replacing traditional graphite anodes because of its high theoretical capacity and low operational voltage. However, during charging and discharging, silicon anodes undergo volume expansion of up to 300%, leading to particle pulverization and electrode structural failure, ultimately affecting battery performance and lifespan. Wang et al. [58] introduced a CA-PAA binder with high adhesion and self-healing properties inspired by the design of natural ivy. This was achieved by incorporating citric acid (CA) into poly(acrylic acid) (PAA), forming a dynamic hydrogen bond network through carboxyl and hydroxyl groups, and realizing the self-healing capability for silicon anodes. CA acts not only as a buffer layer to release internal stress and stabilize the solid electrolyte interphase, but also forms a "soft inside, rigid outside" spatial topology with PAA, enhancing electrode stability during cycling. The silicon anode with CA-PAA binder had an initial areal capacity of 6.5 mAh cm^{-2}, maintaining a capacity retention rate of 78% after 50 cycles, with an ICE of 89.5%. The Si||NCM811 full cell retained a capacity retention rate of 74% after 100 cycles, with a reversible capacity of 1.7 mAh cm^{-2}.

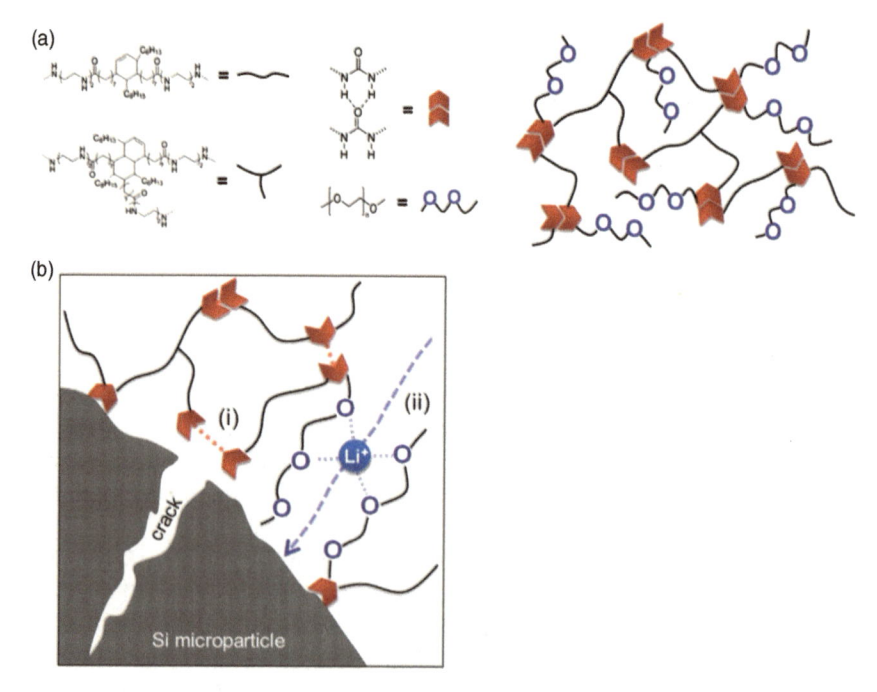

Figure 5.8 (a) Schematic chemical structure of the SHP-PEG binder. (b) Schematic illustration of the Si microparticle electrode with SHP-PEG binder. (i) Self-healing based on dynamic hydrogen bonding close to a crack caused after cycling. (ii) Li-ion conduction facilitated by PEG groups. Source: Munaoka et al. [57]/with permission of John Wiley & Sons.

Zhang et al. [59] tackled this challenge by synthesizing a novel self-healing poly(ether-thiourea) (SHPET) polymer binder. The hydrogen bonds in the SHPET polymer can reform after damage, thus restoring the material's mechanical properties and structural integrity. This mechanism allows cracks generated by volume changes in the silicon anode during charging and discharging to spontaneously heal, significantly enhancing the electrode's cycling stability. The silicon anode using the SHPET binder demonstrated excellent electrochemical performance, with an initial discharge capacity of up to 3744 mAh g^{-1} at a current density of 0.42 A g^{-1}, and it can maintain a capacity of 1917 mAh g^{-1} at a current density as high as 4.2 A g^{-1}. Additionally, after cycling 250 times at a high current density of 4200 mA g^{-1}, the battery could still retain 85.6% of its capacity. These results indicated that the SHPET binder not only possessed excellent self-healing capability but also significantly improved the battery's cycling life and capacity retention rate.

Coskun et al. [60] introduced polymeric binders incorporating Meldrum's acid, which forms ketene intermediates for cross-linked networks, aimed at enhancing binders with self-healing properties. The self-healing capability, facilitated by ion–dipole interactions, ensures recovery of binder-Si and binder-binder connectivity disrupted by silicon's volumetric shifts, stabilizing the electrode structure. Experiments show self-healing binders improve cycling performance, achieving 51% capacity retention over 500 cycles, outperforming traditional types like PVDF.

These binders also exhibit a high ICE of 84%, minimizing initial capacity loss. This self-healing effect is critical for preventing electrode issues like cracks and void formation, essential due to silicon's expansion-contraction.

Lithium-sulfur (Li—S) batteries are considered strong competitors for next-generation energy storage systems because of their high theoretical specific capacity (1675 mAh g^{-1}) and outstanding energy density (2600 Wh kg^{-1}). However, the significant volume change of the sulfur cathode and the shuttle effect of polysulfides severely restrict the practical application of Li—S batteries. To address these challenges, Xiao et al. [61] designed an intelligent self-healing polyurethane binder (SHPUB) that provided self-healing and sulfur-fixing functions through the synergistic effects of aromatic disulfide bonds and hydrogen bonds. The self-healing mechanism is categorized into extrinsic self-healing and intrinsic self-healing. This paper primarily investigates intrinsic self-healing, leveraging the dynamic reversibility of aromatic disulfide bonds and hydrogen bonds to offer self-healing capabilities at room temperature. The breaking and reformation of disulfide bonds, combined with the reorganization of hydrogen bonds, enable the material to self-heal after damage, preserving the structural integrity of the cathode.

The polyurethane binder (PUB) can rapidly self-heal at room temperature, with fractured polyurethane samples almost completely recovering within 40 minutes, and their mechanical strength restoring to >95% of the initial value. The SHPUB-S cathode retained a high reversible electrochemical capacity of 1075.7 mAh g^{-1} after cycling 100 times at 0.2 C, with a capacity retention rate of 84.7%. In comparison, the traditional PVDF-S cathode preserved only 714.0 mAh g^{-1} of reversible electrochemical capacity, with a capacity retention rate of 55.7%. Theoretical calculations and experimental results indicate that SHPUB had higher binding energy and stronger sulfur fixation capability compared to polysulfides (Li$_2$Sx). SHPUB effectively inhibits the shuttle effect of polysulfides by capturing Li$_2$Sx. At a high current density of 2 C, the SHPUB-S cathode retained 80% of its capacity after cycling 500 times, demonstrating excellent long-cycle stability. This study developed a SHPUB based on aromatic disulfide bonds and hydrogen bonds, successfully applied to the sulfur cathode of LSBs. SHPUB not only exhibited excellent self-healing and sulfur-fixation performance but also significantly enhanced the cycling stability and electrochemical performance of batteries, offering a promising solution for the practical application of LSBs.

LSBs have attracted considerable attention due to their ultra-high energy density and environmental friendliness. However, the issues of polysulfide dissolution in the electrolyte and electrode volume expansion result in battery performance that falls significantly short of theoretical expectations. Xu et al. [62] developed a multifunctional polymer binder (LA-GA) with self-healing ability and high adhesion by integrating dynamic disulfide bonds and polar carboxyl and pyrogallol groups. The disulfide bonds provide self-healing capability, allowing the repair of cracks caused by cathode volume expansion. Simultaneously, the polar functional groups (carboxyl and pyrogallol groups) not only strengthen adhesion and maintain electrode structural integrity but also effectively adsorb lithium polysulfides, suppressing the shuttle effect. The sulfur cathode using the LA-GA binder retained a high capacity retention rate of 81.9% after 100 cycles at a 0.2 C rate. Additionally, after 700 cycles at a 1 C rate, the capacity decayed rate is only 0.0469%. In comparison, electrodes using traditional PVDF binders exhibited a capacity decay rate of 0.0852% under the same conditions.

Yang et al. [63] proposed a self-healing binder cross-linked by dynamic hydrogen bonds, introducing a three-dimensional cross-linked polymer network through the formation of abundant dynamic hydrogen bonds between PAA and tannic acid (TA). TA acts as a physical cross-linker, interacting with the PAA main chain through hydrogen bonds, endowing the binder with unique self-healing capability and strong adhesion. This design effectively disperses stress during the lithiation and delithiation process of the Si anode, preventing the pulverization of Si particles. In electrochemical performance tests, the Si@TA-c-PAA electrode showed high reversible specific capacity (3250 mAh g^{-1} at 0.05 C), excellent rate capability (1599 mAh g^{-1} at 2 C), and exceptional cycling stability (1742 mAh g^{-1} at 0.25 C after 450 cycles). In contrast, traditional PAA binders could not provide the same adhesion and self-healing capability, resulting in poorer electrochemical performance. The Si@TA-c-PAA electrode demonstrated stable rate performance at different current densities and maintained a high capacity of 1599 mAh g^{-1} at a high current density (2 C). Additionally, the electrode retained 84.5% capacity retention after 450 cycles, demonstrating its excellent long-term cycling stability.

The limited capacity of graphite anodes in LIBs drives the exploration of silicon anode materials to provide higher theoretical specific capacity. However, during the charge-discharge process, silicon undergoes up to 300% volume change, leading to electrode pulverization and delamination, which further cause a sharp decline in electrochemical performance. Traditional PVDF binders, due to insufficient mechanical properties and weak interaction with silicon, struggle to effectively support the cycling stability of silicon anodes. Jeon et al. [64] introduced a self-healing multifunctional binder aimed at achieving self-healing capabilities through hydrogen bonding and electrostatic interaction. The multifunctional binder provides primary mechanical support through polyacrylic acid (PAA), enhances conductivity with [poly(3,4-ethylenedioxythiophene)] (PEDOT), and employs phytic acid (PA) as a multivalent physical cross-linker to provide dynamic hydrogen bonds and charge interactions, enabling spontaneous healing of cracks in silicon anodes during cycling, thereby restoring structural integrity and mechanical performance (Figure 5.9). Experimental results indicated that silicon anodes using the PEDOT:PAA:PA (PDPP) composite binder retained a reversible capacity of 2312 mAh g^{-1} after 100 cycles at a 0.5 C current density, achieving an ICE of 94%, significantly higher than other reported silicon anodes. Additionally, the PDPP binder significantly enhanced the lithium-ion diffusion coefficient (3.04×10^{-8} cm^2 s^{-1}) and electronic conductivity, demonstrating excellent rate performance and maintaining a capacity of 2084 mAh g^{-1} at a high rate of 5 C.

Due to its high theoretical capacity (4200 mAh g^{-1}), silicon is regarded as the next-generation anode material to replace graphite. However, its substantial volume change can cause pulverization of SiNPs and electrode delamination, leading to an unstable SEI layer, thereby restricting its commercial application. To tackle these

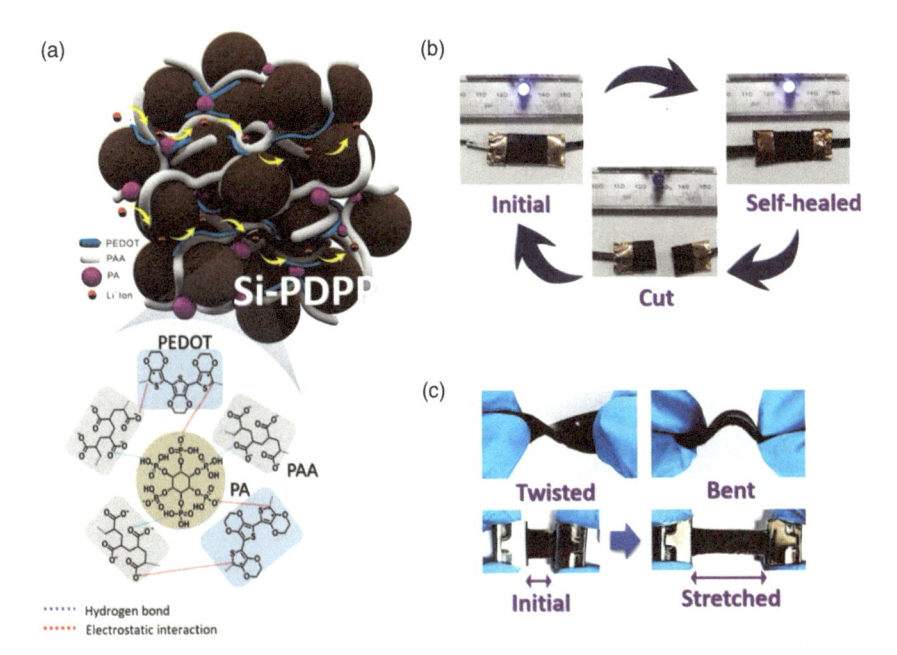

Figure 5.9 (a) Schematic of the self-healing PDPP binder. (b) Self-healability of PDPP. (c) Flexibility and stretchability of PDPP. Source: Reproduced with permission from Chen et al. [63]/John Wiley & Sons.

challenges, Liang et al. [65] developed an inorganic/organic composite binder with both elasticity and self-healing properties, aimed at providing stronger adhesion and improved electrode structural stability. By combining inorganic lithium metasilicate (LS) with organic PVA, a composite binder (PVA@LB) was formed, which is chemically cross-linked by boric acid (BA). LS not only provides sufficient adhesion sites for SiNPs but also forms a protective coating on the SiNPs surface, enhancing Li$^+$ diffusion kinetics. PVA provided elasticity and a self-healing framework due to its low glass transition temperature (72 °C), maintaining electrode structural integrity through hydrogen bond interactions. The Si anode with PVA@LB20 binder maintained a discharge capacity of 2095 mAh g^{-1} after 100 cycles at a current density of 1 A g^{-1} and maintained 1133 mAh g^{-1} capacity at a high current density of 6 A g^{-1}. The Si‖NCM811 full cell maintained a reversible capacity of 1.86 mAh cm^{-2} after 50 cycles at a 0.2 C rate, with a capacity retention rate of 67%.

Kim et al. [66] created a UPy-functionalized PAA grafted with poly(ethylene glycol) (PAU-g-PEG), a self-healing and ion-conducting polymer binder aimed at enhancing the performance of silicon anodes. This new binder addresses the challenges of silicon's significant volumetric expansion during charge/discharge cycles, which often leads to poor cycle life in Si-based electrodes. Key features of the PAU-g-PEG binder include its ability to self-heal through dynamic hydrogen bonding and maintain electronic integrity. The binder effectively accommodates the volume changes in silicon, thereby extending cycle life and improving Coulombic efficiency. The study reported that silicon electrodes using PAU-g-PEG retained high capacities over numerous cycles and performed well at high current densities. The binder's effectiveness is attributed to its dual functionality – self-healing and ion-conducting properties – enabled by the introduction of UPy units and the grafting of PEG. These features resulted in improved electrode adhesion, structural stability, and the ability to form a strong and flexible 3D network. Electrochemical tests demonstrated the binder's ability to facilitate capacity retention and efficiency, making it a promising solution for high-performance, silicon-based lithium-ion batteries.

Hou et al. [67] employed an innovative approach by integrating a self-healing dynamic supramolecular elastomer electrolyte (SHDSE) via in situ polymerization directly onto the silicon anode. This methodology utilizes the dynamic bonds within the SHDSE's molecular structure to achieve superior electrochemical and mechanical performance. SHDSE redefines the interface by acting as both an electrolyte and a binder, thereby enhancing adhesion and enabling continuous lithium-ion transport while drastically reducing silicon particle displacement and electrode swelling.

The all-solid-state lithium-ion battery (ASSLIB) had enhanced cycling stability, achieving over 500 cycles with 68.1% capacity retention and over 99.8% Coulombic efficiency, showcasing the effectiveness of the self-healing SHDSE in maintaining interface integrity despite volumetric changes of silicon. Notably, a 2.0 Ah wave-shaped Si|LiCoO$_2$ soft-pack battery, leveraging in-situ cured SHDSE, maintained a high capacity of 1.68 Ah after 700 cycles, translating to 86.2% retention, even under challenging conditions such as high temperatures up to 100 °C and repeated bending tests. These results are attributed to the self-healing capabilities of SHDSE, which can autonomously repair interface disruptions caused by mechanical stress or silicone expansion.

The SHDSE electrolyte features dynamic supramolecular networks formed by sacrificial hydrogen bonds, which act as self-repairing agents. These bonds help mitigate stress and minimize damage upon cycling, thereby preventing crack propagation and electrode degradation. The polymer's mechanical elasticity ensures robust interfacial contact that accommodates electrode expansion, reinforcing the structural integrity of the silicon anode and facilitating rapid ion transport. The self-healing mechanism was validated via various spectroscopic analyses indicating significant dissociation of lithium salts within the SHDSE, supported by strong intermolecular interactions that maintain the coherence and performance of the composite electrode material. These findings underline the practical potential of SHDSE in sustaining the operational lifespan of silicon-based ASSLIBs under extensive cycling and mechanical deformations.

A persistent challenge persists in integrating robust self-healing and mechanical properties within electrochromic polymeric electrolytes, which are crucial for wearable electronics that endure frequent mechanical stress and potential damage.

Yao et al. [68] designed a self-healing polymeric electrolyte synthesized through a facile, one-pot in-situ polymerization process involving AA, 1-vinylimidazole (VIm), and vinyl hybrid silica nanoparticles (VSN). The self-healing mechanism hinges on the dual cross-linking that combines both physical and covalent network systems, where vinyl hybrid silica nanoparticles serve as cross-linkers. These nanoparticles provide enhanced structural reinforcement, stress distribution, and energy dissipation under strain, which are vital in maintaining structural integrity. The polymer's self-repair ability is particularly attributed to the dynamically cross-linked hydrogen bonds, acting as reversible and non-sacrificial bonds that facilitate the recovery process upon structural damage. Quantitatively, the polymer displayed a notable tensile strength of 11 kPa and achieved a self-healing efficiency of 90.5% within 20 minutes at ambient temperature, showcasing its rapid recovery capability without external stimuli. These healing dynamics are bolstered by hydrogen bonding interactions among carbonyl and imidazole groups, which play a critical role in the polymer chain's intermolecular connectivity.

Electrochemical analysis via electrochemical impedance spectroscopy (EIS) confirmed the self-healing electrolyte's substantial ionic conductivity (1.26×10^{-3} S cm^{-1} at room temperature), demonstrating its efficacy as an electrolyte in EC devices. Moreover, the hybrid self-healing ECD, integrating this novel electrolyte, achieved remarkable electrochromic performances with a coloration efficiency of 406.96 cm^2 C^{-1} and robust cyclic stability over 5000 cycles. The ECD reaches 90% of its maximum color transition in 2.0 seconds and bleached in 1.8 seconds, underscoring its swift optical response. Their study opens possibilities for further developments in flexible and durable electrochromic applications.

5.4 Conclusions and Outlook

Self-repairing materials exhibit significant potential in battery applications, capable of automatically repairing after damage to restore structure and function, thus notably enhancing battery performance, safety, and cyclic lifespan. In batteries, self-healing materials typically achieve functionality through physical bonding, chemical bonding, or composite mechanisms. Physically bonded self-healing materials use dynamic non-covalent bonds, such as hydrogen bonds and electrostatic interactions, to repair damage, making them suitable for applications requiring high flexibility, such as wearable devices. Chemically bonded self-healing materials provide long-lasting performance stability through the breaking and reforming of reversible chemical bonds, such as disulfide bonds, and are widely used in environments demanding high safety.

The application of self-healing materials in gel polymer electrolytes provides batteries with excellent cycling stability and exhibits good electrochemical and mechanical properties, especially in enhancing battery safety. The self-healing binders in electrodes extend the electrode's lifespan and improve the battery's capacity retention rate. The performance improvement of self-healing materials is not limited to enhancing the lifespan and safety of batteries. They effectively suppress dendrite growth, maintain the electrochemical performance of the battery, enhance ionic and electronic conductivity, reduce internal resistance, and improve energy and power densities. Additionally, self-repair materials show outstanding performance in environmental protection and sustainability. By reducing the number of batteries that need frequent replacement due to damage, they significantly reduce the generation of electronic waste, and the renewable nature of many self-healing materials also lessens the environmental impact, aligning with green development strategies.

The technological optimization and large-scale application of self-healing materials will be key to driving their commercialization. Exploring the development of low-cost, high-efficiency production processes, and addressing issues of material performance stability under long-term variable conditions will help maintain its self-healing capabilities. Additionally, research into multifunctional integration and material design will deepen, and composite materials combining different repair mechanisms are expected to quickly address various damage types. Standardization and quality control are vital for ensuring the reliability of materials, necessitating the establishment of international standards and assessment systems. In emerging application fields, as flexible and wearable electronic devices become more prevalent, self-healing technology will occupy an important position in the future of electronics, leading energy storage systems toward greater safety, durability, and environmental friendliness. The advancements in self-repair materials facilitate the creation of more efficient, longer-lasting, and eco-friendlier battery solutions, and as research progresses, self-repair technology is poised to play a more pivotal role in battery applications, contributing to the future of sustainable energy storage.

References

1 Li, Q., Cui, X., and Pan, Q. (2019). Self-healable hydrogel electrolyte toward high-performance and reliable quasi-solid-state Zn-MnO$_2$ batteries. *ACS Applied Materials & Interfaces* 11: 38762–38770.
2 Chen, T., Kong, W., Zhang, Z. et al. (2018). Ionic liquid-immobilized polymer gel electrolyte with self-healing capability, high ionic conductivity and heat resistance for dendrite-free lithium metal batteries. *Nano Energy* 54: 17–25.
3 Ling, W., Mo, F., Wang, J. et al. (2021). Self-healable hydrogel electrolyte for dendrite-free and self-healable zinc-based aqueous batteries. *Materials Today Physics* 20: 100458.
4 Li, H., Xu, F., Li, Y. et al. (2024). Self-healing ionogel-enabled self-healing and wide-temperature flexible zinc-air batteries with ultra-long cycling lives. *Advanced Science* 2402193.
5 Wang, X., Yang, M., Ren, Z. et al. (2024). Mussel-inspired, hydrophobic association-regulated hydrogel electrolytes with super-adhesive and self-healing properties for durable and flexible zinc-ion batteries. *Energy Storage Materials* 70: 103523.
6 Chen, X., Yi, L., Liu, J. et al. (2024). Ionic liquid-based self-healing gel electrolyte for high-performance lithium metal batteries. *Journal of Power Sources* 603: 234433.
7 Davino, S., Callegari, D., Pasini, D. et al. (2022). Cross-linked gel electrolytes with self-healing functionalities for smart lithium batteries. *ACS Applied Materials & Interfaces* 14: 51941–51953.
8 Wang, X., Wang, B., and Cheng, J. (2023). Multi-healable, mechanically durable double cross-linked polyacrylamide electrolyte incorporating hydrophobic interactions for dendrite-free flexible zinc-ion batteries. *Advanced Functional Materials* 33: 2304470.
9 Chen, X., Yi, L., Zou, C. et al. (2023). Boosting the performances of lithium metal batteries through in-situ construction of dual-network self-healing gel polymer electrolytes. *Electrochimica Acta* 446: 142084.

10 Wang, J., Hu, M., Zhu, Y. et al. (2023). Suppression of dendrites by a self-healing elastic interface in a sodium metal battery. *ACS Applied Materials & Interfaces* 15: 16598–16606.

11 Huang, Y., Liu, J., Wang, J. et al. (2018). An intrinsically self-healing NiCo||Zn rechargeable battery with a self-healable ferric-ion-crosslinking sodium polyacrylate hydrogel electrolyte. *Angewandte Chemie International Edition* 57: 9810–9813.

12 Mo, F., Lu, Y., Cui, M. et al. (2023). A self-healable silk fibroin-based hydrogel electrolyte for silver-zinc batteries with high stability. *Journal of Electroanalytical Chemistry* 938: 117466.

13 Guo, P., Su, A., Wei, Y. et al. (2019). Healable, highly conductive, flexible, and nonflammable supramolecular ionogel electrolytes for lithium-ion batteries. *ACS Applied Materials & Interfaces* 11: 19413–19420.

14 Gu, C., Wang, M., Zhang, K. et al. (2023). A full-device autonomous self-healing stretchable soft battery from self-bonded eutectogels. *Advanced Materials* 35: 2208392.

15 Chen, K., Liu, J., Zhang, X. et al. (2024). Three-dimensional cross-linked network deep eutectic gel polymer electrolyte with the self-healing ability enable by hydrogen bonds and dynamic disulfide bonds. *Journal of Colloid and Interface Science* 669: 529–536.

16 Tian, X., Yang, P., Yi, Y. et al. (2020). Self-healing and high stretchable polymer electrolytes based on ionic bonds with high conductivity for lithium batteries. *Journal of Power Sources* 450: 227629.

17 D'Angelo, A.J. and Panzer, M.J. (2019). Design of stretchable and self-healing gel electrolytes via fully zwitterionic polymer networks in solvate ionic liquids for Li-based batteries. *Chemistry of Materials* 31: 2913–2922.

18 Niu, K., Shi, J., Zhang, L. et al. (2024). A self-healing aqueous ammonium-ion micro batteries based on PVA-NH$_4$Cl hydrogel electrolyte and MXene-integrated perylene anode. *Nano Research Energy* 3: e9120127.

19 Huang, S., Wan, F., Bi, S. et al. (2019). A self-healing integrated all-in-one zinc-ion battery. *Angewandte Chemie International Edition* 58: 4313–4317.

20 Tsai, W.-T., Lu, Y.-H., and Liu, Y.-L. (2024). In situ self-healing of gel polymer electrolytes enhancing the cycling stability of lithium ion batteries. *ACS Sustainable Chemistry & Engineering* 12: 7894–7902.

21 Liu, Y., Yin, L., Chen, S. et al. (2024). A hydrogel thermoelectrochemical cell with high self-healability and enhanced thermopower both induced by zwitterions. *Journal of Materials Chemistry A* 12: 18582–18592.

22 Du, X., Tong, Y., Wang, T. et al. (2025). Self-healing and recyclable polyurethane-based solid-state polymer electrolyte via Diels-Alder dynamic network. *Journal of Molecular Structure* 1321: 139793.

23 Wu, N., Shi, Y.-R., Lang, S.-Y. et al. (2019). Self-healable solid polymeric electrolytes for stable and flexible lithium metal batteries. *Angewandte Chemie International Edition* 58: 18146–18149.

24 Wang, Y., Wang, Z., Jin, B. et al. (2023). PolySchiff based self-healing solid-state electrolytes for lithium ion battery. *European Polymer Journal* 193: 112098.

25 Wang, Q., Cui, Z., Zhou, Q. et al. (2020). A supramolecular interaction strategy enabling high-performance all solid state electrolyte of lithium metal batteries. *Energy Storage Materials* 25: 756–763.

26 Chen, K., Sun, Y., Zhang, X. et al. (2023). A self-healing and nonflammable cross-linked network polymer electrolyte with the combination of hydrogen bonds and dynamic disulfide bonds for lithium metal batteries. *Energy & Environmental Materials* 6: e12568.

27 Huang, Y., Shi, Z., Wang, H. et al. (2022). Shape-memory and self-healing polyurethane-based solid polymer electrolytes constructed from polycaprolactone segment and disulfide metathesis. *Energy Storage Materials* 51: 1–10.

28 Zhou, B., Deng, T., Yang, C. et al. (2023). Self-healing and recyclable polymer electrolyte enabled with Boronic Ester transesterification for stabilizing ion deposition. *Advanced Functional Materials* 33: 2212005.

29 Jo, Y.H., Li, S., Zuo, C. et al. (2020). Self-healing solid polymer electrolyte facilitated by a dynamic cross-linked polymer matrix for lithium-ion batteries. *Macromolecules* 53: 1024–1032.

30 Ling, C., Naren, T., Liu, X. et al. (2023). In-situ polymerization induced phase separation to develop high-performance self-healable polymeric electrolytes for lithium metal battery. *Materials Today Energy* 36: 101372.

31 Zhou, B., Zuo, C., Xiao, Z. et al. (2018). Self-healing polymer electrolytes formed via dual-networks: a new strategy for flexible lithium metal batteries. *Chemistry – A European Journal* 24: 19200–19207.

32 Gao, L., Jiang, W., Zhang, X. et al. (2024). A self-healing poly(ionic liquid) block copolymer electrolyte enabled by synergetic dual ion-dipole interactions. *Chemical Engineering Journal* 479: 147822.

33 Liu, D., Tang, Z., Luo, L. et al. (2021). Self-healing solid polymer electrolyte with high ion conductivity and super stretchability for all-solid zinc-ion batteries. *ACS Applied Materials & Interfaces* 13: 36320–36329.

34 Wan, L., Tan, X., Du, X. et al. (2023). Self-healing polymer electrolytes with dynamic-covalent borate for solid-state lithium metal batteries. *European Polymer Journal* 195: 112191.

35 Cao, X., Zhang, P., Guo, N. et al. (2021). Self-healing solid polymer electrolyte based on imine bonds for high safety and stable lithium metal batteries. *RSC Advances* 11: 2985–2994.

36 Le Mong, A. and Kim, D. (2023). Self-healable, super Li-ion conductive, and flexible quasi-solid electrolyte for long-term safe lithium sulfur batteries. *Journal of Materials Chemistry A* 11: 6503–6521.

37 Zhang, L., Zhang, P., Chang, C. et al. (2021). Self-healing solid polymer electrolyte for room-temperature solid-state lithium metal batteries. *ACS Applied Materials & Interfaces* 13: 46794–46802.

38 Deng, K., Zhou, S., Xu, Z. et al. (2022). A high ion-conducting, self-healing and nonflammable polymer electrolyte with dynamic imine bonds for dendrite-free lithium metal batteries. *Chemical Engineering Journal* 428: 131224.

39 Li, R., Fang, Z., Wang, C. et al. (2022). Six-armed and dicationic polymeric ionic liquid for highly stretchable, nonflammable and notch-insensitive intrinsic self-healing solid-state polymer electrolyte for flexible and safe lithium batteries. *Chemical Engineering Journal* 430: 132706.

40 Wang, C., Li, R., Chen, P. et al. (2021). Highly stretchable, non-flammable and notch-insensitive intrinsic self-healing solid-state polymer electrolyte for stable and safe flexible lithium batteries. *Journal of Materials Chemistry A* 9: 4758–4769.

41 Ma, Y., Zhang, R., Wang, L. et al. (2023). Self-healing polymer-based electrolyte induced by amorphous three-dimensional carbon for high-performance solid-state Li metal batteries. *Energy Storage Materials* 61: 102893.

42 Duan, P.-H., Yu, J.-L., Liu, Q.-S. et al. (2024). Dynamic anion enables self-healing single-ion conductor polymer electrolyte for lithium-metal batteries. *Advanced Functional Materials* n/a: 2402065.

43 Ju, Z., Tao, X., Wang, Y. et al. (2024). A self-healing Li-crosslinked elastomer promotes a highly robust and conductive solid-electrolyte interphase. *Energy & Environmental Science* 17: 4703–4713.

44 Whiteley, J.M., Taynton, P., Zhang, W. et al. (2015). Ultra-thin solid-state Li-ion electrolyte membrane facilitated by a self-healing polymer matrix. *Advanced Materials* 27: 6922–6927.

45 Zhang, J., Bai, G., Wang, C. et al. (2024). A self-healing polymer electrolyte based on the Diels-Alder reaction in lithium-ion batteries. *Journal of Applied Polymer Science* 141: e55473.

46 Lee, S., Song, J., Cho, J. et al. (2023). Thermally reprocessable self-healing single-ion conducting polymer electrolytes. *ACS Applied Polymer Materials* 5: 7433–7442.

47 Wan, L., Du, X., Guo, L. et al. (2023). Self-healing polymer electrolytes with nitrogen-boron coordinated boroxine for all-solid-state lithium metal batteries. *Journal of Energy Storage* 74: 109485.

48 Yu, Y., Ling, C., Yang, J. et al. (2024). Self-healing fluorinated polymer deep eutectic electrolytes for stable lithium metal batteries. *Chemical Engineering Journal* 498: 155376.

49 Xia, S., Lopez, J., Liang, C. et al. (2019). High-rate and large-capacity lithium metal anode enabled by volume conformal and self-healable composite polymer electrolyte. *Advanced Science* 6: 1802353.

50 Huang, Y., Zhong, M., Huang, Y. et al. (2015). A self-healable and highly stretchable supercapacitor based on a dual crosslinked polyelectrolyte. *Nature Communications* 6: 10310.

51 Chen, H., Liu, J., Zhou, X. et al. (2021). Rapid leakage responsive and self-healing Li-metal batteries. *Chemical Engineering Journal* 404: 126470.

52 Lin, Y., Li, X., Zheng, W. et al. (2023). Effect of SiO_2 microstructure on ionic transport behavior of self-healing composite electrolytes for sodium metal batteries. *Journal of Membrane Science* 672: 121442.

53 Zhou, B., Jo, Y.H., Wang, R. et al. (2019). Self-healing composite polymer electrolyte formed via supramolecular networks for high-performance lithium-ion batteries. *Journal of Materials Chemistry A* 7: 10354–10362.

54 Zhang, Y., Yu, X., Li, X. et al. (2024). LLZTO crosslinks form a highly stretchable self-healing network for fast healable all-solid lithium metal batteries. *Chemical Engineering Journal* 497: 154397.

55 Li, H., Liu, W., Yang, X. et al. (2021). Fluoroethylene carbonate-Li-ion enabling composite solid-state electrolyte and lithium metal interface self-healing for dendrite-free lithium deposition. *Chemical Engineering Journal* 408: 127254.

56 Lee, Y., Kim, M., Kim, H. et al. (2022). Self-healable and tough polymer electrolyte composites based on associative nanostructural networks. *ACS Applied Polymer Materials* 4: 5821–5830.

57 Munaoka, T., Yan, X., Lopez, J. et al. (2018). Ionically conductive self-healing binder for low cost Si microparticles anodes in Li-ion batteries. *Advanced Energy Materials* 8: 1703138.

58 Wang, Y., Xu, H., Chen, X. et al. (2021). Novel constructive self-healing binder for silicon anodes with high mass loading in lithium-ion batteries. *Energy Storage Materials* 38: 121–129.

59 Chen, H., Wu, Z., Su, Z. et al. (2021). A mechanically robust self-healing binder for silicon anode in lithium ion batteries. *Nano Energy* 81: 105654.

60 Kwon, T.-W., Jeong, Y.K., Lee, I. et al. (2014). Systematic molecular-level design of binders incorporating Meldrum's acid for silicon anodes in lithium rechargeable batteries. *Advanced Materials* 26: 7979–7985.

61 Zhang, X., Chen, P., Zhao, Y. et al. (2021). High-performance self-healing polyurethane binder based on aromatic disulfide bonds and hydrogen bonds for the sulfur cathode of lithium-sulfur batteries. *Industrial & Engineering Chemistry Research* 60: 12011–12020.

62 Wen, Y., Lin, X., Sun, X. et al. (2024). A biomass-rich, self-healable, and high-adhesive polymer binder for advanced lithium-sulfur batteries. *Journal of Colloid and Interface Science* 660: 647–656.

63 Chen, J., Li, Y., Wu, X. et al. (2024). Dynamic hydrogen bond cross-linking binder with self-healing chemistry enables high-performance silicon anode in lithium-ion batteries. *Journal of Colloid and Interface Science* 657: 893–902.

64 Malik, Y.T., Shin, S.-Y., Jang, J.I. et al. (2023). Self-repairable silicon anodes using a multifunctional binder for high-performance lithium-ion batteries. *Small* 19: 2206141.

65 Wang, X., Wang, K., Wan, Z. et al. (2024). Inorganic/organic composite binder with self-healing property for silicon anode in lithium-ion battery. *Materials Today Energy* 43: 101567.

66 Nam, J., Kim, E., Rajeev, K.K. et al. (2020). A conductive self healing polymeric binder using hydrogen bonding for Si anodes in lithium ion batteries. *Scientific Reports* 10: 14966.

67 He, S., Huang, S., Liu, X. et al. (2024). Interfacial self-healing polymer electrolytes for long-cycle silicon anodes in high-performance solid-state lithium batteries. *Journal of Colloid and Interface Science* 665: 299–312.

68 Wang, Y., Zheng, R., Luo, J. et al. (2019). Self-healing dynamically cross linked versatile polymer electrolyte: a novel approach towards high performance, flexible electrochromic devices. *Electrochimica Acta* 320: 134489.

6

Application of Low-Dimensional Materials in Batteries

6.1 Introduction

6.1.1 Lithium-Metal Batteries

With the increasing depletion of fossil fuels and the exacerbation of global warming, the development of renewable and sustainable energy sources has become increasingly important. These include clean energy sources such as solar, wind, geothermal, and tidal energy. In the application of these new energy sources, efficient energy storage and conversion play a critical role. Lithium metal batteries, as one of the leading rechargeable battery technologies today, have attracted widespread attention and research interest. With the deepening research on lithium-ion batteries, significant progress has been made in the three key components: electrode (cathode and anode), separator, and electrolyte materials. However, as the practical energy density of lithium-ion batteries approaches their theoretical limits, alternative lithium battery systems with higher energy density, such as lithium-air batteries, lithium-sulfur (Li—S) batteries, and lithium-sodium batteries, have been proposed and have drawn significant interest.

Taking Li—S batteries as an example, this system utilizes the electrochemical reaction between elemental sulfur and metallic lithium to convert chemical energy into electrical energy. Typically, Li—S batteries employ metallic lithium as the anode and elemental sulfur as the cathode. From Figure 6.1, during discharge, the sulfur cathode generates Li_2S, offering a theoretical specific capacity of up to 1675 mAh g^{-1}, far exceeding that of lithium-ion batteries. The overall reaction of the Li—S battery is as follows:

$$S_8 + 16Li^+ + 16e^- \rightleftharpoons 8Li_2S$$

During the charge and discharge process, the sulfur cathode undergoes a solid-liquid-solid phase transition [1]. Metallic lithium loses electrons, forming lithium-ions that dissolve into the electrolyte, while the electrons are transmitted through the external circuit to the sulfur cathode. Elemental sulfur (solid) accepts electrons and combines with lithium-ions from the electrolyte, forming intermediate lithium polysulfides (LiPSs) (Li_2S_n, where n = 4, 6, 8), which are soluble

Functional Auxiliary Materials in Batteries: Synthesis, Properties, and Applications, First Edition. Wei Hu.
© 2025 WILEY-VCH GmbH. Published 2025 by WILEY-VCH GmbH.

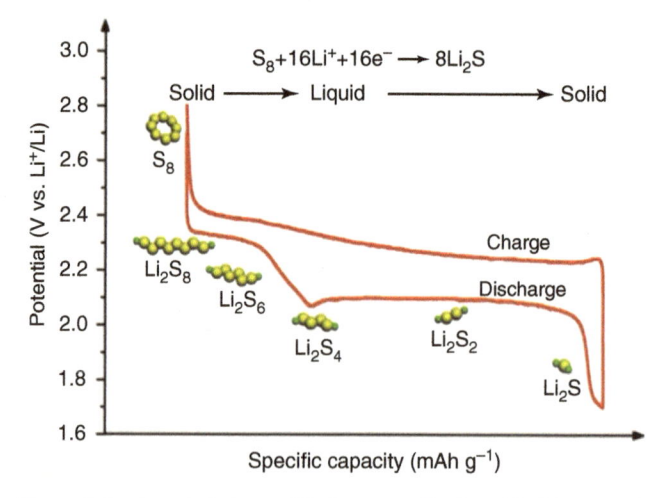

Figure 6.1 A typical charge/discharge profile for a Li–S battery. Source: Fang et al. [1]/with permission of John Wiley & Sons.

in conventional electrolytes. The final discharge products of the polysulfides are insoluble Li_2S_2 or Li_2S.

6.1.2 Low-Dimensional Composite Materials

In the design of high-energy-density batteries, the innovative development of cathode, anode, separator, and electrolyte materials is equally crucial. Low-dimensional materials, with their unique physical and chemical properties, have shown immense potential in these novel battery systems. Low-dimensional materials refer to those whose dimensions are confined or reduced in certain directions, leading to significant differences in their electronic, optical, and mechanical properties compared to traditional three-dimensional (3D) (bulk) materials. Low-dimensional materials often exhibit unique physical phenomena due to their structural characteristics. Based on the extent of dimensional confinement, low-dimensional materials can be categorized into zero-dimensional, one-dimensional (1D), and two-dimensional (2D) materials.

Zero-dimensional materials are confined in all three dimensions, typically with characteristic sizes ranging from 1 to 10 nm, resulting in electron movement being restricted to a very small region. Examples include nanoparticles and quantum dots. 1D materials are confined in two dimensions but extend freely along the third, forming linear structures such as nanowires, nanotubes, and nanorods. 2D materials, on the other hand, have large lateral dimensions but are only a few atomic layers thick in the vertical direction. Notable examples include graphene, transition metal dichalcogenides, and transition metal carbides/nitrides compounds (MXene). For example, graphene, as a prototypical 2D material, exhibits extraordinary mechanical strength, high electrical conductivity, large specific surface area, and excellent planar self-assembly and surface modification capabilities. These characteristics make graphene highly promising for energy storage and conversion, particularly in battery applications.

The size effect of low-dimensional materials significantly influences their physical and chemical properties, including enhanced mechanical strength, quantum confinement effects, and surface effects. By integrating them with other materials, the performance of these low-dimensional materials can be further optimized, enabling the fabrication of more efficient low-dimensional composite materials. In the following sections, this paper will discuss the application of low-dimensional composite materials in Li—S batteries, focusing on the cathode, non-active materials (such as separators and current collectors), and anode materials.

6.2 Low-Dimensional Composite Cathode Materials

During charge and discharge, the sulfur cathode undergoes a phase transition from solid (sulfur) to liquid (Li_2S_n) and back to solid (Li_2S_2 or Li_2S). The main technical challenges faced by Li—S batteries include:

1) Volume expansion of the sulfur cathode. There is a significant density difference between elemental sulfur (2.07 g cm^{-3}, 25 °C) and the discharge product Li_2S (1.66 g cm^{-3}, 25 °C), leading to a volume change of up to 80% during cycling. This substantial volume expansion can cause the cathode structure to collapse over multiple cycles.
2) Poor conductivity of sulfur and Li_2S. Both elemental sulfur and Li_2S have poor electrical conductivity, necessitating the mixing of sulfur with conductive additives (such as conductive carbon black) to improve performance. However, even with such conductive additives, the conductivity remains limited. During the solid-to-solid phase transition, active materials can lose their electronic pathways, resulting in the formation of "dead sulfur".
3) Shuttle effect of polysulfides (Li_2S_n). Soluble polysulfides can easily migrate from the cathode, penetrate the separator, and reach the lithium metal anode, where they react with lithium to form short-chain polysulfides or insoluble Li_2S_2/Li_2S. These short-chain polysulfides may then diffuse back to the cathode, leading to active material loss and a decrease in Coulombic efficiency.

These issues collectively hinder the practical development of Li—S batteries. By thoroughly understanding the reaction mechanisms and failure causes of the sulfur cathode, several key design considerations for sulfur cathodes can be identified:

1) Mechanical stability during cycling.
2) The inherently low electronic and ionic conductivity of the sulfur cathode.
3) Control of the polysulfide shuttle effect while fully utilizing their electrochemical activity.

To address these challenges, the main strategies include:

1) Loading elemental sulfur onto porous conductive scaffold materials, such as carbon materials or conductive polymers, to enhance conductivity and structural stability.
2) Coating a barrier layer on the surface of sulfur particles, the cathode, or the cathode-facing side of a porous polymer separator. This barrier, through physical or chemical means, confines polysulfides within the cathode region and reduces the shuttle effect.

Many low-dimensional materials possess excellent electrical conductivity, abundant pore structures, high specific surface areas, and easily modifiable surfaces. Thus, the application of low-dimensional composite materials in sulfur cathodes can significantly improve the performance of Li—S batteries.

6.2.1 Composite Methods for Low-Dimensional Cathode Materials

In nature, sulfur typically exists in its stable orthorhombic ring structure, α-S_8, with a melting point of approximately 115 °C and a boiling point of around 444.6 °C. Therefore, sulfur cathodes can be prepared through physical methods such as ball milling, precipitation, melt diffusion, vapor infiltration, coating, and melt crystallization.

Ball milling typically refers to the grinding of sulfur and supporting materials in a sealed jar, where weak interactions between the active materials and the support may form sulfur aggregates, leading to reduced active material utilization. This makes it unsuitable for producing high-performance cathode composites. However, ball milling remains a common method for mixing cathode slurry due to its ease of scaling up for mass production. Lin et al. developed a self-supported graphene-sulfur composite using a simple and effective ball milling process with graphite and sulfur [2]. Given the similar electronegativity between sulfur and graphite, their interaction is stronger than the van der Waals forces between graphene layers. During ball milling, sulfur adheres to the surface and edges of graphite, resulting in the mechanical exfoliation of graphite sheets, producing highly conductive graphene with few defects (conductivity of 1820 S cm^{-1}). This promotes rapid electron transport within the sulfur cathode, akin to the Scotch tape method for graphene exfoliation.

Non-polar sulfur can dissolve in certain polar solvents, and by adding other polar solvents, sulfur particles can recrystallize from the solution. Based on this physical precipitation method, Sun et al. fabricated a binder-free nano sulfur-carbon nanotube (CNT) composite material, characterized by sulfur nanocrystals anchored across a superaligned carbon nanotube (SACNT) matrix via a simple solution-based method [3]. The conductive SACNT matrix prevents self-aggregation and ensures the dispersive distribution of sulfur nanocrystals while providing a 3D continuous electron pathway. It also enhances electrolyte infiltration, confines sulfur/polysulfides, and accommodates sulfur's volume changes during cycling. The nanosized sulfur particles shorten the lithium-ion diffusion path, and the confinement within the SACNT network ensures structural and electrochemical stability. The nano S-SACNT composite cathode delivers an initial discharge capacity of 1071 mAh g^{-1}, a peak capacity of 1088 mAh g^{-1}, and retains 85% of its capacity after 100 cycles with a high Coulombic efficiency (~100%) at 1 C. Moreover, at high current rates, the nano S-SACNT composite shows impressive capacities of 1006 mAh g^{-1} at 2 C, 960 mAh g^{-1} at 5 C, and 879 mAh g^{-1} at 10 C.

The viscosity of molten sulfur reaches its lowest point at 155 °C but rapidly increases above 160 °C. The molten diffusion method involves heating pre-ground sulfur and composite materials in a sealed environment at around 155 °C. Molten sulfur infiltrates the pores of the composite via capillary action, and upon cooling, sulfur nanocrystals form, establishing close contact with the composite. The molten diffusion method has become one of the most widely used techniques for preparing composite sulfur cathode materials.

Due to sulfur's low enthalpy of vaporization and sublimation, it is easy to process in the gas phase. At 450 °C, sulfur vapor mainly contains S_8 and S_6 molecules, while above 550 °C, S_8 decomposes into smaller S_4 and S_2 molecules, facilitating the infiltration of sulfur into the host material. The vapor infiltration method not only induces strong carbon-sulfur bonding but also promotes sulfur infiltration into porous structures. Zheng et al. embedded S_2 molecules into a graphene interlayer via vapor infiltration to create a cathode composite [4]. When a mixture of sulfur and graphene oxide (GO) powder was heated under vacuum at 600 °C, GO was reduced to a highly conductive reduced graphene oxide (rGO) network, while S_8 molecules decomposed into S_2 molecules that were inserted between the rGO layers. After cooling, the surface S_2 crystallized back into S_8, which was removed using carbon disulfide (CS_2) to improve cycling stability.

Adding various additives such as conductive agents and binders to sulfur cathodes greatly reduces the energy density of Li—S batteries. When the sulfur content is 54 wt%, the energy density ranges between 283 and 314 Wh l^{-1}, lower than that of commercial $LiCoO_2$-graphite lithium-ion batteries, which diverges from the original high energy density concept. To further improve volumetric energy density, some researchers have developed pure sulfur cathodes. Qie et al. directly coated commercial sulfur powder onto aluminum foil and covered it with a layer of carbon paper [5]. The resulting sulfur cathode achieved areal sulfur loadings between 2.5 and 16.2 mg cm^{-2}. The cathode with a sulfur loading of 16.0 mg cm^{-2} exhibited an initial specific capacity of 1435 mAh g^{-1}, indicating that most sulfur particles were successfully lithiated during discharge.

Sulfur is insoluble in water but has limited solubility in some nonpolar solvents, such as carbon disulfide (CS_2), dimethyl sulfoxide, tetrahydrofuran, and toluene. Porous materials can be added to solvents in which sulfur has been dissolved, and then the solvent can be evaporated, causing sulfur to recrystallize and adhere to the porous material's interior. This method is known as solution crystallization. Zheng et al. prepared a composite with a small amount (10%) of Cu additive in microporous carbon using an ultrasonic-assisted multiple wetness impregnation and synchro-dry technique [6].

By employing chemical methods, such as chemical deposition, copolymerization, oxidation, electrochemical deposition, or reduction reactions, smaller sulfur particles with more uniform distribution can be achieved, thereby enhancing the uniformity of the sulfur cathode's nanostructure and improving its component activity.

Sodium sulfide (Na_2S) and sodium thiosulfate ($Na_2S_2O_3$) are two of the most commonly used raw materials in chemical deposition. Sulfur dissolves in an aqueous solution of sodium sulfide to form a sodium polysulfide solution, and by adding hydrochloric acid or sulfuric acid to the polysulfide or thiosulfate solution, sulfur can be deposited uniformly on the matrix and nucleated. The addition of surfactants and polymers can generate ultra-fine sulfur particles and facilitate the formation of core-shell sulfur/carbon structures, which can effectively suppress the diffusion of polysulfides. Wang et al. synthesized a graphene-sulfur composite using polyethylene glycol (PEG) containing a surfactant through chemical deposition [7]. The PEG chains served as end-capping agents to limit sulfur particle size and as a flexible framework to buffer volume changes during cycling.

Most sulfur cathodes use S_8 as the active material, which undergoes a series of solid–liquid–solid phase transitions during discharge, leading to the formation of soluble polysulfides. Copolymerization, which forms strong covalent bonds between sulfur and carbon, can effectively suppress the dissolution and diffusion of polysulfides. Copolymerization typically involves inverse vulcanization, where liquid-phase S_8 monomers undergo ring-opening polymerization at 159 °C to form linear polysulfane with dual radical segments. These long-chain sulfur radicals can react with functional groups such as vinyl, thiol, cyano, and alkynyl. The resulting organic sulfur polymer's backbone consists of crosslinked long sulfur chains and short polymer chains, giving it electrochemical properties similar to elemental sulfur. For instance, Kim et al. synthesized a triazine-based organic sulfur polymer with interconnected macropores and amine groups, which enhanced the physical and chemical adsorption of polysulfides [8]. The Li—S cells displayed a discharge capacity of 945 mAh g^{-1} after 100 cycles at 0.2 C with a high-capacity retention of 92% and a lifespan of 450 cycles. Particularly, the organized amine groups within the crystals increased the Li$^+$-ion transfer rate, affording a rate performance of 1210 mAh g^{-1} at 0.1 C and 730 mAh g^{-1} at 5 C. Chung et al. copolymerized sulfur with vinyl-containing 1,3-diisopropenylbenzene (DIB) at 185 °C, and then blended the product with conductive carbon and binder through ball milling to produce an organic copolymer sulfur cathode [9]. The cathode retained 823 mAh g^{-1} after 100 cycles at 0.1 C. Li et al. further developed a cathode composite by loading copolymerized sulfur-DIB onto 3D graphene, increasing the sulfur content in the cathode to over 80% [10]. This composite exhibited good cycling performance, with a capacity decay rate of only 0.028% per cycle over 500 cycles. Copolymerization, as a novel method for sulfur composite preparation, constructs covalent C—S bonds that can suppress the polysulfide shuttle effect and promote uniform sulfur dispersion.

Electrochemical deposition has garnered significant attention for synthesizing novel nanomaterials due to its low cost, rapid processing, high purity, and controllable structure. In recent years, this technique has been employed to produce a range of new 2D materials (MXenes) from 3D layered ternary carbides and nitrides (MAX phases). For example, Zhao et al. synthesized carbon/sulfur nanolaminated composites by selectively extracting Ti from the MAX phase Ti$_2$SC using electrochemical deposition [11]. The resulting product consisted of multi-layered C/S sheets, where covalent C—S bonds were present in the nanolaminates. Zhao et al. also employed a two-electrode system to deposit sulfur nanodots onto nickel foam [12]. Due to the conductive and flexible nature of the nickel framework, the electrode exhibited excellent rate performance. The optimized cathode with 0.45 mg cm^{-2} of sulfur on nickel foam demonstrated a high initial discharge capacity (1458 mAh g^{-1} at 0.1 C), superior rate capability (521 mAh g^{-1} at 10 C), and long cycling stability (895 mAh g^{-1} after 300 cycles at 0.5 C and 528 mAh g^{-1} after 1400 cycles at 5 C). Li et al. synthesized vertically aligned sulfur-graphene (S-G) nanowalls on a conductive substrate using electrochemical methods [13]. In this composite, sulfur nanoparticles were uniformly anchored between graphene layers and an orderly graphene array aligned perpendicular to the substrate, enhancing the material's electron/ion conductivity. As a result, the S-G nanowalls as cathodes for Li—S batteries exhibited a high reversible capacity of 1261 mAh g^{-1} in the first cycle and maintained over 1210 mAh g^{-1} after 120 cycles, alongside excellent cycling stability and high-rate performance (over 400 mAh g^{-1} at 8 C, 13.36 A g^{-1}). Electrochemical deposition is an effective method for constructing composite sulfur cathodes, ensuring an orderly nanostructure and uniform sulfur distribution.

Sulfates can also be reduced to sulfur and used for the preparation of composite sulfur cathodes, although this method is associated with lower sulfur content, high cost, and stringent preparation conditions. Zhou et al. employed chemical vapor deposition (CVD) to grow CNTs on an anodic aluminum oxide (AAO) membrane while reducing sulfates to sulfur at 560 °C [14]. This high reaction temperature promoted the formation of strong C—S bonds, similar to those formed in products prepared via conventional CVD methods.

The sulfur loading method has a direct impact on the distribution, particle size, and morphology of sulfur, which in turn affects the electrochemical performance of Li—S batteries. Each preparation method for sulfur cathodes has its unique characteristics and suitable applications. However, most of the reported methods remain limited to laboratory-scale production and are challenging to scale up. Therefore, the development of low-cost, widely applicable, and scalable preparation methods for sulfur composites is still urgently needed.

6.2.2 One-Dimensional Materials in Cathode

6.2.2.1 Carbon Nanotube (CNT) Materials

In 2009, Professor Nazar from the University of Waterloo in Canada first introduced CMK-3 into the sulfur cathode system [15], significantly improving the cycling performance of Li—S batteries. This breakthrough led to the widespread application of carbon-based materials in constructing composite sulfur cathodes. Loading sulfur onto low-dimensional carbon materials not only enhances the electrode's conductivity but also utilizes the unique structure of low-dimensional materials to physically confine polysulfides. Additionally, these materials can effectively mitigate the cathode's volume expansion during battery cycling.

1D CNTs, with their extremely high aspect ratio, not only provide a high specific surface area to anchor sulfur and polysulfides but also form a continuous long-range conductive network that accelerates reaction rates in the cathode and improves sulfur utilization. The hollow structure of CNTs can confine sulfur within their channels, while the porous framework formed by the 1D materials helps alleviate sulfur's volume expansion during cycling, making them ideal for fabricating flexible, binder-free, and self-supporting cathodes. Han et al. first introduced multi-walled carbon nanotubes (MWCNTs) into Li—S battery cathodes [16], which not only improved conductivity but also effectively prevented the dissolution and diffusion of sulfur into the electrolyte. Simple mixing of CNTs as conductive agents with sulfur, or loading sulfur into CNTs, can significantly enhance the electrochemical performance of the cathode. Yuan et al. coated sulfur into MWCNTs through capillary action, preparing a core-shell structured composite cathode material [17]. Compared to the simple mechanical mixture of MWCNT/S cathodes, this special core-shell structure significantly improved the cycling stability of the cathode, with a discharge capacity of 670 mAh g^{-1} remaining after 60 cycles.

However, the bonding interaction between CNTs and sulfur is weak, resulting in a low sulfur loading capacity, which hinders the development of high-loading, high-performance Li—S batteries. Therefore, modifying CNTs and creating synergistic effects with other materials can further enhance the performance of CNT-based composite sulfur cathodes. Zhang et al. hybridized CNTs with graphene [18], arranging aligned CNTs within graphene layers and anchoring them to each other, followed by nitrogen doping, to obtain a sandwich structure N-ACNT/G composite material with efficient 3D electron and ion transport pathways, as shown in Figure 6.2a. Nitrogen doping introduced more defects and electrochemical active sites into the carbon framework, improving the interface's adsorption capacity and electrochemical performance. This material achieved an initial discharge capacity of 1152 mAh g^{-1} at 1 C and a capacity of 770 mAh g^{-1} at 5 C. Wang et al. prepared an MWCNT-based composite cathode grafted with PEG [20]. This composite material has a nest-like structure, with sulfur as the core and the PEG-MWCNT network as the conductive shell layer. PEG grafting enhanced the carbon framework's ability to adsorb polysulfides, reducing their diffusion and shuttle effect. After 200 cycles, this material maintained a capacity of 897 mAh g^{-1} at 0.2 C. Guo et al. introduced a novel strategy to enhance sulfur loading and rate performance in Li—S batteries by coupling a nanostructured cathode with an antifouling separator using electrostatic self-assembly [21]. This method integrates 2D MXene with positively charged 1D CNT-polyethyleneimine, addressing issues of sluggish ionic transport and forming a dynamic crosslinking network in the cathode. The antifouling separator, featuring organized inter-lamellar porosity, dual polarity, and high conductivity, plays a crucial role in low-order polysulfide activation, high-rate cyclability, and Li dendrite inhibition. The design achieves a long-term capacity of 980 mAh g^{-1} at 5 mA cm^{-2} over 500 cycles with a sulfur loading of 2.6 mg cm^{-2}. Additionally, a flexible cathode with high loading (5.8 mg cm^{-2}) and mechanical strength (13 MPa) shows an areal capacity of 7.1 mAh cm^{-2}, performing well at nearly 10 mA cm^{-2}. The cathode's

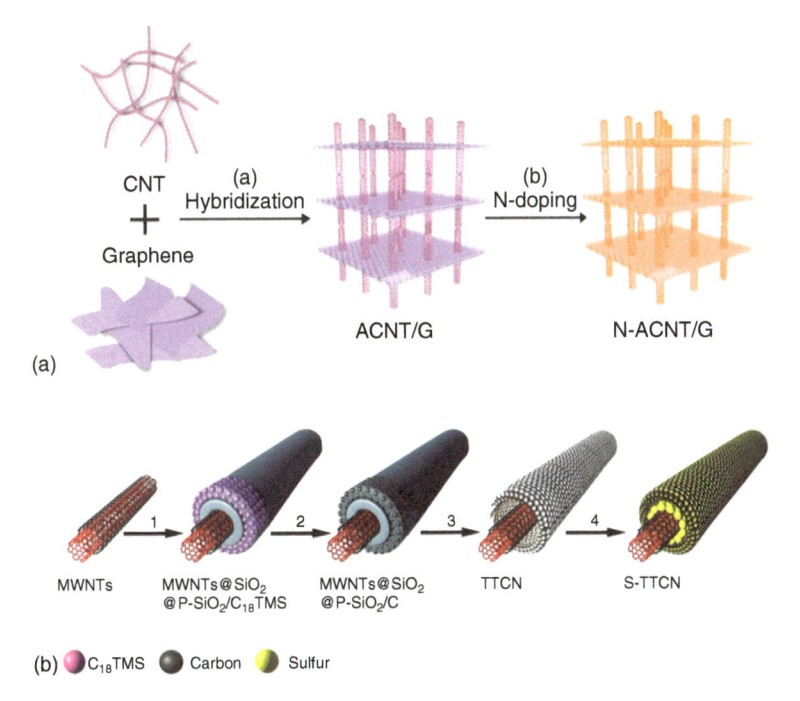

Figure 6.2 (a) The design of N-ACNT/G hybrids with graphene and aligned CNTs as building blocks. Source: Tang et al. [18]/with permission of John Wiley & Sons. (b) The formation of S-TTCN composite. Source: [19]/with permission of John Wiley & Sons.

3D structure, enhanced by $Ti_3C_2T_x$ frameworks and CNT-PEI spacers, improves LiPS adsorption and interfacial reactions. The separator's nanochannels provide antifouling performance and high-rate cyclability, while its artificial interface accelerates LiPS activation. Compared to traditional designs, this integrated approach stabilizes the cathode and suppresses Li dendrites, promising long-term cyclability and capacity for high-rate applications. The T@CP-S|T@CP system demonstrates high loading and rate capacity, bridging the gap between metal anodes and cathodes with rapid ion transport, efficient LiPS activation, and Li anode protection.

The porous structure of CNTs can increase sulfur loading while prolonging battery cycle life. The hollow structure of CNTs not only stores sulfur but also allows for nesting sulfur-loaded CNTs inside, forming a "tube-in-tube" structure. As shown in Figure 6.2b, Guan et al. confined MWCNTs within hollow porous CNTs, preparing a tube-in-tube structured S-TTCN sulfur cathode [19]. This structure not only enhanced conductivity but also effectively prevented the dissolution and diffusion of polysulfides, while the large pore volume created conditions for increasing sulfur loading. At a sulfur content of 71 wt%, S-TTCN still provided a high reversible capacity. Jin et al. filled small-diameter CNTs (approximately 20 nm) and 85.2% sulfur into large-diameter CNTs (approximately 200 nm) [22], forming a composite structure. The high conductivity, flexibility, and structural strength of the small-diameter CNTs worked synergistically with the coating layer of large-diameter CNTs, effectively increasing the utilization of active materials, alleviating the volume expansion of the electrode, and suppressing the dissolution and shuttle effect of polysulfides. After 150 cycles at 5 C, this structure still maintained a capacity of 954 mAh g^{-1}. Additionally, increasing the porosity of CNTs can further enhance sulfur loading capacity. For example, low-density CNT foams, prepared through vapor infiltration and mechanical compression, can yield binder-free, dense sulfur cathodes [23]. This cathode achieved a capacity of 1039 mAh g^{-1} at 0.1 C with a sulfur loading of 19.1 mg cm^{-2} and an areal capacity as high as 19.3 mAh cm^{-2}.

3D printing is a versatile technique for creating robust 3D sulfur cathodes with high porosity and interconnected channels, enabling rapid ionic and electronic transport. Typically, two approaches are used in fabricating sulfur cathodes via direct ink writing (DIW). The first involves infiltrating sulfur vapor into the host material during post-processing, while the second uses 3D-printed electrodes made from sulfur suspension ink. Li—S micro cathodes have been successfully developed using a one-step printing process with aqueous MWCNT inks [24]. This efficient and environmentally friendly method, which eliminates the need for complex microfabrication and toxic organic solvents, produces high-performance electrodes with minimal material waste. The printed microcathodes are capable of high areal discharge capacities (~7 mAh cm^{-2}) with a sulfur content of ~50 wt% and a sulfur loading of ~7 mg cm^{-2}. They demonstrate excellent stability, sustaining high current densities (up to 11.5 mA cm^{-2}) with a Coulombic efficiency of over 95% for >200 cycles. The process allows for easy detachment of the electrodes from the filter membrane and their transfer to various substrates, facilitating versatile application. These MWCNT-based cathodes, with their high sulfur loading and scalable aperiodic electrode architecture, offer promising performance benchmarks for next-generation microelectronics. The system achieves ≥5 mAh cm^{-2} for 50 cycles and can endure over 500 cycles, making it a viable candidate for micro batteries used in powering microdevices and MEMS. Although the current setup uses planar lithium foil as the anode, making it a semi-3D/2.5D battery, this development provides a strategic pathway for integrating printed Li—S batteries into miniature devices, advancing the potential of micro battery technology. A novel 3D-printed sulfur/carbon (S/C) cathode for Li—S batteries was developed using a low-cost commercial carbon black (BP-2000) as the sulfur host [25]. The cathode was fabricated through a simple robocasting 3D printing process, which allows easy control over the areal sulfur loading by adjusting the number of printed layers. The 3D-printed S/C cathode demonstrated excellent electrochemical performance, including high capacity, cycling stability, and rate retention, by enhancing Li$^+$/e$^-$ transport across multiple scales (macro-, micro-, and nano-scale). The sulfur-loaded 3D-printed electrodes (3DP-FDE) were optimized through post-treatment processes like phase inversion and freeze drying, creating a hierarchical porous structure. This structure enables efficient electron and ion transport and mitigates polysulfide shuttling. The Li—S batteries assembled with 3DP-FDE electrodes exhibited a stable discharge capacity of 564 mAh g^{-1} over 200 cycles at 3 C with a sulfur loading of 3 mg cm^{-2}. Cathodes with a higher sulfur loading of 5.5 mg cm^{-2} showed large initial discharge capacities of 1009 and 912 mAh g^{-1}, and retained 87% and 85% of their capacity after 200 cycles at 1 and 2 C, respectively.

6.2.2.2 Carbon Nanofiber (CNF) Materials

Carbon nanofibers (CNFs) have a similar hollow tubular structure to CNTs, but they lack the graphitic characteristics. Due to the permeability and smaller diameter of CNTs, it is difficult to encapsulate a large amount of sulfur, with sulfur mainly being loaded on the surface of CNTs. In contrast, the conductivity and porous structure of CNFs make them more capable of forming conductive interwoven networks, which can effectively capture polysulfides during cycling. The hollow and layered porous structure of CNFs also contributes to increasing sulfur loading.

Cui et al. used AAO as a template to thermally carbonize polystyrene, thereby fabricating a hollow CNF array [26]. The AAO template effectively prevents sulfur from coating the outer walls of CNFs, assisting in the injection of sulfur into the hollow structure of CNFs. After 150 cycles at 0.2 C, this CNF composite cathode still maintained a high specific capacity of 730 mAh g^{-1}. To mitigate capacity fading, Zheng introduced an amphiphilic polymer to modify the surface of CNFs, enhancing the strong interaction between non-polar carbon and polar Li$_x$S clusters [27]. The modified sulfur cathode exhibited excellent cycling performance, with a specific capacity of nearly 1180 mAh g^{-1} at a 0.2 C current rate. After 300 cycles at a 0.5 C rate, the capacity retention reached 80%.

Bian et al. developed a flexible Li—S battery cathode by uniformly depositing sulfur into an electrospun N-rich, Cu-decorated non-woven CNF film, creating a unique sandwich-structured CNF/S-Cu/CNF electrode [28]. This design provides a free-standing cathode that achieves high capacity, rate capability, and long cycling stability. The cathode demonstrates a reversible capacity of 1295 mAh g^{-1} at a current density of 0.1 A g^{-1} and retains over 530 mAh g^{-1} even at a high rate of 1 A g^{-1} after 300 cycles, with nearly 100% Coulombic efficiency. The performance of the CNF/S-Cu/CNF composite is attributed to the synergistic effects of the CNF layers, Cu nanoparticles, and nitrogen doping. The CNF layers act as a flexible and robust framework, enhancing sulfur utilization and promoting electron conductivity. The embedded copper nanoparticles and nitrogen doping further stabilize sulfur, both physically and chemically, and effectively mitigate the polysulfide shuttle effect. This strategy highlights a new approach to fabricating high-performance, flexible, free-standing Li—S battery cathodes, ideal for lightweight and bendable applications. The sandwich structure enhances sulfur loading and minimizes polysulfide migration, showing great potential for next-generation flexible and durable Li—S batteries. The electrode configuration provides robust cycling performance with an 85% capacity retention after 300 cycles at 1 A g^{-1}, making it a strong candidate for future commercial applications.

Additionally, combining CNF with other carbon materials can yield novel structured cathodes. Liu et al. prepared a sandwich core-shell composite cathode, G-S-CNFs, by combining CNFs with graphene [29]. In this structure, graphene sheets encapsulate sulfur-loaded CNFs, with the outer layer of graphene effectively inhibiting the dissolution and diffusion of polysulfides. This composite material exhibited a reversible specific capacity of 694 mAh g^{-1} at 0.1 C, demonstrating significantly higher performance than cathodes without graphene coating. The improved rate capability and cycling stability are attributed to the unique coaxial structure of the nanocomposite, where graphene and CNF significantly enhance conductivity and better capture soluble polysulfides. However, the decline in specific capacity remains unavoidable in such systems. The hollow CNF @ nitrogen-doped porous carbon (HCNF@NPC) core-shell composite, derived from carbonized polyaniline-coated hollow CNF, can serve as a conductive carbon scaffold for sulfur encapsulation, resulting in a composite cathode with a sulfur content of 77.5 wt% [30]. Due to the electron transport and mechanical support provided by the HCNF core, along with the high specific surface area (485.244 $m^2\,g^{-1}$) and large pore volume of the NPC shell, the HCNF@NPC-S composite maintained a discharge capacity of 590 $mAh\,g^{-1}$ after 200 cycles at 0.5 C. Han et al. designed a Li—S battery using a binder-free, self-standing CNF/rGO composite as the host for LiPS-containing liquid active materials [31]. This novel approach leverages the 3D interconnected structure of electrospun CNF combined with uniformly distributed rGO, resulting in a conductive matrix that supports long-range charge transfer and provides sufficient porous space for polysulfide uptake. The flexible nature of the CNF/rGO membrane allows for the construction of thick electrodes with high areal sulfur loading without significant degradation in electrochemical performance. The CNF/rGO composite membrane offers several advantages: it enhances the mechanical flexibility of the electrode, eliminates the need for a polymer binder, and slows down the diffusion of polysulfides by absorbing electrolytes in the porous structure. This design enabled Li—S coin cells with a sulfur content of 49–56 wt% to achieve stable cycling performance and high rate capabilities, with sulfur loadings as high as 20.3 $mg\,cm^{-2}$ and areal specific capacities up to 15.5 $mAh\,cm^{-2}$. Additionally, a Li—S pouch cell based on this structure demonstrated stable performance over extended cycles. Li et al. designed a Janus electrode structure for Li—S batteries to tackle the polysulfide shuttle effect and sluggish reaction kinetics [32]. The electrode features a dual-layer design: one side comprises 1D Fe_3C-decorated N-doped carbon nanofibers (Fe_3C/N-CNFs), while the other side is made up of 2D rGO. This unique configuration provides both chemical immobilization and catalytic abilities, along with physical confinement of polysulfides. The Fe_3C/N-CNF side enhances chemisorption of LiPSs and catalyzes their conversion to Li_2S, while the rGO layer acts as a barrier to prevent LiPS migration. This integrated structure also forms a 3D conductive network that accelerates electron and ion transfer, promoting better electrochemical performance. Li—S batteries with this Janus electrode showed excellent cycling stability with a low decay rate of 0.0089% per cycle at 0.5 C over 300 cycles, along with high rate capabilities, delivering 821.7 $mAh\,g^{-1}$ at 2.0 C. Furthermore, it maintained stable performance at a high sulfur loading of 6.29 $mg\,cm^{-2}$, showcasing the electrode's ability to handle large sulfur content while maintaining efficient redox reactions.

These results indicate that CNF, with its high aspect ratio, is an ideal carrier for sulfur cathode materials. However, to achieve superior performance, CNF-based cathodes should meet the following criteria: (i) large specific surface area and closed structures to physically confine polysulfides, (ii) wide pore size distribution to enhance sulfur loading, and (iii) functional modifications to improve electrical conductivity and chemical adsorption of polysulfides.

6.2.3 Two-Dimensional Materials in Cathode

6.2.3.1 Graphene Materials

2D carbon material graphene shows immense potential in constructing sulfur composite cathodes due to its excellent conductivity, outstanding mechanical strength, and extremely high specific surface area. Sun et al. synthesized a sulfur-reduced graphene oxide composite (SGC) via a one-step hydrothermal method [33]. In this material, sulfur is uniformly dispersed on the rGO sheets, with sulfur content controllable between 20.9% and 72.5%. The SGC-63.6%S cathode maintained a reversible capacity of 804 mAh g^{-1} after 80 cycles at 312 mA g^{-1} (0.186 C) and delivered a reversible capacity of 440 mAh g^{-1} after 500 cycles at approximately 0.75 C. The excellent film-forming ability of graphene enables the construction of 3D self-supporting sulfur cathodes, eliminating the need for current collectors and binders, and thereby increasing the sulfur loading. Jin et al. mixed graphene with nano-sulfur particles and obtained a self-supporting GS/S cathode through vacuum filtration [34]. However, during vacuum filtration, graphene sheets self-assemble into a parallel stacked structure, leading to good in-plane conductivity and ion transport but poor interlayer conductivity and ion transport, limiting the overall performance of the battery. To address this issue, Wang et al. prepared a self-supporting sulfur-reduced graphene oxide (S-rGO) cathode via freeze-drying and low-temperature heat treatment [35]. The sulfur content of this cathode reached as high as 71%, with the 3D rGO framework providing multi-dimensional electronic and ionic transport pathways, effectively mitigating volume changes in the cathode during cycling. The discharge capacity reached 800 mAh g^{-1} after 200 cycles at a current density of 300 mA g^{-1}. Moreover, by altering the composite structure of graphene and sulfur, various graphene-sulfur composite cathodes can be prepared, which help further improve rate performance, cycle stability, and electrochemical properties.

Introducing functional groups on the graphene surface can enhance the adsorption of polysulfides by graphene sheets, effectively mitigating the shuttle effect. GO contains a large number of oxygen-containing functional groups, such as hydroxyl, carbonyl, and epoxy groups, which can interact with polysulfides. Jin et al. used a simple chemical deposition strategy followed by low-temperature heat treatment to obtain a uniform sulfur coating (approximately tens of nanometers thick) on GO sheets [36]. However, GO has poor conductivity, and electrolyte penetration is difficult, leading to severe polarization, requiring an activation process to promote electrolyte infiltration of sulfur particles. To address this, Fang et al. combined GO with graphene [37], using high-conductivity graphene as the current collector, high-porosity graphene as the sulfur host, and partially reduced graphene oxide as the polysulfide adsorption layer, creating an all-graphene sulfur cathode. This cathode achieved a sulfur loading of 5 mg cm^{-2}, with an initial specific capacity as high as 1500 mAh g^{-1}, and after 400 cycles, the discharge capacity remained at 841 mAh g^{-1}.

Doping graphene with heteroatoms introduces polar, electroactive sites that enhance the adsorption of polysulfides. Xie et al. synthesized a porous interconnected 3D boron-doped graphene aerogel (BGA) as a scaffold material for sulfur cathodes via a one-pot hydrothermal method [38]. Boron doping induced negative charges on the surrounding carbon atoms, enhancing the adsorption of polysulfides. The BGA-S cathode maintained a discharge capacity of 994 mAh g^{-1} after 100 cycles at 0.2 C. Nitrogen doping, on the other hand, imparts positive charges to surrounding carbon atoms. Li et al. fabricated a nitrogen-doped graphene cathode scaffold [39], where the introduction of nitrogen atoms improved the electrochemical stability of the battery and accelerated the electrochemical reaction kinetics. This cathode achieved an initial specific capacity of 1200 mAh g^{-1} at 0.3 A g^{-1}, with a capacity decay rate of only 0.05% per cycle after 300 cycles at 0.75 A g^{-1}.

Additionally, Yin et al. used density functional theory (DFT) calculations to study the adsorption effects of pyridinic nitrogen, pyrrolic nitrogen, and graphitic nitrogen-doped graphene on polysulfides [40]. The results showed that localized pyridinic nitrogen exhibited the strongest adsorption of polysulfides, effectively anchoring soluble polysulfides. Hou et al. further conducted systematic DFT calculations on different heteroatom-doped nanocarbon materials [41]. The results in Figure 6.3a indicated that N and O doping significantly enhanced dipole–dipole electrostatic interactions between carbon materials and polysulfides, effectively

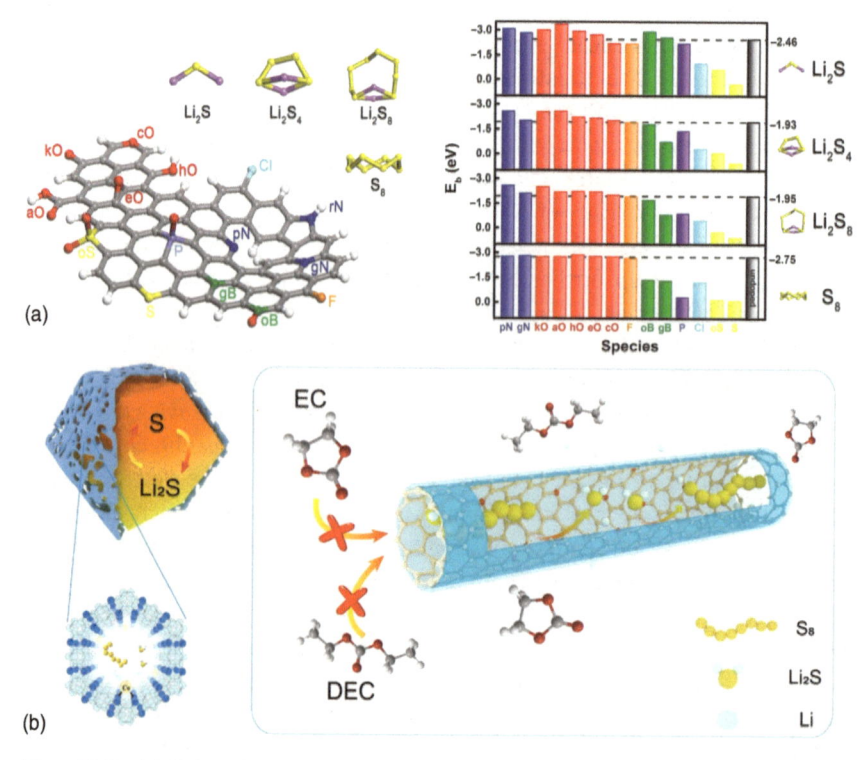

Figure 6.3 (a) X-doped nanocarbon materials (X = N, O, F, B, P, S, and Cl) and PBE level optimized structure of Li$_2$S, Li$_2$S$_4$, Li$_2$S$_8$, S$_8$ molecules. Right plot is the binding energy of Li$_2$S, Li$_2$S$_4$, Li$_2$S$_8$, and S$_8$ interacting with X-doped GNRs. Source: Hou et al. [41]/with permission of John Wiley & Sons. (b) Solid-state conversion of sulfur confined in the micropore channels. Source: Yang et al. [42]/with permission of John Wiley & Sons.

preventing the shuttle effect, and thereby achieving high capacity and high Coulombic efficiency. However, doping with single atoms of B, F, S, P, and Cl was relatively less effective. The study also proposed a volcano-shaped relationship between the binding energy of doped carbon materials and polysulfides and the electronegativity of the doping element, providing a theoretical basis for optimizing the design of doped materials.

Graphene can also combine with other low-dimensional carbon materials to further enhance the conductivity of the electrode and design a rich pore structure. Zhao et al. used CVD to simultaneously catalyze the deposition of graphene and single-walled carbon nanotubes (SWCNTs) on layered double hydroxides [43]. SWCNTs were embedded between graphene layers, forming a conductive 3D network framework. The interlayer spaces of graphene and the hollow structure of the CNTs can both load sulfur. The resulting G/SWCNT-S cathode retained a discharge capacity of 650 mAh g^{-1} after 100 cycles at a rate of 5 C. Yang et al. uniformly covered porous carbon on both sides of graphene [44], utilizing the high specific surface area and porosity of the porous carbon to achieve a high dispersion of sulfur, while the graphene layers served as electron transport layers, significantly improving conductivity.

Rationally designing surface defects and heterostructures on low-dimensional materials can impart electrocatalytic functionality to the materials, thereby catalyzing the conversion of polysulfides and enhancing the redox reaction kinetics of Li—S batteries. As shown in Figure 6.3b, Yang et al. introduced single-atom cobalt (Co) into a microporous carbon host with ultra-large pore size and loaded sulfur into this composite material [42]. The ultra-large pore size effectively prevented contact between the electrolyte and internal sulfur, avoiding the formation and dissolution of polysulfides, while the catalytic sites of single-atom Co accelerated the solid–solid conversion of sulfur. When combined with a liquid carbonate electrolyte, the Coulombic efficiency reached 99.88%, and after 1000 cycles at 0.5C, the capacity decay per cycle was only 0.016%. When combined with a solid electrolyte, the capacity reached 1100 mAh g^{-1}, and after 200 cycles, the Coulombic efficiency was as high as 99.83%. Tian et al. modified porous graphene aerogels with monodisperse polar nickel cobalt oxide nanoparticles (NCO-GA) [45]. This aerogel composite exhibited high conductivity, a layered porous structure, high chemical adsorption capacity, and excellent electrocatalytic performance. The sulfur-loaded cathode prepared using NCO-GA achieved a discharge capacity of 1214.1 mAh g^{-1} at 0.1 C, and after 1000 cycles at a 2 C rate, the capacity decay per cycle was only 0.031%. Li et al. used an immersion method followed by high-temperature heat treatment to prepare 3D porous rGO composites containing various metal nanoparticles (Fe, Co, Ni, Cu, Zn, and CoZn) [46], which were used to load sulfur and serve as cathode materials. The study found that CoZn-rGO composites had the best catalytic effect. The resulting Li—S batteries demonstrated stable cycling under conditions of high sulfur loading (5 mg cm^{-2}), lean electrolyte (E/S = 4.8), and high current density (8.0 C), with a power density of 26 120 W kg^{-1}, enabling the battery to complete charging and discharging in just five minutes.

Wang et al. introduced a groundbreaking approach by using single-atom catalysts (SACs) to enhance the electrochemical conversion kinetics in Li$_2$S-based cathodes [47]. This strategy addresses the inherent kinetic limitations of Li—S batteries, especially during the conversion of Li$_2$S to sulfur, which involves multi-phase transformations. By dispersing single iron atoms (SAFe) onto nitrogen-rich carbon matrices (NC), the authors demonstrated that SACs can effectively reduce the activation energy required for the delithiation of Li$_2$S, accelerating the electrochemical reactions without compromising the current rate. Both spectroscopic and electrochemical analyses, supported by theoretical simulations, confirmed that the presence of SAFe improved catalytic activity, lowering the activation voltage while maintaining ultra-fast charge/discharge rates. The resulting Li$_2$S/Li batteries exhibited exceptional performance, with a discharge capacity of 588 mAh g^{-1} at 12 C and a remarkably low capacity fading rate of 0.06% per cycle over 1000 cycles at 5 C. This study highlights the potential of single-atom catalysis in overcoming the kinetic barriers in conversion-based battery systems, enabling faster reaction rates and longer cycle life. The findings not only validate the use of SACs in enhancing Li$_2$S cathodes but also open new avenues for future high-energy-density batteries designed for electric vehicle applications. Du et al. explored the use of SACs comprising cobalt atoms embedded in nitrogen-doped graphene (Co-N/G) to improve the performance of Li—S batteries [48]. Through a combination of advanced imaging techniques such as high-angle annular dark field (HAADF) imaging, X-ray absorption spectroscopy (XAS), and theoretical calculations, they demonstrated that the Co-N-C coordination centers in the Co-N/G composite act as bifunctional electrocatalysts. These centers facilitate both the formation of Li$_2$S during discharge and its decomposition during charging, thereby improving the overall sulfur utilization and cycle stability of the battery. The S@Co-N/G composite, with an impressive sulfur content of 90 wt%, delivered a gravimetric capacity of 1210 mAh g^{-1} and an areal capacity of 5.1 mAh cm^{-2} at 0.2 C with a minimal capacity decay rate of 0.029% per cycle over 100 cycles. The study highlights how Co-N/G effectively addresses the challenges of polysulfide conversion in Li—S batteries, enabling high sulfur loading and excellent electrochemical performance. This work underscores the potential of SACs in designing advanced conductive hosts for high-performance Li—S batteries and other electrochemical energy storage systems.

Qin et al. developed a hierarchically porous graphene/carbon aerogel (HPGCA) as a cost-effective sulfur host for lithium-sulfur batteries [49]. They used NaCl to assist in the hydrothermal self-assembly process, which enabled the formation of a highly porous structure without the need for special drying techniques. The NaCl reagent was crucial in controlling the flexibility of the carbon aerogel and preventing the collapse of pores during conventional drying, thereby preserving a rich network of mesopores and macropores. This porous structure provided sufficient space to accommodate sulfur, leading to a high sulfur loading of 69 wt% and over 6 mg cm^{-2} on the cathode. The HPGCA-S cathode, fabricated without the addition of extra conductive agents, demonstrated impressive electrochemical performance. It achieved an initial capacity of 1121 mAh g^{-1} and retained a reversible capacity of 797 mAh g^{-1} after 100 cycles at a current density of 0.2 C. This stable cycling performance was attributed to the well-preserved porous structure, which enhanced sulfur utilization and mitigated the polysulfide shuttle effect. He et al. developed a 3D CNT/graphene-Li$_2$S (3DCG-Li$_2$S) hybrid aerogel as a high-performance cathode for Li—S batteries [50]. By employing a solvothermal reaction followed by liquid infiltration and evaporation coating, they achieved a record 81.4 wt% Li$_2$S loading in the cathode. The 3D mesoporous structure, composed of interconnected 2D graphene nanosheets and 1D CNTs, facilitates efficient electron transfer and ionic diffusion while suppressing polysulfide solubility during charge and discharge cycles. The unique architecture of the 3DCG-Li$_2$S aerogel allows it to function as a freestanding, binder-free cathode, eliminating the need for polymeric binders or conductive additives. This design results in outstanding electrochemical performance, including a high reversible discharge capacity of 1123.6 mAh g^{-1}, a capacity decay as low as 0.02% per cycle, and a high-rate capacity of 514 mAh g^{-1} at 4 C. These features demonstrate the potential of the 3DCG-Li$_2$S aerogel in achieving stable long-term cycling and high-rate performance in Li—S batteries. Cavallo et al. developed a novel free-standing rGO aerogel to act as a supporting electrode in catholyte-based Li—S batteries [51]. This mesoporous rGO aerogel offers an interconnected 3D conductive framework that facilitates electrochemical reactions and improves electron transport pathways. The oxygen-containing groups on the surface of the rGO structure help stabilize the cycling performance by reducing the polysulfide shuttle effect, where Li$_2$S$_n$ species migrate to the anode. The use of a fluorine-free catholyte, where polysulfides act as both the active material and the conducting agent, further enhances the sustainability of the system by eliminating traditional fluorine-based Li-salts such as LiN(SO$_2$CF$_3$)$_2$ or LiCF$_3$SO$_3$. The resulting Li—S cell with this aerogel structure demonstrated impressive electrochemical performance, achieving a high areal capacity of 3.4 mAh cm^{-2} with a sulfur loading of 6.4 mg cm^{-2}. Additionally, the cell exhibited remarkable cycling stability, retaining 85% of its capacity after 350 cycles. This approach offers a sustainable and cost-effective solution, utilizing environmentally friendly materials such as carbon and sulfur while achieving high energy densities without the need for expensive fluorine-based salts.

A novel wearable Li—S bracelet battery has been developed using advanced 3D printing technology. This innovative approach allows for the precise control of electrode thickness while simplifying the manufacturing process in a cost-effective way [52]. The 3D-printed battery features a highly conductive skeleton that facilitates efficient electron and ion transport, achieving a specific capacity of 505.4 mAh g^{-1} after 500 cycles with a high active material loading of 10.2 mg cm^{-2}. The practical applicability of the battery is demonstrated by its ability to power a red light-emitting diode when worn as a bracelet. The construction of the 3D Li—S battery involved using graphene/phenol formaldehyde resin as a spot of printing ink, which provided the cathode skeleton with mechanical strength. A porous structure was created using a SiO$_2$-template method to store sulfur, further enhancing performance. This battery design integrates two types of 3D printing technologies: DIW for the electrode and fused deposition for the battery casing. The 3D-printed battery offers a promising solution for wearable energy storage devices due to its scalability, environmental friendliness, and low cost. Additionally, this technology could inspire new developments in high-energy, high-power Li—S batteries and other energy storage systems, such as lithium-ion and metal-air batteries.

6.2.3.2 MXene Materials

2D transition metal carbides/nitrides, known as MXenes, represent a novel class of 2D materials. MXenes are typically produced by selectively etching the "A" layers in laminated MAX phases ($M_{n+1}AX_n$, where $n = 1-4$), where "M" is an early transition metal (e.g. Ti, V, Zr, and Nb), "A" is an IIIA or IVA element (e.g. Al, Ga and, Si), and "X" is carbon and/or nitrogen. Since the first discovery of $Ti_3C_2T_x$ MXene in 2011, various methods have been developed for MXene synthesis, including HF etching, in situ HF-forming etching, alkali etching, electrochemical etching, and molten salt etching. These processes remove the "A" atoms and introduce surface terminations such as —O, —OH, —F, and —Cl, which can be tailored by post-treatment. To date, over 30 types of MXenes have been synthesized, with ongoing research continuing to expand the family.

MXenes exhibit key features such as a 2D structure, metallic conductivity, high surface area, and water dispersibility. Consequently, they are extensively studied for applications in energy storage, electrocatalysis, biomedicine, and electromagnetic interference shielding. In energy storage devices, MXenes have shown significant potential in lithium-ion, sodium-ion, Li—S, zinc-ion batteries, and supercapacitors [53]. They can serve either as high-rate electrode materials or conductive substrates that enhance electrochemical performance. The metallic conductivity of MXenes facilitates rapid electron transfer, improving sulfur utilization and capacity. Surface terminations on MXenes strongly adsorb LiPSs, forming intense metal-S bonds that inhibit the shuttle effect. Additionally, MXenes exhibit effective catalytic activity for the conversion between LiPSs and Li_2S, enhancing redox kinetics during cycling. MXenes also aid in the uniform nucleation and growth of lithium, mitigating dendrite formation and accommodating volume expansion. These characteristics have led to substantial improvements in Li—S battery performance.

By simply loading sulfur on the MXene nanosheets, the electrochemical performance of the Li—S batteries can be significantly improved. Zhao et al. developed a sulfur-loaded Ti_3C_2 ($S/L-Ti_3C_2$) composite to enhance the long-term stability and performance of sulfur cathodes in Li—S batteries [54]. The layered Ti_3C_2 material, featuring an accordion-like structure, was prepared by exfoliating Ti_3AlC_2 in a 40% hydrofluoric acid solution. Sulfur was then incorporated into the $L-Ti_3C_2$, with a sulfur content of 57.6 wt%. The resulting $S/L-Ti_3C_2$ composite was tested as a cathode material for Li—S batteries, demonstrating an impressive initial discharge capacity of 1291 mAh g^{-1} and retaining 970 mAh g^{-1} after 100 cycles at a current density of 200 mA g^{-1}. These promising results highlight the potential of 2D carbides, similar to graphene or graphite, as effective materials for high-performance Li—S batteries. The high electrical conductivity and excellent cycle stability of the $S/L-Ti_3C_2$ composite, along with its ability to maintain a high capacity and coulombic efficiency of 99%, underscore the exciting opportunities for developing sulfur-loaded 2D carbides as electrode materials in future Li—S battery systems and other electrochemical energy storage devices. Zhang et al. developed a sulfur-host composite using titanium carbide (Ti$_3$C$_2$T$_x$) Mxene as the sulfur host to improve the performance of Li—S batteries [55]. The S@Mxene composite, synthesized via a hydrothermal method, features sulfur nanoparticles uniformly decorated onto the surface of the Mxene sheets. The layered, stacked structure of the Ti$_3$C$_2$T$_x$ Mxene allows it to accommodate the volume expansion of sulfur during cycling, while its high electrical conductivity facilitates efficient electron transfer. This unique architecture not only suppresses the LiPS shuttle effect but also enhances the electrochemical performance of the Li—S batteries. With a high areal sulfur loading (~4.0 mg cm^{-2}), the S@Mxene composite demonstrates an initial reversible capacity of 1477.2 mAh g^{-1} and retains a specific reversible capacity of 1212.2 mAh g^{-1} after 100 cycles at 0.2 C, exhibiting a low capacity loss of only 0.18% per cycle. The simplicity and scalability of the hydrothermal method used in this study suggest a promising approach for the large-scale production of high-performance cathodes for Li—S batteries.

The single-layered MXene nanosheets prepared by the one-step LiF/HCl etching strategy are also excellent candidates as conductive sulfur hosts. Liang et al. designed a sulfur host material by leveraging the high conductivity and polar surface properties of MXene phases [58]. MXenes, which are 2D early-transition-metal carbides, have shown great promise as sulfur hosts due to their conductive layered structure and strong interaction with polysulfide species. In this study, sulfur nanoparticles were uniformly embedded onto Ti$_2$C MXene sheets through a hydrothermal method, forming a composite cathode for Li—S batteries. This S/Ti$_2$C composite exhibited an impressive sulfur content of 70 wt% and delivered a specific capacity of nearly 1200 mAh g^{-1} at a C/5 rate. The sulfur was strongly anchored to the MXene surface through metal-sulfur interactions, which were confirmed via X-ray photoelectron spectroscopy (XPS). This strong bonding interaction effectively suppressed the polysulfide shuttle effect while maintaining excellent cycling stability, with a capacity retention of 80% over 400 cycles at a C/2 rate. The results indicate that MXene materials, particularly Ti$_2$C, are highly effective as conductive and sulfurphilic hosts for Li—S batteries, providing both high capacity and long-term cycling stability. As shown in Figure 6.4a, Yao et al. developed a dual-confinement strategy for LiPs by integrating polydopamine-coated MXene nanosheets (S@Mxe@PDA) as a high-performance sulfur host for Li—S batteries [56]. This design leverages MXene's high conductivity and its ability to chemically bond with intermediate polysulfides, while the polydopamine (PDA) coating serves to spatially confine sulfur and prevents its direct contact with the electrolyte. The Mxe component plays a key role in trapping LiPs through stable Ti—S bonding, while the PDA layer provides mechanical flexibility and maintains the structural integrity of the cathode during cycling. This dual-confinement approach was confirmed through DFT calculations, which demonstrated the strong interaction between S@Mxe@PDA and LiPs. The S@Mxe@PDA cathode exhibited exceptional electrochemical properties, including a high reversible capacity of 1044 mAh g^{-1} after 150 cycles at 0.2 C and excellent rate

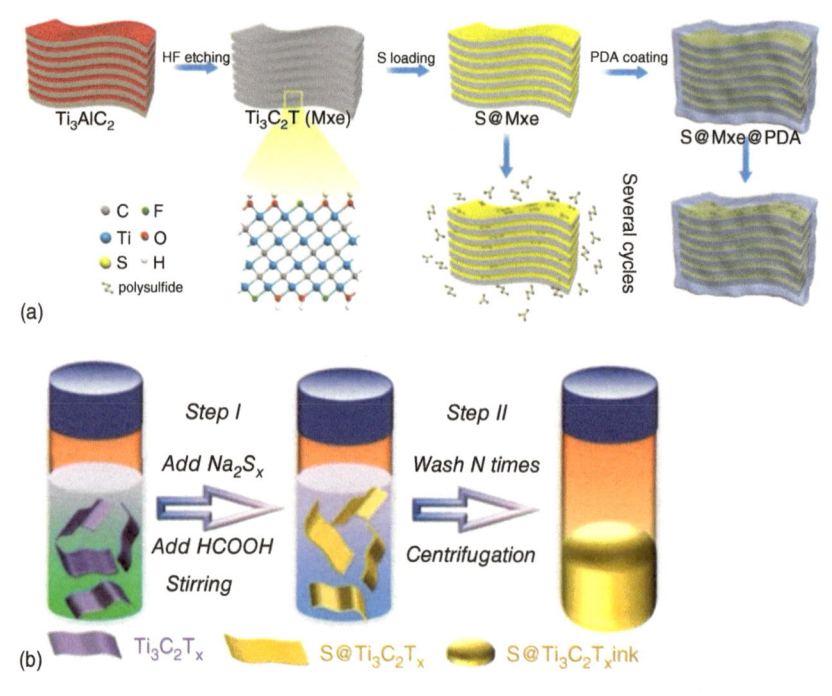

Figure 6.4 (a) Synthesis procedure of the S@Mxe@PDA hybrid electrode. Source: Yao et al. [56]/with permission of John Wiley & Sons. (b) Schematic of the synthesis process of the S@Ti$_3$C$_2$T$_x$ ink. Source: Tang et al. [57]/with permission of John Wiley & Sons.

performance with 624 mAh g^{-1} at 6 C. Furthermore, the cathode maintained a high capacity of 556 mAh g^{-1} after 330 cycles at 0.5 C, even with a high sulfur loading of 4.4 mg cm^{-2}. These results highlight the effectiveness of this dual-confinement strategy in enhancing both the cycling stability and rate performance of Li—S batteries.

Due to their 2D morphology and excellent mechanical properties, MXenes are well-suited as flexible substrates for sulfur electrodes [57]. The negatively charged single-layered MXene nanosheets, with their abundant surface terminations, disperse uniformly in aqueous solutions, forming stable colloidal suspensions, as shown in Figure 6.4b. By introducing Na$_2$S$_x$ and HCOOH into a Ti$_3$C$_2$T$_x$ colloidal suspension, sulfur nanoparticles can be synthesized in situ, resulting in a homogeneous Ti$_3$C$_2$T$_x$/S ink. This ink can then be processed via vacuum filtration to produce a freestanding S@Ti$_3$C$_2$T$_x$ film electrode. The resulting electrode, containing 70 wt% sulfur, delivers capacities of 1184 mAh g^{-1} at 0.2 C and 1044 mAh g^{-1} at 2 C, and retains a capacity of 724 mAh g^{-1} after 800 cycles at 0.2 C. Furthermore, a pouch cell assembled with this S@Ti$_3$C$_2$T$_x$ cathode and a lithium ribbon anode demonstrated an initial capacity of 1263 mAh g^{-1} at 0.5 C. Under bending conditions, the cell retained 1119 mAh g^{-1} after five cycles, outperforming the cell's performance in a flat state, which exhibited an initial capacity of 1124 mAh g^{-1} with 903 mAh g^{-1} retained after five cycles. Additionally, a flexible Ti$_3$C$_2$T$_x$/S film electrode was fabricated using a vapor deposition method to load sulfur onto a freestanding, robust $Ti_3C_2T_x$ paper film. This electrode, with 30 wt% sulfur content, achieved capacities of 1383 mAh g^{-1} at 0.1 C and 1075 mAh g^{-1} at 2 C, and exhibited a low capacity decay rate of 0.014% per cycle over 1500 cycles at 1 C.

These findings highlight the significant enhancement in electrochemical performance achieved by using MXene nanosheets directly as sulfur hosts in Li—S batteries. However, the electrodes still face challenges related to the restacking tendency of 2D MXene nanosheets, which restricts Li$^+$ diffusion and limits the full utilization of MXenes. To address this issue and fully capitalize on the advantages of MXenes in Li—S batteries, various strategies have been explored, including chemical modification, structural control of MXenes, and the development of MXene-based composites.

Chemical modification is a widely used strategy to enhance the performance of MXenes. Studies have shown that doping heteroatoms such as nitrogen, sulfur, or metal atoms into the conductive matrix can significantly improve its catalytic properties. Song et al. prepared a porous N-doped Ti_3C_2 MXene (P-NTC) using a sacrificial templating method to address the polysulfide shuttle effect and sluggish sulfur reaction kinetics in Li—S batteries [59]. This P-NTC structure, with its high surface area, excellent conductivity, and strong interaction with LiPSs, significantly enhanced the redox reactions during cycling. Nitrogen doping further improved the interfacial interaction with lithium and lowered the dissociation barrier for Li_2S. The resulting S/P-NTC composite cathode exhibited a low capacity decay of 0.033% per cycle at 2.0 C over 1200 cycles. Additionally, with a sulfur mass loading of 8.2 mg cm^{-2}, it achieved an areal capacity of 9.0 mAh cm^{-2}. Zhang et al. introduced single-atom zinc implanted MXene (SA-Zn-MXene) into a sulfur cathode to address the sluggish reaction kinetics and severe shuttle effect in Li—S batteries [60]. The high electronegativity of zinc atoms on MXene facilitates strong interactions with polysulfides, while simultaneously catalyzing the conversion reactions by reducing the energy barriers from Li_2S_4 to Li_2S_2/Li_2S. Additionally, the uniformly dispersed zinc atoms accelerate the nucleation of Li_2S_2/Li_2S on the 2D MXene layers during cycling. As a result, the SA-Zn-MXene sulfur cathode delivers a high reversible capacity of 1136 mAh g^{-1}. After electrode optimization, it achieves a high areal capacity of 5.3 mAh cm^{-2}, with a rate capability of 640 mAh g^{-1} at 6 C, and 80% capacity retention after 200 cycles at 4 C. These findings suggest that the SA-Zn-MXene layers not only enhance polysulfide adsorption but also accelerate the conversion of polysulfides, leading to improved redox kinetics and high electrochemical performance.

Structural control, particularly in designing open structures for MXenes, is crucial for preventing restacking and increasing the exposed surface area, thereby maximizing the utilization of MXenes. Carbon materials and transition metal compounds are conventional and effective sulfur hosts. Porous carbon materials can physically confine LiPSs, provide ample space for enhanced sulfur loading, and mitigate volume expansion, while transition metal compounds chemically adsorb LiPSs to suppress the shuttle effect. Dong et al. developed an all-MXene-based flexible and integrated sulfur cathode by employing a 3D alkalized Ti_3C_2 MXene nanoribbon (a-Ti$_3$C$_2$ MNR) framework as a sulfur host and a 2D delaminated Ti$_3$C$_2$ MXene (d-Ti$_3$C$_2$) interlayer on a polypropylene (PP) separator [61]. This design significantly improved the performance of Li—S batteries by addressing the polysulfide shuttle effect and promoting fast lithiation/delithiation kinetics. The a-Ti$_3$C$_2$ MNR framework, with its open macropores and large surface area, enabled high sulfur loading and facilitated ionic diffusion, while the d-Ti$_3$C$_2$ nanosheet-based interlayer provided effective chemical and physical barriers against the polysulfide shuttle. As a result, the all-MXene electrode achieved a high reversible capacity of 1062 mAh g^{-1} at 0.2 C, with enhanced capacity retention of 632 mAh g^{-1} after 50 cycles at 0.5 C. Additionally, the electrode exhibited excellent rate capability, delivering 288 mAh g^{-1} at 10 C, and demonstrated long-term cyclability. This integrated MXene-based cathode outperformed traditional configurations using aluminum current collectors, offering a promising strategy for the development of flexible, high-performance Li—S batteries. The approach of utilizing MXene nanostructures can be extended to other MXene-derived materials, suggesting broad applicability for energy storage devices such as batteries and supercapacitors. Chen et al. developed a flexible free-standing S@lithium-ion-intercalated V$_2$C MXene/rGO-CNT (S@V$_2$C–Li/C) electrode to address the challenge of lithium-ion transport in high-rate Li—S batteries [62]. In this unique nanoarchitecture, rGO and CNTs provide a flexible conductive skeleton, while V$_2$C-Li MXene serves as a sulfur host with high polarity and an enlarged interlayer distance, enhancing both lithium-ion migration and chemical absorption of polysulfides. The synergistic effects of strong polysulfide absorption and improved lithium-ion transport resulted in long-term cycling stability, with low capacity decay rates of 0.053% and 0.051% per cycle over 500 cycles at 1 and 2 C, respectively. The electrode was prepared through vacuum filtration, utilizing the high conductivity of rGO and CNTs while leveraging V$_2$C-Li MXene to promote electron transport and ion exchange. The intercalation of lithium-ions in V$_2$C MXene expanded its interlayer spacing, facilitating faster ion migration and better electrolyte infiltration, especially under high sulfur loading conditions.

Given their unique 2D structure and high electrical conductivity, MXenes are promising candidates for combining with carbon materials or transition metal compounds to form composite materials with synergistic effects for Li—S batteries. In these composites, MXenes create a continuous conductive network that enables fast charge transport, while offering strong adsorption and catalytic properties for LiPSs. Additionally, the other components not only retain their inherent functions but also serve as spacers to prevent MXene restacking and promote Li$^+$ diffusion. Consequently, MXene-based composites often demonstrate excellent electrochemical performance as sulfur hosts, making them a promising direction for Li—S battery applications. Song et al. developed a unique 3D porous Ti$_3$C$_2$T$_x$ MXene/rGO (MX/G) hybrid aerogel as a free-standing polysulfide reservoir to enhance the performance of Li—S batteries [63]. In this design, highly conductive MXene and rGO are combined into a 3D interconnected porous structure, which provides efficient Li$^+$/electron transport and strong chemical anchoring of LiPSs. The 2D polar interfaces of the MXene enable fast redox reaction kinetics and help mitigate the shuttle effect. As a result, the MX/G hybrid aerogel electrodes exhibit excellent electrochemical performance, including a high capacity of 1270 mAh g^{-1} at 0.1 C and an extended cycling life of up to 500 cycles with a low capacity decay rate of 0.07% per cycle. Additionally, the aerogel demonstrates a high areal capacity of 5.27 mAh m^{-2}, making it a promising material for next-generation energy storage systems. This innovative design allows the MX/G aerogel to serve as an efficient 3D scaffold for accommodating Li$_2$S$_6$ catholyte, improving sulfur utilization and redox kinetics. The strong Ti—S interaction on the polar surface of Ti$_3$C$_2$T$_x$ MXene effectively anchors sulfur species and restricts the shuttle effect.

6.3 Low-Dimensional Composite Materials in Separators

As one of the key components in a battery system, the primary function of the separator is to prevent direct contact between the cathode and anode, thereby avoiding internal short circuits in the battery. Although the separator is an inactive material that does not directly participate in electrode reactions, it has a significant impact on internal ion transport, internal resistance, rate performance, and cycle life of the battery. An ideal separator should possess good electronic insulation and mechanical stability. Additionally, to mitigate the effects of electrode volume changes and prevent lithium dendrite penetration, the separator needs to have a certain level of tensile strength and puncture resistance. Under extreme conditions (such as high temperatures and shocks), the separator must also maintain structural and chemical stability. Furthermore, the separator should have appropriate porosity to ensure smooth ionic conduction within the electrolyte, enhancing the rate performance of the battery. It also needs to exhibit good wettability with the electrolyte to reduce internal resistance.

Traditional separator materials are mainly polyolefins, such as polyethylene (PE) and PP separators. These separators have mature manufacturing processes, low impedance, excellent chemical stability, and are widely used in lithium-ion secondary batteries. However, in Li—S battery systems, due to the polysulfide shuttle effect, traditional polyolefin separators often lead to rapid capacity decay and low Coulombic efficiency, limiting the application of high-capacity Li—S batteries. Therefore, for separators in Li—S batteries, functionalized design and modification have become key. By optimizing the pore structure and introducing electrostatic repulsion to achieve selective ion conduction, enhancing the ability to adsorb polysulfides, and catalyzing the redox reactions of active materials, the overall performance of Li—S batteries can be effectively improved, facilitating the realization of high-energy-density batteries.

To achieve high energy density and output in Li—S batteries, the functional layer loading of modified separators should be as light as possible to match the high sulfur-loading cathode and achieve high cycling capacity at lower rates. For high-capacity Li—S batteries, functionalized separators can be designed through charge repulsion effects, steric hindrance effects, and adsorption effects to effectively suppress the polysulfide shuttle phenomenon.

The introduction of functional separators plays a crucial role in improving the capacity, Coulombic efficiency, cycle life, and safety of Li—S batteries. Functional layers enhance battery performance mainly by inhibiting the polysulfide shuttle effect, regulating the cathode and anode interfaces, and activating "dead sulfur." However, there is still ample room for further improvement.

6.3.1 Zero-Dimensional Materials in Separators

The ionic radius of polysulfide anions is larger than that of lithium-ions, and this difference can be leveraged to design functionally modified separators with size-screening or blocking capabilities. These separators use physical confinement effects to mitigate the polysulfide shuttle effect. As shown in Figure 6.5a, Balach et al. designed a mesoporous carbon (MPC)-coated separator, where the large surface area of the MPC acts as a physical barrier to the diffusion of polysulfides [64]. At a rate of 0.2 C, the initial capacity reached 1378 mAh g^{-1}, and after 500 cycles at 0.5 C, a remaining capacity of 723 mAh g^{-1} was maintained, with a per-cycle decay rate of 0.081%.

By modifying low-dimensional carbon materials, negatively charged groups can be introduced. Zeng et al. introduced sulfonic acid groups into acetylene black, creating a sulfonated acetylene black (AB-SO$_3^-$) modification layer [66]. The porous structure of sulfonated acetylene black provides a more accessible path for lithium-ions, resulting in an initial specific capacity of 1262 mAh g^{-1}, and after 100 cycles at 0.1 C, a remaining capacity of 955 mAh g^{-1} was maintained.

The physical adsorption of separators is primarily governed by van der Waals forces. However, if the adsorption is too weak, LiPSs may not be effectively trapped, or the adsorbed LiPSs could easily detach from the adsorbent. Materials modified to enhance physical adsorption generally rely on their high porosity and large surface areas. Chung et al. developed a composite separator featuring a PEG-supported MPC coating, designed to improve the electrochemical performance of Li—S batteries by utilizing pure sulfur cathodes [67]. The separator, attached to a Celgard PP membrane, incorporates an MPC/PEG coating on the side facing the pure sulfur cathode, which serves two critical functions: acting as an upper current collector

Figure 6.5 (a) SEM image of the mesoC-coating separator. Source: Reproduced with permission from Balach et al. [64]/John Wiley & Sons. (b) The flexibility and mechanical strength of the C-coated separator. Source: Reproduced with permission from Chung and Manthiram [65]/John Wiley & Sons.

to enhance sulfur utilization and functioning as a polysulfide trap to suppress the diffusion of LiPSs. This approach effectively addresses the issues of polysulfide shuttling and sulfur loss during cycling. The unique structure of the MPC/PEG-coated separator contributes to its high mechanical strength, lightweight design, and ability to maintain the flexibility of the separator. This configuration allows for the efficient reactivation of immobilized polysulfides, resulting in improved sulfur utilization and extended battery life. As a result, Li—S cells using this separator achieved a high discharge capacity of 1307 mAh g^{-1}, exceptional cycle stability with only 0.11% capacity fade per cycle over 500 cycles, and remarkable reversibility.

Despite this, physical adsorption alone is insufficient to meet the practical demands of Li—S batteries, as the dissolution and diffusion of LiPSs cannot be fundamentally prevented through simple blocking mechanisms. To enhance the overall trapping effect, chemical adsorption has also been introduced in separators. Chemical trapping mechanisms can be categorized into two types: chemical adsorption and electrochemical catalysis. Chemical adsorption occurs when the coating materials immobilize LiPSs through electron transfer, exchange, or chemical bonding – interactions that are significantly stronger than van der Waals forces. As a result, the adsorbed LiPSs remain securely attached to the coating materials, aided by polar–polar interactions or Lewis acid–base interactions, preventing their easy release. Zheng et al. developed innovative ultralight carbon flakes (CFs)-modified separators (CFs@PP) to tackle the polysulfide shuttle effect in Li—S batteries [68]. The CFs are synthesized by the direct carbonization of sodium citrate, which provides a scalable and cost-effective approach for large-scale production. When applied to a PP separator, the CFs form a highly conductive, ultralight coating (0.16 mg cm^{-2}), which enhances electron transportation and effectively blocks LiPS migration. This CFs-modified separator significantly improves the electrochemical performance of Li—S batteries. It offers an initial discharge capacity of 1063 mAh g^{-1} at 0.5 C and retains 683 mAh g^{-1} after 500 cycles, with an impressively low capacity decay rate of 0.071% per cycle. Additionally, the CFs@PP separator exhibits excellent rate performance, delivering a discharge capacity of 725 mAh g^{-1} at a high current density of 2 C. The separator also mitigates self-discharge, with only 66 mAh g^{-1} capacity loss after 72 hours of shelving.

By functionalizing the separator to enhance its chemical adsorption capacity and strengthen the chemical anchoring of polysulfides, the diffusion of polysulfides toward the anode can be effectively reduced, retaining them on the cathode side as much as possible. This significantly mitigates the polysulfide shuttle effect. Heteroatom doping and functionalization are effective methods to improve the chemical adsorption ability of carbon materials for polysulfides. Balach et al. developed an N, S co-doped mesoporous carbon functional coating (NSMPC-HS), where the co-doping structure provides stronger binding energy with polysulfides [69]. The Li—S battery using this functional separator maintained a specific capacity of 740 mAh g^{-1} after 500 cycles at a 0.5 C rate, with a per-cycle capacity decay rate of about 0.041%, which is significantly better than that of a singly doped MPC functional separator. Pei et al. designed a high-performance functional separator for Li—S batteries by coating nitrogen-doped porous carbon nanosheets onto a commercial PP separator [70]. The lightweight barrier layer (0.075 mg cm^{-2}) of 2D porous carbon nanosheets exhibited excellent polysulfide-entrapping capabilities, which effectively suppressed the shuttle effect that typically hampers Li—S battery performance. This modified separator improved the cycling stability and capacity retention of sulfur cathodes made from commercial carbon materials. With the G@PC/PP separator, a CB/S cathode containing 64 wt% sulfur and a sulfur loading of 3.5 mg cm^{-2} delivered a high capacity of 754 mAh g^{-1} and retained 88.6% of its capacity after 500 cycles at 1 C. Additionally, a self-supporting CNT/S cathode with 70 wt% sulfur exhibited excellent cycling stability, retaining 793 mAh g^{-1} after 400 cycles at 0.5 C with a sulfur loading of 6.0 mg cm^{-2}. At a higher sulfur loading of 12.0 mg cm^{-2}, the cell achieved a high areal capacity of 12.1 mAh cm^{-2} after 100 cycles at 0.2 C.

In Li—S batteries, the poor conductivity of elemental sulfur and its final discharge product, lithium sulfide (Li$_2$S), as well as the continuous changes in the electrode surface structure during cycling, cause the cathode interface resistance to gradually increase, severely limiting the battery's power output density. Therefore, introducing a functional layer to reduce cathode interface resistance is crucial for improving electrode reaction kinetics. As shown in Figure 6.5b, Chung et al. extensively studied conductive carbon material coatings for separators, such as coating conductive carbon black on one side of a PP separator via vacuum filtration [65]. Conductive carbon materials can not only physically block or adsorb polysulfides but also serve as an "upper current collector," providing an electronic transmission network for the active material attached to the separator, significantly reducing the electronic transmission resistance of the cathode. With this carbon material-coated separator, pure sulfur can be used directly as the cathode, achieving an initial specific capacity of 1389 mAh g^{-1} at a 0.2 C rate and 1045 mAh g^{-1} at a 2 C rate. After 50 cycles, the capacity remained at 920 mAh g^{-1}, with a noticeable reduction in the self-discharge phenomenon.

In addition to suppressing dendrite growth, the functional layer of the separator can also regulate the growth behavior of metallic lithium. Liu et al. coated the separator with functionalized nanocarbon (FNC) material that has immobilized lithium-ions [71]. During cycling, lithium dendrites grow simultaneously from the FNC layer on the separator and the lithium metal anode toward each other. When the dendrites meet, under the action of mechanical stress, the growth direction of lithium changes from axial to radial. Therefore, the dendrites do not penetrate the separator but form a dense lithium layer between the separator and the lithium anode. This controlled growth alleviates the formation of the solid electrolyte interphase (SEI) and reduces electrolyte decomposition. The assembled Li/LiFePO$_4$ battery can stably cycle for over 800 cycles, with a capacity retention rate of 80% and a Coulombic efficiency of over 97%.

6.3.2 One-Dimensional Materials in Separators

The ionic radius of polysulfide anions is larger than that of lithium-ions, and this difference can be leveraged to design functionally modified separators with size-screening or blocking capabilities. These separators use physical confinement effects to mitigate the polysulfide shuttle effect. Chung et al. coated MWCNTs onto a commercial separator [72], where the porous network of MWCNTs facilitated the rapid transport of lithium-ions while acting as a second current collector to enhance electron transport on the cathode side, and simultaneously blocked the shuttle of polysulfides. The coating had a surface mass of only 0.17 mg cm^{-2}, and after 300 cycles at 1 C, the remaining capacity was 621 mAh g^{-1}, with a per-cycle decay rate of 0.14%. Chang et al. developed a SWCNT-modulated separator to address the challenges of polysulfide migration in Li—S batteries [73]. The SWCNT coating on the separator not only directly suppresses the migration of polysulfides but also indirectly protects the lithium-metal anode from contamination, thereby enhancing the stability of both the cathode and anode. The conductive sp^2-carbon scaffold in the SWCNT network continuously reactivates and reutilizes trapped active materials, promoting high sulfur content and loading in the cathode. This SWCNT-modulated separator enables Li—S cells to achieve high sulfur loadings of up to 6.3 mg cm^{-2} and sulfur content of 78 wt%, leading to a high sulfur/carbon (S/C) ratio ranging from 3.5 to 7.0. The SWCNT-modulated separator enhances the electrochemical utilization of active materials by limiting polysulfide migration and improving lithium-ion diffusion. As a result, Li—S cells with this separator demonstrated a high discharge capacity of 1132 mAh g^{-1} and an excellent capacity retention rate, with a low fade rate of 0.18% per cycle over 300 cycles. Zhu et al. coated glass fiber (GF) separators with CNTs [74]. Due to the high porosity of GF separators, they can absorb more electrolytes, while the electron pathways provided by CNTs facilitate charge transfer and reduce interface resistance. After 200 cycles at a 1 C rate, the battery maintained a specific capacity of 721 mAh g^{-1}, and its performance at a 4 C rate was also significantly better than that of a pure PP separator. Additionally, the introduction of carbon materials improved the electrolyte wettability of the functional separator, enabling fast lithium-ion transport and uniform distribution.

The role of functional separators on the cathode side mainly involves blocking polysulfides and improving cathode interface resistance, while modifications on the anode side focus on two main aspects: enhancing the "lithiophilicity" of the separator on the anode side to ensure uniform lithium-ion distribution at the anode interface and regulating the deposition behavior and growth direction of metallic lithium to reduce dendrite formation. As shown in Figure 6.6, Cheng et al. utilized GFs with strong polar functional groups as an intermediate layer between the lithium metal anode and the separator [75]. The polar groups can adsorb a large number of lithium-ions, weakening the electrostatic interactions and concentration gradients between lithium-ions and protrusions on the copper foil of the anode, preventing lithium-ions from accumulating around the protrusions. This promotes uniform lithium-ion distribution and smooth deposition on the anode, inhibiting the tip effect that leads to lithium-ion accumulation and dendrite growth, significantly improving the safety and long-cycle stability of the lithium metal anode.

Furthermore, Bai et al. designed a metal-organic framework (MOF) material with a pore size of only 9 Å [76]. The MOF-modified separator could selectively screen lithium-ions while inhibiting the shuttle of large-radius polysulfide anions.

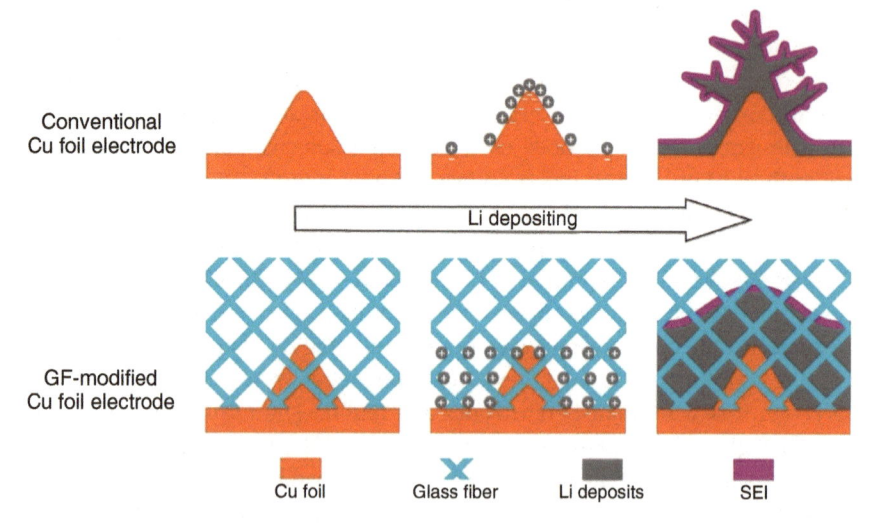

Figure 6.6 Li deposition of the routine 2D Cu foil electrode and GF-modified Cu foil electrode. Source: Cheng et al. [75]/with permission of John Wiley & Sons.

The Li—S battery using the MOF functional separator showed a per-cycle decay rate of only 0.019% over 1500 cycles, with almost no capacity loss during the first 100 cycles.

Dai et al. designed a hierarchical fibrous CNF-CNT membrane interlayer decorated with cobalt-doped nickel disulfide nanoparticles (Co-NiS$_2$@CNF-CNT) to enhance the performance of Li—S batteries [77]. Inspired by the root system of grass, where root hairs increase surface area and absorption efficiency, the CNF-CNT membrane mimics this structure, with CNF acting as the "root" and CNT as the "root hairs." This structure improves sulfur utilization and polysulfide absorption. Cobalt doping in NiS$_2$ introduces electron-deficient regions, which enhances the chemical adsorption of LiPSs and increases catalytic activity, as demonstrated by DFT calculations. The conductive CNF-CNT skeleton decreases cell polarization and enhances the catalytic conversion of polysulfides, leading to improved electrochemical performance. The Li—S battery paired with the Co-NiS$_2$@CNF-CNT interlayer exhibited an impressive rate capability of 951.4 mAh g^{-1} at 3 C, a reversible capacity of 944.1 mAh g^{-1} after 500 cycles at 0.2 C, and a long cycle life of 3000 cycles at high rates (3 and 5 C). Additionally, the battery achieved an areal capacity of 7.96 mAh cm^{-2} with a high sulfur loading of 9.6 mg cm^{-2} and maintained 6.61 mAh cm^{-2} after 200 cycles.

Lin et al. introduced a novel separator design using nickel cobaltite/carbon nanofiber (NiCo$_2$O$_4$/CNF) composites to address the shuttle effect and slow conversion kinetics of LiPSs in Li—S batteries [78]. NiCo$_2$O$_4$, with its polar nature, effectively traps polysulfides through Ni—S and Co—S bonds, while the conductive CNF network enhances electron transport and provides structural stability. This dual-functional separator not only adsorbs LiPSs but also accelerates the redox conversion between sulfur and Li$_2$S, protecting the lithium anode from degradation. Experimental studies and DFT calculations confirmed the catalytic role of $NiCo_2O_4$ in improving sulfur redox kinetics, while the CNF network further acts as a secondary barrier to suppress polysulfide diffusion. As a result, Li—S batteries with $NiCo_2O_4$/CNF-modified separators achieved a high rate capacity of 870 mAh g^{-1} at 5 C and retained 95% capacity after 200 cycles at 0.1 C, with a sulfur loading of 6.3 mg cm^{-2}. Even at a higher sulfur loading of 7.9 mg cm^{-2} and a lean electrolyte-to-sulfur ratio of 7.2 μL mg^{-1}, the battery delivered an impressive areal capacity of 8.1 mAh cm^{-2}.

6.3.3 Two-Dimensional Materials in Separators

In Li—S batteries, the dissolution and diffusion of soluble polysulfides are key factors leading to the loss of sulfur active material from the cathode and corrosion of the lithium anode. Since polysulfides carry negative charges, modifying the separator with a functional layer containing negatively charged groups can utilize electrostatic repulsion to confine polysulfides to the cathode side, thereby reducing their shuttle diffusion to the anode. Huang et al. developed an ultra-thin Nafion modification layer rich in sulfonic acid groups [79]. This modification layer not only effectively repels polysulfide anions through electrostatic repulsion but also provides pathways for lithium-ion transport, enabling efficient selective ion passage. The Li—S battery using this modification layer showed a capacity fade rate of only 0.08% per cycle after 500 cycles. Nafion can also be combined with other materials to form a separator modification layer, utilizing its electrostatic repulsion effect. For instance, Nafion-Super P [80] and Nafion-Super P-PEO [81] composite systems effectively suppress the shuttle effect.

Lu et al. designed a sulfonated reduced graphene oxide (SRGO) interlayer [82]. The SRGO interlayer serves two primary functions: acting as an ion-selective layer that physically blocks polysulfide migration and chemically anchoring LiPSs through the electronegative sulfonic groups. Based on Lewis acid–base theory and DFT calculations, it was demonstrated that the sulfonic groups strongly interact with lithium-ions in polysulfides, thereby stabilizing the electrochemical process. Additionally, the sulfonic groups induce uniform sulfur distribution, reducing excessive sulfur growth and enhancing sulfur utilization. The Li—S battery equipped with the SRGO interlayer exhibited impressive electrochemical performance, achieving a high reversible discharge capacity of over 1300 mAh g^{-1} and retaining 802 mAh g^{-1} after 250 cycles at 0.5 C. The cell maintained 1100 mAh g^{-1} at 0.2 C after varying rates from 0.2 to 4 C over 60 cycles, showcasing robust cycling stability and high-rate capability. Lei et al. have designed a negatively charged graphene composite membrane to effectively suppress the polysulfide shuttle effect [83]. By directly coating a layer of rGO/sodium lignosulfonate (SL) composite onto a standard PP separator, they fabricated a rGO@SL/PP membrane. The abundant negatively charged sulfonate groups in this membrane can effectively inhibit the transfer of negatively charged polysulfide ions while not affecting the transport of positively charged lithium-ions. Li—S batteries using this membrane retained 74% of their capacity after 1000 cycles, demonstrating extremely high cycling stability.

Moreover, SL and rGO are covalently bonded, forming a negatively charged rGO@SL composite material that not only effectively prevents polysulfide shuttling but also ensures excellent lithium-ion transport properties. This membrane design enables the Li—S battery to exhibit a capacity decay rate of <0.026% per cycle after 1000 cycles at a sulfur loading of 1.5 mg cm^{-2}, and a capacity decay rate of <0.074% per cycle after 670 cycles at a sulfur loading of 3.8 mg cm^{-2}. In situ Raman spectroscopy and polysulfide permeability tests confirmed that the rGO@SL/PP membrane strongly suppresses polysulfide shuttling through the charge repulsion effect of its abundant sulfonate groups, thereby achieving excellent battery performance. Notably, this negatively charged membrane can be easily integrated into other optimized electrode structures simply by replacing the traditional separator, without the need for modifying complex electrode structures, thus potentially impacting practical technology rapidly.

Additionally, other materials containing negatively charged groups can also be used as modification layers. Jin et al. utilized the negatively charged —SO$_2$C(CN)$_2$Li group in Li-PFSD to achieve selective lithium-ion transport [84]. Conder et al. modified the surface of a PP membrane with lithium polystyrene sulfonate [85], effectively suppressing the polysulfide shuttle effect. Yim et al. coated barium titanate (BaTiO$_3$) nanoparticles on the surface of a PE separator [86], leveraging its ferroelectric properties to generate permanent dipoles under an external electric field, which exerted strong repulsion against polysulfides, as shown in Figure 6.7a. Moreover, the addition of BaTiO$_3$ ceramic particles reduced the thermal shrinkage of the separator, thereby enhancing battery safety.

Huang et al. modified a PP separator with GO, where the abundant negatively charged oxygen-containing functional groups on GO repelled polysulfide anions [90]. The dense structure formed by the self-assembled GO layers blocked the shuttle of polysulfides, while the stacked pores allowed smooth lithium-ion transport. Huang later combined GO with Nafion to create a PP/GO/Nafion three-layer functional separator [91]. This modification layer was only 130 nm thick, combining the electrostatic repulsion of Nafion with the spatial hindrance effect of GO, effectively blocking the transport of polysulfides. The Li—S battery using this separator achieved an initial specific capacity of 1225 mAh g^{-1} at 0.2 C, with a per-cycle decay rate of just 0.18%.

Zhang et al. mixed GO and oxidized carbon nanotubes (o-CNTs) and coated them onto a PP separator [92]. The abundant oxygen-containing functional groups enhanced the chemical adsorption capacity for polysulfides, while the nanotube channels effectively restricted the diffusion of polysulfides. The Li—S battery with this functional separator maintained a capacity of 750 mAh g^{-1} after 100 cycles at a 1 C rate.

As shown in Figure 6.7c, Sun et al. used vacuum filtration to deposit a layer of 2D black phosphorus (BP) material onto a conventional separator [88]. The strong chemical interaction between BP and polysulfides allowed effective chemical adsorption. The high conductivity of BP also acted as a second current collector, activating the anchored polysulfides, resulting in highly efficient cycling with a high sulfur-loaded cathode. After 100 cycles at a current density of 0.4 A g^{-1}, the capacity

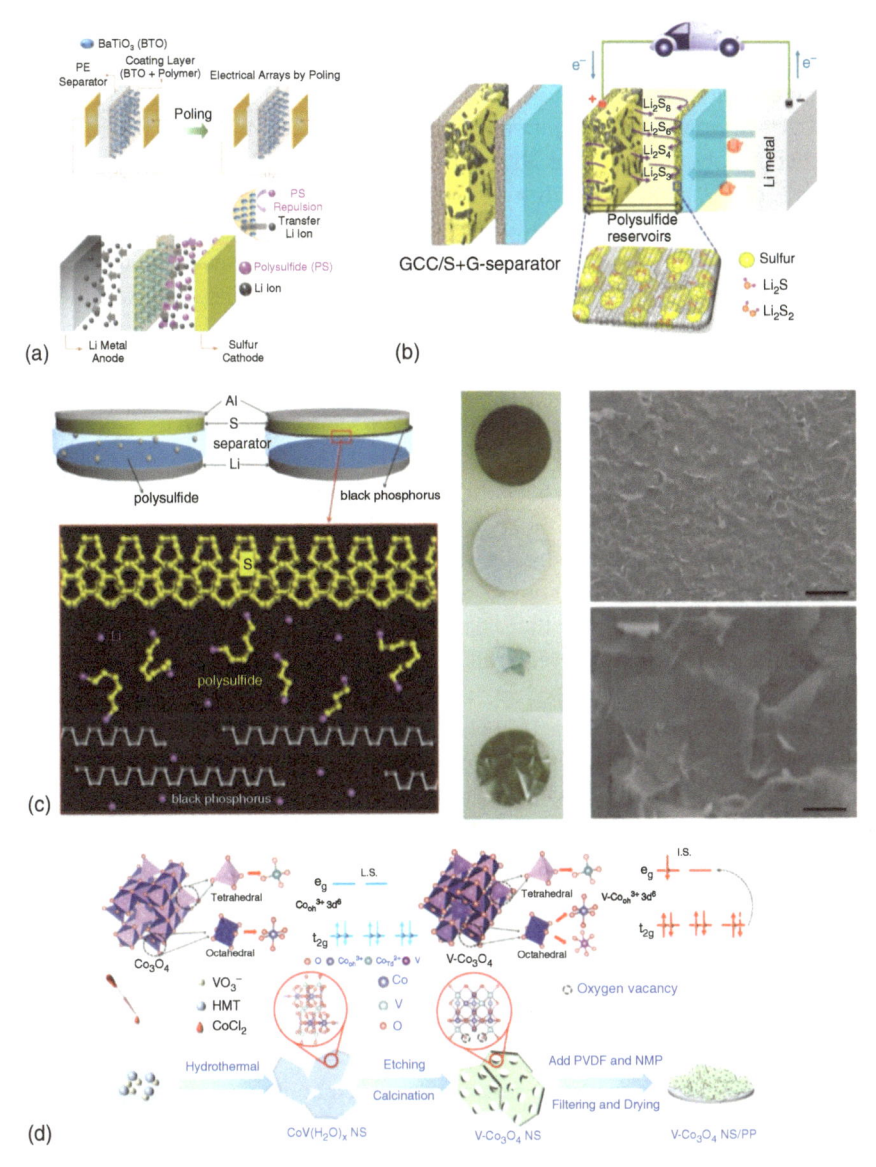

6.3 Low-Dimensional Composite Materials in Separators | **317**

Figure 6.7 (a) The poling process of the BTO-coated PE separator and the effect of the poled BTO toward polysulfide rejection. Source: Yim et al. [86]/with Permission of John Wiley & Sons. (b) Li–S cell with a GCC, sulfur cathode, and G-separator. Source: Zhou et al. [87]/with permission of John Wiley & Sons. (c) Li–S cell with black phosphorus (BP) coated separator. Photographs of the as-prepared BP-coated separator: cathode-facing side, anode-facing side, folded, unfolded. Right is SEM images of the as-prepared BP-coated separator. Source: Reproduced with permission from Sun et al. [88]/John Wiley & Sons. (d) Crystal structure and Co_{Oh}^{3+} of spinel Co_3O_4 NS and spinel V-Co_3O_4 NS with the tetrahedral and octahedral sites, respectively. And the preparation of spinel V-Co_3O_4 NS/PP separator. Source: Zhou et al. [89]/with permission of John Wiley & Sons.

remained at 800 mAh g^{-1}. Song et al. deposited MXene material Ti$_3$C$_2$T$_x$ on one side of the separator using vacuum filtration [93]. The coating was only 522 nm thick. The high conductivity of MXene and the strong adsorption between Ti atoms and polysulfides improved the utilization of active material and enhanced the effective anchoring of polysulfides. After 500 cycles at a 0.5 C rate, the battery maintained a capacity of 550 mAh g^{-1}, with a per-cycle decay rate of 0.062%. Additionally, some composite modification coatings combine physical blocking and chemical adsorption effects, significantly enhancing the cycling performance and stability of Li—S batteries. For example, the TiO$_2$/graphene composite coating prepared by Xiao et al., [94] and the PEG/MPC composite coating developed by Chung et al., [67] both demonstrated effective suppression of the polysulfide shuttle effect and significant improvement in battery performance.

As shown in Figure 6.7b, Zhou et al. modified the separator by exfoliating graphene using an intercalation method, preparing a flexible, integrated graphene/sulfur/graphene composite cathode [87]. Due to the 2D layered structure of graphene, it can tightly contact the separator while also acting as a physical barrier to block polysulfides. The conductive graphene also enhances the utilization of active sulfur, reduces interface contact resistance, and lowers battery polarization. Later, they proposed a design that coats sulfur and graphene on the separator surface in an integrated structure, significantly improving the battery's gravimetric energy density without the need for metal current collectors or sulfur matrix materials [95].

Song et al. designed a graphene/polypropylene/Al$_2$O$_3$ (GPA) three-layer structured separator, where graphene is positioned on the side near the cathode, serving as both a conductive layer and an electrolyte storage layer, enabling fast transmission of both electrons and ions [96]. Al$_2$O$_3$ is placed on the side near the anode, where its high stability and mechanical strength improve the thermal stability and safety of the separator, while also effectively suppressing dendrite growth. Yu et al. replaced the commercial PP separator with a porous nanocellulose membrane, which exhibits higher electrolyte retention, better thermal stability, and good wettability with lithium metal, effectively inhibiting dendrite formation, with a cycle life exceeding 1000 cycles [97].

Ye et al. designed a polymer microporous membrane/GO/polymer microporous membrane three-layer structured separator [98]. This design endows the LMO/GO/Li battery with both anti-puncture and dendrite-eliminating advantages. When the separator is pierced by lithium dendrites, the middle GO layer chemically etches the dendrites through a spontaneous redox reaction, significantly reducing the risk of short circuits caused by dendrite penetration through the separator. The LMO/GO/Li battery can cycle 6000 times, 48 times the life of traditional LMO/Li batteries, with a Coulombic efficiency of up to 93%. Based on a similar principle, Liu et al. designed a PE/SiO$_2$/PE separator, where SiO$_2$ reacts with lithium dendrites to prevent them from penetrating the separator [99].

In the research on low-dimensional composite cathode materials, the electrocatalytic effects of low-dimensional carbon materials have gained considerable attention. In fact, directly coating these electrocatalytic low-dimensional materials onto the surface of the separator can also exert catalytic effects, promoting the conversion of polysulfides and thereby reducing the loss of active materials. As shown in Figure 6.7d, Zhou et al. designed the in-spin state of spinel oxides and prepared V doping-induced hexagonal nanosheets (V-Co_3O_4 NS) catalysts with an intermediate spin state [89]. The V-substituted Co^{3+} sites optimize the spin state and the d-band center of spinel Co_3O_4 NS, which maximizes the interaction between LiPSs and the catalyst. The V-Co_3O_4 NS/PP separator delivers outstanding rate performance (600.1 mAh g^{-1} at 10.0 C) and exhibits negligible capacity decay, with only 0.015% loss per cycle over 1900 cycles at 3.0 C. Catalytic coatings can also combine synergistic effects such as spatial confinement and charge repulsion to further enhance the cycling performance of Li—S batteries.

Lei et al. developed a novel metallic oxide composite, $NiCo_2O_4$@rGO, to address challenges in Li—S batteries, which are known for high energy density and low cost [100]. Traditional carbon materials hinder commercialization due to low lithium-ion transport rates. The $NiCo_2O_4$@rGO composite, used as a separator, facilitates robust cycling stability and areal capacity by providing abundant active sites for catalytic conversion of polysulfides and reducing the energy barrier for lithium-ion diffusion. With a high sulfur loading of 6 mg cm^{-2}, the batteries achieve an initial capacity of 5.04 mAh cm^{-2} and retain 92% capacity after 400 cycles. The $NiCo_2O_4$@rGO/PP separator enhances ion diffusion and suppresses polysulfide dissolution. DFT calculations indicate a low energy barrier for lithium-ion diffusion (0.15 eV), leading to improved cycling performance. The batteries, using a sulfur-carbon mixture cathode, show an areal capacity of 7.1 mAh cm^{-2}, surpassing commercial standards. The separator simplifies integration into existing systems, playing a catalytic role in redox reactions and preventing Li_2S deposition, thus enhancing stability and extending battery life.

Another effective strategy is to add transition metal or precious metal atoms. Various SACs have been prepared for cathodes and separators for Li—S batteries. Zhang et al. developed a multi-functional separator by coating commercial PP separators with nitrogen-doped graphene (NG) impregnated with SACs such as Fe, Co, and Ni [101]. The SAC-modified separator was designed to effectively block the crossover of polysulfides and enhance their redox kinetics. Among the SACs tested, Fe showed the best performance due to its higher binding energy with polysulfides, which was confirmed through theoretical calculations and in situ Raman measurements. This modification not only improves polysulfide adsorption but also catalyzes their conversion during cycling, addressing the polysulfide shuttle effect in Li—S batteries. With an extremely low metal loading of approximately 2 μg in the entire cell, the Fe/NG-modified separator delivered an impressive performance, retaining 83.7% of its capacity after 750 cycles at 0.5 C with a high sulfur loading of 4.5 mg cm^{-2}. The battery also demonstrated excellent rate capability, achieving a discharge capacity of 673 mAh g^{-1} at 5 C and significantly reducing the voltage gap from 0.48 V (commercial PP separator) to 0.24 V. This study highlights the potential of SAC-modified separators in enhancing both the cycle life and rate performance of high-energy Li—S batteries by mitigating the shuttle effect and promoting polysulfide redox reactions.

Functional separators are crucial in mitigating the shuttle effect and improving redox reaction kinetics, thereby enhancing the overall electrochemical performance [102]. The design principles for electrocatalytic materials used in separator modification can be summarized as follows:

1) The synthesis methods for catalytic materials should be straightforward and scalable for commercial production. Techniques such as hydrothermal processes and high-temperature pyrolysis are preferred for their feasibility in mass production.
2) Catalytic materials should possess hierarchical porous structures, appropriate pore sizes, and a large specific surface area. The combination of micropores and mesopores offers strong adsorption capacity and high pore volume, which can physically block polysulfides while storing electrolytes. Proper pore sizes are essential to retain soluble polysulfides without hindering ion and electron diffusion, and a large surface area ensures sufficient active sites for polysulfide adsorption and uniform Li_2S deposition.
3) Stability in both physicochemical and mechanical properties is essential. Catalytic materials must be conductive, have good wettability, and maintain flexibility. These properties promote long-term cycling stability by improving electrolyte contact without unwanted reactions, while high conductivity enhances ion and electron transport to accelerate redox reactions. For materials with inherently low conductivity, introducing defects or vacancies can improve electrical performance.

Moreover, cost efficiency is a key consideration in material selection, and the catalytic materials should be lightweight to maximize sulfur utilization and increase the battery's energy density.

6.4 Low-Dimensional Composite Current Collectors

6.4.1 Design of Current Collector

In lithium batteries, the active materials in the electrodes often cannot conduct electricity efficiently on their own, requiring a conductive medium to assist. The current collector serves as this medium by contacting the active materials and transferring electrons from the electrode to the external circuit or back to the electrode from the external circuit. In traditional lithium batteries, aluminum foil is typically used as the current collector for the cathode, while copper foil is used for the anode. However, Li—S batteries differ from conventional lithium-ion batteries based on intercalation reactions. Li—S batteries not only require the rapid transfer of lithium-ions but also demand quick electron transfer, meaning that the active materials must have excellent electronic conduction pathways. Unfortunately, the dendritic growth and instability of the interface at the lithium metal anode can cause the active material to detach from the electronic conduction network, reducing the utilization of the active material and potentially leading to battery failure. This issue is particularly pronounced in high sulfur-loading and high-current charge–discharge cycles.

When preparing sulfur cathodes, sulfur-containing active materials, conductive agents, and binders are mixed into a slurry and coated onto aluminum foil. Although adjusting the sulfur host structure can improve contact between sulfur and conductive agents, cathodes that rely solely on aluminum foil current collectors struggle to achieve sulfur loadings exceeding 6 mg cm^{-2}. Therefore, ideal current collector materials for Li—S batteries should possess the following characteristics:

1) Good mechanical strength and adhesion: The current collector provides physical support to the electrode, ensuring the uniform distribution of active materials, maintaining the stability of the battery structure, and promoting uniform electrochemical reactions.
2) Long-range electronic pathways and low ionic diffusion resistance: Common conductive agents for sulfur cathodes, such as carbon black or acetylene black, tend to disperse as particles, making it difficult to form continuous electronic and ionic transport channels, thus increasing internal resistance. During cycling, repeated redox reactions of sulfur can cause the redistribution of active materials, which may hinder lithium-ion migration. Nano current collectors typically incorporate long-range electronic conduction frameworks that ensure rapid electron transfer, while their nanochannels provide continuous ionic transport paths, speeding up reaction kinetics and improving performance under high sulfur-loading and high-current conditions.
3) Surface modification to regulate polysulfide behavior: The hierarchical porous structure of nano current collectors can regulate the adsorption sites for polysulfides and prevent their migration. By introducing heteroatom doping and other techniques, polysulfides can be anchored onto the nano current collector, preventing the formation of "dead sulfur" (unreacted sulfur on the conductive framework) and improving the utilization of active materials.
4) Mitigation of volume expansion and maintenance of cathode structural integrity: In Li—S batteries, the redox processes of active materials are accompanied by volume changes, which may destabilize the contact with the current collector or cause active material detachment. The flexible and porous structure of nano current collectors can effectively alleviate such expansion, maintaining good contact with the active material.
5) Inhibition of side reactions and enhancement of electrode chemical stability: Aluminum foil can react with TFSI$^-$ ions in the electrolyte to form Al(TFSI)$_3$, which is particularly problematic at high sulfur loadings. In contrast, nano current collectors do not react with the electrolyte, helping to extend the cycle life of the battery.

Such current collector designs provide more efficient electronic and ionic transport pathways for Li—S batteries, improving their stability and efficiency under challenging conditions, such as high sulfur loading and high current densities.

The anode in Li—S batteries typically uses lithium metal, but a major challenge with lithium metal is the instability of the SEI and the growth of lithium dendrites. Therefore, the primary goals in designing the anode current collector are to suppress dendrite formation and stabilize the SEI. The design strategies for Li—S battery anode current collectors include the following:

1) Increased surface area: By increasing the surface area of the current collector, the local current density is reduced, which helps mitigate the growth of lithium dendrites.
2) Lithiophilic sites and uniform electric field distribution: The use of Lithiophilic sites or uniform electric field distribution can guide lithium-ion deposition, creating abundant lithium nucleation sites. This promotes uniform lithium deposition and avoids irregular dendrite growth.
3) Provision of buffer space: During charge and discharge, the volume of the lithium anode undergoes significant changes, which may lead to SEI rupture. The current collector should offer some degree of free space to buffer this volume change and ensure structural stability.

6.4.2 Nanocomposite Current Collectors

Based on these design requirements, researchers have developed various nano current collector materials, including low-dimensional composite current collectors based on CNT, CNF, graphene, and other carbon-based composites. Barchasz et al. constructed a nano-scale current collector for the sulfur cathode using vertically aligned carbon nanotubes (VACNT) [103]. VACNT features a high specific surface area, excellent conductivity, mechanical strength, and anisotropy, which not only enhances sulfur loading but also suppresses changes in sulfur morphology during cycling. At 0.01 C, this current collector achieved a first discharge-specific capacity of 1300 mAh g^{-1}. However, the drawback of VACNT is its poor wettability by the electrolyte, necessitating further surface modification of the CNT. Peng et al. designed a three-dimensional carbon nanotube (3D-CNT) current collector [104], which significantly improved the electrode's chemical stability and effectively suppressed the formation of passivation layers, as shown in Figure 6.8. With a sulfur loading of >3.7 mAh cm^{-2}, the electrode could cycle 950 times with a capacity decay rate of only 0.029% per cycle. Amir et al. developed a composite current collector composed of sulfurized polyacrylonitrile (SPAN) and a CNT conductive framework through electrospinning and sulfurization processes [105]. By co-spinning PAN and CNT into nanofibers, a 3D porous structure with conductive pathways was formed, and after sulfurization in a high-temperature sulfur atmosphere, the SPAN-CNT composite material was obtained. At a rate of 0.2 C, the initial discharge-specific capacity of the material reached 1610 mAh g^{-1}.

CNFs, as another class of 1D nanocarbon materials, exhibit excellent mechanical properties and conductive and thermal characteristics. Through interface modification and pore structure design, CNF can achieve a high specific surface area, heteroatom doping, and core-shell structures. Chung et al. proposed using a porous carbon of interwoven CNF as the current collector to immobilize polysulfides [106]. Due to the good conductivity and interconnected porous structure of CNF, the sulfur loading and utilization were significantly improved. Zhang et al. developed a new type of 3D structured material by incorporating uniformly distributed

6.4 Low-Dimensional Composite Current Collectors

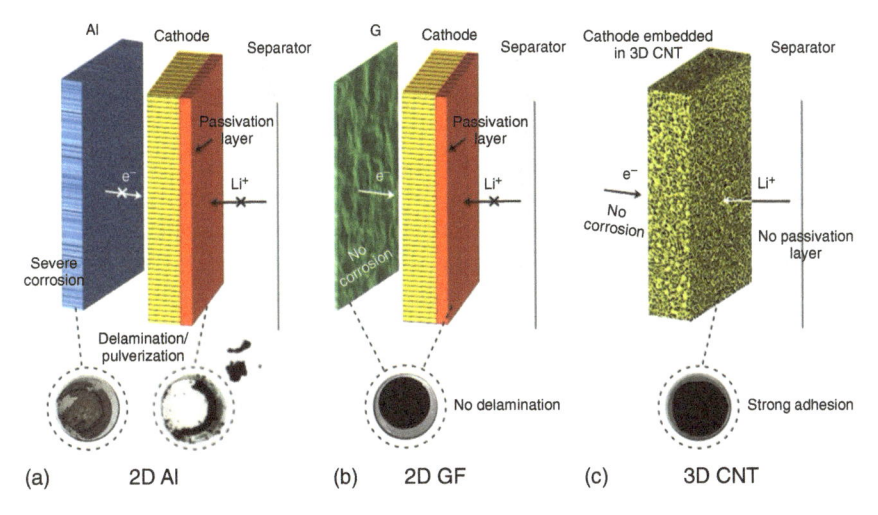

Figure 6.8 (a) 2D-Al, (b) 2D-GF and (c) 3D-CNT current collectors for Li–S batteries. Source: Reproduced with permission from Peng et al. [104]/John Wiley & Sons.

cobalt (Co) nanoparticles and nitrogen (N)-doped CNF onto carbon cloth (CC) [107]. This material not only has excellent conductivity but also mitigates volume expansion during charge-discharge cycles. N-doping and Co nanoparticles provided adsorption sites for polysulfides, accelerating their conversion reactions. Wang et al. fabricated a 3D porous conductive carbon fiber current collector (CCP) for lithium anodes [108]. The porous structure of CCP facilitated non-dendritic lithium deposition, enhanced material transport within the electrode, and effectively mitigated volume changes during the lithium anode's cycling. After 800 cycles at a 2 C rate, the capacity retention rate of this current collector remained as high as 91%.

As a representative 2D carbon material, graphene has promising applications in Li—S batteries due to its ultra-high specific surface area, excellent conductivity, superior chemical stability, and mechanical strength. Lu et al. prepared a 3D randomly dual-continuous micron-porous graphene foam (3D-MPGF) as a binder-free current collector [109]. This 3D-MPGF material possesses a unique micron-scale porous structure, featuring interconnected tubular pores and non-tubular micropores. This porous structure can effectively suppress the polysulfide shuttle effect while supporting sulfur without the need for a binder, significantly enhancing battery performance. Additionally, Zhang et al. developed a nitrogen-doped graphene as the current collector for lithium metal anodes [110]. This current collector can regulate the lithium metal nucleation process to suppress dendrite formation. Nitrogen doping introduces lithiophilic functional groups that induce uniform lithium metal nucleation, leading to uniform lithium deposition on the anode surface and reducing the risk of dendrite growth.

In addition to being directly used as current collectors for both the cathode and anode, low-dimensional carbon materials can also be combined with traditional metal current collectors to form composite current collectors, enhancing overall performance. Huang et al. significantly improved the electronic conductivity and adhesion of the current collector to the active materials by coating a mixture of graphene and carbon nanotubes on the surface of conventional aluminum foil [111]. This hybrid structure not only provides abundant lithium-ion pathways for redox reactions but also reduces the overpotential during the reaction process. At a rate of 0.5 C, the assembled Li—S battery achieved an initial specific capacity of 1113 mAh g^{-1}, with a capacity retention rate of 73% after 300 cycles. Cheng et al. used a 3D foam aluminum/CNT composite as the current collector [112]. This design established long-range and short-range electronic conduction pathways and provided sufficient space for high sulfur loading. For batteries with a sulfur loading of 12 mg cm^{-2}, excellent performance was observed, while for those with a sulfur loading of 7 mg cm^{-2}, the initial specific capacity reached 860 mAh g^{-1}.

Despite the significant advantages of low-dimensional materials in the design of current collectors for Li—S batteries, several challenges remain. Currently, metal current collectors remain the mainstream choice, with a focus on lightweight design and corrosion resistance. In contrast, while carbon-based current collectors offer many benefits, their affinity for lithium is relatively weak. Composite current collectors, which combine the advantages of multiple materials, are more effective than single-material systems, but their design is complex, and large-scale production faces several technical bottlenecks. In the future, research on low-dimensional composite current collectors must address issues such as the polysulfide shuttle effect and interface stability while considering their feasibility in large-scale applications to promote further advancements in Li—S battery technology.

6.5 Low-Dimensional Composite Anode Materials

6.5.1 Formation of SEI and Failure Mechanism

Due to the high reactivity of lithium metal, a passivation film composed of Li$_2$CO$_3$, LiOH, and Li$_2$O forms on its surface even before it comes into contact with the electrolyte. When lithium metal is exposed to the electrolyte, a new layer called the SEI forms on its surface. Currently, there are three main models describing the structure of the SEI: the Peled model, the mosaic model, and the Coulombic interaction model. The main inorganic components of SEI include Li$_2$O, LiOH, LiF, Li$_2$CO$_3$, and Li$_3$N. In Li—S batteries, the influence of polysulfides can also lead to the formation of Li$_2$S and Li$_2$S$_2$ within the SEI. The organic components of SEI mainly consist of ROLi, RCOOLi, ROCOLi, RCOO$_2$Li, and ROCO$_2$Li (where R represents an alkyl group). Generally, the organic components of SEI are more stable than the inorganic ones, but their lithium-ion transport rate is relatively slower. Additionally, the chemical composition of SEI varies significantly depending on the type of electrolyte used.

In Li—S batteries, the failure mechanisms of lithium metal anodes can be broadly categorized into two types: one is related to the intrinsic kinetics of lithium deposition/dissolution, and the other stems from the redox reactions in the Li—S battery, particularly the side reactions triggered by polysulfides migrating to the lithium metal anode. Over time, porous materials, including the SEI layer and dead lithium, accumulate on the surface of the lithium metal anode, increasing interfacial resistance and causing electrode volume expansion. As the electrolyte is gradually depleted, internal resistance also rises. Furthermore, the growth of lithium dendrites may pierce the separator, resulting in a short circuit and further accelerating the anode's failure [113]. In addition, the polysulfides generated at the cathode shuttle to the anode surface and react with the lithium metal, which can lead to electrolyte consumption and pulverization and deactivation of the anode, exacerbating its failure [114]. These issues highlight the importance of SEI stability and its ability to suppress polysulfides, which are critical research directions for achieving the long-term stability of the anode in Li—S batteries.

6.5.2 Nanocomposite Lithium Metal Anodes

The use of pure lithium metal anodes in Li—S batteries is severely limited by issues such as lithium dendrite growth and volumetric expansion. Although lithium-ion batteries can alleviate the issue of volume expansion through intercalation techniques, Li—S batteries experience more severe expansion of the lithium anode, leading to frequent rupture of the SEI during charge and discharge cycles, accompanied by low coulombic efficiency. Additionally, the pure lithium metal anode has a 2D planar structure, resulting in a high local current density that promotes dendrite growth. Therefore, designing the lithium anode into a 3D porous structure not only effectively mitigates volume expansion but also reduces local current density and suppresses dendrite growth. As a result, nanostructured lithium anodes and nanocomposite anodes have emerged.

In nanostructured anodes, the scaffold typically does not contain metallic lithium, and lithium needs to be deposited onto this scaffold. These scaffolds can be classified as either conductive or non-conductive but lithiophilic. During the charge and discharge process, lithium undergoes deposition and dissolution on these scaffolds. In conductive scaffold-based nanolithium anodes, lithium deposits not only on the surface of the scaffold but also within its internal structure. However, lithium deposited on the surface can block the internal channels of the scaffold, making it difficult for subsequent lithium-ions to deposit further inside. In nonconductive lithiophilic scaffolds, the lithiophilic nature of the scaffold surface promotes the adsorption and reduction of lithium-ions, preventing the blockage of channels. However, due to the non-conductive nature of the scaffold, lithium tends to preferentially deposit near the bottom close to the current collector, making it difficult for the lithium farther from the current collector to dissolve during discharge, resulting in dead lithium and reduced Coulombic efficiency.

Carbon materials are one of the most widely used conductive scaffolds. Nanocarbon scaffolds, with their large specific surface area and high porosity, can effectively suppress dendrite growth, stabilize the anode/electrolyte interface, and alleviate volume changes. Mukherjee et al. fabricated a porous graphene scaffold, using defects in the scaffold to induce lithium deposition, while the porous structure restricted dendrite growth [115]. This design enabled the anode to maintain a

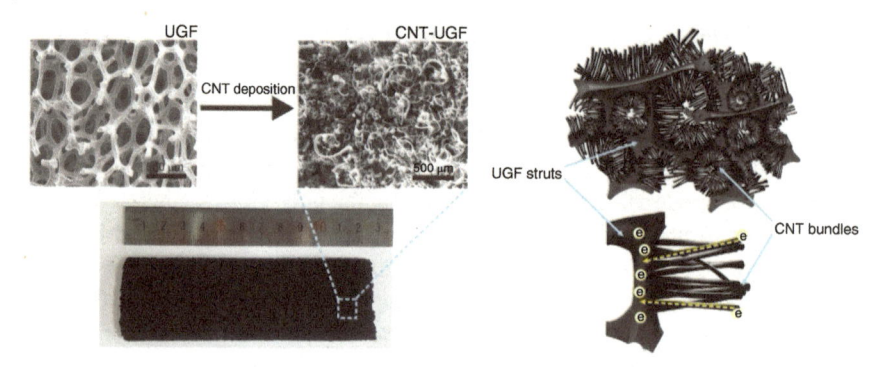

Figure 6.9 SEM images of UGF and CNT-UGF. Right is a schematic of the 3D interconnected network of the CNT-UGF hybrid which allows free electron transfer between the UGF and each CNT. Source: Reproduced with permission from Jin et al. [117]/John Wiley & Sons.

Coulombic efficiency >99% after over 1000 cycles. Nonconductive scaffolds, on the other hand, typically contain polar functional groups that adsorb lithium-ions from the electrolyte, thereby accelerating lithium-ion transport and suppressing dendrite formation. Cheng et al. used GFs with polar functional groups as scaffold materials, enabling uniform lithium-ion deposition and significantly reducing dendrite formation [75].

In comparison, nanocomposite anodes, which integrate scaffold materials with metallic lithium, exhibit superior comprehensive performance. Lin et al. prepared a Li-rGO composite anode by injecting molten lithium into rGO [116]. The polar functional groups on the rGO surface enhanced the material's lithiophilicity, resulting in only 20% volume change during cycling, excellent mechanical flexibility, lower overpotential, and a high capacity of 3390 mAh g^{-1}. As shown in Figure 6.9, Jin et al. grew CNTs on ultrathin graphite foam (UGF) to fabricate a covalently bonded carbon nanostructure (CNT-UGF) [117]. They used an electrochemical pre-deposition technique to prepare a Li/CNT-UGF anode, and the assembled Li—S battery exhibited a capacity decay rate of only 0.057% per cycle after 400 cycles at a 2 C rate.

To address the dendrite issue in lithium anodes, designing a suitable host for lithium metal is an effective strategy. MXene materials are considered potential candidates for suppressing the formation and growth of lithium dendrites, contributing to more stable lithium anodes. The surface terminations on MXenes exhibit a strong affinity for lithium, guiding its nucleation. Their metallic conductivity and fast lithium-ion diffusion ensure rapid electrochemical kinetics. Additionally, the diverse structures and robust mechanical properties of MXenes offer ample opportunities for creating advanced MXene/lithium composites. As a result, MXenes have garnered increasing attention for enhancing the performance of lithium anodes, particularly in high-performance Li—S batteries. Li et al. proposed a novel approach to address the issues of uncontrollable lithium dendrite growth and large volume changes in lithium-metal anodes, which are major obstacles to the practical use of lithium-based batteries [118]. By creating a lamellar structured Ti$_3$C$_2$ MXene (graphene, BN)-lithium film anode, they utilized the unique ductility of metallic lithium and the lubricity of these nanosheets to effectively control dendrite formation. The lithiophobic nature of the atomic layers helps confine lithium plating within nanoscaled gaps, preventing vertical dendrite growth. This lamellar structure acts as an "artificial solid electrolyte interphase," enabling stable lithium plating and stripping during cycling. The fabricated Ti_3C_2 MXene-lithium anode demonstrated excellent electrochemical performance, including a low overpotential of 32 mV at 1.0 mA cm^{-2} and stable voltage profiles over 200 cycles. The anode also exhibited impressive high-rate performance with minimal overpotential increase, and in a full-cell configuration with a sulfur-carbon cathode, it achieved a high energy density of 656 Wh kg^{-1}.

Compared to lamellar MXenes, perpendicular MXene structures are more favorable for lithium-ion transport. A simple rolling-cutting method was used to create perpendicular MXene-lithium arrays with dual periodic interspaces, significantly promoting lithium-ion transport and effectively mitigating dendrite formation during lithium stripping and plating. Zhao et al. developed a novel perpendicular MXene-Li array with dual periodic interspaces as an anode material for lithium metal batteries [119]. The array consists of nanometer-scale interspaces within the MXene walls, which promote fast lithium-ion transport during stripping and plating, and micrometer-scale interspaces between the MXene walls, which help homogenize the electric field and prevent the formation of lithium dendrites. This unique structure effectively addresses two major challenges in lithium metal batteries: dendrite growth and large volume changes. The design leads to a dendrite-free lithium anode with low potential (25 mV) and a high specific capacity of 2056 mAh g^{-1}. The MXene-Li arrays exhibit excellent electrochemical performance, including a long cycle life of up to 1700 hours and impressive high-rate capabilities of 2500 cycles at a current density of 20 mA cm^{-2}. The perpendicular structure also mitigates the "lightning rod effect," ensuring homogeneous lithium deposition and deep stripping/plating capacity of up to 20 mAh cm^{-2}.

The construction of a 3D MXene framework as a lithium host further suppresses dendrite growth and minimizes volume changes during the lithium stripping/plating process. Zhang et al. proposed 3D porous MXene aerogels as scaffolds for high-rate lithium metal anodes to address challenges such as dendrite growth, short lifespan, and infinite volume expansion during Li plating/stripping, particularly under high current densities [120]. Using Ti_3C_2 MXene as a scaffold, they demonstrated that the material's high metallic conductivity, rapid lithium-ion transport, and abundant nucleation sites significantly improve cycling stability and lower overpotential at current densities up to 10 mA cm^{-2}. The interconnected structure of the MXene aerogel effectively mitigates volume fluctuations during Li deposition, enabling uniform Li growth and high efficiency. This approach also showed promising results in full-cell configurations with $LiFePO_4$ as the cathode, suggesting the potential for high-rate performance in practical applications. By leveraging the inherent properties of MXene, such as functional groups that enhance Li nucleation and fast electron/ion transport, this strategy opens up new avenues for the use of non-graphene 2D materials as scaffolds for high-energy-density Li metal batteries.

6.5.3 Low-Dimensional Materials in 3D-Printing Anodes

Recent advancements in 3D-printing technology have provided new approaches for designing complex energy storage devices, particularly for lithium anodes. A dendrite-free lithium anode with remarkably low overpotential (~10 mV), long cycle life (up to 1200 hours), and high areal capacities have been developed through extrusion-based 3D printing. The printed anode utilizes MXene (Ti$_3$C$_2$Tx) arrays with large interspaces, which serve dual purposes: guiding uniform lithium nucleation and dispersing lithium-ion flux and electric fields, thereby effectively inhibiting lithium dendrite growth [121]. The interspaces also accommodate cobblestone-like lithium deposits, contributing to high areal capacities (10–20 mAh cm^{-2}). When paired with a LiFePO$_4$ cathode, the full cell demonstrates high-rate performance (up to 30 C) and long cycle life (over 300 cycles). The MXene arrays, derived from highly concentrated MXene ink (~300 mg ml^{-1}), possess favorable rheological properties, making them ideal for 3D printing applications. These arrays guide parallel lithium growth and prevent dendrite formation during repeated plating/stripping. Even after 1200 hours of cycling, lithium dendrites were not observed. This development highlights the potential of 3D printing for producing advanced lithium-based batteries with high areal capacities and long-term cycling stability.

Lithium metal anodes hold immense potential for high-energy-density batteries but are hindered by safety concerns, primarily due to uncontrolled dendrite growth. To address this, a novel nitrogen-doped carbon (N-doped carbon) framework has been developed using extrusion-based 3D printing, derived from a zinc-based metal-organic framework (Zn-MOF) precursor [122]. This framework features a hierarchically porous structure with micro-, meso-, and macro-pores, providing a large specific surface area of 869 m^2 g^{-1}. These structural characteristics effectively suppress dendrite formation, accommodate large lithium deposits, and stabilize the Li/electrolyte interface, while also dissipating the high current densities that often exacerbate dendrite growth. The 3D-printed N-doped carbon framework (3DP-NC) enables an ultrahigh areal capacity of 30 mAh cm^{-2} at a current density of 10 mA cm^{-2} and demonstrates a high Coulombic efficiency of 97.9% over a long lifespan (~2000 hours) at 1 mA cm^{-2}. The high surface area helps dissipate the effects of high current densities, while the N-doped carbon surface ensures uniform lithium nucleation, preventing dendrite growth. Additionally, in symmetric cells, the Li-plated 3DP-NC anode operates efficiently at ultrahigh current densities of up to 20 mA cm^{-2} with low overpotentials. Coupled with a 3D-printed LiFePO$_4$ cathode, this setup demonstrates improved rate capabilities in full cells. This work highlights the potential of 3D printing as a powerful tool for fabricating high-performance lithium-metal batteries, leveraging its ability to create highly structured, porous frameworks that enhance lithium growth behavior and battery stability.

Therefore, 3D-printing technology has introduced a new approach to synthesizing effective lithium metal hosts. With a deeper understanding of the fabrication process, further improvements can be made to meet practical application requirements.

6.6 Conclusion and Outlook

The application of low-dimensional composite materials in the cathode, inactive components (separator, current collector), and anode of Li—S batteries has demonstrated significant advantages, playing a crucial role in addressing the performance bottlenecks of Li—S batteries.

In terms of the cathode, low-dimensional materials, particularly carbon-based materials such as graphene and CNTs, provide a stable host for sulfur by constructing porous and highly conductive framework structures. This not only enhances the sulfur loading but also significantly improves the interfacial stability of the sulfur cathode, which is often compromised by volume expansion during the charge and discharge process. Additionally, the multi-functionality of low-dimensional composite materials allows for doping or surface modification to introduce lithiophilic sites that enhance the adsorption and immobilization of polysulfides, thereby suppressing the shuttle effect and prolonging the cycle life of batteries. However, the long-term stability of high-loading sulfur cathodes under high current densities remains a challenge. Future research should focus more on the structural design of low-dimensional composite materials to optimize electronic and ionic transport channels, further improving sulfur utilization and increasing the energy density of the battery.

In the field of inactive materials, low-dimensional materials also play a crucial role in the design of separators and current collectors. Modified separators made with low-dimensional materials can effectively block the migration of polysulfides to the anode, suppressing the shuttle effect while maintaining good lithium-ion transport performance. For example, composite separators made of graphene and CNTs, with their unique pore structures and surface chemistry, can simultaneously allow ion penetration and adsorb and immobilize polysulfides, significantly enhancing the battery's long-term cycling stability. In the future, the design of separators should further focus on optimizing pore size distribution and surface functionalization to balance polysulfide isolation with lithium-ion transport efficiency. As for current collectors, nanocarbon materials and 3D graphene foams, among other low-dimensional materials, have improved the mechanical strength of electrodes by constructing highly conductive, lightweight structures with excellent interfaces with active materials. These materials also enhance the conductivity and electrochemical kinetics of sulfur cathodes. Future designs for composite current collectors will need to focus on maintaining low cost and high performance in large-scale production while improving corrosion resistance.

For the anode, low-dimensional composite materials offer new approaches for lithium metal anodes. By constructing 3D porous nanoframeworks, low-dimensional materials can effectively alleviate the issue of SEI film rupture caused by volume changes during charge and discharge cycles, while also reducing local current density and suppressing dendrite growth. At the same time, lithiophilic low-dimensional material frameworks can induce uniform lithium deposition, reducing the formation of "dead lithium." In the future, the application of low-dimensional materials in lithium anodes will need further optimization of their mechanical stability and electrochemical interface to ensure stability under high rates and long cycling conditions.

The multi-functionality of low-dimensional composite materials presents a broad prospect for improving the performance of Li—S batteries. However, more research is needed on material cost, scalability of manufacturing processes, and interface stability. Furthermore, the integrated design of low-dimensional materials will be a key focus of future research. By applying these materials to the coordinated optimization of cathodes, separators, current collectors, and anodes, Li—S batteries can be further advanced toward higher energy density, longer cycle life, and lower cost.

References

1 Fang, R., Zhao, S., Sun, Z. et al. (2017). More reliable lithium-sulfur batteries: status, solutions and prospects. *Advanced Materials* 29: 1606823.
2 Lin, T.Q., Tang, Y.F., Wang, Y.M. et al. (2013). Scotch-tape-like exfoliation of graphite assisted with elemental sulfur and graphene-sulfur composites for high-performance lithium-sulfur batteries. *Energy & Environmental Science* 6: 1283–1290.
3 Sun, L., Li, M., Jiang, Y. et al. (2014). Sulfur nanocrystals confined in carbon nanotube network as a binder-free electrode for high-performance lithium sulfur batteries. *Nano Letters* 14: 4044–4049.
4 Zheng, S.Y., Wen, Y., Zhu, Y.J. et al. (2014). In situ sulfur reduction and intercalation of graphite oxides for Li-S battery cathodes. *Advanced Energy Materials* 4: 1400482.
5 Qie, L. and Manthiram, A. (2016). High-energy-density lithium-sulfur batteries based on blade-cast pure sulfur electrodes. *ACS Energy Letters* 1: 46–51.
6 Zheng, S.Y., Yi, F., Li, Z.P. et al. (2014). Copper-stabilized sulfur-microporous carbon cathodes for Li-S batteries. *Advanced Functional Materials* 24: 4156–4163.
7 Wang, H., Yang, Y., Liang, Y. et al. (2011). Graphene-wrapped sulfur particles as a rechargeable lithium-sulfur battery cathode material with high capacity and cycling stability. *Nano Letters* 11: 2644–2647.
8 Kim, H., Lee, J., Ahn, H. et al. (2015). Synthesis of three-dimensionally interconnected sulfur-rich polymers for cathode materials of high-rate lithium-sulfur batteries. *Nature Communications* 6: 7278.
9 Chung, W.J., Griebel, J.J., Kim, E.T. et al. (2013). The use of elemental sulfur as an alternative feedstock for polymeric materials. *Nature Chemistry* 5: 518–524.
10 Li, B., Li, S.M., Xu, J.J. et al. (2016). A new configured lithiated silicon-sulfur battery built on 3D graphene with superior electrochemical performances. *Energy & Environmental Science* 9: 2025–2030.

11 Zhao, M.Q., Sedran, M., Ling, Z. et al. (2015). Synthesis of carbon/sulfur nanolaminates by electrochemical extraction of titanium from Ti_2SC. *Angewandte Chemie* 127: 4892–4896.

12 Zhao, Q., Hu, X., Zhang, K. et al. (2015). Sulfur nanodots electrodeposited on Ni foam as high-performance cathode for Li-S batteries. *Nano Letters* 15: 721–726.

13 Li, B., Li, S., Liu, J. et al. (2015). Vertically aligned sulfur-graphene nanowalls on substrates for ultrafast lithium-sulfur batteries. *Nano Letters* 15: 3073–3079.

14 Zhou, G.M., Wang, D.W., Li, F. et al. (2012). A flexible nanostructured sulphur-carbon nanotube cathode with high rate performance for Li-S batteries. *Energy & Environmental Science* 5: 8901–8906.

15 Ji, X., Lee, K.T., and Nazar, L.F. (2009). A highly ordered nanostructured carbon-sulphur cathode for lithium-sulphur batteries. *Nature Materials* 8: 500–506.

16 Han, S.C., Song, M.S., Lee, H. et al. (2003). Effect of multiwalled carbon nanotubes on electrochemical properties of lithium sulfur rechargeable batteries. *Journal of the Electrochemical Society* 150: A889–A893.

17 Yuan, L.X., Yuan, H.P., Qiu, X.P. et al. (2009). Improvement of cycle property of sulfur-coated multi-walled carbon nanotubes composite cathode for lithium/sulfur batteries. *Journal of Power Sources* 189: 1141–1146.

18 Tang, C., Zhang, Q., Zhao, M.Q. et al. (2014). Nitrogen-doped aligned carbon nanotube/graphene sandwiches: facile catalytic growth on bifunctional natural catalysts and their applications as scaffolds for high-rate lithium-sulfur batteries. *Advanced Materials* 26: 6100–6105.

19 Zhao, Y., Wu, W., Li, J. et al. (2014). Encapsulating MWNTs into hollow porous carbon nanotubes: a tube-in-tube carbon nanostructure for high-performance lithium-sulfur batteries. *Advanced Materials* 26: 5113–5118.

20 Li, H., Sun, L., and Wang, G. (2016). Self-assembly of polyethylene glycol-grafted carbon nanotube/sulfur composite with Nest-like structure for high-performance lithium-sulfur batteries. *ACS Applied Materials & Interfaces* 8: 6061–6071.

21 Guo, D., Ming, F., Su, H. et al. (2019). MXene based self-assembled cathode and antifouling separator for high-rate and dendrite-inhibited Li-S battery. *Nano Energy* 61: 478–485.

22 Jin, F., Xiao, S., Lu, L. et al. (2015). Efficient activation of high-loading sulfur by small CNTs confined inside a large CNT for high-capacity and high-rate lithium-sulfur batteries. *Nano Letters* 16: 440–447.

23 Li, M., Carter, R., Douglas, A. et al. (2017). Sulfur vapor-infiltrated 3D carbon nanotube foam for binder-free high areal capacity lithium-sulfur battery composite cathodes. *ACS Nano* 11: 4877–4884.

24 Milroy, C., Manthiram, A., and Milroy, C. (2016). Printed microelectrodes for scalable, high-areal-capacity lithium-sulfur batteries. *Chemical Communications* 52: 4282–4285.

25 Gao, X., Sun, Q., Yang, X. et al. (2019). Toward a remarkable Li-S battery via 3D printing. *Nano Energy* 56: 595–603.

26 Zheng, G., Yang, Y., Cha, J.J. et al. (2011). Hollow carbon nanofiber-encapsulated sulfur cathodes for high specific capacity rechargeable lithium batteries. *Nano Letters* 11: 4462–4467.

27 Zheng, G., Zhang, Q., Cha, J.J. et al. (2013). Amphiphilic surface modification of hollow carbon nanofibers for improved cycle life of lithium sulfur batteries. *Nano Letters* 13: 1265–1270.

28 Bian, Z.H., Xu, Y., Yuan, T. et al. (2019). Hierarchically designed CNF/S-Cu/CNF nonwoven electrode as free-standing cathode for lithium-sulfur batteries. *Batteries & Supercaps* 2: 560–567.

29 Lu, S., Cheng, Y., Wu, X. et al. (2013). Significantly improved long-cycle stability in high-rate Li-S batteries enabled by coaxial graphene wrapping over sulfur-coated carbon nanofibers. *Nano Letters* 13: 2485–2489.

30 Li, Q., Zhang, Z., Guo, Z. et al. (2014). Improved cyclability of lithium-sulfur battery cathode using encapsulated sulfur in hollow carbon nanofiber@nitrogen-doped porous carbon core-shell composite. *Carbon* 78: 1–9.

31 Han, S., Pu, X., Li, X. et al. (2017). High areal capacity of Li-S batteries enabled by freestanding CNF/rGO electrode with high loading of lithium polysulfide. *Electrochimica Acta* 241: 406–413.

32 Li, J., Zhang, H., Luo, L. et al. (2021). Blocking polysulfides with a Janus Fe_3C/N-CNF@RGO electrode via physiochemical confinement and catalytic conversion for high-performance lithium-sulfur batteries. *Journal of Materials Chemistry A* 9: 2205–2213.

33 Sun, H., Xu, G.L., Xu, Y.F. et al. (2012). A composite material of uniformly dispersed sulfur on reduced graphene oxide: aqueous one-pot synthesis, characterization and excellent performance as the cathode in rechargeable lithium-sulfur batteries. *Nano Research* 5: 726–738.

34 Jin, J., Wen, Z.Y., Ma, G.Q. et al. (2013). Flexible self-supporting graphene-sulfur paper for lithium sulfur batteries. *RSC Advances* 3: 2558–2560.

35 Wang, C., Wang, X.S., Wang, Y.J. et al. (2015). Macroporous free-standing nano-sulfur/reduced graphene oxide paper as stable cathode for lithium-sulfur battery. *Nano Energy* 11: 678–686.

36 Ji, L., Rao, M., Zheng, H. et al. (2011). Graphene oxide as a sulfur immobilizer in high performance lithium/sulfur cells. *Journal of the American Chemical Society* 133: 18522–18525.

37 Fang, R., Zhao, S., Pei, S. et al. (2016). Toward more reliable lithium-sulfur batteries: an all-graphene cathode structure. *ACS Nano* 10: 8676–8682.

38 Xie, Y., Meng, Z., Cai, T. et al. (2015). Effect of boron-doping on the graphene aerogel used as cathode for the lithium-sulfur battery. *ACS Applied Materials & Interfaces* 7: 25202–25210.

39 Li, L., Zhou, G., Yin, L. et al. (2016). Stabilizing sulfur cathodes using nitrogen-doped graphene as a chemical immobilizer for Li S batteries. *Carbon* 108: 120–126.

40 Yin, L.C., Liang, J., Zhou, G.M. et al. (2016). Understanding the interactions between lithium polysulfides and N-doped graphene using density functional theory calculations. *Nano Energy* 25: 203–210.

41 Hou, T.Z., Chen, X., Peng, H.J. et al. (2016). Design principles for heteroatom-doped nanocarbon to achieve strong anchoring of Polysulfides for lithium-sulfur batteries. *Small* 12: 3283–3291.

42 Yang, H.T., Wang, L., Geng, C.N. et al. (2024). Catalytic solid-state sulfur conversion confined in micropores toward superhigh coulombic efficiency lithium-sulfur batteries. *Advanced Energy Materials* 2400249.

43 Zhao, M.Q., Liu, X.F., Zhang, Q. et al. (2012). Graphene/single-walled carbon nanotube hybrids: one-step catalytic growth and applications for high-rate Li-S batteries. *ACS Nano* 6: 10759–10769.

44 Yang, X., Zhang, L., Zhang, F. et al. (2014). Sulfur-infiltrated graphene-based layered porous carbon cathodes for high-performance lithium-sulfur batteries. *ACS Nano* 8: 5208–5215.

45 Tian, X., Zhou, Y., Zhang, B. et al. (2022). Monodisperse polar $NiCo_2O_4$ nanoparticles decorated porous graphene aerogel for high-performance lithium sulfur battery. *Journal of Energy Chemistry* 74: 239–251.

46 Li, H., Meng, R., Ye, C. et al. (2024). Developing high-power Li||S batteries via transition metal/carbon nanocomposite electrocatalyst engineering. *Nature Nanotechnology* 19: 792–799.

47 Wang, J., Jia, L., Zhong, J. et al. (2019). Single-atom catalyst boosts electrochemical conversion reactions in batteries. *Energy Storage Materials* 18: 246–252.

48 Du, Z., Chen, X., Hu, W. et al. (2019). Cobalt in nitrogen-doped graphene as single-atom catalyst for high-sulfur content lithium-sulfur batteries. *Journal of the American Chemical Society* 141: 3977–3985.

49 Qin, F., Zhang, K., Zhang, Z. et al. (2019). Graphene/carbon aerogel for high areal capacity sulfur cathode of Li-S batteries. *Ionics* 25: 4615–4624.

50 He, J., Chen, Y., Lv, W. et al. (2016). Three-dimensional CNT/graphene-Li2S aerogel as freestanding cathode for high-performance Li-S batteries. *ACS Energy Letters* 1: 820–826.

51 Cavallo, C., Agostini, M., Genders, J.P. et al. (2019). A free-standing reduced graphene oxide aerogel as supporting electrode in a fluorine-free Li_2S_8 catholyte Li-S battery. *Journal of Power Sources* 416: 111–117.

52 Chen, C.L., Jiang, J.M., He, W.J. et al. (2020). 3D printed high-loading lithium-sulfur battery toward wearable energy storage. *Advanced Functional Materials* 30: 1909469.

53 Zhao, Q., Zhu, Q.Z., Liu, Y. et al. (2021). Status and prospects of MXene-based lithium-sulfur batteries. *Advanced Functional Materials* 31: 2100457.

54 Zhao, X., Liu, M., Chen, Y. et al. (2015). Fabrication of layered Ti_3C_2 with an accordion-like structure as a potential cathode material for high performance lithium-sulfur batteries. *Journal of Materials Chemistry A* 3: 7870–7876.

55 Zhang, F., Zhou, Y., Zhang, Y. et al. (2020). Facile synthesis of sulfur@titanium carbide Mxene as high performance cathode for lithium-sulfur batteries. *Nanophotonics* 9: 2025–2032.

56 Yao, Y., Feng, W., Chen, M. et al. (2018). Boosting the electrochemical performance of Li-S batteries with a dual polysulfides confinement strategy. *Small* 14: e1802516.

57 Tang, H., Li, W.L., Pan, L.M. et al. (2018). In situ formed protective barrier enabled by sulfur@titanium carbide (MXene) ink for achieving high-capacity, long lifetime Li-S batteries. *Advanced Science* 5: 1800502.

58 Liang, X., Garsuch, A., and Nazar, L.F. (2015). Sulfur cathodes based on conductive MXene nanosheets for high-performance lithium-sulfur batteries. *Angewandte Chemie (International Edition in English)* 54: 3907–3911.

59 Song, Y., Sun, Z., Fan, Z. et al. (2020). Rational design of porous nitrogen-doped Ti_3C_2 MXene as a multifunctional electrocatalyst for Li-S chemistry. *Nano Energy* 70: 104555.

60 Zhang, D., Wang, S., Hu, R.M. et al. (2020). Catalytic conversion of polysulfides on single atom zinc implanted MXene toward high-rate lithium-sulfur batteries. *Advanced Functional Materials* 30: 2002471.

61 Dong, Y., Zheng, S., Qin, J. et al. (2018). All-MXene-based integrated electrode constructed by Ti(3)C(2) nanoribbon framework host and Nanosheet interlayer for high-energy-density Li-S batteries. *ACS Nano* 12: 2381–2388.

62 Chen, Z., Yang, X., Qiao, X. et al. (2020). Lithium-ion-engineered interlayers of V_2C MXene as advanced host for flexible sulfur cathode with enhanced rate performance. *Journal of Physical Chemistry Letters* 11: 885–890.

63 Song, J., Guo, X., Zhang, J. et al. (2019). Rational design of free-standing 3D porous MXene/rGO hybrid aerogels as polysulfide reservoirs for high-energy lithium-sulfur batteries. *Journal of Materials Chemistry A* 7: 6507–6513.

64 Balach, J., Jaumann, T., Klose, M. et al. (2015). Functional mesoporous carbon-coated separator for long-life, high-energy lithium-sulfur batteries. *Advanced Functional Materials* 25: 5285–5291.

65 Chung, S.H. and Manthiram, A. (2014). Bifunctional separator with a light-weight carbon-coating for dynamically and statically stable lithium-sulfur batteries. *Advanced Functional Materials* 24: 5299–5306.

66 Zeng, F.L., Jin, Z.Q., Yuan, K.G. et al. (2016). High performance lithium-sulfur batteries with a permselective sulfonated acetylene black modified separator. *Journal of Materials Chemistry A* 4: 12319–12327.

67 Chung, S.H. and Manthiram, A. (2014). A polyethylene glycol-supported microporous carbon coating as a polysulfide trap for utilizing pure sulfur cathodes in lithium-sulfur batteries. *Advanced Materials* 26: 7352–7357.

68 Zheng, B.B., Yu, L.W., Zhao, Y. et al. (2019). Ultralight carbon flakes modified separator as an effective polysulfide barrier for lithium-sulfur batteries. *Electrochimica Acta* 295: 910–917.

69 Balach, J., Singh, H.K., Gomoll, S. et al. (2016). Synergistically enhanced polysulfide chemisorption using a flexible hybrid separator with N and S dual-doped mesoporous carbon coating for advanced lithium-sulfur batteries. *ACS Applied Materials & Interfaces* 8: 14586–14595.

70 Pei, F., Lin, L.L., Fu, A. et al. (2018). A two-dimensional porous carbon-modified separator for high-energy-density Li-S batteries. *Joule* 2: 323–336.

71 Liu, Y.D., Liu, Q., Xin, L. et al. (2017). Making Li-metal electrodes rechargeable by controlling the dendrite growth direction. *Nature Energy* 2.

72 Chung, S.H. and Manthiram, A. (2014). High-performance Li-S batteries with an ultra-lightweight MWCNT-coated separator. *Journal of Physical Chemistry Letters* 5: 1978–1983.

73 Chang, C.-H., Chung, S.-H., and Manthiram, A. (2016). Effective stabilization of a high-loading sulfur cathode and a lithium-metal anode in Li-S batteries utilizing SWCNT-modulated separators. *Small* 12: 174–179.

74 Zhu, J.D., Ge, Y.Q., Kim, D. et al. (2016). A novel separator coated by carbon for achieving exceptional high performance lithium-sulfur batteries. *Nano Energy* 20: 176–184.

75 Cheng, X.B., Hou, T.Z., Zhang, R. et al. (2016). Dendrite-free lithium deposition induced by uniformly distributed lithium ions for efficient lithium metal batteries. *Advanced Materials* 28: 2888–2895.

76 Bai, S.Y., Liu, X.Z., Zhu, K. et al. (2016). Metal-organic framework-based separator for lithium-sulfur batteries. *Nature Energy* 1: 16094.

77 Dai, X., Lv, G., Wu, Z. et al. (2023). Flexible hierarchical co-doped NiS$_2$@CNF-CNT electron deficient interlayer with grass-roots structure for Li-S batteries. *Advanced Energy Materials* 13: 2300452.

78 Lin, J.-X., Qu, X.-M., Wu, X.-H. et al. (2021). NiCo$_2$O$_4$/CNF separator modifiers for trapping and catalyzing polysulfides for high-performance lithium-sulfur batteries with high sulfur loadings and lean electrolytes. *ACS Sustainable Chemistry & Engineering* 9: 1804–1813.

79 Huang, J.Q., Zhang, Q., Peng, H.J. et al. (2014). Ionic shield for polysulfides towards highly-stable lithium-sulfur batteries. *Energy & Environmental Science* 7: 347–353.

80 Hao, Z.X., Yuan, L.X., Li, Z. et al. (2016). High performance lithium-sulfur batteries with a facile and effective dual functional separator. *Electrochimica Acta* 200: 197–203.

81 Cai, W.L., Li, G.R., He, F. et al. (2015). A novel laminated separator with multi functions for high-rate dischargeable lithium-sulfur batteries. *Journal of Power Sources* 283: 524–529.

82 Lu, Y., Gu, S., Guo, J. et al. (2017). Sulfonic groups originated dual-functional interlayer for high performance lithium-sulfur battery. *ACS Applied Materials & Interfaces* 9: 14878–14888.

83 Lei, T.Y., Chen, W., Lv, W.Q. et al. (2018). Inhibiting polysulfide shuttling with a graphene composite separator for highly robust lithium-sulfur batteries. *Joule* 2: 2091–2104.

84 Jin, Z.Q., Xie, K., and Hong, X.B. (2013). Electrochemical performance of lithium/sulfur batteries using perfluorinated ionomer electrolyte with lithium sulfonyl dicyanomethide functional groups as functional separator. *RSC Advances* 3: 8889–8898.

85 Conder, J., Urbonaite, S., Streich, D. et al. (2015). Taming the polysulphide shuttle in Li-S batteries by plasma-induced asymmetric functionalisation of the separator. *RSC Advances* 5: 79654–79660.

86 Yim, T., Han, S.H., Park, N.H. et al. (2016). Effective polysulfide rejection by dipole-aligned BaTiO$_3$ coated separator in lithium-sulfur batteries. *Advanced Functional Materials* 26: 7817–7823.

87 Zhou, G., Pei, S., Li, L. et al. (2014). A graphene-pure-sulfur sandwich structure for ultrafast, long-life lithium-sulfur batteries. *Advanced Materials* 26: 625–631. 664.

88 Sun, J., Sun, Y., Pasta, M. et al. (2016). Entrapment of Polysulfides by a black-phosphorus-modified separator for lithium-sulfur batteries. *Advanced Materials* 28: 9797–9803.

89 Zhou, W., Chen, M., Zhao, D. et al. (2024). Engineering spin state in spinel Co$_3$O$_4$ nanosheets by V-doping for bidirectional catalysis of polysulfides in lithium-sulfur batteries. *Advanced Functional Materials* 34: 2402114.

90 Huang, J.Q., Zhuang, T.Z., Zhang, Q. et al. (2015). Permselective graphene oxide membrane for highly stable and anti-self-discharge lithium-sulfur batteries. *ACS Nano* 9: 3002–3011.

91 Zhuang, T.Z., Huang, J.Q., Peng, H.J. et al. (2016). Rational integration of polypropylene/graphene oxide/nafion as ternary-layered separator to retard the shuttle of polysulfides for lithium-sulfur batteries. *Small* 12: 381–389.

92 Zhang, Y.B., Miao, L.X., Ning, J. et al. (2015). A graphene-oxide-based thin coating on the separator: an efficient barrier towards high-stable lithium-sulfur batteries. *2D Materials* 2: 024013.

93 Song, J., Su, D., Xie, X. et al. (2016). Immobilizing polysulfides with MXene-functionalized separators for stable lithium-sulfur batteries. *ACS Applied Materials & Interfaces* 8: 29427–29433.

94 Xiao, Z., Yang, Z., Wang, L. et al. (2015). A lightweight TiO(2)/graphene interlayer, applied as a highly effective polysulfide absorbent for fast, long-life lithium-sulfur batteries. *Advanced Materials* 27: 2891–2898.

95 Zhou, G., Li, L., Wang, D.W. et al. (2015). A flexible sulfur-graphene-polypropylene separator integrated electrode for advanced Li-S batteries. *Advanced Materials* 27: 641–647.

96 Song, R.S., Fang, R.P., Wen, L. et al. (2016). A trilayer separator with dual function for high performance lithium-sulfur batteries. *Journal of Power Sources* 301: 179–186.

97 Yu, B.C., Park, K., Jang, J.H. et al. (2016). Cellulose-based porous membrane for suppressing Li dendrite formation in lithium-sulfur battery. *ACS Energy Letters* 1: 633–637.

98 Ye, M.H., Xiao, Y.K., Cheng, Z.H. et al. (2018). A smart, anti-piercing and eliminating-dendrite lithium metal battery. *Nano Energy* 49: 403–410.

99 Liu, K., Zhuo, D., Lee, H.W. et al. (2017). Extending the life of lithium-based rechargeable batteries by reaction of lithium dendrites with a novel silica nanoparticle sandwiched separator. *Advanced Materials* 29: 1603987.

100 Lv, X., Lei, T., Wang, B. et al. (2019). An efficient separator with low Li-ion diffusion energy barrier resolving feeble conductivity for practical lithium-sulfur batteries. *Advanced Energy Materials* 9: 1901800.

101 Zhang, K., Chen, Z., Ning, R. et al. (2019). Single-atom coated separator for robust lithium-sulfur batteries. *ACS Applied Materials & Interfaces* 11: 25147–25154.

102 Yuan, C., Yang, X., Zeng, P. et al. (2021). Recent progress of functional separators with catalytic effects for high-performance lithium-sulfur batteries. *Nano Energy* 84: 105928.

103 Barchasz, C., Mesguich, F., Dijon, J. et al. (2012). Novel positive electrode architecture for rechargeable lithium/sulfur batteries. *Journal of Power Sources* 211: 19–26.

104 Peng, H.J., Xu, W.T., Zhu, L. et al. (2016). 3D carbonaceous current collectors: the origin of enhanced cycling stability for high-sulfur-loading lithium-sulfur batteries. *Advanced Functional Materials* 26: 6351–6358.

105 Razzaq, A.A., Yao, Y.Z., Shah, R. et al. (2019). High-performance lithium sulfur batteries enabled by a synergy between sulfur and carbon nanotubes. *Energy Storage Materials* 16: 194–202.

106 Chung, S.H. and Manthiram, A. (2013). Nano-cellular carbon current collectors with stable cyclability for Li-S batteries. *Journal of Materials Chemistry A* 1: 9590–9596.

107 Zhang, W., Zhang, J.F., Zhao, Y. et al. (2019). Multi-functional carbon cloth infused with N-doped and co-coated carbon nanofibers as a current collector for ultra-stable lithium-sulfur batteries. *Materials Letters* 255: 126595.

108 Wang, Z.H., Pan, R.J., Sun, R. et al. (2018). Nanocellulose structured paper-based lithium metal batteries. *ACS Applied Energy Materials* 1: 4341–4350.

109 Lu, L.Q., De Hosson, J.T.M., and Pei, Y.T. (2019). Three-dimensional micron-porous graphene foams for lightweight current collectors of lithium-sulfur batteries. *Carbon* 144: 713–723.

110 Zhang, R., Chen, X.R., Chen, X. et al. (2017). Lithiophilic sites in doped graphene guide uniform lithium nucleation for dendrite-free lithium metal anodes. *Angewandte Chemie (International Edition in English)* 56: 7764–7768.

111 Huang, J.Q., Zhai, P.Y., Peng, H.J. et al. (2017). Metal/nanocarbon layer current collectors enhanced energy efficiency in lithium-sulfur batteries. *Science Bulletin (Beijing)* 62: 1267–1274.

112 Cheng, X.B., Peng, H.J., Huang, J.Q. et al. (2014). Three-dimensional aluminum foam/carbon nanotube scaffolds as long- and short-range electron pathways with improved sulfur loading for high energy density lithium sulfur batteries. *Journal of Power Sources* 261: 264–270.

113 Wood, K.N., Noked, M., and Dasgupta, N.P. (2017). Lithium metal anodes: toward an improved understanding of coupled morphological, electrochemical, and mechanical behavior. *ACS Energy Letters* 2: 664–672.

114 Qie, L., Zu, C.X., and Manthiram, A. (2016). A high energy lithium-sulfur battery with ultrahigh-loading lithium polysulfide cathode and its failure mechanism. *Advanced Energy Materials* 6: 1502459.

115 Mukherjee, R., Thomas, A.V., Datta, D. et al. (2014). Defect-induced plating of lithium metal within porous graphene networks. *Nature Communications* 5: 3710.

116 Lin, D., Liu, Y., Liang, Z. et al. (2016). Layered reduced graphene oxide with nanoscale interlayer gaps as a stable host for lithium metal anodes. *Nature Nanotechnology* 11: 626–632.

117 Jin, S., Xin, S., Wang, L. et al. (2016). Covalently connected carbon nanostructures for current collectors in both the cathode and anode of Li-S batteries. *Advanced Materials* 28: 9094–9102.

118 Li, B., Zhang, D., Liu, Y. et al. (2017). Flexible Ti_3C_2 MXene-lithium film with lamellar structure for ultrastable metallic lithium anodes. *Nano Energy* 39: 654–661.

119 Cao, Z., Zhu, Q., Wang, S. et al. (2019). Perpendicular MXene arrays with periodic interspaces toward dendrite-free lithium metal anodes with high-rate capabilities. *Advanced Functional Materials* 30: 1908075.

120 Zhang, X., Lv, R., Wang, A. et al. (2018). MXene aerogel scaffolds for high-rate lithium metal anodes. *Angewandte Chemie (International Edition in English)* 57: 15028–15033.

121 Shen, K., Li, B., and Yang, S. (2020). 3D printing dendrite-free lithium anodes based on the nucleated MXene arrays. *Energy Storage Materials* 24: 670–675.

122 Lyu, Z., Lim, G.J.H., Guo, R. et al. (2020). 3D-printed electrodes for lithium metal batteries with high areal capacity and high-rate capability. *Energy Storage Materials* 24: 336–342.

7

Applications of Porous Organic Framework Materials in Batteries

7.1 Introduction

7.1.1 Overview of Energy Demand and Battery Technologies

Energy is the foundation of modern society's survival and development. With the growth of the global economy and population, the demand for energy is continuously increasing. Traditional fossil fuels (such as coal, oil, and natural gas) have long been the main sources of energy. However, their limited reserves and environmental damage make seeking renewable and clean energy an inevitable choice. In recent years, renewable energy technologies such as solar, wind, and hydro have developed rapidly, but these energies have intermittency and instability, requiring efficient energy-storage systems to ensure a stable energy supply. Batteries, as an important energy-storage technology, are not only widely used in portable electronic devices, transportation (such as electric vehicles), and grid energy-storage systems, but also play a significant role in enhancing renewable energy utilization.

Battery technology has undergone a long development journey, from early lead-acid and nickel-cadmium batteries to the widely used lithium-ion batteries (LIBs) today. Each technological innovation has greatly advanced electronic devices and energy-storage systems. To meet the ever-growing energy demands and the limitations of traditional battery technology, scientists are actively exploring and developing new battery technologies, including solid-state batteries (SSBs), lithium-sulfur (Li-S) batteries, and lithium-air batteries. In recent years, framework materials, due to their unique structure and excellent properties, have shown broad application prospects in battery technology, providing new opportunities for further enhancement. This article will summarize the recent applications of framework materials in batteries and forecast future development trends.

7.1.2 Limitations of Traditional Battery Material

Although traditional battery materials, such as graphite, metal oxides, and sulfides, have been widely applied in the energy-storage field and achieved significant accomplishments in certain areas, their inherent limitations restrict further enhancement of battery performance. Firstly, traditional materials have low energy

Functional Auxiliary Materials in Batteries: Synthesis, Properties, and Applications, First Edition. Wei Hu.
© 2025 WILEY-VCH GmbH. Published 2025 by WILEY-VCH GmbH.

density, making it difficult to meet the increasingly growing energy demands. With the rapid development of electric vehicles and portable electronic devices, the demand for high-energy-density batteries has become particularly urgent. However, existing traditional materials, due to their structural properties, limit the energy density of batteries, making it difficult to overcome this bottleneck.

Secondly, the cycle life and stability issues of traditional battery materials also become important factors limiting their application. As the battery undergoes repeated charging and discharging in practical applications, electrode materials are prone to structural changes and performance degradation, leading to a shortened battery lifespan. Additionally, traditional materials may form dendrites or undergo volume expansion during battery charging and discharging, further affecting battery safety and stability.

Furthermore, traditional battery materials are costly, and involve complex manufacturing processes, and some key materials, like cobalt, face supply chain issues. This not only increases the production cost of batteries but also adversely affects the environment. To achieve broader application and lower costs, developing new battery materials to replace traditional ones has become a focal point of current research.

7.1.3 Potential of Porous Organic Framework Materials for Energy Storage

Porous Organic Frameworks (POFs) have shown great application potential, becoming an emerging research direction in the field of energy storage. POFs perform exceptionally well in the energy-storage field owing to their unique structural advantages. Firstly, POFs possess highly tunable porous structures and large surface areas, enabling them to offer more active sites to effectively enhance battery energy density. Additionally, the organic molecular framework of POF materials imparts excellent flexibility and lightweight characteristics, enabling the design of lighter batteries without sacrificing energy density, which is particularly crucial for applications like portable electronic devices and electric vehicles.

Secondly, the conductivity of POF materials can be optimized through chemical modification and structural design, thereby significantly enhancing the electrochemical performance of electrode materials. By introducing conductive functional groups or doping with other functional materials, POF materials can achieve excellent charge transport capabilities, thus improving battery power density and cycle stability. These characteristics make POFs show great application potential in energy-storage devices such as supercapacitors, LIBs, and sodium-ion batteries (SIBs).

Additionally, the synthesis methods for POF materials are diverse and environmentally friendly, allowing for functionalization through molecular design to meet the needs of different energy-storage devices. This not only helps enhance the overall performance of batteries but also provides possibilities for efficient, low-cost, large-scale production in the future. Compared to traditional inorganic materials, POF materials also have significant advantages in terms of environmental friendliness and sustainability, giving them an important and undeniable position in the development of future energy-storage technologies.

7.2 Types of Porous Organic Framework Materials

Framework materials are a class of materials with highly ordered and tunable porous structures, mainly divided into Metal-Organic Frameworks (MOFs), Covalent Organic Frameworks (COFs), Hydrogen-Bonded Organic Frameworks (HOFs), and coordination polymers based on their composition and construction.

7.2.1 Metal-Organic Frameworks (MOFs)

MOFs are a class of porous crystalline materials formed by metal ions or metal clusters connected to organic ligands through coordination bonds [1]. They have garnered widespread attention in the energy-storage field due to their tunable structures, high porosity, and large specific surface area. MOF materials have enormous potential for application in batteries, particularly demonstrating excellent performance in energy-storage devices such as LIBs, SIBs, and supercapacitors.

Firstly, the tunable structure of MOF materials allows for precise control over the pore size and shape of the material by selecting different metal ions and organic ligands. This characteristic allows MOF materials to optimize the ion transport paths in electrode materials, reduce ion diffusion resistance, and thereby enhance the rate performance and energy density of batteries. For example, in LIBs, the high porosity of MOF materials can effectively alleviate the volume changes of electrode materials during the charge and discharge processes, preventing pulverization and structural collapse, and thereby extending the cycle life of the battery.

Secondly, MOF materials have excellent conductivity and electrochemical activity, and through reasonable structural design and functional modification, their performance in batteries can be further enhanced. For example, by compositing conductive polymers or carbon-based materials with MOF materials, their conductivity and cycling stability can be significantly improved. These composite materials not only demonstrate higher energy density in batteries but also improve the power density and cycling performance of the batteries.

Additionally, the multifunctionality and structural diversity of MOF materials make them highly promising for a wide range of applications in the energy-storage field. Besides serving as electrode materials, MOF materials can also be used in the design of electrolytes and separator materials, further enhancing the overall performance of batteries. For example, the high porosity and tunability of MOF materials make them excellent electrolyte carriers, capable of increasing ion conductivity, reducing the battery's internal resistance, and thereby improving energy efficiency.

In summary, MOF materials have broad application prospects in batteries. With their unique structural advantages and excellent electrochemical performance, MOF materials can enhance the performance of existing battery technologies and offer new insights for the development of novel energy-storage devices. With continuous advancements in the synthesis technology of MOF materials and in-depth application research, the status of MOF materials in future energy-storage technologies will further improve, propelling battery technology to new heights.

7.2.1.1 Types of MOFs

Based on the types of organic ligands used to construct MOFs, these materials mainly include MOFs constructed with carboxylic acid ligands and MOFs constructed with nitrogen-containing heterocyclic ligands.

Carboxylate MOFs Carboxylate MOFs are the earliest researched, fastest developed, and most diverse type of MOF materials. The organic ligands used in constructing carboxylate MOF materials are usually various types of aromatic carboxylic acid ligands, such as terephthalic acid and trimesic acid. These ligands can combine with transition metal (TM) ions or ion clusters to form various types of MOFs, mainly including the IRMOF series such as MOF-5, HKUST-1, UiO series, CAU series, and MIL series.

Nitrogen-Containing Heterocyclic MOFs Nitrogen-containing heterocyclic ligands are another common type of organic ligands used for MOF construction. Deprotonated azole molecules form anionic ligands that can coordinate with metals, with nitrogen atoms in the ligand participating in coordination via sp^2 hybridization. The number of nitrogen atoms in the ligands, their arrangement in the five-membered ring, and the hydrogen atom sites on the heterocycle that can be substituted with functional groups provide rich tunability. In 2006, Yaghi's research group [2] reported 12 types of porous crystalline materials formed by coordinating imidazole-based ligands with metals such as Zn or Co, modified with different functional groups. Because the coordination bond angle between the imidazole ligands and metal ion centers was very similar to the Al—O—Al or Si—O—Si bond angles in aluminosilicate zeolites, Yaghi's group named these materials ZIFs (zeolitic imidazolate frameworks).

7.2.2 Covalent Organic Frameworks (COFs)

COF is a type of organic porous material formed by covalent bonds, characterized by a highly ordered channel structure and adjustable framework composition [3]. Due to the synthetic flexibility and structural diversity of COF materials, their application potential in the energy-storage field, especially in battery technology, is gradually emerging. Compared to MOF, the organic composition of COF materials gives them higher chemical stability and structural rigidity, making them perform better under high-pressure conditions.

The greatest feature of COF materials is their completely organic framework, which endows them with excellent tunability and functionalization potential. By precisely designing the structure of organic monomers, researchers can synthesize COF materials with specific pore channel structures, sizes, and surface functional groups. This customization capability enables COF materials to more effectively enhance the ion conductivity and charge storage capacity of electrode materials in energy-storage devices. For example, in SIBs, COF materials can achieve higher energy-storage efficiency and longer cycle life by optimizing their pore size and surface chemical properties.

7.2.2.1 Types of COFs

Currently, common linkage types in COFs include boronate esters, imine bonds, thiazoles, imidazoles, C—C bonds, and sp^2-carbon linkages. The elements that mainly constitute COFs are C, H, O, N, B, and S. Based on spatial configuration, COFs can be categorized into two-dimensional covalent organic frameworks (2D COFs) and three-dimensional covalent organic frameworks (3D COFs). 2D COFs have a honeycomb-like planar structure similar to that of graphite, while 3D COFs possess a spatial network structure akin to the tetrahedral shape of the diamond. By replacing the carbon atoms in graphite and diamond with molecules of geometric configurations connected via covalent bonds, one can obtain typical 2D COF and 3D COF structures.

7.2.3 Hydrogen-Bonded Organic Frameworks (HOFs)

Additionally, COF materials, due to the flexibility of their organic frameworks, can effectively buffer the volumetric changes of electrode materials during charge and discharge cycles, thus reducing mechanical stress and structural degradation. This characteristic gives COF significant advantages in enhancing battery life and stability, especially in energy-storage devices that require long-term high-performance operation.

Besides MOF and COF materials, there are other types of organic framework materials within the POFs family that also show broad application prospects in battery technology and the energy-storage field. These materials not only compensate for certain limitations of MOF and COF but also offer more diverse properties and functions, providing new insights for innovations in battery technology.

HOFs are a noteworthy type of material, characterized as crystalline molecular frameworks with permanent porosity assembled via hydrogen bonding. This class of porous materials offers unique advantages. HOFs form stable 3D network structures through intermolecular hydrogen bonds, and this weak interaction endows HOF materials with enhanced processability and tunability, as exemplified in Figure 7.1 [4]. HOFs' porous structures can be adjusted through simple post-processing methods, making them promising candidates in the field of energy storage, particularly in supercapacitors and SIBs. Compared to MOFs and COFs, HOF materials also exhibit lower synthesis costs and higher environmental friendliness, providing them with a greater advantage for large-scale applications.

7.2.3.1 Types of HOF

HOFs are a type of crystalline POFs material assembled by hydrogen bonds connecting organic–organic or metal–organic building units, often supported by other weak interactions such as π–π interactions, electrostatic forces, and van der Waals forces. HOFs can be simply divided into two parts: the first part is the intermediary

344 | *7 Applications of Porous Organic Framework Materials in Batteries*

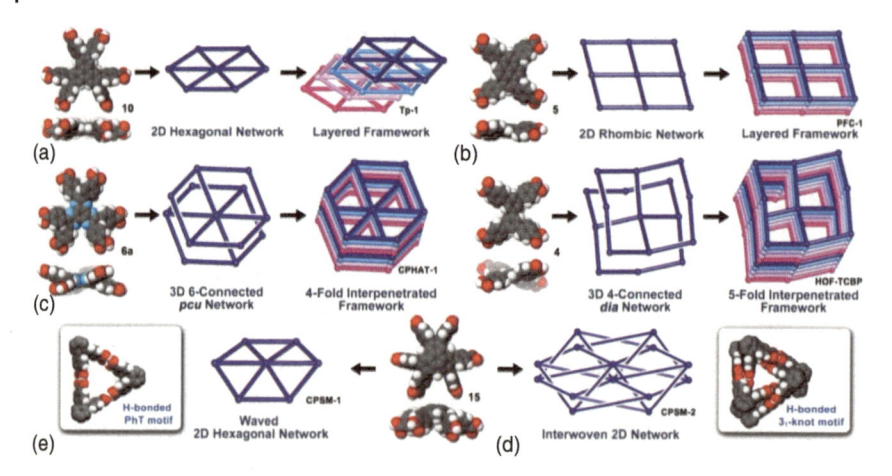

Figure 7.1 Representative examples of the relationship between molecular conformation and network topology of HOFs. Planar (a) C_3- and (b) C_2-symmetric tectons gave layered frameworks. Twisted (c) C_3- and (d) C_2-symmetric tectons gave interpenetrated frameworks. (e) C_3-symmetric bowl-shaped tecton yielded both waved 2D-hexagonal and 2D-interwoven networks via H-bonded PhT or 3_1-knot motif. Source: Hisaki et al. [4]/with permission of John Wiley & Sons.

structural framework (the core part), and the second part is the hydrogen-bond donor (the pendant part). The core structure and the peripheral hydrogen-bond donor significantly influence the stability of the resulting HOF. Different combinations not only affect the stability of the HOFs but also their overall performance, highlighting the essential structure-performance relationship in these materials.

Diaminotriazine Series Before 2011, structures of HOFs with permanent porosity had not been established. It wasn't until Chen Banglin's research group [5] successfully synthesized a stable HOF, HOF-1, using tetraphenylmethane as the intermediary structural framework and diaminotriazine as the peripheral hydrogen-bond donor building units. This was achieved by allowing the vapor diffusion of 1,4-dioxane solution into a formic acid solution containing the ligands.

Through studying the activation methods, HOF-1 was first vacuum dried at room temperature for 24 hours, followed by vacuum drying at 100 °C for another 24 hours, resulting in the first instance of a HOF, HOF-1, with permanent porosity.

Carboxylic Acid Dimers Carboxyl ligands are among the most common ligands used in constructing framework materials, due to their strong hydrogen bonding capability as well as their strong directional nature. As a result, they are frequently used in building porous HOF materials. Currently, carboxyl-based HOF materials dominate the reported HOFs and have garnered considerable attention from researchers.

Other Hydrogen Bond Donors Pyrazole monomers can form pyrazole trimers through N—H···N hydrogen bonding, creating planar triangular structures. HOFs constructed in this manner can also form stable framework structures. Additionally, amide groups are widely used as hydrogen bond donors in the construction of HOF materials.

In summary, besides MOF, COF, and HOF organic framework materials also hold significant application potential in the energy-storage field. These materials offer diverse choices for enhancing battery technologies through their unique structures and properties. Not only do these materials exhibit outstanding performance in energy density, cycle life, and stability, but they also have clear advantages in terms of environmental friendliness and low-cost fabrication. In the future, as material science further progresses, these diverse organic framework materials are expected to play more crucial roles in the energy-storage field, strongly supporting the realization of more efficient and environmentally friendly energy-storage technologies.

7.3 Applications of Porous Organic Framework Materials in Batteries

With the continuous rise in energy demand and the advancement of energy-storage technologies, POFs, owing to their unique structural advantages and superior electrochemical performance, display extensive application potential in the battery industry. In the previous sections, we discussed the unique characteristics of POFs and their potential in the energy-storage field. In this chapter, we will further explore the practical applications of organic framework materials in specific battery components. Whether used as electrode materials or in electrolytes, catalyst supports, and battery separators, these materials not only play a crucial role in enhancing battery performance but also drive the innovation and development of next-generation efficient and environmentally friendly energy-storage technologies.

7.3.1 Applications in Electrode Materials

Electrode materials are critical to battery performance, directly affecting the energy density, cycle life, and rate performance of the battery. POFs, with their high specific surface area, adjustable pore structures, and excellent conductivity, are gradually becoming the ideal choice for next-generation electrode materials. In the field of electrode materials, POFs can provide more active sites and faster ion transport pathways while effectively mitigating the volume expansion and structural degradation problems faced by traditional electrode materials during charge-discharge cycles. Therefore, exploring and optimizing the application of POFs in electrode materials has become one of the key directions for enhancing overall battery performance.

7.3.1.1 MOF as Electrode Materials

As a conventional anode material, graphite has inherent drawbacks of low theoretical specific capacity and operating voltage, which can easily lead to the formation of lithium dendrites and the risk of electrolyte combustion. To address this issue, Sarakonsri et al. [6] prepared a series of isostructural dual-ligand MOFs using the microwave heating method, where Co-MOF exhibited higher efficiency compared to its Mn and Zn counterparts, with an excellent specific capacity (732 mAh g^{-1} after 200 cycles). Moreover, through first-principle calculations, it was revealed that lithium-ions diffuse along the metal centers rather than through the pores in MOF, highlighting the importance of metal centers in MOF materials in enhancing LIB performance.

Correspondingly, the uneven magnesium electrodeposition in hybrid lithium batteries poses a significant barrier to practical applications. In addressing this issue, Fu et al. [7] utilized the prominently exposed magnesium-philic centers in the MgMOF substrate and the very low lattice mismatch rate with Mg (0.93%) to achieve epitaxial electrocrystallization driven by the MgMOF@PPy@CC electrode. By employing a periodic electric field and the spatial confinement effect exerted by the nanoscale channels within the MgMOF, they successfully induced uniform nucleation of Mg at the molecular scale and restricted its growth. As a result, stability was achieved for 1200 hours at 10 mA cm^{-2}. The most groundbreaking aspect is that the battery assembled with the MgMOF@PPy@CC anode and Mo$_6$S$_8$ cathode demonstrated an ultra-long operational lifespan with a capacity retention rate of 96.23% after over 10,000 cycles.

Another important strategy to solve the electrochemical kinetics problems of electrode materials is to introduce heterostructures with interfacial effects. Chen et al. [8] proposed a multi-template synthesis strategy to design and fabricate a NiS/SnO$_2$/MOF (NSM) heterostructure as a high-performance electrode for LIBs, where the SnO$_2$-NiS heterostructure creates a strain buffer and enhances Li$^+$ storage activity (Figure 7.2). Additionally, introducing the ion channels and framework structures of MOF into the electrode can enhance reaction kinetics and suppress electrode degradation compared to NS electrodes without MOF. By introducing MOF prepared with hydroxyl-containing organic ligands, strong hydrogen bonds can be formed with poly-oxyethylene, and the two exhibit good compatibility. The NSM electrode shows highly repeatable charge-discharge curves and achieves a high specific capacity of over 1000 mAh g^{-1}, whereas the NS electrode experiences irreversible damage within a limited operating time, indicating that hydrogen bonds enhance the interfacial stability between NSM electrodes and polymer electrolytes, and increase the Li$^+$ migration rate.

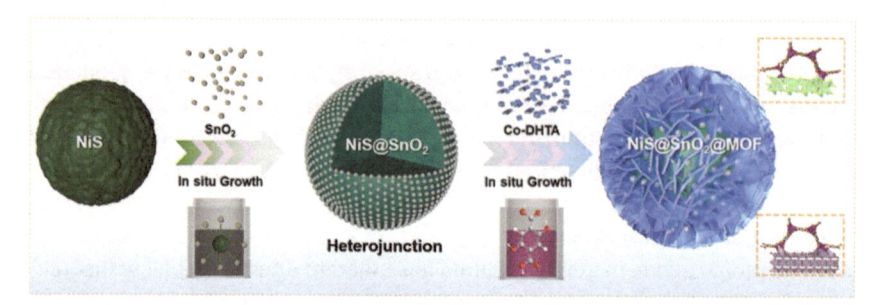

Figure 7.2 Schematic illustration of the synthetic route of NSM composite. Source: Zhang et al. [8]/with permission of John Wiley & Sons.

Zhang et al. [9] investigated the application of MOFs in LIBs, particularly focusing on their composite materials' role and mechanisms in enhancing battery performance. They developed nitrogen-doped porous carbon nanofibers (PCNFs) derived from polyacrylonitrile (PAN) using electrospinning and subsequent hydrothermal treatment, which were then combined with molybdenum disulfide (MoS$_2$) nanosheets to form the composite material (PCNF@MoS$_2$). The incorporation of MOF, particularly the assembly of MOF nanoparticles (such as ZIF-8) within the PAN fibers during the electrospinning process, not only created a porous nanostructure but also preserved a high specific surface area and good conductivity during the subsequent thermal treatment. The PCNF@MoS$_2$ composite exhibited outstanding lithium storage performance. The structural characteristics of the MOF provided an ideal growth template for MoS$_2$, ensuring a uniform and vertical alignment of the MoS$_2$ nanosheets on the PCNF surface, which enhanced the battery's reaction kinetics and conductivity. By optimizing the precursor amount of MoS$_2$, the researchers effectively controlled the density of the MoS$_2$ nanosheets, preventing material agglomeration and thereby enhancing the battery's cycling stability. Furthermore, PCNF, serving as a supporting nanostructure, mitigated reorganization and stacking phenomena, ensuring that active sites remained fully exposed, thus facilitating rapid lithium-ion diffusion and effective electron transfer. Additionally, the nitrogen doping of PCNF improved the material's conductivity and electrochemical activity, further enhancing battery performance. The excellent electrochemical performance of PCNF@MoS$_2$ was validated through cyclic voltammetry (CV), galvanostatic charge-discharge tests, and electrochemical impedance spectroscopy. The composite demonstrated a high initial reversible capacity during charge-discharge cycles and maintained a capacity of 1116.2 mAh g^{-1} after 450 cycles, demonstrating exceptional cycling stability. In contrast, the electrochemical performance of MoS$_2$ alone was significantly weaker, which underscores the synergistic interaction between PCNF and MoS$_2$ as a key factor in enhancing battery performance. The application of MOF in LIBs, through the formation of a supportive framework and the provision of active sites, effectively enhances the energy-storage capabilities of MoS$_2$. This design concept for MOF-derived materials offers new insights into battery technology development, particularly in improving cycling life and charge-discharge rates, showing promising practical applications. This research suggests that the potential of MOFs extends beyond LIBs, possibly influencing other energy storage and conversion fields, thereby showcasing their promise for the development of high-performance batteries.

Potassium-ion batteries (KIBs) experience severe volume expansion at the insertion point due to the larger diameter (2.76 Å) and greater mass of potassium-ions compared to lithium-ions (1.52 Å). At high temperatures, the chemical and mechanical instability of the electrodes is further amplified, often leading to severe volume expansion and interfacial instability. Li et al. [10] developed a HAN-Cu-MOF (HAN = hexaazatrinaphthalene) composed of nitrogen-rich aromatic molecules and π-d conjugated CuO$_4$, using it as a KIB anode. Calculations have shown that the large π-conjugated C=N groups in HAN can serve as redox-active sites for K ion storage, and the hybridization of its π orbitals with the d orbitals of metal nodes is expected to enhance the delocalized electron distribution of the structure. This not only enhances the electronic conductivity of the electrode but also achieves chemical stability of the MOF electrode by controlling the arrangement of the electron cloud. A porous networked HAN-Cu-MOF anode was obtained, characterized by numerous redox-active sites that facilitate ion transport and ensure the stability of KIB anodes at high temperatures. When assembled into a battery and operated at 60°C, this MOF anode delivers a reversible capacity of 455 mAh g^{-1} at 50 mA g^{-1}. After 1600 cycles (initial 1000 mA g^{-1}), the capacity retention remains at approximately 96.7% (with a capacity decay of about 0.0021% per cycle), demonstrating ultra-stable cycling performance and exceptional rate capability (161 mAh g^{-1} at 2000 mA g^{-1}), proving the material's excellent potassium storage capability at high temperatures, indicating the great potential of MOFs to support stable electrochemical performance under high-temperature conditions.

In their study, Park et al. [11] successfully synthesized an electrode composed of nitrogen-doped porous carbon nanofibers (N-PCNFs) and ZnSe nanocrystals. Utilizing electrospinning techniques, this composite was crafted by employing zinc acetate dihydrate and 2-methylimidazole as precursors, followed by a high-temperature pyrolysis process. This led to a uniform integration of ZnSe nanoparticles within a porous carbon framework. The unique properties of MOFs were leveraged, offering a high surface area that facilitates rapid electrolyte diffusion within the electrode and effectively curbing the growth and agglomeration of ZnSe particles. In the material design, nitrogen doping not only significantly enhanced the electrical conductivity of the carbon nanofibers but also bolstered structural stability through the formation of a porous architecture. The porous carbon fiber network improved potassium-ion transport efficiency, reduced ion diffusion resistance, and increased the contact area between the electrode and the electrolyte, thereby boosting overall battery performance. Remarkably, this composite material demonstrated superior electrochemical performance, achieving an initial discharge-specific capacity of 512 mAh/g and retaining approximately 87% capacity after 200 cycles at a current density of 100 mA/g, underscoring its exceptional cycling stability. Moreover, at a higher current density of 500 mA/g, the composite maintained a commendable capacity retention rate, revealing its excellent rapid charge-discharge capability. These performance enhancements are attributed to the ZnSe@N-PCNFs structure's enhanced conductivity and the introduction of reactive sites that effectively buffer the volumetric changes of potassium-ions during charge-discharge cycles. In conclusion, this study introduces an effective and innovative approach to enhance the performance of KIBs through the design of MOF-based composite materials. The findings not only highlight the potential of ZnSe@N-PCNFs composites in improving cycling and rate performances but also provide new strategies and inspiration for the design and optimization of materials for KIBs in the future. The further development of KIBs could benefit from the application of such composite materials, advancing green energy-storage technologies.

In recent years, emerging two-dimensional metal-organic frameworks (2D-MOFs) not only possess advantages such as high porosity but also can have larger π planes through selective combinations of ligands, allowing more effective utilization of electron delocalization to achieve lower bandgaps and higher electrical conductivity. Bu et al. [12] successfully synthesized 2D c-MOF, Cu-TAC by introducing hexahydroxy-substituted ligand trialzine (6OH-TAC) into 2D-COF. This strategy improved the density of redox-active sites and conductivity, significantly enhancing the electrochemical performance of LIB anodes. Fourier transform infrared (FT-IR) spectroscopy and X-ray photoelectron spectroscopy (XPS) analytical techniques were used to study the compositional changes of Cu-TAC electrodes at different discharge/charge states, clarifying the Li^+ ion storage mechanism of Cu-TAC. It was found that during discharge, the C=N group combines with Li^+ ions to form C—N—Li. Additionally, changes in the CuO_4 part of Cu-TAC during discharge/charge were explored, and it was found that a reversible redox reaction occurred between Cu^+ and Cu^{2+} during the process. A synergistic electron storage mechanism between CuO_4 and TAC units was proposed, enabling Cu-TAC to exhibit excellent Li^+ storage capacity.

Besides using directly modified MOFs as battery electrodes, utilizing carbonized frameworks as supports for battery electrodes is also a research hotspot. Andrey et al. [13] synthesized two precursors, Ni&Zn-MOF and Zn-MOF, using nitrogen and oxygen-containing ligands, and carbonized them to assemble and study Na—Se batteries. For the cathode, a nitrogen and oxygen co-doped carbon matrix modified with Ni single atoms (Ni SA) was prepared from Ni&Zn-MOF for Se storage (denoted as Ni SA/NOC). Studies show that batteries with Ni SA/NOC/Se composites as cathodes and NOC/Na composites as anodes exhibit excellent electrochemical performance. The Ni SA/NOC/Se cathode maintains a high reversible specific capacity of up to 315 mAh g^{-1} after 100 cycles (0.1C), and shows excellent rate performance of 195 mAh g^{-1} at a 5C rate. This superior performance is attributed to the Ni SA catalyst weakening the Se—Se bonds in Se_8 rings, inducing the formation of chain-like Se_x molecules, ensuring direct conversion between Se and Na_2Se, and preventing the formation of soluble intermediates; simultaneously, the Ni SA catalyst enhanced the kinetics of electrochemical reactions, improving the utilization and rate capability of Se. Moreover, dendrite-free anodes made from N, O codoped carbon derived from Zn-MOF efficiently store sodium, enabling Na—Se full cells to maintain a capacity of 213 mAh g^{-1} after 1000 cycles at a 1C rate, demonstrating excellent long-cycle performance and high stability.

Due to the increasing scarcity and rising costs of lithium resources, SIBs have gained attention as promising alternatives, leveraging the abundance and environmental advantages of sodium. In a pioneering study, Shen et al. [14] investigated the use of a novel flower-structured cobalt 2,5-thiophenedicarboxylate coordination polymer (Co-TDC) as an anode material for SIBs, reporting significant electrochemical performance achievements. The Co-TDC was synthesized via a solvothermal reaction at 120 °C, involving cobalt nitrate and 2,5-thiophenedicarboxylic acid in mixed solvents. This method yielded a flower-like, porous material that optimizes ion diffusion paths and electrolyte storage, thereby enhancing electrode functionality. Electron microscopy analyses (SEM and TEM) confirmed an orderly flower-like microstructure, composed of multiple layers that significantly improve ion diffusion channels, thus aiding electrochemical performance. In electrochemical tests, Co-TDC as an anode material exhibited excellent charge cycles, achieving an initial discharge capacity of 737.5 mAh g^{-1} and maintaining a capacity of 328 mAh g^{-1} after 100 cycles, surpassing the durability of many carbon-based materials. Notably, the material demonstrated superior high-rate charge-discharge capabilities under various current densities, underscoring its adaptability to rapid current changes and efficient sodium-ion storage and release. Soft X-ray absorption spectroscopy analysis revealed that the valence state of Co ions remained stable as Co^{2+} during cycling, suggesting sodium-ions predominantly occupy the organic components without significant cobalt interaction. This research underscores the potential of MOFs in enhancing SIB technology, offering new avenues for the exploration of similar materials in next-generation rechargeable batteries and energy-storage devices.

Due to the larger ionic radius of Na$^+$, traditional battery materials face challenges of structural instability during Na$^+$ intercalation. In a significant advancement, Li et al. [15] introduced 2D conjugated metal-organic frameworks (c-MOFs) based on divalent copper, which demonstrated excellent electrochemical performance in SIBs. The materials studied, HATN-XCu (X = O or S), were designed with dual redox sites to ensure high-capacity sodium storage performance. Their layered stacked structure offers rapid transport and diffusion pathways, significantly enhancing charge transfer rates and ion diffusion capabilities. FT-IR and XPS revealed that the dual redox sites within these c-MOFs are derived from the electron-rich C=N groups and [CuX_4] units, facilitating multi-electron reactions and substantial Na$^+$ accumulation. HATN-OCu, when used as an anode material, exhibited a reversible capacity of 500 mAh g^{-1} at a current density of 0.1 A g^{-1}, and maintained a capacity of 151 mAh g^{-1} even at higher rates (5 A g^{-1}). Furthermore, a sodium-ion full cell assembled with HATN-OCu as the anode and Na$_3$V$_2$(PO$_4$)$_2$O$_2$F (NVPOF) with voltage platforms of 4.0 and 3.6 V as the cathode showed excellent rate performance (117 mAh g^{-1} at 5 A g^{-1}) and long cycle life (capacity retention of 80% after 500 cycles at 2 A g^{-1}). This innovative c-MOFs design addresses the capacity limitations faced by traditional MOFs due to single redox sites, and the introduction of dual redox sites significantly enhances the material's electrochemical performance. During electrochemical cycling, the HATN-XCu electrode materials displayed excellent reversibility and stability with an initial discharge capacity of 626 mAh g^{-1}, maintaining high capacity and Coulombic efficiency after extensive dynamic electrochemical testing. Electrochemical kinetics analysis indicated that the sodium-ion storage behavior of HATN-OCu is influenced by both capacitive behavior and diffusion response, characterized by higher carrier mobility and lower polarization. This research significantly advances the design philosophy of high-capacity sodium batteries through the synthesis of novel c-MOFs with dual redox sites, providing crucial theoretical and experimental foundations for their potential future application in energy-storage devices.

Zhao et al. [16] have successfully synthesized metal selenide (e.g., (CoFe)Se$_2$) @carbon nanofiber (CNS) composites using MOFs as templates through various methods, including liquid-phase anion exchange, electrospinning, and selenization. This composite exhibits exceptional electrochemical performance, including a high specific capacity of 1170 mAh g^{-1} and outstanding rate capability at high current densities, maintaining a capacity of 563 mAh g^{-1} even after 5630 cycles. This approach not only aids in controlling the morphology of the final product but also ensures the uniformity and consistency of the materials. The hierarchical and hollow nanostructures exhibit exceptional performance under high-rate discharge conditions, effectively mitigating volume changes during cycling and preventing structural collapse. Furthermore, the conduction characteristics of MOFs enhance the electrochemical activity, while CNS provides a conductive backbone that improves the overall conductivity of the electrodes. By rationally designing the metal composition in (CoFe)Se$_2$, synergistic effects among different metals can enhance sodium-ion storage capacity. Studies indicate that these composites maintain outstanding discharge capacity after extensive cycling, outperforming many conventional anode materials. Ultimately, the high specific surface area and porosity of MOFs improve their permeability in the electrolyte, optimizing the interface reactions and facilitating sodium-ion transfer. This significant improvement in electrochemical performance and stability highlights the potential of MOFs as crucial materials in advancing the next generation of high-performance SIBs.

In the field of SIBs, to overcome challenges such as low energy density, poor rate capability, and poor cycling stability, Xiong et al. [17] study focuses on utilizing MOFs to enhance battery performance. They synthesized a 3D framework superstructure (Bi@C ⊂ CFs) with arrays of carbon nanoribbons and a coating of metal bismuth (Bi) nanospheres, significantly enhancing the sodium storage capacity and cycling life of the battery. Specifically, the study involves pyrolyzing one-dimensional MOF nanorods at 600 °C in an inert atmosphere, transforming them into a carbon dioxide framework and Bi@C nanospheres to serve as SIB anode materials. This material exhibits excellent electrochemical performance, maintaining a capacity of 308.8 mAh g^{-1} even at a high current density of 80 A g^{-1}. Moreover, after 200 cycles at 1 A g^{-1}, the full cell still provides a high capacity of 272 mAh g^{-1}. This exceptional performance is primarily attributed to two factors: first, the hybrid structure formed during the MOF transformation process; second, the uniform solid electrolyte interphase (SEI) layer and porous nanostructure formed during cycling in the ether-based electrolyte. During charging and discharging, the homogeneous and thin SEI layer formed on the surface of the active material creates a nanoporous structure, improving the utilization of active material and shortening the ion/electron diffusion path. Using techniques such as XPS and high-resolution transmission electron microscopy (HRTEM), the study further uncovers the advantages of dimethyl ether (DME) electrolytes in enhancing sodium storage capability. Compared to carbonate mixtures (EC/DEC), the SEI layer formed by DME electrolyte not only maintains the structural stability of the active material but also provides excellent rate capability and cycling stability. Additionally, the low electronic density and high sodium-ion diffusivity of DME electrolytes significantly reduce the internal resistance of the battery. Ultimately, the double-layer carbon protective layer derived from the MOF structure allows the Bi@C ⊂ CFs composite as SIB anode material to exhibit remarkable electrochemical performance, making it a promising sodium storage material.

7.3.1.2 COF as Electrode Materials

COFs are considered promising electrode materials due to their inherent insolubility in electrolytes, abundant porosity, orderly controllable channels to facilitate ion transport, and π-conjugated frameworks to enhance charge transport. In recent years, 2D materials have attracted extensive research interest and have been the subject of numerous studies.

Regarding the redox properties of organic cathode materials, organic electroactive groups are typically classified as p-type or n-type. Due to the diversity of n-type structural units, most reported COF-based cathode materials are composed of n-type active centers consisting of groups like C=O and C=N, generally exhibiting high theoretical specific capacity. However, n-type COF-based cathodes have relatively low redox potentials. In contrast, COFs with p-type active centers (such as phenothiazine, aromatic amines, and triazines) have more "holes," which can exhibit high electrode potentials (>3.0 V vs. Li$^+$/Li). However, p-type structural units generally have large molecular weights and low content of redox-active centers, leading to significantly lower theoretical specific capacities for corresponding p-type COFs compared to n-type COFs, hindering the improvement of energy density.

Jiang et al. [18] introduced p-type phenazine (Pz) molecules into COFs, using Pz-based building block 5,10-dimethyl-5,10-dihydrophenazine-2,7-dicarboxaldehyde (DMPZD) along with another p-type semiconductor, tris-(4-aminophenyl)-amine (TAPA), and 4,4′,4″-(1,3,5-triazine-2,4,6-triyl)trainline (TATTA) through a solvothermal synthesis reaction to prepare TAPA-Pz-COF and TATTA-Pz-COF. TATTA-Pz-COF exhibited a high specific capacity of 324 mAh g^{-1} at 100 mA g^{-1}, and a record-high energy density of 737 W h kg^{-1} among all organic polymer and COF electrodes. The structure and evolution of redox energy during the electron transport process were further explored through density functional theory (DFT) calculations. Based on the calculation results, a three-step discharge scheme for an asymmetric TATTA-Pz-COF unit was proposed: TATTA-Pz-COF + 3PF$_6^-$ → TATTA-Pz-COF + 2PF$_6^-$ → STATTA-Pz-COF → TATTA-Pz-COF + 2Li$^+$. Simulations showed that ion storage in TATTA-Pz-COF can facilitate electron transport and improve rate performance.

Besides introducing p-type and n-type active groups into COFs to obtain monopolar COFs, researchers have found that bipolar organic/polymer cathode materials, which contain both p-type and n-type electroactive groups, hold great potential for achieving both high voltage and high specific capacity. COFs possess high crystallinity and porosity, allowing for the insertion of electroactive groups at specific sites as needed, resulting in COF materials with bipolar characteristics (combining both p-type and n-type). Zhang et al. [19] used 4,4′,4″,4‴-(1,4-phenyl-enebis(azanetriyl))tetrabenzaldehyde (PTB) and 2,5-bis(hexyloxy)terephthalohydrazide (DHZ) as structural units to design and construct a COF material with p/n bipolar characteristics for use as a cathode in LIBs. The PTB unit contains two sp^3 N (C—N group) redox-active sites, while the C=N bond formed via Schiff base condensation between the PTB and DHZ units serves as another type of active site. Consequently, the prepared p/n bipolar PTB-DHZ-COF contains multiple C—N and C=N redox-active groups, which can effectively achieve high voltage and large theoretical capacity. Combined with highly conductive CNTs, the prepared PTB-DHZ-COF40, used as cathode material, achieved a high capacity of 114.24 mAh g^{-1} at 1000 mA g^{-1}. After 5000 cycles, the capacity retention rate remained at 86.3%, demonstrating excellent cycling stability. Meanwhile, the battery achieved a maximum energy density of 486 Wh kg^{-1} and a power density of 3.185 kW kg^{-1}, outperforming currently commercialized batteries.

COF-based electrode materials offer advantages such as low cost, designable structures, and sustainability. However, reported COF-based electrode materials exhibit poor rate performance due to their relatively poor electronic and ionic conductivity, resulting in cathodes with capacity and rate performance still far from commercial viability. The reported method to enhance the rate performance of COFs is by combining them with conductive carbon materials like graphene. However, adding too much graphene and similar materials can reduce the battery's energy density. To achieve high capacity and good rate performance, Chen et al. [20] designed TQBQ-COF by removing inactive linker groups from the framework and doping heteroatoms, composing them of multiple carbonyl and pyrazine groups. Since both carbonyl and pyrazine groups are designed as redox sites of the TQBQ-COF electrode, a theoretical capacity of 515 mAh g^{-1} can be achieved. Electrochemical performance characterization revealed that the constant current charge-discharge curve of the TQBQ-COF electrode in sodium batteries showed an initial discharge capacity of 452.0 mAh g^{-1} (0.02 A g^{-1}, 1–3.6 V) and maintained a capacity of 352.3 mAh g^{-1} after 100 cycles. The excellent rate performance can be attributed to the honeycomb structure of TQBQ-COF and its high electronic conductivity (1.973×10^{-9} S cm^{-1}). Additionally, after the battery was left to rest for 24 hours after the fifth cycle, it still retained 92% of its original capacity. The good sodium storage performance can be attributed to the stability of the TQBQ-COF material and fewer side reactions with the electrolyte. TQBQ-COF electrodes maintained low impedance and showed no significant cracks after 500 cycles (169.4 Ω) and 1000 cycles (220.2 Ω). The TQBQ-COF electrode displayed high energy density and higher power density. The porous and nitrogen-doped structure of TQBQ-COF facilitates rapid ion/electron transport among multiple active sites, thereby endowing TQBQ-COF electrodes with excellent cycle stability and rate performance in sodium batteries.

Wang et al. [21] synthesized a COF@carbon nanotube (COF@CNTs) composite material using a simple room temperature method, and applied it as an anode material for LIBs. Based on electrode characterization, DFT calculations, and electrochemical analysis at different voltages during various cycles, the lithium-ion storage mechanism of COF@CNTs indicates that the few-layered conjugated COF structure on conductive CNTs, as well as the gradual increase in interlayer spacing due to repeated lithium insertion, facilitates efficient storage of 14 lithium-ions per COF monomer. COF@CNTs exhibited excellent electrochemical performance, with reversible capacity gradually increasing before approximately 320 cycles, reaching a high and stable reversible capacity of 1021 mAh g^{-1} between approximately 320–500 cycles. This capacity value can be attributed to the capacity contribution of 1536 mAh g^{-1} from the COF in the COF@CNTs composite material.

COF electrodes often form a closely packed 2D layer due to π–π stacking interactions, which usually prevent ions from penetrating interlayer internal active sites, requiring ions to travel along longer paths to reach active centers. This results in reduced utilization of active centers, especially under fast charging/discharging conditions. Wang et al. [22] successfully converted micron-sized 2,6-diaminoanthraquinone-2,4,6-triformylphloroglucinol (DAAQ-TFP)-COF into 2D nanosheets DAAQ-ECOF using a room-temperature vibratory ball milling method. DAAQ-ECOF exhibits highly reversible lithiation/delithiation reactions at 2.34 and 2.48 V, providing a practical voltage range comparable to emerging cathode materials. Its reduction potential is 0.4 V higher than that of monomer DAAQ, attributed to the extended conjugated structure within the polymer layer. In contrast, the reaction kinetics of multi-layer DAAQ-TFP-COF are primarily determined by the lithium-ion diffusion in the bulk material. Few-layer DAAQ-ECOF exhibits ultra-fast charge and discharge capability. At a current density of 20 mA g^{-1}, the capacity of DAAQ-ECOF gradually increases before 320 cycles, reaching 1021 mAh g^{-1} between 320 and 500 cycles, which is approximately 96% of its theoretical capacity. In contrast, DAAQ-TFP-COF only reaches 73% of its theoretical value. Due to dissolution issues, the initial capacity of DAAQ-based cathodes is 197 mAh g^{-1}, but quickly fades to 75 mAh g^{-1} within 20 cycles. DAAQ-TFP-COF and DAAQ-ECOF possess good chemical and thermal stability, are insoluble in the electrolyte, and ensure excellent cycling performance.

COF materials possess ordered nanoporous structures that facilitate fast ion transport and abundant internal redox sites, while reducing volume expansion, showcasing remarkable advantages. However, the large surface area of COF-based electrodes leads to an expanded electrode-electrolyte interface (EEI), forming an unstable SEI layer, severely impacting battery performance. Despite optimizations through methods such as molecular modification and chemical doping, enhancing the performance of COF-based electrodes remains a significant challenge. Stabilizing the COF-based EEI to achieve practical application of KIB remains a huge challenge. The introduction of fluorine can lead to a stable EEI, suppress side reactions, and enhance ion diffusion and intercalation kinetics.

##Kim et al. [23] proposed a fluorine-rich Covalent Organic Framework (F-COF), synthesized through a solvothermal method, under conditions mixing an organic solvent with acetic acid aqueous solution, where the carbonyl groups of 1,3,5-triformylphloroglucinol (TFP) reacted with the fluorinated amino group of 2,3,5,6-tetrafluoro-1,4-phenylenediamine (F-DA) (Figure 7.3). High-performance and chemically stable KIB electrodes were fabricated using F-COF as a base material. F atoms on the COF framework enhance K$^+$ ion insertion kinetics and act as electroactive sites, facilitating the reversible transition of C—F bonds from semi-ionic to fully ionic (as shown in Figure 7.3). Thus, the SEI layer can reversibly form and degrade during ion insertion and extraction steps. F-COF maintains about 99.7% capacity after 5000 cycles and achieves 95 mAh g^{-1} in just 12 minutes at a 5C rate, demonstrating fast-charging capability, while the original COF-based electrodes without F atoms exhibit a sharp drop in coulombic efficiency after 4000 cycles. Computational analysis elucidates the important mechanistic role of F

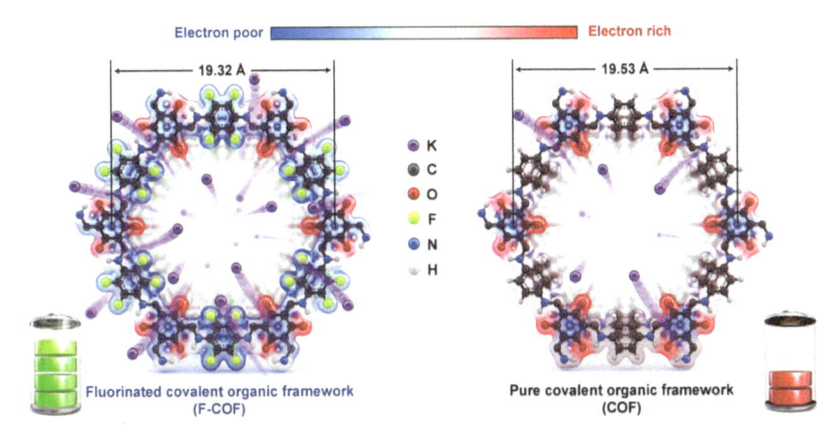

Figure 7.3 The illustration of significant enhancement of K⁺ ion storages and kinetics of F-COF than that of COF in the KIB system during the charging and discharging. Source: Lee et al. [23]/with permission of John Wiley & Sons.

atoms in enhancing the electrode's electron affinity and conductivity, both of which help stabilize K⁺ ion binding.

Traditional zinc-ion battery (ZIB) cathode materials, such as MnO_2 and V_2O_5, suffer from issues like easy hydrolysis/redox reactions and structural transformations, leading to poor cycling stability and rate performance. To overcome the shortcomings of traditional cathode materials, Vaidhyanathan et al. [24] introduced a COF with redox-active ZnI_2 as the cathode material for ZIBs. They synthesized, for the first time, a COF based on tannic acid and hydroquinone (IISERP-COF22), which exhibits high crystallinity and a large specific surface area, with a pore size of 30.9 Å. This COF is linked by Schiff base bonds and contains abundant nitrogen and oxygen functional groups in its structure, allowing it to chelate cations during charge and discharge processes, thus enhancing the performance of ZIBs. Introducing ZnI_2 into $ZnSO_4$ aqueous electrolyte significantly increased the battery capacity from 208 to 690 mAh g⁻¹ by generating various polyiodides (such as I_3^- and IO_3^-). The high performance of IISERP-COF22 is mainly attributed to its unique structural design and electrochemical properties. The diverse coordination sites of the COF can form stable chelates with Zn^{2+} ions, facilitating rapid ion migration and uniform charge distribution on the electrode surface. The introduction of polyiodides not only enhances the conductivity of the electrolyte but also improves the rate performance and cycling stability of the battery by reducing charge transfer impedance and overall resistance. Electrochemical tests show that IISERP-COF22 achieves a discharge capacity of up to 690 mAh g⁻¹ at a current density of 1.5 A g⁻¹, and maintains a capacity retention rate of over 80% after 6000 cycles at 5 A g⁻¹. In situ UV-visible spectroscopy and XPS reveal that during charge-discharge processes, polyiodides like I_2 and I_3^- form and are stored on the COF electrode surface, significantly enhancing the battery's energy and power density. Theoretical calculations and molecular dynamics simulations also support that the pore structure of the COF facilitates the rapid diffusion and transport of Zn^{2+} and I_3^-, contributing significantly to the overall electrochemical performance.

7.3.1.3 HOF as Electrode Materials

In the design of high-performance battery materials, HOFs have attracted researchers' attention due to their unique hydrogen-bond structure and excellent electrochemical properties. Wu et al. [25] designed 2D organic framework materials, G_2NDI and G_2PDI, using the G-quartet as the electron donor and naphthalene-1,4 : 5,8-bis(dicarboximide) (NDI) or 2,5,8,11-tetrahexylperylene-3,4 : 9,10-bis(dicarboximide) (PDI) as the electron acceptor, respectively. The guanine at both ends of the G_2NDI and G_2PDI monomer molecules hydrogen bonds with guanine from adjacent molecules to form a square grid structure. This orderly framework, formed by multiple hydrogen bonds and intermolecular π-π stacking, remains highly chemically stable in organic solvents. Subsequently, they tested G_2PDI as a cathode material for LIBs, finding that the material retained >93% capacity after 300 constant current charge-discharge cycles. This corresponds to a capacity retention rate of 99.97% per cycle, significantly higher than the 72.4% retention rate when using one of its monomers, diacetylene-functionalized perylene diimide, as a cathode material for LIBs. This demonstrates that the unique grid structure and multiple hydrogen bonds confer excellent electrochemical stability. Furthermore, transient absorption spectroscopy and electron paramagnetic resonance tests confirm that the orderly PDI stacking provides ideal electron conduction channels, thereby partially overcoming the poor conductivity of organic materials, and allowing Li^+ transport through the porous framework's electrolyte.

Meanwhile, in the positive research on SIBs, Wu et al. [26] used 3,5-diaminotriazole (DAT) as a monomer to synthesize 2D HOF materials (HOF-DAT) with excellent chemical stability and studied their application in SIBs. The N—H bonds in DAT can form multiple hydrogen bonds with adjacent structural units, providing the HOF-DAT material with good structural stability, and allowing it to maintain its chemical structure in various solvents of different polarities. Therefore, HOF-DAT, when used as a cathode material for SIBs, exhibited high specific capacity and excellent cycling stability, capable of stable cycling for 10,000 rounds at a current density of $1 A g^{-1}$, surpassing the cycling performance of most organic electrode materials. DFT calculations indicate that the amino and oxygen atoms within the DAT structure can stably adsorb Na^+, achieving effective Na^+ adsorption and storage during charge and discharge processes (Figure 7.4). Concurrently, the ultrathin layered structure provides sodium-ions with a shorter diffusion distance, with sodium-ion diffusion within and between layers having extremely low activation barriers, lower than in other organic or inorganic electrode materials.

Luo et al. [27] designed and synthesized a structurally stable hydrogen-bonded organic framework (HOFs-8) regulated by aldehydes, which contains an array of negative potential sites for sodium-ion storage. Thanks to the flexible hydrogen bonds and unique structural symmetry, HOFs-8 enables efficient utilization of active sites and rapid transport of sodium-ions and electrons. First, at 0 °C using tetrahydrofuran (THF) as the reaction solvent, the N1, N3, N5-tris(pyridin-4-yl)-benzene-1,3,5-tricarboxamide (TPBTC) molecule was synthesized by reacting acetyl chloride (1,3,5-benzenetricarbonyl chloride) with 4-aminopyridine organic amine monomer. Subsequently, TPBTC self-assembled into HOFs-8 in a mixed solution

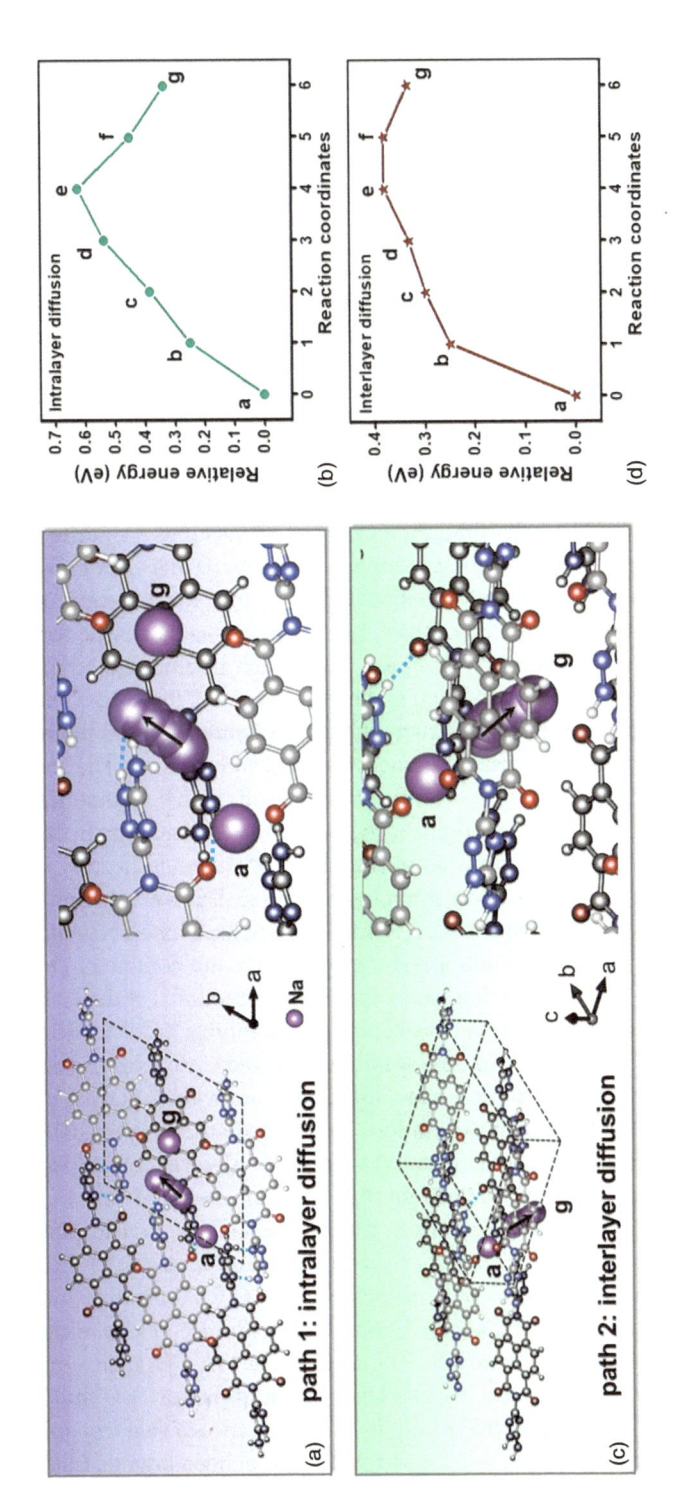

Figure 7.4 Simulated Na$^+$ ion diffusion within HOF-DAT. (a) Schematic showing the P1 diffusion path and (b) corresponding activation energy profile; (c) schematic showing the P2 diffusion path and (d) corresponding activation energy profile. Source: Wu et al. [26]/with permission of John Wiley & Sons.

of chloroform and methanol. The HOFs-8 electrode exhibits remarkable cycling stability under high current density, maintaining good reversible capacity even after 5000 cycles. The elastic hydrogen bonds in HOFs-8 inhibit the dissolution of organic molecules, enhance structural integrity, and reduce volume expansion, thus ensuring long-term stability. The abundant C=O active sites in these framework structures have been proven to be crucial for sodium-ion storage. HOFs not only demonstrate excellent thermal and chemical stability but also provide abundant active sites through the transfer of hydrogen bonds and π–π interactions, favoring enhanced sodium-ion storage performance. This study demonstrates the potential of HOFs in the development of high-performance green organic electrode materials, especially highlighting their broad prospects in the design and application of rechargeable SIBs.

Aqueous zinc metal batteries are considered strong contenders for next-generation energy-storage systems due to their high theoretical capacity, environmentally friendly low cost, and high safety. However, the dendrite growth and severe side reactions faced by zinc metal anodes limit their practical applications. Dendrite growth exacerbates surface exposure, leading to more side reactions and thus making the surface rougher. Designing a multifunctional coating that can promote uniform zinc deposition and inhibit side reactions is key to improving battery performance and lifespan. The Wang team [28] designed a HOF rich in zinc-philic functional groups, which can serve as a multifunctional interface coating for ZMBs. This HOF is self-assembled into a framework structure through hydrogen bond interactions with N- and O-rich functional groups (carboxyl, amino, and triazine). The abundant N and O active sites in the HOF restrict the 2D diffusion of Zn^{2+}, promoting rapid and uniform zinc deposition. Simultaneously, the hydrogen-bonding environment in HOF helps regulate the desolvation of $[Zn(H_2O)_6]^{2+}$ and significantly inhibits side reactions on the zinc electrode by trapping coordinated active water molecules. Contact angle tests show that the MA-BTA@Zn electrode has higher wettability and a more zinc-philic surface, facilitating the transportation and conduction of Zn^{2+}. AFM images of the MA-BTA HOF indicate that Young's modulus of the HOF layer is approximately 7.8 GPa, which is considered to help inhibit Zn dendrite growth and improve battery safety. In situ optical microscopy and CA curves indicate that the coating can regulate Zn^{2+} diffusion and nucleation behavior, achieving uniform distribution of Zn^{2+}. Theoretical calculations demonstrate that the MA-BTA HOF with abundant hydrogen bonds facilitates the desolvation process of $[Zn(H_2O)_6]^{2+}$ ions. SEM and XRD observations indicate that the HOF coating enables uniform and dense deposition of Zn^{2+}, helping to expose the Zn(002) crystal plane and inhibit the growth of zinc dendrites and by-product generation. In situ FT-IR spectra show increased transmittance of H_2O molecules and SO_4^{2-}, indicating increased consumption of H_2O and SO_4^{2-}, which reflects the occurrence of side reactions in bare zinc symmetric cells. In contrast, FT-IR curves of MA-BTA@Zn symmetric cells show no significant change in the characteristic peaks of H_2O and SO_4^{2-}, suggesting that the coating effectively inhibits the occurrence of side reactions.

Wang et al. [29] constructed a redox-active hydrogen-bonded organic framework (HOF-HATN) based on hexaazatrinaphthylene and diaminotriazine groups and confirmed its benefit in facilitating hydrogen ion storage when used as a cathode material for aqueous ZIBs. Its unique hydrogen-bonding network and strong π-π stacking endow HOF-HATN with a fast Grotthuss proton conduction mechanism, stable supramolecular structure, and enhanced hydrogen proton storage capability. The redox-active hydrogen-bonded organic framework (HOF-HATN) was synthesized by reacting tricyanohexaazatrinaphthylene (HATN-3CN) with dicyandiamide (DCD). The DAT fragment is a common hydrogen-bond building block, allowing the DCD-HATN molecules to spontaneously self-assemble into a 2D supramolecular structure. This bottom-up molecular engineering strategy gives these electrode materials periodically arranged redox-active centers, a supramolecular hydrogen-bonding network, and a layered π-π stacking structure. Theoretical calculations confirm the presence of its hydrogen bond and strong π–π interactions. Thanks to its unique hydrogen bond network and strong π-π stacking, HOF-HATN exhibits a high specific capacity (320 mAh g^{-1} at a current density of 0.05 A g^{-1}), excellent cycling stability (almost no capacity decay over 10,000 cycles at 5 A g^{-1}), and good rate performance (130 mAh g^{-1} at 2 A g^{-1}). Furthermore, a pouch cell was assembled using HOF-HATN, confirming its feasibility for practical applications. Additionally, HOF-HATN was applied in ammonium ion batteries, demonstrating its widespread applicability as an electrode material. A series of spectral characterizations and various electrolyte experiments confirmed the Zn^{2+}/H$^+$ costorage mechanism. To further verify the H$^+$ storage mechanism, a covalent triazine framework (HATN-CTF) containing hexaazatrinaphthylene structural units was also synthesized and tested. By comparing the electrochemical test results of HOF-HATN and HATN-CTF, it was found that the HATN-CTF without a hydrogen-bond network exhibited lower capacity and weaker redox peaks contributing to H$^+$ storage. Experimental and theoretical calculations revealed that HOF-HATN can achieve 18e$^-$ storage, with 23% coming from Zn^{2+} and 77% from H$^+$, confirming the H$^+$ diffusion path and storage mechanism along hydrogen bonds, thereby verifying its unique Grotthuss proton-hopping mechanism.

7.3.2 Applications in Electrolytes and Electrolyte Additives

Electrolytes, as the medium for ion conduction in batteries, directly affect the energy efficiency, safety, and cycle life of the battery. Traditional liquid and solid electrolytes, despite having certain advantages in practical applications, still face issues such as low conductivity, poor interfacial stability, and insufficient safety. POFs can not only improve the ionic conductivity of electrolytes but also enhance the interfacial stability between electrolytes and electrode materials by tuning pore structure and surface chemistry, thereby enhancing the overall performance of the battery. In future energy-storage technologies, the application potential of POFs in electrolytes will further drive the development of efficient and safe batteries.

7.3.2.1 MOF as Electrolytes and Electrolyte Additives

Traditional liquid electrolytes have safety hazards such as flammability, leakage, and dendrite growth, which limit their application. As a result, solid electrolytes have gained widespread attention due to their excellent mechanical strength and thermal stability. However, solid electrolytes currently face issues such as low ionic conductivity and poor electrochemical stability. Chou et al. [30] confined polymer electrolyte molecules within MOFs to achieve safe and high-performance all-solid-state sodium metal batteries. Specifically, the researchers synthesized a polymer-MOF single-ion conducting solid polymer electrolyte (SICSPE). By confining the polymer molecules within the nanoporous structure of the MOF and leveraging the Lewis acid–base interactions with ZIF-8 particles, this innovative design effectively enhanced the safety and performance of sodium metal batteries. Through Lewis acid–base interactions, sodium-ions are in a weak coordination state, promoting the dissociation of sodium-ions, which enhances their transference number and conductivity. The orderly migration channels provided by nanopores facilitate rapid sodium-ion transport, significantly enhancing the ionic conductivity of the electrolyte (reaching 4.01×10^{-4} S cm^{-1}) and the sodium-ion transference number ($t_{Na^+} = 0.87$). This design also expands the electrochemical stability window of the electrolyte, reaching up to 4.89 V vs. Na/Na$^+$. In terms of electrochemical performance, the polymer-MOF SICSPE exhibited excellent cycling stability and rate performance. The assembled all-solid-state sodium metal batteries showed no dendrite formation on the sodium metal surface, exhibiting excellent rate capability and stable cycling performance, maintaining a 96% capacity retention rate after 300 cycles. Moreover, this design significantly improved the compatibility and stability of the electrolyte–electrode interface, further ensuring the long cycle life and safety of the batteries. By confining polymer electrolytes within the MOF nanopores, this innovative design not only boosts the ionic conductivity and sodium-ion transference number but also significantly enhances the battery's cycling stability and safety.

Inspired by ion channel mechanisms in biological systems, Wang et al. [31] have innovatively combined MOFs to enhance the conductivity and safety of LIBs. By utilizing open metal sites (OMSs) within MOFs to complex with electrolyte anions, a biomimetic ion channel is formed. This design facilitates the smooth passage of lithium-ions through the electrolyte channels with reduced activation energy, thereby increasing ionic conductivity. The MOFs provide a stable structure that immobilizes electrolyte anions, diminishing direct interactions between lithium-ions and anions and expediting lithium-ion mobility. Specifically, studies using HKUST-1 as a model material demonstrate that post-activation and LiClO$_4$ solution incorporation, these MOF channels form biomimetic structures akin to ion channels in biological systems. The binding of electrolyte anions with OMSs generates a negative charge distribution which optimizes the conduction pathways for lithium-ions and minimizes migration resistance within the electrolyte.

Electrochemical characterization reveals that this MOF electrolyte achieves an ionic conductivity of 0.38 mS cm^{-1} at room temperature. While slightly lower than that of pure liquid electrolytes, it offers superior structural stability and safety benefits. Furthermore, LiFePO$_4$ | Li lithium metal batteries (LMBs), utilizing this MOF electrolyte as a prototype component, exhibit high capacity retention and long-term cycle stability. Notably, under charge-discharge rates from 0.2 to 2C, the batteries maintain excellent performance, retaining 75% of their initial capacity even after 500 cycles. This study pioneers the use of MOFs as efficient lithium-ion conductors, leveraging their structural design flexibility and ion channel mechanisms, offering new insights and practical application potential for next-generation battery development.

SSBs have garnered significant attention due to their high safety and stability. However, existing solid-state electrolytes (SEs) face challenges in ionic conductivity and ion transport across electrode/electrolyte interfaces. Traditional SEs, including ion-conducting ceramics and polymers, have not succeeded in simultaneously facilitating rapid lithium-ion transport within the SE and across the SE-electrode interface. Addressing these challenges, Guo et al. [32] explored nanostructured solid-state electrolytes derived from MOFs. They developed an innovative UIO/Li-IL solid-state electrolyte composed of lithium-ion liquid (Li-IL) adsorbed within Uio-66 MOFs, achieving an ionic conductivity of 3.2×10^{-4} S cm^{-1} at 25 °C. Due to the nanostructured UIO/Li-IL's high surface tension and excellent electrode contact, the impedances at the Li/SE and LiFePO$_4$/SE interfaces were reduced to 44 Ω cm^2 and 206 Ω cm^2 at 60 °C, respectively, enabling stable lithium plating/stripping behavior.

This novel UIO/Li-IL SE demonstrated exceptional cycle stability and high discharge capacity, with the SSB retaining 100% capacity after 100 cycles at 0.2C and 94% capacity after 380 cycles at 1C, indicating significant application potential. MOFs in these batteries provide high-area channels for efficient movement of EMIM$^+$, TFSI$^-$, and Li$^+$, enhancing the overall ionic conductivity of the electrolyte and reducing impedance at the electrode-SE interface. MOFs exhibit a dual mechanism of rapid conduction both in the bulk and at the interfaces, rendering them an ideal choice for improving SSB performance. The study's focus on MOF-derived SEs not only highlights their potential to facilitate rapid lithium-ion transport but also provides novel insights into their physicochemical characteristics. Overall, this work outlines new opportunities for enhancing LIB performance through MOF-derived nanostructured SEs, promising advancements in safe and long-lasting energy-storage systems.

Meanwhile, solid polymer electrolytes (SPEs) are considered strong candidates for SSBs due to their high flexibility and good interfacial contact with electrodes. To develop SPEs with high ionic conductivity and good compatibility with high-voltage electrodes, Guo et al. [33] proposed a novel "host-guest recognition" gel polymer electrolyte (GPE) strategy and constructed a MOFs-GPE system using in-situ polymerization technology. Specifically, a Ti-based MOF with various synergistic sites was used as the "host" platform of the GPE to modulate electrolyte performance. These MOFs not only serve as efficient lithium-ion conduction accelerators but also exhibit excellent performance in mechanical strength and high voltage tolerance. Ti-based MOFs synergistically enhance the transport dynamics of lithium-ions by providing diverse redox-inactive Ti metal centers and oxygen sites. Their ordered channels and microporosity assist in uniform Li$^+$ ion plating during charge-discharge processes, promoting the formation of stable interfacial layers and improving cycling performance. The MOFs-GPE structure exhibited an electrochemical stability window of 5.05 V, a transference number of 0.71, and an improved ionic conductivity of 1.36×10^{-3} S cm^{-1}. When combined with a LiFePO$_4$ cathode, the assembled battery retained 98.1% capacity after 500 cycles at a 1C current density. In situ measurements and DFT calculations indicate that the introduced MOF host materials can accelerate Li$^+$ ion transport dynamics within composite electrolytes and promote uniform lithium-ion deposition on the electrode interface, significantly enhancing the battery's electrochemical performance. This study provides crucial guidance for improving the Li$^+$ ion migration capability in host-guest recognition SPEs and offers new insights and methods for developing high energy density SSBs.

SIBs are considered strong candidates for large-scale electrochemical energy-storage devices due to their low cost, abundant resources, and compatibility with LIB manufacturing processes. However, highly reactive free solvent molecules in liquid electrolytes can induce continuous interfacial side reactions between the electrodes and electrolyte, leading to degradation of SIBs' cycling performance. Zhang et al. [34] addressed this issue by designing and synthesizing a copper-based metal-organic framework (Cu-MOF-74) with uniform 1.1 nm nanopores to enhance electrolyte interface compatibility and construct long-life and high-safety SIBs. Cu-MOF-74 is composed of Cu^{2+} nodes and 2,5-dihydroxyterephthalic acid ligands. By providing OMSs and nanopores, the MOF physically confines solvent molecules and chemically coordinates with Na salts. This allows the formation of highly aggregated solvated structures, effectively suppressing side reactions at the EEI and enhancing the thermal stability of the electrolyte. Through physical confinement in nanopores and chemical coordination, Cu-MOF-74 significantly reduces the presence of free solvent molecules, thereby decreasing the likelihood of solvent decomposition and improving the thermal and interfacial stability of the electrolyte. Experimental results show that the MOF-constrained electrolyte-based battery retains 93% capacity after 3000 cycles and maintains 90% capacity after 600 cycles at 60 °C. MOF-constrained electrolytes effectively reduce EEI side reactions, and form stable SEI and cathode electrolyte interface (CEI), thereby enhancing the cycle life and safety of the battery. DFT calculation results indicate that the migration energy barrier of Na$^+$ ions in MOF pores is significantly reduced, owing to the anion-enhanced migration pathway. The OMSs in the MOF and chemically coordinated ClO$_4^-$ anions provide rapid ion transport pathways. The nanoconfinement effect of Cu-MOF-74 not only significantly enhances the electrochemical stability of the electrolyte but also effectively improves the overall performance of SIBs by reducing free solvent molecules, optimizing ion migration paths, and enhancing interfacial compatibility.

The direct application of lithium metal anodes faces many safety issues, such as dendrite growth leading to short circuits, thermal runaway, and even explosions. To address these issues, researchers have proposed various strategies, including structured electrodes, functionalized separators, the development of solid electrolytes, and interfacial engineering. Among these, the non-fluidity and high modulus of quasi-solid electrolytes can mitigate safety issues caused by traditional liquid electrolytes while maintaining good interfacial contact.

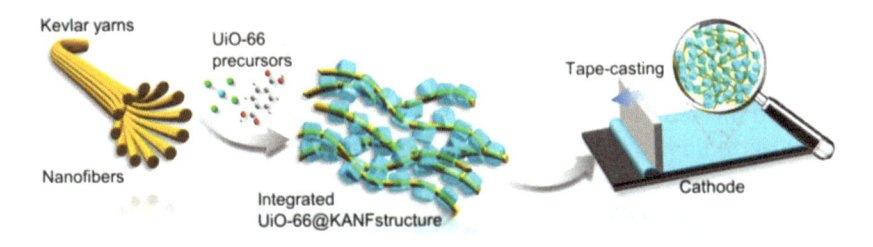

Figure 7.5 Schematic illustration for the configuration of cathode-supported UiO-66@KANF layer. Source: Liu et al. [35]/with permission of John Wiley & Sons.

Wang et al. [35] proposed an innovative ultrathin rigid composite electrolyte solution designed to provide superior safety and performance for LMBs. This design utilizes Kevlar nanofibers (KANFs) as a framework to fabricate interwoven necklace-like nanostructures of MOFs, which are built on the cathode plate using a coating method, forming a cathode-supporting functional layer with a thickness of 5.1 μm (Figure 7.5). This highly cross-linked structure exhibits a high elastic modulus (5.4 GPa) at room temperature, and through the encapsulation of a deep eutectic solvent within the MOF channels, combined with in situ polymerization, it forms a quasi-solid electrolyte (QSE) with an interconnected 3D network, achieving an ionic conductivity of 0.73 mS cm^{-1}. The abundant porous structure and Lewis acid sites of MOF particles facilitate lithium-ion transport and effectively constrain anion mobility. KANFs can bridge between MOF pores through functional groups, further promoting the free state of lithium-ions and accelerating transport kinetics, thus reducing activation energy. DFT calculations indicate that KANFs as MOF hosts play a crucial role in the free state and rapid diffusion kinetics of lithium-ions. Electrochemical tests show that the assembled Li | LiFePO$_4$ battery retains a high reversible capacity of 128.7 mAh g^{-1} after 500 cycles at 3C. This innovative electrolyte design, by enhancing ionic conductivity, reducing interfacial impedance, and improving interfacial compatibility, not only significantly increases the battery's charging rate but also extends its lifespan.

Huang et al. [36] proposed a nano-confinement in situ curing strategy, creating a new melt guest-mediated metal-organic framework (MGM-MOF). By embedding the newly developed melt crystal organic electrolyte (ML20) into the nano-cages of anionic MOF-OH, an MGM-MOF-OH with multimodal supramolecular action sites and a continuous negative charge environment was obtained. These nanochannels promote lithium-ion transport by continuous transitions between adjacent negative charge environments and chemically restrict anion movement through hydroxyl-functionalized pore walls, significantly improving lithium-ion conductivity (up to 7.1×10^{-4} S cm^{-1}) and lithium-ion transference number (0.81). The nanochannels of MGM-MOF significantly enhance ion transport performance by providing multimodal supramolecular interaction sites. The re-solidification of ML20 guest molecules within the MOF nanopores forms diverse polar networks, significantly reducing the activation energy for ion migration (to 0.29 eV), thereby increasing ion conductivity. Additionally, due to the chemical restriction of anions by hydroxyl-functionalized pore walls, the diffusion coefficient of TFSI⁻ anions is greatly reduced, further enhancing the selective conduction capability of lithium-ions. Verified by one-dimensional and 2D nuclear magnetic resonance (NMR) and theoretical calculations, these multimodal supramolecular interaction sites enable lithium-ions to quickly transition between multiple negative charge environments, thereby suppressing anion movement.

7.3.2.2 COF as Electrolytes and Electrolyte Additives

Solid electrolytes suffer from low ionic conductivity at room temperature, and the unclear electrochemical mechanisms of inorganic/organic polymer electrolytes hinder their development. Single lithium-ion conductor electrolytes can provide a high lithium-ion transference number (>0.7) by fixing anions, reducing ion concentration gradient formation and lithium dendrite growth, thus greatly improving energy efficiency and battery life. Guo et al. [37] designed three lithium carboxylate COFs with diverse skeletal structures as single lithium-ion conductor electrolytes. LiOOC-COF1 was obtained by the reaction of 1,3,5-triformylbenzene (Tf) and 2,5-diaminobenzoic acid (DAA). LiOOC-COF2 and LiOOC-COF3 were prepared by the reaction of Tf or 1,3,5-triformylphloroglucinol (Tp) with DAA or 4,4′-diamino-[1,1′-biphenyl]-3,3′-dicarboxylic acid (DBA), respectively. By covalently anchoring active groups within the COF structure, the migration of single lithium-ions is ensured. Through this design, these anionic frameworks exhibit directional ion channels that can eliminate interfacial side reactions and dendrite growth. LiOOC-COF materials exhibit excellent electrochemical properties, particularly their high ionic conductivity and high transference number. Through NMR and electrochemical impedance spectroscopy tests, the ionic conductivity of LiOOC-COF3 at 30 °C is 1.36×10^{-5} S cm^{-1}, with a high transference number of 0.91. This is attributed to its ordered directional ion channels and stable structure. Additionally, it exhibits a wide electrochemical stability window (4.2 V) and low activation energy (0.17 eV). Experimental results indicate that it demonstrates stable lithium plating/stripping performance on lithium metal electrodes and significantly inhibits the growth of lithium dendrites. In a 320-hour cycling test at a current density of 50 μA cm^{-2}, LiOOC-COF3 exhibits uniform lithium-ion migration, and the lithium metal surface remains smooth and clean. Additionally, it was used to assemble a quasi-solid-state organic full battery, where the C$_6$O$_6$ | LiOOC-COF3 | Li battery exhibited a discharge-specific capacity of 420 mAh g^{-1} at a current density of 50 mA g^{-1}, with excellent cycling stability and rate performance. DFT calculations further reveal the lithium-ion migration mechanism in LiOOC-COF3, showing preferential lithium-ion transport in the channels, benefiting from the pore structure and the role of oxygen atoms. The unique C=O sites of LiOOC-COF3 provide abundant π-electron delocalization resources, enhancing lithium-ion migration kinetics.

Zheng et al. [38] synthesized COF materials characterized by specific pore size distribution and surface chemical features using solution-based and template synthesis techniques. These materials feature a highly ordered porous structure that facilitates ion conduction within batteries and offers more active sites for electrochemical reactions, thereby significantly boosting the battery's specific energy and capacity. In the context of battery applications, COFs demonstrate significant potential as electrode materials. The porous architecture of COFs helps mitigate the volume fluctuations experienced by conventional electrode materials during charge-discharge processes, thereby improving their mechanical stability and cycle lifespan. Using structural functionalization, COFs can improve their conductivity and ionic transference, optimizing charge transport pathways to effectively enhance the overall battery performance. In experimental investigations, researchers thoroughly tested COF materials using electrochemical methods including CV, constant-current charge-discharge tests, and electrochemical impedance spectroscopy. The results indicated that the COF materials demonstrated outstanding electrochemical performance in multi-cycle testing conditions. Under constant current cycling conditions, the initial discharge capacity was approximately 750 mAh g^{-1}, maintaining over 90% capacity retention (675 mAh g^{-1}) after 200 cycles, indicating exceptional cycling stability. Additionally, rate performance tests demonstrated that COFs could stably deliver 500 mAh g^{-1} at 1C and retain about 350 mAh g^{-1} at 2C, highlighting their strong charge exchange capabilities and rapid energy support at high rates.

Electrochemical impedance spectroscopy analysis revealed a significant reduction in the charge transfer resistance of COFs, from 300 Ω to about 100 Ω, after the initial full cycle, further indicating improved conductivity with structural optimization. These metrics demonstrate that COFs excel in long-term battery stability and offer robust performance under high power demands. The study achieved significant improvements in the electrochemical performance of COF materials for LIBs through optimized synthesis and functionalization. The performance enhancements contribute to not only improving the total energy efficiency and cycle lifespan of LIBs but also provide crucial technical reference and support for the future development of high-performance, durable energy-storage devices. Due to their superior chemical stability and remarkable electrochemical performance, COFs are anticipated to serve as crucial components in the future of energy-storage systems.

Manthiram et al. [39] dissolved 2 mol/l lithium bis(trifluoromethanesulfonyl)-imide (LiTFSI) in dimethylacrylamide (DMA) to obtain a novel liquid electrolyte, DMA@LiTFSI. The electrolyte was then introduced into lithium-treated COF-SO3Li (LiCOF) by immersing the DMA@LiTFSI solution into LiCOF, using low-pressure driven methods and vacuum treatment to fill the liquid electrolyte into the COF channels, and performing in situ polymerization at room temperature to form DLC electrolyte. At room temperature, the DLC electrolyte achieved a lithium-ion conductivity of 1.7×10^{-4} S cm^{-1}, about 100 times higher than other COF electrolytes, with a lithium-ion transference number of 0.85. The functional groups on the DMA chain can effectively release lithium-ions and promote orderly lithium-ion migration by altering the channel environment. Additionally, the DLC electrolyte showed excellent mechanical properties, and it can be fabricated into ultrathin and flexible films, aiding in the creation of foldable SSBs. XPS and peakforce quantitative nanomechanics (PFQNM) results further confirmed that the

DLC electrolyte formed a double-layer SEI on the lithium metal surface, consisting of an inorganic inner layer and an organic outer layer. This structure effectively inhibits dendrite growth and maintains contact at the electrode/electrolyte interface. This structure endows the full battery with excellent cycling stability and rate performance and supports foldable SSBs that can operate stably in various bent states. The framework material plays a crucial role in enhancing electrolyte performance and battery stability.

Kim et al. [40] proposed a covalent organic framework with zwitterionic functional groups (Zwitt-COF) for use in all-solid-state LMBs. The Zwitt-COF was prepared by introducing 2-(pyridinium-1-yl)acetate groups into a pyridine-rich non-ionic COF (NCOF). The Zwitt-COF retained the stacked crystalline structure of the original NCOF, but its crystallinity decreased upon the introduction of zwitterionic functional groups, aiding the improvement of ion transport. BET analysis revealed that Zwitt-COF has 1.4 nm micropores and several mesopores, facilitating rapid ion migration. As a solid-state electrolyte, Zwitt-COF demonstrated significantly enhanced electrochemical performance. At room temperature, its ionic conductivity reached 1.65×10^{-4} S cm^{-1}, far surpassing that of NCOF electrolyte. The conductivity remained stable at high temperatures, showing excellent thermal stability. Activation energy calculations indicated that the ion transport activation energy of the Zwitt-COF solid-state electrolyte is 0.149 eV, significantly lower than that of NCOF, suggesting more effective ion pair dissociation and ion transport. Theoretical simulation results further revealed the migration mechanism of lithium-ions in Zwitt-COF. The introduction of zwitterionic functional groups facilitated the dissociation of strong ion pairs and reopened linear hexagonal ion channels by adopting an AA-stacking structure, significantly improving ion conductivity.

ZIBs are considered important candidates for future energy-storage technologies due to their high theoretical capacity, low cost, and environmental friendliness. However, the development of ZIBs faces numerous challenges, including the growth of zinc dendrites, side reactions (such as hydrogen evolution reaction), and electrolyte leakage. These issues significantly restrict the performance of ZIBs in high energy density and long lifespan applications. Additionally, the strong interaction between Zn^{2+} ions and surrounding solvent water molecules severely limits the mobility of charge carriers in the electrolyte, leading to slow reaction kinetics. Although some strategies like "water-salt coexistence" and hydrated molten electrolytes have been proposed, these methods have drawbacks such as high cost and complex preparation processes.

To overcome the aforementioned challenges, Tan et al. [41] proposed a rapidly synthesized single-ion conductive hydrogel electrolyte (TCOF-S-Gel) for high-performance quasi-solid-state ZIBs (Figure 7.6a). Researchers used light-responsive sulfonate group modified COFs as the initiator for acrylamide polymerization, utilizing the solvating effect to reduce electrostatic interactions between Zn^{2+} ions and sulfonate groups, thereby preparing TCOF-S-Gel electrolyte with high ionic conductivity and Zn^{2+} transference number (Figure 7.6b). This COF-based hydrogel electrolyte is formed through instantaneous in situ photopolymerization of internal acrylamide within COF pores, initiated by light-responsive sulfonate

Figure 7.6 Design philosophy of the quasi-solid electrolyte (a) Operation perspective diagram of Zn||MnO$_2$ full battery based on TCOF-S-Gel quasi-solid electrolyte. TCOF-S-Gel electrolyte design in this work: (b) top view and sectional view show $-SO_3^-$ and PAM in the COF pore channels provide a "stair" for the rapid migration of Zn^{2+} ions; (c) Zn^{2+} ions are uniformly and reversibly deposited/stripped under the guidance of COF, and almost no dendrites are produced. Source: Qiu et al. [41]/with permission of John Wiley & Sons.

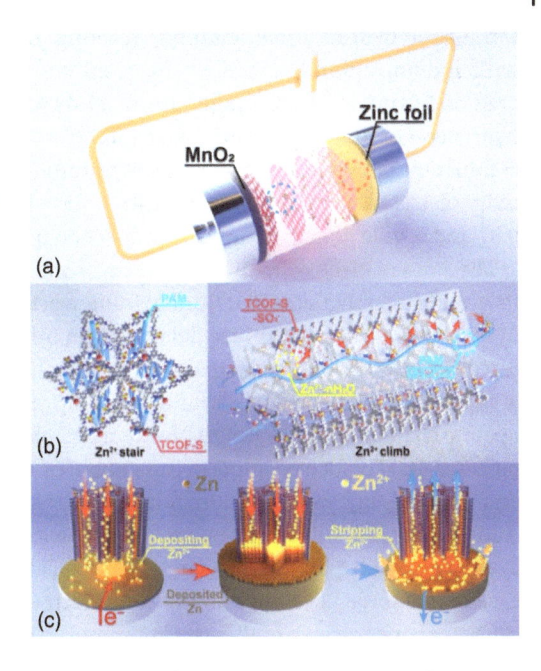

group modified COFs (TCOF-S). TCOF-S-Gel electrolyte demonstrated significantly enhanced electrochemical performance. At room temperature, its ionic conductivity reached 27.2 mS cm^{-1}, and the Zn^{2+} transference number was as high as 0.89, which is much higher than traditional hydrogel electrolytes (such as PAM-Gel). This high performance is attributed to the solvating effect between sulfonate groups in TCOF-S and acrylamide chains, which significantly reduces the interaction between Zn^{2+} ions and COF pore walls, enhancing ion migration efficiency. Molecular dynamics simulations and DFT calculations further revealed the zinc-ion migration mechanism in TCOF-S-Gel. Simulation results indicated that the diffusion rate of Zn^{2+} in TCOF-S-Gel is much higher than in traditional hydrogel electrolytes, while the migration of SO$_4^{2-}$ anions is significantly restricted, confirming the single-ion conductivity characteristics of TCOF-S-Gel. Additionally, TCOF-S-Gel forms a uniform zinc-ion deposition on the zinc foil surface, significantly inhibiting the growth of zinc dendrites and enhancing battery safety (Figure 7.6c).

7.3.2.3 HOF as Electrolytes and Electrolyte Additives

Traditional LMBs are considered an ideal choice for energy storage due to their high energy density. However, they pose safety risks as lithium dendrites easily form in liquid electrolytes. Solid-state electrolytes are viewed as effective alternatives, capable of eliminating this issue, yet common inorganic and polymer solid-state electrolytes are limited by their conductivity and interfacial compatibility. Shi et al. [42] investigated a novel solid-state electrolyte in LIBs, focusing on using a hydrogen-bonding framework based on porous coordination chains (NKU-1000) to enhance battery performance. The NKU-1000 material, composed of ((CH$_3$)$_2$NH$_2$)$_{0.5}$[In$_{0.5}$(HTATAB)]·1.5H$_2$O·0.5CH$_3$OH·DMF, was synthesized through a hydrothermal method, reacting triaryl amine tetraacid (H$_3$TATAB) with indium chloride (InCl$_3$). The resulting material showcases a BET surface area of 540 m^2 g^{-1} and a pore size of 11.48 Å, forming a one-dimensional chain hydrogen-bonding framework within its structure, providing excellent lithium-ion conduction pathways. Notably, this structure demonstrates a lithium-ion conductivity of 1.13×10^{-3} S cm at 25 °C, with a high lithium-ion transference number of 0.87 and a wide electrochemical window of 5.0 V.

The unique electrochemical behavior of the NKU-1000 material is attributed to its orderly negative sites and flexible framework, maintaining efficient and uniform Li$^+$ transport without lithium dendrite formation. The structural design facilitates rapid state transitions, particularly by guiding linear lithium-ion transit to maintain a consistent, high-speed Li$^+$ flow, resulting in excellent cycling life and a wide operational temperature range. In experiments, solid-state lithium batteries based on NKU-1000 displayed a low overpotential of 45 mV under 0.5 mA cm^2 conditions and showed no short-circuiting in prolonged cycling tests lasting 1200 hours. Additionally, no lithium dendrite growth was observed in the stripping/plating cycling tests at 0.5 and 1.0 mA cm^2. This indicates that NKU-1000 not only features excellent interfacial compatibility and electrochemical stability but also significantly enhances the overall performance of LIBs through its unique electrode and electrolyte interface structure. Furthermore, quasi-SSBs paired with LiFePO$_4$ and LiNi$_{0.8}$Co$_{0.1}$Mn$_{0.1}$O$_2$ (NCM-811) cathode materials demonstrated high capacity retention and good rate performance, highlighting NKU-1000 as an innovative solution to improve battery electrochemical stability and lifespan. This strongly suggests NKU-1000's valuable application prospects and research significance in enhancing the safety and long-term stability of LMBs.

7.3.3 Applications in Catalysts and Catalyst Supports

In battery technology, catalysts play a crucial role, particularly in energy conversion devices such as fuel cells and metal-air batteries, where catalyst performance directly determines the reaction efficiency and overall battery performance. Traditional catalyst materials, although they have improved battery efficiency to some extent, still face numerous challenges in terms of activity, selectivity, and stability. POFs, due to their unique pore structure, high specific surface area, and tunable chemical environment, exhibit excellent catalytic performance. These materials can not only provide abundant active sites for electrochemical reactions but also achieve fine control over the catalytic process by adjusting their pore structure and surface functionalization. Moreover, as catalyst supports, POFs can effectively enhance catalyst dispersibility and stability, extend catalyst lifespan, and further boost battery performance. Therefore, the application of POFs in catalysts and catalyst supports offers new directions and possibilities for the development of efficient batteries.

7.3.3.1 MOF as Catalysts and Catalyst Supports

Li et al. [43] synthesized a class of MOF-74(M), composed of metal cations (such as Mg, Ca, Co, Ni, Zn, Fe, Mn, and Cu) and 2,5-dihydroxy-1,4-benzenedicarboxylic

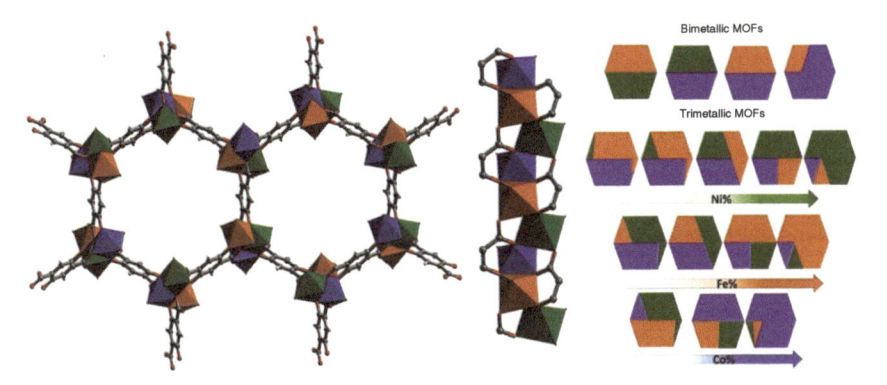

Figure 7.7 The left diagram shows a tri-metallic (Fe, Ni, and Co) MOF-74 structure. The schematic on the right shows the different combinations of the synthesized MOF-74 families: binary MOFs, and the three families of ternary MOFs. Ni, Fe, and Co are represented by green, brown, and purple colors, respectively. Source: Abdelhafiz et al. [43]/CC BY 4.0/with permission of John Wiley & Sons.

acid (DOBDC) ligands (Figure 7.7). The type and composition of its metal nodes can be easily adjusted; the large one-dimensional hexagonal channels of MOF-74 can accelerate the diffusion of reactants, ensuring rapid structural conversion under harsh conditions; its highly accessible unsaturated metal centers can maximize the density of active sites, thereby facilitating the binding of reaction intermediates. By varying the proportions of the three metals Ni, Co, and Fe, the effect on oxygen evolution reaction (OER) activity was systematically studied, and the composition ratio with the best catalytic performance was determined. Electrochemical performance tests indicated that this trimetallic MOF exhibited remarkable activity and stability under OER conditions. Using characterization techniques such as in situ X-ray absorption spectroscopy, the study revealed reasons for enhanced OER activity, including increased metal oxidation states, formation of oxygen vacancies, and enhanced metal-oxygen covalency. These factors synergistically improved the catalyst's electronic conductivity and OER activity, thereby significantly enhancing the electrochemical performance of the electrolyte.

Li-O_2 batteries have garnered attention due to their exceptionally high theoretical specific capacity, but their electrochemical performance is limited by the slow kinetics of redox reactions and the irreversibility of the discharge product Li_2O_2. To address these issues, Shu et al. [44] introduced an electron-rich ferrocene acid (FcA) as a ligand into Mn-MOF-74, forming Mn-MOF-74-FcA to optimize the oxygen electrode reactions in Li-O_2 batteries. Specifically, the introduction of FcA ligands enhanced the metal-ligand interaction at Mn sites, resulting in the rearrangement of 3d orbitals of the [MnO_x] coordination sphere and electron transfer from FcA to the Mn sites. This regulation of electronic structure and orbital arrangement optimized the orbital coupling between Mn sites and oxygen species, enhanced electron exchange, and reduced the energy barrier of the oxygen electrode reactions. By introducing FcA, the synthesized Mn-MOF-74-FcA displayed significant performance enhancement. This material not only has a high specific surface area and good pore size distribution, facilitating the transport of oxygen and lithium-ions, but also performs excellently in electrochemical performance. When Mn-MOF-74-FcA is used as the oxygen electrode in Li-O_2 batteries, it exhibits a low overpotential (0.82 V) and a long cycle life (276 cycles), attributed to its excellent electronic conductivity and fast oxygen electrode reaction kinetics. Additionally, Mn-MOF-74-FcA forms a uniform film-like Li_2O_2 during discharge, which has good interface contact with the electrode surface, facilitating complete decomposition during charging and achieving the reversible formation and decomposition of Li_2O_2. The study demonstrates that regulating the coordination environment of MOFs by introducing electron-rich ligands can significantly enhance the electrochemical performance of Li-O_2 batteries, providing new ideas for the design of high-performance electrocatalysts.

Li-S batteries have garnered attention due to their high energy density but face challenges such as sulfur insulation, volume changes, and the polysulfide (LiPS) shuttle effect. To overcome these issues, researchers have developed quasi-MOF nanostructures. This structure partially disrupts the connection between metal nodes and organic ligands through a controlled ligand exchange strategy, exposing a wealth of active sites. Fransaer et al. [45] developed quasi-MOF nanospheres through a solvent-assisted ligand exchange method, possessing a hierarchical structure and abundant zinc nodes. These provide numerous active sites to enhance sulfur fixation and LiPS chemical interactions, while their dynamic structural evolution can effectively catalyze the LiPS conversion, thereby accelerating reaction kinetics. The application of quasi-MOF nanospheres in Li-S batteries shows significant improvements in electrochemical performance. Through electrochemical performance testing with sulfur loading and low electrolyte/sulfur ratio, S/quasi-MOF nanospheres exhibit an areal capacity of up to 6.0 mAh cm^{-2} and maintain good capacity after 100 cycles. This excellent electrochemical performance is attributed to the morphological and structural advantages of the quasi-MOF nanospheres. The uniform sulfur distribution and excellent conductivity within its porous architecture ensure high sulfur utilization and rapid reaction kinetics, while strong physical sulfur fixation and superior simultaneous removal reaction (SRR) catalytic activity further suppress LiPS dissolution and shuttle behavior, significantly extending cycle life.

Huang et al. [46] used 2-aminoterephthalic acid and titanium tetraisopropoxide as precursors to synthesize the initial Ti-MOF-NH_2. By mixing the synthesized Ti-MOF-NH_2 with tannic acid and controlling different etching times, they obtained a hierarchically porous catalytic metal-organic framework (HPC-MOF). This MOF features a hierarchically porous structure that combines the advantages of macropores and micropores, where macropores ensure rapid mass transport and micropores provide a high density of catalytic active sites. In batteries, the HPC-MOF effectively suppresses the shuttle effect of LiPSs through its porous structure and accelerates redox reactions through abundant catalytic sites, thereby enhancing the electrochemical performance of the battery. The discharge capacity of Li-S batteries with the introduction of HPC-MOF increased by 164.6% at 1.0C, and the decay rate in long-term cycling decreased by 83.3%. Even under high sulfur loading (7.1 mg cm^{-2}) and dilute electrolyte conditions, the battery still exhibits stable cycling performance, showing an initial areal capacity of 6.34 mAh cm^{-2} and maintaining at 5.85 mAh cm^{-2} after 100 cycles under high sulfur loading. In summary, the introduction of HPC-MOF effectively enhances the electrolyte performance and electrochemical stability of Li-S batteries by optimizing mass transport and catalytic site density, providing new insights and methods for the development of high-performance energy-storage devices.

7.3.3.2 COF as Catalysts and Catalyst Supports

Sun et al. [47] designed a zwitterionic covalent organic framework (zwitterionic COF) by reacting 2,4,6-triformylphloroglucinol (TFP) with 2,5-diaminobenzene-sulfonic acid (Pa-SO$_3$H) and ethidium bromide, resulting in a COF with cationic and anionic functional groups. The synthesized COF was then stirred with a 15 wt% aqueous solution of LiTFSI at 50 °C for two days to enhance its electrolyte performance. This zwitterionic COF possesses a bidirectional electrostatic catalytic function, which helps improve the conversion efficiency of LiPSs, suppress the shuttle effect, and promote rapid Li$^+$ conduction. Experimental results show that the Li-S battery, with the introduction of the zwitterionic COF, improves its discharge capacity to 1426.6 mAh g^{-1} at 1.0C, with a capacity retention rate of 83.5% during long-term cycling. Under high sulfur loading (5.0 mg c m^{-2}) and dilute electrolyte conditions, the battery exhibits excellent cycling stability, with an initial areal capacity of 696.8 mAh g^{-1} and maintains high efficiency after 50 cycles.

Zinc-air batteries are severely limited in energy efficiency and lifespan by the slow kinetics of the oxygen reduction reaction (ORR) and OER. However, the slow reaction kinetics of ORR and OER leads to high overpotential and low efficiency during the charge-discharge process of the battery, thus necessitating effective electrocatalysts to promote these reactions. Kim et al. [48] synthesized PTCOF via a nucleophilic substitution reaction between cyanuric chloride and 2,6-diaminopyridine under mild conditions (Figure 7.8). They then introduced cobalt nanoparticles (CoNP) into the PTCOF, forming a pyridine-linked triazine COF with bifunctional electrocatalytic activity, known as CoNP-PTCOF (Figure 7.8). This COF, through its unique structural design, serves as a bifunctional electrocatalyst in the battery. The electronic structure of CoNP-PTCOF, after regulation, can effectively catalyze ORR and OER, significantly reducing the overpotential of these reactions and demonstrating excellent stability. CoNP-PTCOF shows a half-wave potential (E1/2) of 0.85 V (relative to RHE) in ORR, with a diffusion-limited current density of −6.0 mA cm^{-2}, comparable to commercial Pt/C ($E_{1/2}$ of 0.87 V). Additionally, CoNP-PTCOF has an OER overpotential of 0.45 V (relative to RHE, 10 mA cm^{-2}), close to the level of commercial RuO$_2$ (0.4 V). Moreover, zinc-air batteries assembled with CoNP-PTCOF exhibit excellent electrochemical performance, with charge-discharge curves showing a peak power density of 53 mW cm^{-2}, comparable to batteries using a Pt/C and RuO$_2$ mixture (72.2 mW cm^{-2}). The zinc-air battery assembled with CoNP-PTCOF shows a specific capacity of 796.9 mAh g^{-1} at 10 mA cm^{-2}, with a voltage efficiency of 60.4% after 50 cycles, significantly better than the 52.1% of commercial Pt/C and RuO$_2$ mixture batteries.

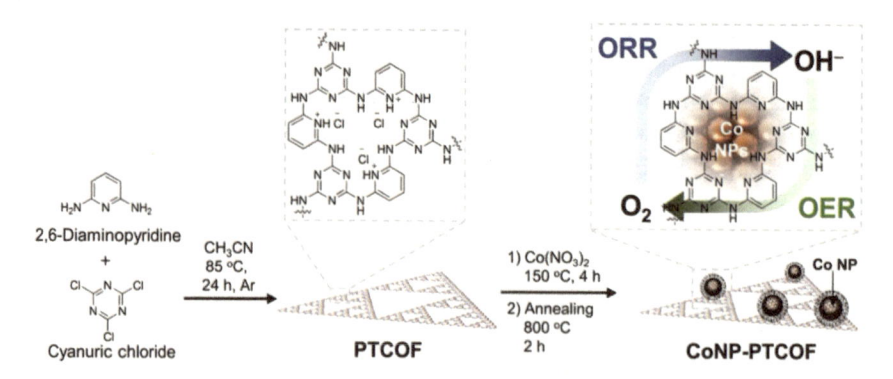

Figure 7.8 Schematic illustration of the synthesis of pyridine-linked triazine covalent organic framework (PTCOF) and cobalt nanoparticle-embedded PTCOF oxygen electrocatalysts (CoNP-PTCOF). Source: Park et al. [48]/with permission of John Wiley & Sons.

Fang et al. [49] designed and synthesized two unique triazine-cyclotriphosphazene-based COFs (Q3CTP-COFs, named JUC-610 and JUC-611, respectively) and applied their ultrathin nanosheets (JUC-610-CON) as ORR electrocatalysts. Q3CTP-COFs are synthesized through the condensation reaction of hexa(4-formylphenoxy)cyclotriphosphazene (CTP-6-CHO) with 2,4,6-tris(4-aminophenyl)triazine (TAPT). Due to their unique bilayer stacking structure, these COFs can expose more active carbon sites and accelerate mass diffusion during the ORR process. Weak interlayer π-π interactions make these COFs easily exfoliate into ultrathin nanosheets, further enhancing electrocatalytic performance. In alkaline electrolyte, JUC-610-CON exhibited a half-wave potential of 0.72 V (relative to RHE), and under 300 mA cm^{-2} conditions, the power density of the JUC-610-CON based zinc-air battery reached 156 mW cm^{-2}, surpassing most reported COF materials. Additionally, JUC-610-CON showed excellent stability, with almost no voltage loss over 200 hours of operation. DFT calculations confirmed that the electrophilic structures in Q3CTP-COFs induce high-density positively charged carbon active sites, thereby facilitating the oxygen adsorption and reduction process.

To address the energy density and stability issues of zinc-air batteries, Chen et al. [50] designed a one-dimensional van der Waals heterostructure by depositing thiophenyl groups on a graphene substrate to construct TAPTt-COF material, which was used as an efficient metal-free oxygen reduction electrocatalyst. The superior performance of the COF material was validated through a series of electrochemical tests. In 0.1 M KOH electrolyte, the zinc-air battery based on CC-3 material showed excellent ORR catalytic activity, with a half-wave potential ($E_{1/2}$) of 0.828 V relative to RHE, surpassing the commercial Pt/C's 0.865 V. In OER tests, CC-3 material exhibited an overpotential (η10) of 389 mV at a current density of 10 mA cm^{-2}, comparable to the performance of commercial IrO$_2$ catalyst. At a current density of 10 mA cm^{-2}, the specific capacity of the zinc-air battery based on CC-3 reached 714 mAh g^{-1}, and at a high discharge current density of 40 mA cm^{-2}, its specific capacity was 696 mAh g^{-1}, outperforming Pt/Ir-C based batteries. After 120 charge-discharge cycles, the battery using CC-3 material maintained 94% of its initial current density, demonstrating excellent cycling stability.

7.3.3.3 HOF as Catalysts and Catalyst Supports

With societal development, traditional LIBs no longer meet our energy-storage demands. Lithium-carbon dioxide (Li-CO$_2$) batteries are considered one of the most promising high-capacity storage systems due to their high theoretical energy density (1876 Wh kg^{-1}), low cost, and environmental friendliness. However, the slow redox kinetics of CO$_2$ in the Li-CO$_2$ battery system results in poor rate capability and low cycle stability, making the practical application of Li-CO$_2$ batteries extremely challenging. Zhang's team [51] designed a stable HOF-based multifunctional cathode catalyst, achieving long-cycle stable operation of lithium-carbon dioxide batteries under high-rate conditions (Figure 7.9). This HOF-based multifunctional catalyst is composed of a highly stable hydrogen-bonded organic framework (HOF-FJU-1) combined with ruthenium nanoparticle-modified carbon nanotubes. The introduction of HOF-FJU-1 effectively enhances the catalyst's ability to transport CO$_2$ and Li$^+$, enabling efficient catalytic conversion of CO$_2$. Specifically, the introduction of HOF-FJU-1 into the catalyst offers the following advantages: The porous HOF-FJU-1 has the ability to adsorb CO$_2$, improving CO$_2$ conversion efficiency; The porous HOF-FJU-1 containing cyano functional groups induces uniform deposition of discharge products and alleviates volume expansion during charge-discharge processes to some extent; HOF-FJU-1 causes a positive shift in the ruthenium d-band, significantly lowering the reaction energy barriers during lithium carbonate formation/decomposition, forming uniform and efficient catalytic sites; HOF-FJU-1 has regularly arranged pore channels that facilitate efficient transport of CO$_2$ and Li$^+$, enhancing CO$_2$ redox kinetics. Consequently, a Li-CO$_2$ battery based on HOF-FJU-1 can operate stably for 1800 hours at 400 mA g^{-1} and maintain a low overpotential of 1.96 V even at a high current density of up to 5 A g^{-1}. This work provides significant insights into the development of HOF-based lithium-carbon dioxide battery catalysts and broadens the application range of HOF in the energy-storage field.

Luo et al. [52] proposed a hydrogen-bonded cobalt porphyrin framework that can effectively accommodate iodine and serve as an electrocatalyst for aqueous zinc-iodine (Zn-I$_2$) organic batteries. FT-IR, XPS, and DFT results indicate that

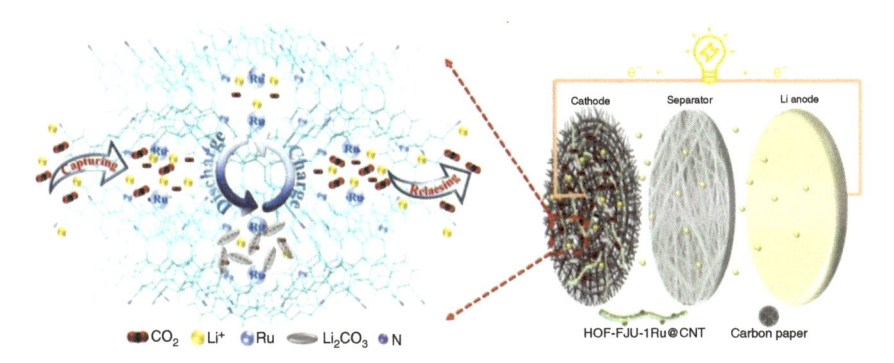

Figure 7.9 The schematic diagram of a Li-CO$_2$ battery. Source: Cheng et al. [51]/with permission of John Wiley & Sons.

the HOF exhibits excellent iodine species adsorption capability. In situ Raman spectroscopy showed that the redox mechanism of the Zn-I$_2$ battery depends on the I/I$^-$ redox reaction, with I$_3^-$/I$_5^-$ serving as intermediate products. In situ Ultraviolet-visible (UV-vis) spectroscopy further revealed that HOF restricts polyiodide dissolution. The aqueous Zn-I$_2$ organic battery with an I$_2$@PFC-72-Co cathode exhibited excellent rate performance, achieving 134.9 mAh g^{-1} at 20C. Moreover, these batteries demonstrated long-term cycling stability, enduring over 5000 cycles at 20C. The impressive electrochemical performance of I$_2$@PFC-72-Co can be attributed to the synergistic Co single-atom electrocatalyst within the HOF-Co structure. Additionally, the benzene ring structure and carboxyl functional groups of HOFs have a strong capability to adsorb iodine and iodide. Due to these synergistic effects, aqueous Zn-I$_2$ batteries with an I$_2$@PFC-72-Co cathode exhibit exceptional electrochemical performance.

7.3.4 Applications in Battery Separators

Battery separators, as crucial components of batteries, can effectively prevent internal short circuits, enhance battery safety, and facilitate improved battery performance. Currently, the development of battery separators has entered a new stage, with continuous advancements and innovations in material selection, preparation processes, and structural design. Especially in the field of new energy, research and progress on battery separators have become significant research directions. Traditional polyolefin and glass fiber separators are prone to dendrite puncture, exhibit poor cycling stability, and cannot regulate ion transport, which to some extent impairs the electrochemical performance of batteries. Even when cellulose is used to make separators, uneven porosity cannot mitigate these issues. POFs, as a new class of materials with excellent ion transport properties, chemical stability, and mechanical toughness, can replace traditional polymer separators in areas such as LIBs, Li-S batteries, aqueous ZIBs, and supercapacitors.

7.3.4.1 MOF as Battery Separator

To address the challenges of Li-S batteries, such as the dissolution and shuttle effect of lithium polysulfides (LiPS), Deng et al. [53] designed an efficient separator by introducing doped metal sites (Fe) to synthesize a flower-like bimetallic MOF, namely Fe-ZIF-8, based on the original ZIF-8. These MOFs were then coated onto a polypropylene (PP) separator to form a modified separator. The role of Fe-ZIF-8 as a separator material is mainly reflected in two aspects: first, it blocks the migration of LiPS through selective channels; second, it enhances the conversion of LiPS via Fe electrocatalytic sites. Additionally, the Fe-ZIF-8 modified separator prevents the formation of lithium dendrites due to its uniform pore size (approximately 3.4 and 10 Å), ensuring even transmission and deposition of lithium-ions. Li-S batteries using the Fe-ZIF-8/PP separator exhibited an initial discharge capacity of 863 mAh g^{-1} at 0.5C and 746 mAh g^{-1} at 3C. Under conditions of high sulfur loading (5.0 mg cm^{-2}) and a low electrolyte-to-sulfur ratio (5 μL mg^{-1}), the capacity retention rate was 92% after 1000 cycles. Moreover, Li||Li symmetric cells demonstrated stable cycling performance for over 4000 hours at a high current density of 10 mA cm^{-2}, highlighting the significant advantages of the Fe-ZIF-8/PP separator in enhancing the long cycle life and capacity retention of the battery.

Liu et al. [54] synthesized MOF-808 by reacting ZrCl$_4$ with 2-amino terephthalic acid. Subsequently, the MOF was mixed with glucose and graphene oxide and subjected to pyrolysis, resulting in the formation of the MOF@CC composite material. The MOF@CC composite was then coated onto a commercial PP separator, creating the MOF@CC@PP separator. This advanced structure provides specific transport channels that accelerate the diffusion of Li$^+$ through the aggregation of anionic and cationic clusters while enhancing electron exchange via C—N bridge bonds. Consequently, this configuration improves the electrocatalytic efficiency of polysulfides (LiPS) and effectively suppresses their accumulation and deposition. The Li-S battery utilizing the MOF@CC@PP separator achieved a remarkable reversible capacity of 1063 mAh g^{-1} at 0.5C, with a capacity retention rate of 88% after 100 cycles and a capacity of 765 mAh g^{-1} at a high discharge rate of 5C. Furthermore, the large-area pouch cell demonstrated stable performance, maintaining a capacity of 855 mAh g^{-1} after 70 cycles.

Traditional MOF particle modification layers often fail to achieve true "ion sieving" effects due to particle gap influences, but constructing continuous, ultrathin, crack-free MOF membranes can effectively shield polysulfides and homogenize lithium-ion deposition. Xu et al. [55] explored the application and mechanism of MOF membranes in Li-S batteries, emphasizing the key role of an ultrathin, crack-free MOF film in suppressing polysulfides. MOFs are used to construct modification layers to block polysulfide migration in batteries due to their unique microporous structure that can serve as an ion sieve. Researchers have successfully fabricated a 20 nm thick MOF film without cracks on the battery separator using atomic layer deposition (ALD) technology for the first time. The ultrathin MOF film made from ZIF-8 not only blocks polysulfides but also synergistically promotes uniform lithium-ion deposition. The pore size design of the MOF membrane is critical, with pore sizes between 0.152 nm for lithium-ions and 1.2 to 1.7 nm for polysulfides, allowing effective screening of both ions. The working mechanism of this MOF layer was validated through phase field simulations and in situ Raman spectroscopy, showing it significantly mitigates polysulfide shuttling, thus enhancing the battery's cycle stability. The study results indicate that Li-S batteries with MOF membranes exhibit excellent performance during continuous cycling, maintaining a capacity retention rate of 95.6% after 500 cycles. Even at high sulfur loadings, the Li-S batteries maintained good electrochemical performance. Electrochemical impedance spectroscopy testing showed that batteries with MOF membranes have lower charge transfer resistance than those with traditional separators, further indicating superior ionic conductivity and interface stability. Additionally, no polysulfide signals were observed when monitoring battery performance with in situ Raman spectroscopy, further proving the effective blocking action of the MOF membrane against polysulfides. Comparing different types of separators, the study also showed that the uniformity and continuity of MOF membranes significantly surpass those of traditional MOF particle layers, effectively enhancing Li-S battery performance and stability. In Li-S batteries, MOFs not only enhance cycle performance by blocking polysulfide shuttling but also promote uniform lithium-ion deposition, offering new strategies and methods for developing efficient, stable Li-S batteries. These findings lay the groundwork for future improvements and applications of Li-S batteries, showcasing the tremendous potential and broad application prospects of MOF materials.

To address the challenges of LMB cycling stability, which are affected by lithium dendrite growth and the instability of the SEI, Chen et al. [56] proposed a lithium-friendly separator made of a fluorinated polymer (LS). This innovative separator was developed through electrospinning and hydrothermal methods, allowing for the in situ growth of MOF-801 on the surface of the fluorinated polymer fibers, resulting in a uniformly distributed multilayer cross-structured separator. The lithium-friendly groups (C—F) provided by the fluorinated polymer fibers effectively guide the uniform deposition of lithium. Moreover, the MOF-801 enhances the transport rate of Li$^+$ through the adsorption of PF$_6^-$ ions. The Li||Li symmetrical cells utilizing the LS separator demonstrated an impressive cycling life exceeding 1000 hours at a current density of 1 mA cm^{-2}, with a remarkably low overpotential of 34 mV. Under high current density conditions of 5 mA cm^{-2}, the LS separator exhibited an average overpotential of 180 mV, whereas cells using the Celgard separator experienced short-circuiting at high currents (as shown in Figure 7.10). Additionally, in NCM-811||Li batteries, the discharge capacity reached 159.7 mAh g^{-1} at 2C using the LS separator, exceeding the capacity by 13.7 mAh g^{-1} compared to those with the Celgard separator. The capacity retention rate increased by 11% after 400 cycles at 0.5C. For the LiFePO$_4$||Li batteries, a capacity retention

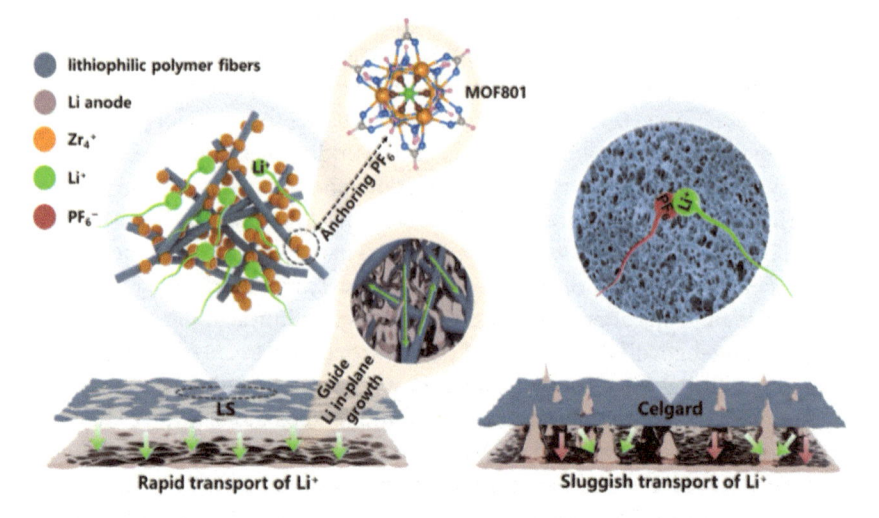

Figure 7.10 Schematic illustration of the role and Li deposition behaviors with (left) LS and (right) Celgard separators. Source: Zuo et al. [56]/with permission of John Wiley & Sons.

rate of 92.9% was achieved after 500 cycles at 1C, significantly outperforming the batteries equipped with the Celgard separator, which retained only 75.2% of their initial capacity.

High-voltage LMBs present significant potential for next-generation electric vehicle applications due to their high energy density. However, their electrochemical stability under high voltage and elevated temperature conditions is threatened by the dissolution of TM ions and the shuttle effect. Zhang et al. [57] introduced ethylenediaminetetraacetic acid (EDTA)-modified MOF-808 (M-E) as a multifunctional ion-selective separator coating, creating M-E with natural ion channels and uniform chelating sites that accurately capture TM ions, thereby preventing their electro-deposition at the anode. Structural characterization revealed that the crystal structure of MOF-808 remained intact after EDTA modification, with a slight reduction in pore size (from 13.8 to 11.8 Å), while still effectively facilitating lithium-ion transport. The NCM622//Li batteries utilizing the M-E@Celgard separator achieved a coulombic efficiency of 99.68% and maintained over 70% capacity retention after 1000 cycles. Moreover, the separator exhibited excellent cycling stability at elevated temperatures, specifically at 55 °C, preventing battery failure. The NCM622//Li batteries equipped with the M-E@Celgard separator delivered a discharge capacity of 139.1 mAh g^{-1} at a rate of 5C, with a capacity retention of 116.1 mAh g^{-1} after 1000 cycles, demonstrating significantly enhanced capacity retention and cycling life compared to unmodified batteries. Additionally, NCM811//Li batteries showed notable improvements in cycling stability at a rate of 1C, achieving a capacity retention rate of 88.7%.

Kang et al. [58] developed a functional ultrathin separator (FUS) with a thickness of 23 µm, achieving excellent electrochemical stability of the zinc anode and long-term durability of the ultrathin separator by modulating the interfacial chemistry of zinc deposition. The Cu-BTC MOF precursor was pyrolyzed in an H$_2$/Ar atmosphere to produce C/Cu nanocomposites, which featured a high specific surface area and a uniform distribution of nanoparticles. These nanocomposites were then coated onto a mechanically robust cellulose nanofiber membrane substrate, resulting in the FUS. The uniform nanopore structure and ion-selective capabilities enhanced the transport of Zn^{2+} ions, while the abundant zinc-affinity sites reduced local current density, promoting uniform zinc nucleation and deposition while suppressing the growth of zinc dendrites. The Zn//Zn symmetrical batteries utilizing the FUSs operated stably for over 600 hours at a current density of 5 mA cm^{-2}, in stark contrast to batteries with glass fiber separators, which failed within 15 hours. Furthermore, as the current density increased from 0.2 to 10 mA cm^{-2}, the batteries with FUSs exhibited superior charge-discharge performance, with an exchange current density reaching 26.9 mA cm^{-2}. Additionally, in the Zn//AC zinc-ion hybrid supercapacitor, the discharge capacity with the FUS was 117 mAh g^{-1} at 0.1 A g^{-1} and 75 mAh g^{-1} at 1 A g^{-1}, significantly outperforming the batteries with glass fiber separators, which exhibited capacities of 97 and 53 mAh g^{-1}, respectively. Cycling lifespan tests demonstrated that the supercapacitors utilizing FUSs maintained near 100% coulombic efficiency after 5000 cycles at 1 A g^{-1}.

7.3.4.2 COF as Battery Separator

Existing commercial separators, such as polyethylene (PE) membranes, are inadequate in preventing lithium dendrite penetration and exhibit limited lithium-ion conductivity. Therefore, there is a pressing need to develop high-performance functional separators to enhance the electrochemical performance and stability of LMBs. COFs present unique characteristics that allow for effective lithium-ion conduction pathways, improved mechanical strength of the separator, and suppression of lithium dendrite growth. Wu et al. [59] synthesized sulfonic acid-functionalized COF nanosheets (TpPa-SO$_3$H) through a heterogeneous polymerization reaction between 2,5-diaminobenzenesulfonic acid and 1,3,5-trihydroxybenzaldehyde, followed by an ion exchange reaction to obtain lithium-intercalated COF nanosheets (TpPa-SO$_3$Li). Ultimately, a super-thin composite separator (TpPa-SO$_3$Li@PE) was fabricated by vacuum self-assembly of TpPa-SO$_3$Li onto the surface of a commercial PE separator. The COF provides efficient lithium-ion conduction pathways, enhances the mechanical strength of the separator, and facilitates the uniform flow of lithium-ions, effectively inhibiting dendrite growth.

The TpPa-SO$_3$Li@PE separator exhibited excellent lithium-ion conductivity (0.96 mS cm^{-1}) and a high lithium-ion transference number (0.83) at room temperature. Under a high current density of 5 mA cm^{-2}, Li/Li symmetrical batteries employing the TpPa-SO$_3$Li@PE separator demonstrated stable lithium deposition/stripping cycles exceeding 2600 hours. In Li/LiFePO$_4$ full batteries, a capacity retention rate of 94.9% was achieved after 300 cycles at 1C, while a high specific capacity of 113.6 mAh g^{-1} was observed at a rate of 5C.

Existing Li-S batteries face significant challenges during charge-discharge cycles due to the migration of LiPS, leading to irreversible loss of active materials and self-discharge phenomena. Additionally, the growth of lithium dendrites poses risks of short circuits and safety concerns. Wang et al. [60] addressed these issues by designing and synthesizing a sulfonic acid-rich covalent organic framework (SCOF-2) to modify the separator in Li-S batteries. The framework was created by condensing 2,5-diaminobenzenesulfonic acid with 2,4,6-trihydroxybenzaldehyde through a Schiff base reaction, resulting in a COF featuring abundant sulfonic acid groups (as shown in Figure 7.11a). Compared to single-sulfonic and nonsulfonic COFs, SCOF-2 exhibited enhanced electronegativity and a larger interlayer spacing, which effectively hindered the migration of LiPS and suppressed the formation of lithium dendrites. As illustrated in Figure 7.11b, the modification with SCOF-2 improved the conductivity of lithium-ions and inhibited dendrite growth, thereby enhancing the cycling stability and capacity retention of the batteries.

The SCOF-2-modified batteries demonstrated an ultra-low capacity decay rate of only 0.047% per cycle after 800 cycles at a current density of 1C. Furthermore, the modified batteries exhibited only a 6.0% capacity decay during one week of rest, indicating outstanding anti-self-discharge performance. Under conditions of high sulfur loading (3.2–8.2 mg cm^{-2}) and a sparse electrolyte (5 μl mg^{-1}), the SCOF-2-modified batteries retained approximately 80% of their capacity after 100 cycles, highlighting their significant potential for practical applications.

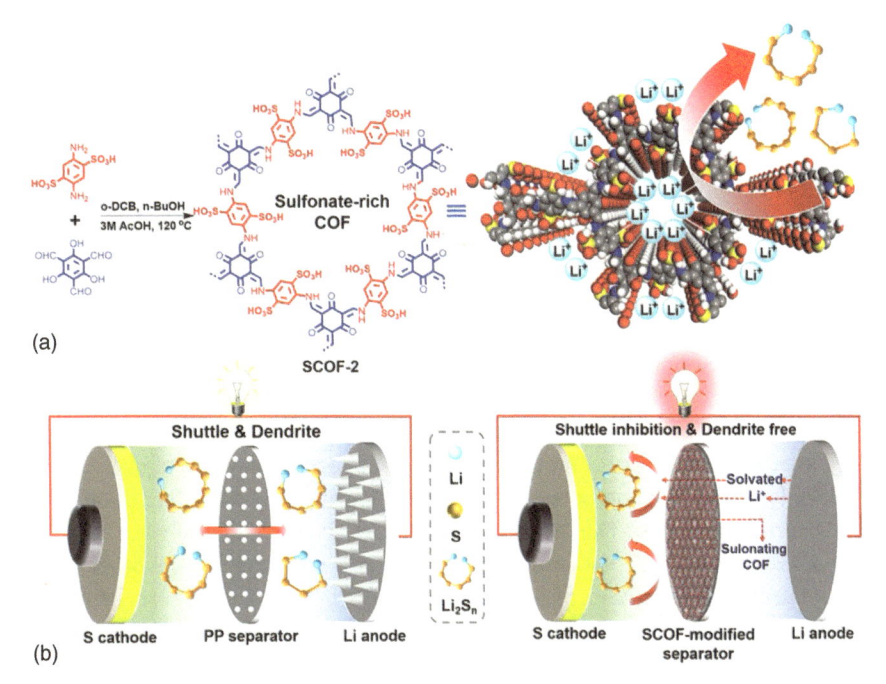

Figure 7.11 (a) Schematic synthesis of the sulfonated COFs, selective permeability of Li$^+$, and the blocking of polysulfides in Li-S batteries. (b) Graphic comparison of the batteries with different separators: the battery with PP separator shows shuttle effect and dendrite growth (left), while the SCOF-modified battery shows shuttle inhibition and dendrite-free (right). Source: Xu et al. [60]/with permission of John Wiley & Sons.

Crown ether groups can catalyze the diffusion and transformation of LiPS and facilitate the deposition and decomposition of Li$_2$S through strong interactions with Li$^+$ ions. Fu et al. [61] synthesized a crown-ether-functionalized COF (BPTA-CE-COF) via a Schiff base reaction between 5′,5″-bis(4-aminophenyl)-[1,1′ : 3′,1″ : 3″, 1‴-quaterphenyl] (BPTA) and 4,4′,4″,4‴-(6,7,9,10,17,18,20,21-octahydrodibenzo[b,k][1,4,7,10,13,16]hexaoxacyclo-octadecane-2,3,13,14-tetrayl)tetrabenzaldehyde (CEs). This COF was then in situ grown on the surface of CNTs to form a CE-COF@CNT composite.

The structural properties of CE-COF@CNT include good conductivity and an abundance of crown ether functional groups that form supramolecular channels. These channels effectively catalyze the diffusion and conversion of LiPS, enhancing the rates of Li$_2$S formation and decomposition, thereby significantly improving battery performance. Li-S batteries with separators modified by CE-COF@CNT exhibited a remarkable initial discharge capacity of 1618.1 mAh g^{-1} at 0.1C, demonstrating excellent rate performance and a capacity decay rate of only 0.040% after 1000 cycles at 1C. Furthermore, even under conditions of high sulfur loading (8.04 mg cm^{-2}) and low electrolyte volume (4.0 μl mg^{-1}), the batteries maintained a high capacity retention of 92.9% after 100 cycles.

Similar to Li-S batteries, sodium-sulfur (Na-S) batteries face challenges like rapid capacity decay and short lifespan. To enhance the performance of Na-S batteries,

Wang et al. [62] introduced a COF film containing azobenzene side groups as an active separator to address these issues. The COF film was synthesized via the condensation reaction between 1,3,5-triformylbenzene (Tb) and azobenzene-modified hydrazine (Azo-Th), forming Azo-TbTh. This film was fabricated on a glass fiber (GF) substrate through interfacial polymerization, resulting in a continuous film approximately 70 nm thick. The azobenzene side groups help narrow pore sizes, suppress the shuttle effect of polysulfides, and provide ion hopping sites to facilitate sodium-ion migration.

This COF functions in dual roles within the battery: firstly, the azobenzene side groups effectively narrow the pore sizes to block polysulfide migration; secondly, they provide ion hopping sites that reduce the transport resistance of sodium-ions, thereby facilitating their migration. Na-S batteries utilizing the Azo-TbTh film as a separator achieved a capacity of 1295 mAh g^{-1} at 0.2C, with a capacity retention rate of 90% after 1000 cycles at 1C, corresponding to a capacity decay rate of just 0.036% per cycle. They also demonstrated excellent rate performance and cycling stability at various current densities. The discharge capacities at 0.1C, 0.2C, 0.5C, 1C, 2C, and 3C were 1540, 1336, 1057, 840, 646, and 465 mAh g^{-1}, respectively. At a 3C current density, the discharge capacity recovered to 1314 mAh g^{-1}, showcasing excellent rate performance and cycling stability. After 1000 cycles at 1C, the battery maintained a capacity of 524 mAh g^{-1}, with a capacity decay rate of just 0.036% per cycle.

Uncontrolled side reactions between sodium metal and the electrolyte can lead to sodium dendrite growth and instability in battery performance. Lee et al. [63] addressed these challenges by reacting dihydroxybenzophenone with terephthalaldehyde to synthesize a COF with a conjugated π-system. This COF plays a critical role in regulating sodium-ion transport and suppressing sodium dendrite growth in batteries. The COF separator features an open nanoporous structure that significantly enhances the sodium-ion transference number (t_{Na^+} = 0.78) and induces the formation of sodium fluoride (NaF)-rich SEI. At current densities as high as 20 mA cm^{-2}, the COF separator effectively inhibited sodium dendrite growth, demonstrating excellent electrochemical performance.

In symmetric cells using the COF separator, a cycle life exceeding 1500 hours was achieved at a current density of 20 mA cm^{-2}. Additionally, a full cell configuration of Na|COF|NaTi$_2$(PO$_4$)$_3$(NTPO) exhibited over 5000 cycles at high rates of 5C and 10C, maintaining excellent capacity retention.

7.3.4.3 HOF as Battery Separator

HOFs offer a promising solution for battery separators due to their lack of TM ions, resulting in more stable, lighter, and thinner membranes with higher energy densities. Liu et al. [64] synthesized an HOF material named cyanuric acid-melamine (CAM) by self-assembling melamine and cyanuric acid in an aqueous solution. This HOF was then mixed with polytetrafluoroethylene and rolled into a uniformly thick separator. During charge-discharge cycles, the triazine rings within CAM serve as excellent lithium-ion conductors. Li$^+$ ions preferentially form Li—N bonds with the nitrogen atoms in the triazine rings, facilitating effective migration within CAM. This mechanism allows Li$^+$ ions to preferentially deposit on the surface of lithium metal, while solvent molecules are released back into the electrolyte, reducing side reactions. In contrast, conventional Celgard separators in LMBs allow both Li⁺ deposition and solvent molecule decomposition.

Liu's group assembled full cells using CAM and Celgard separators, lithium metal, and NCM 622 cathodes for electrochemical performance testing. Under the condition where the anode-to-cathode capacity ratio (N/P) was 5, the batteries with CAM separators demonstrated superior cycling stability and rate performance compared to those with Celgard separators. This exceptional performance is attributed to the excellent desolvation capability of the CAM separator, significantly reducing side reactions between lithium metal and solvent molecules, and highlighting its potential to enhance the performance of LMBs.

LIBs have become the dominant energy-storage devices in recent years due to their maturity and widespread application compared to other energy-storage technologies. However, as the pursuit for high energy density LIB systems continues, safety concerns have emerged. $LiNi_xCo_yMn1-x-yO_2$ (NCM) is considered the ideal cathode material to meet the needs of high-energy-density batteries. Nevertheless, issues such as structural instability and poor thermal stability of NCM pose significant challenges to cycling stability and safety, restricting the large-scale commercial application of NCM-based LIBs.

Sun et al. [65] addressed these challenges by modifying PP separators with varying thicknesses of HOF material, specifically cyanuric acid melamine (MCA). MCA features a 0.9 nm pore channel structure that enables selective lithium-ion transport through physical size sieving. Additionally, MCA contains various polar groups such as C=N and C=O, which exhibit strong adsorption toward TM elements leaching from NCM, preventing the cross-membrane shuttling of these elements that degrades long-cycle battery performance. Moreover, MCA's high thermal stability and flame-retardant properties enhance battery safety via a coverage-based flame-retardant mechanism.

The MCA functional layer modification strategy effectively addresses the issues of poor lithium-ion conductivity and low safety of commercial polyolefin separators. This advancement improves both the electrochemical and safety performance of NCM-based LIBs, offering a new direction for the development of novel battery separators.

7.4 Conclusion and Outlook

7.4.1 Conclusion

In the research domain of POFs, recent years have witnessed significant advancements in battery applications. POFs, encompassing MOFs, COFs, and HOFs, have emerged as pivotal materials driving innovation in next-generation battery technologies due to their finely tunable pore structures and exceptional chemical and thermal stability. These materials have demonstrated remarkable potential in a variety of energy-storage devices including LIBs, SIBs, and zinc-air batteries. POFs are poised to reshape the design and development of battery materials, offering a transformative approach to enhancing battery performance and safety.

1) Innovative Applications in Electrode Materials: The high surface area and customizable pore structures of POFs offer abundant electrochemical active sites, significantly enhancing the specific capacity and cycle life of electrode materials. Notably, in Li-S and lithium-oxygen batteries, POFs exhibit superior performance in accelerating the kinetics of redox reactions and inhibiting the shuttle effect of LiPS.
2) Electrolyte Design and Optimization: The nano-confinement effect of POF frameworks boosts ion transport dynamics, effectively increasing ionic conductivity and electrochemical stability of electrolytes, showcasing broad application prospects in the development of all-solid-state electrolytes. Additionally, flexible chemical modification through pore channel design can enhance the stability of the electrolyte-electrode interface, improving overall battery performance.
3) Advantages as Catalysts and Supports: As catalysts or catalyst supports, POFs leverage their unique pore structures to significantly enhance catalytic activity and stability, improving the efficiency of key reactions in batteries such as the ORR and OER, while demonstrating prolonged electrochemical stability.
4) Innovative Battery Separators: POF-based membrane materials particularly excel in preventing the diffusion of LiPS in electrolytes and inhibiting lithium dendrite growth. Their high ion selectivity and excellent structural stability provide reliable support for enhancing the safety and cycling capability of batteries.

7.4.2 Outlook

The future development direction of the application of porous frame materials in batteries is mainly concentrated on the following aspects:

1) Structural Functionalization and Performance Optimization: Future research should focus on optimizing the structural and functional properties of POFs through chemical synthesis and functionalization techniques. This effort aims to achieve higher specific capacities, excellent conductivity, and chemical stability to meet specific application demands.
2) From Laboratory to Large-Scale Applications: Despite POFs showing impressive results in laboratory settings, overcoming the barriers of production costs and preparation processes is essential for scaling up to industrial applications. Simplifying efficient synthesis processes and cost-effective material design will be critical factors.
3) Environmental Friendliness and Sustainable Development: Developing environmentally friendly synthetic routes to reduce the use of organic solvents and harmful chemical reagents represents a significant challenge for the sustainable development of POFs. Enhancing the recyclability of materials will also lay the foundation for durable, eco-friendly green energy-storage solutions.
4) Interdisciplinary Collaborative Innovation: Combining materials science, electrochemical engineering, and computational simulation techniques to

deeply understand the performance mechanisms of POFs in batteries is crucial. Promoting interdisciplinary collaborative innovation will accelerate the application and development of emerging technologies.

As research continues to progress, POFs are expected to transcend the limitations of current materials, becoming central to reshaping energy-storage technologies. Their broad application prospects in energy-storage devices will drive the development and widespread adoption of high-performance, long-lasting, and environmentally friendly battery systems.

References

1 Luo, D., Li, M., Ma, Q. et al. (2022). Porous organic polymers for Li-chemistry-based batteries: functionalities and characterization studies. *Chemical Society Reviews* 51: 2917–2938.
2 Park, K.S., Ni, Z., Cote, A.P. et al. (2006). Exceptional chemical and thermal stability of zeolitic imidazolate frameworks. *Proceedings of the National Academy of Sciences of the United States of America* 103: 10186–10191.
3 Côté, A.P., Benin, A.I., Ockwig, N.W. et al. (2005). Porous, crystalline, covalent organic frameworks. *Science* 310: 1166–1170.
4 Hisaki, I., Xin, C., Takahashi, K. et al. (2019). Designing hydrogen-bonded organic frameworks (HOFs) with permanent porosity. *Angewandte Chemie International Edition* 58: 11160–11170.
5 He, Y., Xiang, S., and Chen, B. (2011). A microporous hydrogen-bonded organic framework for highly selective C2H2/C2H4 separation at ambient temperature. *Journal of the American Chemical Society* 133: 14570–14573.
6 Sammawipawekul, N., Kaeosamut, N., Autthawong, T. et al. (2024). Isostructural dual-ligand-based MOFs with different metal centers in response to diverse capacity lithium-ion battery anode. *Chemical Engineering Journal* 482: 148904.
7 Wang, Y.Q., Cheng, F.L., Ji, J.W. et al. (2024). Reshaping Li-Mg hybrid batteries: epitaxial electrodeposition and spatial confinement on MgMOF substrates via the lattice-matching strategy. *Carbon Energy* 6: e520.
8 Zhang, N., Meng, Q., Wu, H. et al. (2023). Co-MOF as stress-buffered architecture: an engineering for improving the performance of NiS/SnO2 heterojunction in lithium storage. *Advanced Energy Materials* 13: 2300413.
9 Zhang, C.-L., Jiang, Z.-H., Lu, B.-R. et al. (2019). MoS2 nanoplates assembled on electrospun polyacrylonitrile-metal organic framework-derived carbon fibers for lithium storage. *Nano Energy* 61: 104–110.
10 Yang, M., Zeng, X., Xie, M. et al. (2024). Conductive metal–organic framework with superior redox activity as a stable high-capacity anode for high-temperature K-ion batteries. *Journal of the American Chemical Society* 146: 6753–6762.
11 Ho Na, J., Chan Kang, Y., and Park, S.-K. (2021). Electrospun MOF-based ZnSe nanocrystals confined in N-doped mesoporous carbon fibers as anode materials for potassium ion batteries with long-term cycling stability. *Chemical Engineering Journal* 425: 131651.

12 Yin, J.-C., Lian, X., Li, Z.-G. et al. (2024). Triazacoronene-based 2D conductive metal–organic framework for high-capacity lithium storage. *Advanced Functional Materials* 34: 2403656.

13 Hu, P., Xiao, F., Wang, H. et al. (2022). Dual-functional hosts derived from metal-organic frameworks reduce dissolution of polyselenides and inhibit dendrite growth in a sodium-selenium battery. *Energy Storage Materials* 51: 249–258.

14 Ning, Y., Lou, X., Shen, M. et al. (2017). Mesoporous cobalt 2,5-thiophenedicarboxylic coordination polymer for high performance Na-ion batteries. *Materials Letters* 197: 245–248.

15 Wang, B., Li, J., Ye, M. et al. (2022). Dual-redox sites guarantee high-capacity sodium storage in two-dimension conjugated metal–organic frameworks. *Advanced Functional Materials* 32: 2112072.

16 Ma, X., Chen, J., and Zhao, W. (2020). Construction of series-wound architectures composed of metal-organic framework-derived hetero-(CoFe)Se$_2$ hollow nanocubes confined into a flexible carbon skeleton as a durable sodium storage anode. *Nanoscale* 12: 22161–22172.

17 Liang, Y., Song, N., Zhang, Z. et al. (2022). Integrating Bi@C nanospheres in porous hard carbon frameworks for ultrafast sodium storage. *Advanced Materials* 34: 2202673.

18 Liu, Z., Yang, X., Wang, K. et al. (2024). P-type semiconducting covalent organic frameworks for Li-ion battery cathodes with high-energy density. *Energy Storage Materials* 68: 103337.

19 Li, W., Huang, Q., Shi, H. et al. (2024). Integrated high voltage, large capacity, and long-time cyclic stability of lithium organic battery based on a bipolar-type cathode of triphenylamine-hydrazone COF material. *Advanced Functional Materials* 34: 2310668.

20 Shi, R., Liu, L., Lu, Y. et al. (2020). Nitrogen-rich covalent organic frameworks with multiple carbonyls for high-performance sodium batteries. *Nature Communications* 11: 178.

21 Lei, Z., Yang, Q., Xu, Y. et al. (2018). Boosting lithium storage in covalent organic framework via activation of 14-electron redox chemistry. *Nature Communications* 9: 576.

22 Wang, S., Wang, Q., Shao, P. et al. (2017). Exfoliation of covalent organic frameworks into few-layer redox-active nanosheets as cathode materials for lithium-ion batteries. *Journal of the American Chemical Society* 139: 4258–4261.

23 Lee, J., Lim, H., Park, J. et al. (2023). Fluorine-rich covalent organic framework to boost electrochemical kinetics and storages of K+ ions for potassium-ion battery. *Advanced Energy Materials* 13: 2300442.

24 Kushwaha, R., Jain, C., Shekhar, P. et al. (2023). Made to measure squaramide COF cathode for zinc dual-ion battery with enriched storage via redox electrolyte. *Advanced Energy Materials* 13: 2301049.

25 Wu, Y.-L., Horwitz, N.E., Chen, K.-S. et al. (2017). G-quadruplex organic frameworks. *Nature Chemistry* 9: 466–472.

26 Wu, Y., Mao, X., Zhang, M. et al. (2021). 2D molecular sheets of hydrogen-bonded organic frameworks for ultrastable sodium-ion storage. *Advanced Materials* 33: 2106079.

27 Guo, C., Gao, Y., Li, S.-Q. et al. (2024). Chemical-stabilized aldehyde-tuned hydrogen-bonded organic frameworks for long-cycle and high-rate sodium-ion organic batteries. *Advanced Functional Materials* 34: 2314851.

28 Ding, J., He, J., Chen, L. et al. (2024). Zincophilic sites enriched hydrogen-bonded organic framework as multifunctional regulating interfacial layers for stable zinc metal batteries. *Angewandte Chemie International Edition* e202416271. https://doi.org/10.1002/anie.202416271.

29 Chu, J., Liu, Z., Yu, J. et al. (2024). Boosting H+ storage in aqueous zinc ion batteries via integrating redox-active sites into hydrogen-bonded organic frameworks with strong π-π stacking. *Angewandte Chemie International Edition* 63: e202314411.

30 Zhang, J., Wang, Y., Xia, Q. et al. (2024). Confining polymer electrolyte in MOF for safe and high-performance all-solid-state sodium metal batteries. *Angewandte Chemie International Edition* 63: e202318822.

31 Shen, L., Wu, H.B., Liu, F. et al. (2018). Creating lithium-ion electrolytes with biomimetic ionic channels in metal–organic frameworks. *Advanced Materials* 30: 1707476.

32 Wu, J.-F. and Guo, X. (2019). Nanostructured metal–organic framework (MOF)-derived solid electrolytes realizing fast lithium ion transportation kinetics in solid-state batteries. *Small* 15: 1804413.

33 Liu, Q., Yang, L., Mei, Z. et al. (2024). Constructing host–guest recognition electrolytes promotes the Li+ kinetics in solid-state batteries. *Energy & Environmental Science* 17: 780–790.

34 Fan, Y., Chang, Z., Wu, Z. et al. (2024). Nano-confined electrolyte for sustainable sodium-ion batteries. *Advanced Functional Materials* 34: 2314288.

35 Liu, Y., Wang, S., Chen, W. et al. (2024). 5.1 μm ion-regulated rigid quasi-solid electrolyte constructed by bridging fast li-ion transfer channels for lithium metal batteries. *Advanced Materials* 36: 2401837.

36 Xie, Y., Xu, L., Tong, Y. et al. (2024). Molten guest-mediated metal–organic frameworks featuring multi-modal supramolecular interaction sites for flame-retardant superionic conductor in all-solid-state batteries. *Advanced Materials* 36: 2401284.

37 Zhao, G., Mei, Z., Duan, L. et al. (2022). COF-based single Li+ solid electrolyte accelerates the ion diffusion and restrains dendrite growth in quasi-solid-state organic batteries. *Carbon Energy* 5: e248.

38 Han, X.-B., Tang, X.-Y., Lin, Y. et al. (2019). Ultrasmall abundant metal-based clusters as oxygen-evolving catalysts. *Journal of the American Chemical Society* 141: 232–239.

39 Guo, D., Shinde, D.B., Shin, W. et al. (2022). Foldable solid-state batteries enabled by electrolyte mediation in covalent organic frameworks. *Advanced Materials* 34: 2201410.

40 Kang, T.W., Lee, J.-H., Lee, J. et al. (2023). An ion-channel-restructured zwitterionic covalent organic framework solid electrolyte for all-solid-state lithium-metal batteries. *Advanced Materials* 35: 2301308.

41 Qiu, T., Wang, T., Tang, W. et al. (2023). Rapidly synthesized single-ion conductive hydrogel electrolyte for high-performance quasi-solid-state zinc-ion batteries. *Angewandte Chemie International Edition* 62: e202312020.

42 Han, Z., Zhang, R., Jiang, J. et al. (2023). High-efficiency lithium-ion transport in a porous coordination chain-based hydrogen-bonded framework. *Journal of the American Chemical Society* 145: 10149–10158.

43 Abdelhafiz, A., Mohammed, M.H., Abed, J. et al. (2024). Tri-metallic catalyst for oxygen evolution reaction enables continuous operation of anion exchange membrane electrolyzer at 1A cm^{-2} for hundreds of hours. *Advanced Energy Materials* 14: 2303350.

44 Wang, X., Xu, H., Wen, X. et al. (2024). Coordination environment modulation to optimize d-orbit arrangement of Mn-based MOF electrocatalyst for lithium-oxygen battery. *Energy Storage Materials* 70: 103519.

45 Zhang, X., Luo, J., Lin, H.-F. et al. (2019). Tailor-made metal-nitrogen-carbon bifunctional electrocatalysts for rechargeable Zn-air batteries via controllable MOF units. *Energy Storage Materials* 17: 46–61.

46 Xie, L., Xiao, Y., Zeng, Q. et al. (2024). Balanced mass transfer and active sites density in hierarchical porous catalytic metal–organic framework for enhancing redox reaction in lithium–sulfur batteries. *ACS Nano* 18: 12820–12829.

47 Cao, Y., Zhang, Y., Han, C. et al. (2023). Zwitterionic covalent organic framework based electrostatic field electrocatalysts for durable lithium–sulfur batteries. *ACS Nano* 17: 22632–22641.

48 Park, J.H., Lee, C.H., Ju, J.-M. et al. (2021). Bifunctional covalent organic framework-derived electrocatalysts with modulated p-band centers for rechargeable Zn–Air batteries. *Advanced Functional Materials* 31: 2101727.

49 Chang, J., Li, C., Wang, X. et al. (2023). Quasi-three-dimensional cyclotriphosphazene-based covalent organic framework nanosheet for efficient oxygen reduction. *Nano-Micro Letters* 15: 159.

50 Liu, C., Liu, F., Li, H. et al. (2021). One-dimensional van der waals heterostructures as efficient metal-free oxygen electrocatalysts. *ACS Nano* 15: 3309–3319.

51 Cheng, Z., Fang, Y., Yang, Y. et al. (2023). Hydrogen-bonded organic framework to upgrade cycling stability and rate capability of Li-CO$_2$ batteries. *Angewandte Chemie International Edition* 62: e202311480.

52 Guo, C., Cao, Y., Gao, Y. et al. (2024). Cobalt single-atom electrocatalysts enhanced by hydrogen-bonded organic frameworks for long-lasting zinc-iodine batteries. *Advanced Functional Materials* 34: 2314189.

53 Razaq, R., Din, M.M.U., Småbråten, D.R. et al. (2024). Synergistic effect of bimetallic MOF modified separator for long cycle life lithium-sulfur batteries. *Advanced Energy Materials* 14: 2302897.

54 Li, L., Tu, H., Wang, J. et al. (2023). Electrocatalytic MOF-carbon bridged network accelerates Li$^+$-solvents desolvation for high Li$^+$ diffusion toward rapid sulfur redox kinetics. *Advanced Functional Materials* 33: 2212499.

55 Zhou, C., Dong, C., Wang, W. et al. (2024). An ultrathin and crack-free metal-organic framework film for effective polysulfide inhibition in lithium–sulfur batteries. *Interdisciplinary Materials* 3: 306–315.

56 Zuo, L., Ma, Q., Xiao, P. et al. (2024). Upgrading the separators integrated with desolvation and selective deposition toward the stable lithium metal batteries. *Advanced Materials* 36: 2311529.

57 Feng, Y., Zhong, B., Zhang, R. et al. (2023). Taming active-ion crosstalk by targeted ion sifter toward high-voltage lithium metal batteries. *Advanced Energy Materials* 13: 2302295.

58 Li, Y., Peng, X., Li, X. et al. (2023). Functional ultrathin separators proactively stabilizing zinc anodes for zinc-based energy storage. *Advanced Materials* 35: 2300019.

59 Yu, G., Cui, Y., Lin, S. et al. (2024). Ultrathin composite separator based on lithiated COF nanosheet for high stability lithium metal batteries. *Advanced Functional Materials* 34: 2314935.

60 Xu, J., An, S., Song, X. et al. (2021). Towards high performance Li–S batteries via sulfonate-rich COF-modified separator. *Advanced Materials* 33: 2105178.

61 Han, D., Sun, L., Li, Z. et al. (2024). Supramolecular channels via crown ether functionalized covalent organic frameworks for boosting polysulfides conversion in Li–S batteries. *Energy Storage Materials* 65: 103143.

62 Yin, C., Li, Z., Zhao, D. et al. (2022). Azo-branched covalent organic framework thin films as active separators for superior sodium–sulfur batteries. *ACS Nano* 16: 14178–14187.

63 Kang, T., Sun, C., Li, Y. et al. (2023). Dendrite-free sodium metal anodes via solid electrolyte interphase engineering with a covalent organic framework separator. *Advanced Energy Materials* 13: 2204083.

64 Yang, Y., Yao, S., Wu, Y. et al. (2023). Hydrogen-bonded organic framework as superior separator with high lithium affinity C=N bond for low N/P ratio lithium metal batteries. *Nano Letters* 23: 5061–5069.

65 Han, C., Cao, Y., Yang, M. et al. (2024). Hydrogen-bonded organic framework modified separator for simultaneously enhancing the safety and electrochemical performance of Ni-rich lithium-ion battery. *Journal of Energy Chemistry* 96: 72–78.

Index

a

addition polymerization 114, 158
all-deep eutectic soft battery (AESB) 252, 253
all-solid-state batteries (ASSBs) 116, 120, 164, 168
all-solid-state lithium ion battery (ASSLIB) 280
aluminum current collectors 308
amide groups as hydrogen bond donors 345
anion-rich solvation structure 5–6, 10, 11, 31, 36, 39
anodes stabilization 30–31
anodic aluminum oxide (AAO) 293, 297
aqueous zinc metal batteries (AZMBs) 358
Arrhenius curve 20
Azo-TbTh film as a separator 380

b

ball milling 290, 292, 354
battery innovation 239
battery safety 177, 210, 217, 224, 239, 240, 244, 266, 272, 281, 316, 340, 358, 367, 374, 381
battery separators 374
  COF 378–380
  HOF 380–381
  MOF 374–377
Bi@C ⊂ CFs composite as SIB anode material 351
biopolymers, in anode functional layers

cellulose 96
chitosan 96–98
sodium alginate 96–98
biopolymers, in binders
  carboxymethyl cellulose (CMC) 53–54, 67–68
  carrageenan 66
  cellulose (CLS) 63–64
  chitosan (CS) 54–56, 68–70
  citrulline (Cit) 64–65
  gelatin 61–62, 76–78
  guar arabic (GG) 59
  gum arabic (GA) 57–58, 74–75
  lignin 57, 72–74
  pectin 65
  sodium alginate (SA) 56–57, 70–72
  starch 60–61, 75–76
  tragacanth gum (TG) 62–63, 78–79
  trehalose (THL) 64
  xanthan gum (XG) 59–60, 75
biopolymers, in electrolyte additives
  cellulose 88
  citrulline 89–90
  pectin 90
  trehalose 88–89
biopolymers in electrolytes
  cellulose 79–81
  chitosan 81–82
  gelatin 85–87
  lignin 82–85
biopolymers, in separators
  carrageenan 95
  cellulose 91–93

Functional Auxiliary Materials in Batteries: Synthesis, Properties, and Applications, First Edition. Wei Hu.
© 2025 WILEY-VCH GmbH. Published 2025 by WILEY-VCH GmbH.

biopolymers, in separators (*contd.*)
    starch   94
bis(trifluoromethylsulfonyl)imide (TFSI)
    anions   264
bivalent ionic conductors   215
borate groups function   265

c

carbon disulfide (CS_2)   291
carbon nanofibers (CNF) materials   92, 93, 297, 298, 315, 323
carbon nanotube (CNT) materials   293–296
carboxylate-based SICs   216
carboxylate MOFs   342
carboxyl-based HOF materials   344
carboxymethyl cellulose (CMC)   53–54, 67–68, 161, 246
    biopolymers in binders   53–54, 67–68
carrageenan
    biopolymers in binders   66–79
    biopolymers in separators   95
catalyst and catalyst supports
    COF as   371–373
    HOF as   373–374
    MOF as   368–371
cathode electrolyte interface (CEI) layer   5, 7–9, 12–13, 22, 30, 31, 37–39, 118, 119, 131, 174, 166, 197, 199, 362
cathode interface modification   23–24
cathodes stabilization   30–31
cathode-supported UiO-66@KANF layer   363
cationic single-ion conductor (SIC)   216, 217, 219, 221–227, 229, 230
CE-COF@CNT composite   379
CEI-forming additives   12–13, 36–38
cellulose
    biopolymers in anode functional layers   96
    biopolymers in binders   63–64
    biopolymers in electrolyte additives   88
    biopolymers in electrolytes   79–81

    biopolymers in separators   91–93
cellulose nanocrystals (CNC)   52, 80, 81, 88, 129
cellulose separators   80, 92
chitosan
    biopolymers in anode functional layers   96–98
    biopolymers in binders   54–56, 68–70
    biopolymers in electrolytes   81–82
citrulline (Cit)
    biopolymers in binders   64–65
    biopolymers in electrolyte additives   89–90
classical molecular dynamics (CMD) simulation   2
cobalt nanoparticle-embedded PTCOF oxygen electrocatalysts (CoNP-PTCOF)   371, 372
cobalt 2,5-thiophenedicarboxylate coordination polymer (Co-TDC) anode   349
COF@carbon nanotube (COF@CNTs) composite material   353
complex nitrate-based additives   14–15
complex nitro-based additives   15
composite electrolytes   128, 168, 214, 225, 228, 272–275, 362
CoNP-PTCOF   371, 372
copolymerization   19, 73, 132, 166, 167, 174, 248, 254, 291, 292
copper-based metal-organic framework (Cu-MOF-74)   362
Coulombic efficiency   14, 16, 69, 81, 92, 199, 245, 252, 253, 255, 262, 264, 266–268, 274, 275, 279, 280, 289, 291, 296, 297, 301, 304, 309, 310, 312, 318, 325, 326, 328, 350, 354, 377
covalent organic frameworks (COFs)   201, 205, 206, 225, 341–343, 352, 353, 364–367, 372, 378, 379, 381
    as battery separator   378–380
covalent-organic frameworks (COF) materials

as catalyst and catalyst supports 371–372
as electrode materials 352–355
as electrolytes and electrolyte additives 364–367
feature 342
types 343–344
CPAU 77
cross-linked network gel polymer electrolyte (CNGPE) 253
cyanuric acid melamine (MCA) functional layer modification strategy 381

d

DAAQ-ECOF 354
DA cycloaddition reaction 270
dead lithium 3, 4, 11, 164, 325, 330
deep eutectic solvent (DES) 27, 36, 85, 132, 252, 262, 263, 271, 363
density functional theory (DFT) 28, 117, 193, 264, 300, 352
desolvation activation energy 65, 90
desolvation barrier, reduction of 4, 6
diaminotriazine (DAT) 344, 356, 358
differential scanning calorimetry (DSC) 2, 74, 194, 257, 270
diglycidyl ether of bisphenol A (DGEBA) 265
direct method 193
DMA@LiTFSI electrolyte 365
donor number (DN) solvent 7
dual-ion batteries (DIBs) 24
dual-network structured self-healing gel polymer electrolyte (DN-SHGPE) 249, 250
dual-salt strategy 25

e

electric double layer (EDL) 12, 32, 89
electrochemical deposition 24, 291–293
electrochemical impedance spectroscopy (EIS) 266, 275, 281, 347, 364, 365, 375
electrochemical stability, PA-based electrolyte 131

electrochemical window 22, 95, 113, 126, 171, 173, 176, 177, 190, 193, 197, 202, 207, 211, 217, 222, 223, 226, 263, 267, 270, 271, 274, 368
electrode binders 275–281
electrode materials 345
  COF as 352–356, 364–367
  HOF as 367–368
  MOF as 345–351
electrolyte additives 1, 11, 30, 39, 52, 87–90, 359–364, 367–368
electrolyte decomposition 22, 117, 119, 312
electron-deficient Lewis acid 16
electrostatic self-assembly 294
emulsion styrene-butadiene rubber (ESBR) 110
energy storage, POF materials for 340–341
entropy-driven solubilization strategy 16
epoxy resin (EPR) 156–157, 161–162, 167–170, 223
  synthetic polymers in electrolytes 167–170
  synthetic polymers in binders 156–157, 161–162
ethoxylated trimethylpropane triacrylate (ETPTA) 28
ethyl α-cyanoacrylate (ECA) 273
ethylenediaminetetraacetic acid (EDTA)-modified MOF-808 377

f

Fe-ZIF-8 374, 375
Fe-ZIF-8/PP separator 374, 375
fine-tuning agent 1
flame retardant 24, 29–32, 35, 39, 107, 125, 129, 134, 135, 139, 145, 155, 161, 164, 171, 176–179, 181, 199, 202, 207, 220, 271, 381
fluorinated additives
  anion-rich solvation structure 5–6
  high oxidation stability 4–5
  improvement of safety performance 2
  reduction of desolvation barrier 4, 6

fluorinated additives (*contd.*)
  SEI-forming additives 2–4
fluorine-rich Covalent Organic
    Framework (F-COF) 354, 355
fluoroethylene carbonate (FEC) 25, 33,
    34, 132, 177, 208, 274
  and glyme 7–8
  HF gas generation 8
  incompatibility with other electrodes
    8, 9
  and Lewis base 7, 8
  loss of impedance 8
  and other fluorinated electrolytes 6–7
  recycling issues 9
Fourier transform infrared (FTIR) 10,
    22, 74, 97, 163, 194, 222, 349
freeze-drying 141, 179, 299
functional ultrathin separator (FUS) 377

g

γ-butyrolactone (GBL) solvent 29
Garnet electrolytes 212
gelatin 60–62, 76–78, 85–87
  biopolymers in binders 61–62, 76–78
  biopolymers in electrolytes 85–87
gel polymer electrolytes 79, 198, 244,
    256, 281
glyme, FEC 7–8
graphene materials 299–304
graphene/phenol formaldehyde resin
    303
guar arabic (GG) 59
  biopolymers in binders 59, 75
gum arabic (GA)
  biopolymers in binders 57–58, 74–75

h

HATN-XCu electrode materials 350
HCPE 273, 274
HF gas generation 8
hierarchically porous catalytic
    metal-organic framework
    (HPC-MOF) 370, 371
high concentration electrolyte (HCE)
    10, 203

high-conductivity graphene (HCG) 299
high decomposition activation energy
    18–19
high DN value 13, 15, 16
high oxidation stability 3–5, 22, 24
HOF-based multifunctional cathode
    catalyst 373
HOF-DAT material 356
HOF-FJU-1 373
HOFs-8 356, 358
"host-guest recognition" gel polymer
    electrolyte (GPE) strategy 361
hydrogel electrolyte 80–83, 85–87, 245,
    246, 248, 249, 254, 255, 366, 367
hydrogen-bonded organic frameworks
    (HOF) materials
  advantages 343
  as battery separator 380–381
  as catalyst and catalyst supports 373
  as electrolytes and electrolyte additives
    367–368
  types 343–344
hydrothermal methods 86, 299, 300, 305,
    368, 376

i

IISERP-COF22 355
inorganic-rich SEI formation 10–12
in situ ion exchange method 193
in situ sol-gel method 224
in-situ synthesis 86
intercalation method 318
ionic concentration polarization 215
ionic gel electrolyte membranes (IGEMs)
    203
ionic liquids
  advantages of 193
  catalog of 191–192
  catalyst 196
  chemical synthesis 194
  development of 190–191
  drug delivery 196
  electrochemistry 194
  electrolyte 198–206
  electrolyte additive 197–198

Index | 393

lubricants 195–196
organic-inorganic composite 210–215
organic solvent electrolyte 206–210
separation technology 195
synthesis and characterization method of 193, 194
I$_2$@PFC-72-Co cathode 374
isosorbide nitrate (ISDN) 14

j

JUC-610-CON 372

k

Kevlar nanofibers (KANFs) as MOFs 363

l

layer-by-layer self-assembly 96
Lewis base, FEC 7, 8
lignin
   biopolymers in binders 57, 72–74
   biopolymers in electrolytes 82–85
lithiophilicity 313, 326
lithium battery electrolyte 176, 196, 197
lithium bis(trifluoromethanesulfonyl) imide (LiTFSI) 11, 20, 21, 25, 26, 131, 132, 159, 173, 174, 176, 177, 208, 209, 211, 214, 222, 245, 253, 258, 259, 262, 267, 365, 371
lithium-carbon dioxide (Li-CO$_2$) batteries 373
lithium diallylborate 223
lithium garnet 211
lithium-ion batteries (LIBs) 1, 339
lithium metal
   compatibility of 32–33
   protective layer 25, 27
lithium-metal batteries 16, 113, 287–288, 328
lithium nitrate 15, 16
lithium oxynitride 19
lithium plating/stripping 11, 361, 364
lithium polysulfide (LiPS/LPs) 13, 78, 162, 277, 287, 374
   migration 311

lithium salt 5, 9, 22, 28, 29, 32, 33, 36, 208
   dissolution 21, 22
   lithium salt + solvent 215
lithium-sulfur (Li-S) batteries 13, 54, 161, 277, 302, 304, 305, 312, 314, 320, 370, 374
localized high-concentration electrolytes (LHCE) 10
long term durability 53, 75, 114, 377
low-dimensional cathode materials 290–293
   composite methods for 290–293
low-dimensional composite anode materials 324–328
   nanocomposite lithium metal anodes 325–327
   SEI and failure mechanism 324–325
   3D-printing anodes 328
low-dimensional composite current collectors
   design of 320–322
   nanocomposite 322–324
low-dimensional composite materials 288–289
   in separators
      one-dimensional materials 312–315
      two-dimensional materials 315–320
      zero-dimensional materials 310–312
lysine-modified catecholin lignin (AL-Lys-D) 72

m

MA-BTA@Zn electrode 358
mass spectrometry titration (MST) 4
M-E@Celgard separator 377
mechanical kneading method 17
melt guest-mediated metal-organic framework (MGM-MOF) 363
metal-organic frameworks (MOFs) 201, 212, 341–342, 350
   as battery separator 374–377
   materials 341

metal-organic frameworks (MOFs) (contd.)
    as catalyst and catalyst supports 368–371
    as electrode materials 345–351
    as electrolytes and electrolyte additives 359–368
    multifunctionality 341
    structural advantage 341
    types 342
metal selenide (e.g. (CoFe)Se$_2$)@carbon nanofiber (CNS) composites 350
2-methacryloyloxyethyl phosphorylcholine (MPC) 254
methoxy polyethylene glycol maleimide (mPEG-MAL) 256–257
methylene methanedisulfonate (MMDS) 38
MgMOF@PPy@CC electrode 346
Michael addition reaction 162
Mn-MOF-74-FcA 369, 370
MOF@CC@PP separator 375
molecular transporter 220
multi-walled carbon nanotubes (MWCNTs) 313
MXene materials 304–309, 318, 326

n

nanocomposite current collectors 322–324
nanocomposite lithium metal anodes 325–327
NCM622//Li batteries with M-E@Celgard separator 377
nickel cobaltite/carbon nanofiber (NiCo$_2$O$_4$/CNF) composites 314
Ni SA/NOC/Se cathode 349
NiS/SnO$_2$/MOF (NSM) electrode 346
nitrile additives
    cathode interface modification 23–24
    crystallinity of 28
    electrochemical window 22
    electrolyte decomposition 22
    ionic association, weakening of 24
    ion transport, facilitation of 20–21
    lithium salt dissolution 21, 22
    low flammability 22–23
    low mechanical strength 28
    plasticization 19–20
    polymer flexibility 23
    prone to polymerization 28
    SEI formation 24
    Zn^{2+} structure 24
nitrile and lithium metal
    compatibility of 25–27
    incompatibility of 25
nitro additive
    CEI-forming additives 12–13
    inorganic-rich SEI formation 10–12
    lithium-sulfur batteries 13
    in solvation and desolvation structures 9–10
    water molecules, stabilization of 14
nitro-C60 derivative 15
nitrogen-containing heterocyclic MOFs 342
NKU-1000 material 367, 368
N-methylacetamide (NMA) 253
nonpolar sulfur 290
non-solvent induced phase separation (NIPS) 224
nuclear magnetic resonance (NMR) spectroscopy 4, 194
N-vinyl pyrrolidone (NVP) 113, 114

o

one-dimensional materials
  in cathode
    carbon nanofibers (CNF) 297–298
    carbon nanotube (CNT) 293–296
one-step printing process 296
organic borate-based SIC 217
organic-inorganic composite 210–215, 225–228
organic nitro additive
    complex nitrate-based additives 14–15
    complex nitro-based additives 15
    high decomposition activation energy 18–19
    low solubility 15–16

sacrificial additives 17–18
organic solvent electrolyte 197, 206–210
oxygen evolution reaction (OER) activity 369, 371

p

PBPE 266, 267
PCNF@MoS$_2$ composite 347
pectin
   biopolymers in binders 65
   biopolymers in electrolyte additives 90
PGC 77
phosphate ester additives
   cathodes and anodes stabilization 30–31
   compatibility of 32–33
   flame retardant 29–30
   incompatibility with anodes 32
   solvation structure regulation 31–32
physical confinement effects 310, 313
plastic-crystalline electrolyte (PCE) layer 23, 28
polyacrylics (PA)
   synthetic polymers in anodes 143
   synthetic polymers in binders 111–112, 122–124
   synthetic polymers in electrolytes 131–133
polyacrylonitrile (PAN) 20, 38, 80, 112–113, 129–131, 139–140, 142, 322, 347
   synthetic polymers in anodes 142–143
   synthetic polymers in battery separators 139–140
   synthetic polymers in binders 112–113
   synthetic polymers in electrolytes 129–131
polycarbonate propylene glycol ether (PPCAGE) 219
polydopamine (PDA) 305
polyethylene glycol (PEG) 22, 56, 132, 133, 159, 171, 220, 224, 225, 246, 247, 257, 263, 274, 292

poly(ethylene glycol) methyl ether acrylate (PEGA) 262
polyethylene oxide (PEO)
   synthetic polymers in binders 158–159
   synthetic polymers in electrolytes 173–176
polyethylene terephthalate (PET)
   synthetic polymers in battery separators 178–179
   synthetic polymers in binders 159–160
polyethylenimine (PEI) 161
   synthetic polymers in binders 157–158, 162–164
polyformaldehyde (POM) 27
polyimide (PI)
   synthetic polymers in battery separators 179–180
   synthetic polymers in binders 160–161, 166–167
   synthetic polymers in electrolytes 176–177
polylactic acid precursor (PAP) 268
polymer flexibility 23
polymer-MOF single-ion conducting solid polymer electrolyte (SICSPE) 360
polyolefin (PO)
   synthetic polymers in battery separators 136–138
   synthetic polymers in binders 114–115
polypropylene (PP) separator 308, 312, 374
polysulfide (PS) 13
   redox reactions 319
   shuttle effect 57–61, 310
polytetrafluoroethylene (PTFE) 109–110, 117–118, 380
   synthetic polymers, in binders 109–110, 117–118
polyurethane (PU) 260
   synthetic polymers in binders 158, 164–165

polyurethane (PU) (*contd.*)
  synthetic polymers in electrolytes 170–173
polyurethane binder (PUB) 277
polyvinyl alcohol (PVA)
  synthetic polymers in battery separators 140–141
  synthetic polymers in binders 111, 119–122
  synthetic polymers in electrolytes 133–135
polyvinylidene difluoride (PVDF)
  synthetic polymers in battery separators 138–139
  synthetic polymers in binders 108–109, 115–117
  synthetic polymers in electrolytes 125–129
polyvinyl pyrrolidone (PVP)
  synthetic polymers in binders 113–114
porous networked HAN-Cu-MOF anode 348
porous organic framework (POF) materials
  advantages 340–341
  application in electrode materials 345–359
  covalent organic frameworks 342
  hydrogen-bonded organic frameworks 343
  metal-organic frameworks 341–343
PTB-DHZ-COF40 353

q
quasi-MOF nanospheres 370

r
radial distribution function (RDF) 22
Raman spectroscopy 10, 263, 316, 374, 375
reactive ion exchange method 193
redox-active hydrogen-bonded organic framework (HOF-HATN) 358, 359

reversible addition-fragmentation chain transfer (RAFT) polymerization 261, 262
ring-opening polymerization 85, 158, 160, 168, 292

s
Seebeck coefficient 256
self-extinguishing time (SET) tests 2
self-healing hydrogel polymer electrolyte (SHGPE) 245, 246
self-healing materials
  benefits of 240–242
  chemically bonded 243–244
  composite electrolytes 272–275
  electrode binders 275–281
  gel polymer electrolytes 244–256
  with multiple repair mechanisms 244
  overview of 239–240
  physically bonded 243
  scaling and commercializing 242–243
  solid polymer electrolytes 256–271
  stability and durability of 242
self-healing mechanism 124, 244, 245, 250, 252, 256, 261, 263–267, 270, 271, 274, 277, 280
self-healing poly(ether-thiourea) (SHPET) 276
single-atom catalysts (SACs) 302, 319
single-ion conductive 215, 216
  catalog of 216–217
  organic 217–225
  organic-inorganic composite 225–228
single-wall carbon nanotube (SWCNT) 301, 313
SiO_2-template method 303
sodium alginate (SA)
  biopolymers in anode functional layers 96–98
  biopolymers in binders 56–57, 70–72
sodium dendrite 250, 380
sodium-ion batteries (SIBs) 20, 56, 57, 139, 340, 362
sodium sulfide (Na_2S) 292
sodium thiosulfate ($Na_2S_2O_3$) 292

soft X-ray absorption spectroscopy (sXAS) 22
solid-electrolyte interphase (SEI)
　and failure mechanism　324–325
　forming additives　2–4, 34–36
solid polymer electrolytes (SPEs)　19, 112, 201, 256–271, 361
　additives　19–20
solid state battery electrolyte　197, 258
soluble styrene-butadiene rubber (SSBR)　110
solution polymerization　110, 112–114
solvation structure regulation　31–32
starch
　biopolymers in binders　60–61, 75–76
　biopolymers in separators　94
styrene-butadiene rubber (SBR)
　synthetic polymers in binders　110, 118–119
sulfate ester additives
　CEI-forming additives　36–38
　SEI-forming additives　34–36
sulfobetaine vinylimidazole (SBVI)　254
sulfonate-based SICs　216
sulfonated reduced graphene oxide (SRGO)　315
sulfonic acid-rich covalent organic framework (SCOF-2) modified batteries　378
sulfonimide-based SICs　216
sulfur-carbon mixture cathode　319
sulfur-reduced graphene oxide (S-rGO) cathode　299
synchro-dry technique　291
synthetic polymers, in anodes
　polyacrylics　143
　polyacrylonitrile　142–143
synthetic polymers, in battery separators
　polyacrylonitrile　139–140
　polyethylene terephthalate　178–179
　polyimide　179–180
　polyolefin　136–138
　polyvinyl alcohol　140–141
　polyvinylidene difluoride　138–139

synthetic polymers, in binders
　epoxy resin (EPR)　156–157, 161–162
　polyacrylics (PA)　111–112, 122–124
　polyacrylonitrile (PAN)　112–113
　polyethylene oxide (PEO)　158–159
　polyethylene terephthalate (PET)　159–160
　polyethylenimine (PEI)　157–158, 162–164
　polyimide (PI)　160–161, 166–167
　polyvinylidene difluoride (PVDF)　108–109, 115–117
　polyolefin (PO)　114–115
　polytetrafluoroethylene (PTFE)　109–110, 117–118
　polyurethane (PU)　158, 164–165
　polyvinyl alcohol (PVA)　111
　polyvinyl pyrrolidone (PVP)　113–114
　styrene-butadiene rubber (SBR)　110, 118–119
synthetic polymers, in electrolytes
　epoxy resin　167–170
　polyacrylics　131–133
　polyacrylonitrile　129–131
　polyethylene oxide　173–176
　polyimide　176–177
　polyurethane　170–173
　polyvinyl alcohol　133–135
　polyvinylidene difluoride (PVDF)　125–129

t

TAPTt-COF material　372
TCOF-S-Gel electrolyte　366, 367
terephthalaldehyde (TPA)　257, 265, 380
thermopower enhancement　256
three-dimensional covalent organic frameworks (3D COFs)　343
3D-printing anodes　328
Ti-based MOFs　361
titanium carbide　305
TpPa-SO$_3$Li@PE separator　378
TQBQ-COF electrode　353
traditional battery materials, limitations of　339–340

tragacanth gum (TG)   62–63, 78, 79
  biopolymers, in binders   62–63, 78–79
transmission electron microscope (TEM)   5, 12, 15, 38, 128, 349
trehalose (THL)
  biopolymers in binders   64
  biopolymers in electrolyte additives   88–89
triazine-cyclotriphosphazene-based COFs (Q3CTP-COFs)   372
triethylene glycol dinitrate (TEGDN)   14
trifluoroacetate anion (TFA$^-$)   16
trifunctional furan (TF)   255
trifunctional maleimide (TMI)   255
triglycidyl isocyanurate (TGIC)   161
tris(trimethylsilyl) phosphite (TMSP)   7, 30, 31
two-dimensional covalent organic frameworks (2D COFs)   343
two-dimensional materials
  in cathode
    graphene materials   299–304
    MXene materials   304–309
two-dimensional metal-organic frameworks (2D-MOFs)   348

u

UIO/Li-IL solid-state electrolyte   361
ultrasonic-assisted multiple wetness impregnation   291
UPy dimers   249, 263
ureidopyrimidinone   248

v

vacuum filtration   299, 306, 308, 312, 316, 318

vapor deposition method   307
vertically aligned carbon nanotubes (VACNT)   322
Vienna ab initio simulation package (VASP)   22
vinyl hybrid silica nanoparticles (VSN)   280
vinylene carbonate (VC)   9
Vogel–Fulcher–Tammann (VFT) curve   20

x

xanthan gum (XG)
  biopolymers in binders   59–60, 75
X-ray photoelectron spectroscopy (XPS)   2, 5, 123, 253, 305, 349
  analysis   2

z

zinc-air batteries (ZABs)   80, 134, 246, 371, 372, 381
  with CoNP-PTCOF   371
zinc-iodine (Zn-I$_2$) batteries
  with I$_2$@PFC-72-Co cathode   373–374
zinc-ion batteries (ZIBs)   24, 31, 52, 56, 62, 63, 65, 79–81, 85–92, 96, 99, 118, 140, 244, 255, 264, 304, 359, 366, 374
  cathode materials   355
ZnSe@N-PCNFs composites   348
Zn^{2+} structure   24, 32, 90
Zwitt-COF   366
zwitterionic covalent organic framework   371
zwitterionic polymer networks   254